Lecture Notes in Computer Science 6189

Commenced Publication in 1973
Founding and Former Series Editors:
Gerhard Goos, Juris Hartmanis, and Jan van Leeuwen

Boualem Benatallah   Fabio Casati
Gerti Kappel   Gustavo Rossi (Eds.)

Web Engineering

10th International Conference, ICWE 2010
Vienna, Austria, July 5-9, 2010
Proceedings

 Springer

Volume Editors

Boualem Benatallah
CSE, University of New South Wales, Sydney, Australia
E-mail: boualem@cse.unsw.edu.au

Fabio Casati
University of Trento
Department of Information Engineering and Computer Science, Italy
E-mail: fabio.casati@unitn.it

Gerti Kappel
Vienna University of Technology, Business Informatics Group, Austria
E-mail: gerti@big.tuwien.ac.at

Gustavo Rossi
Universidad Nacional de La Plata. LIFIA, Facultad de Informática, Argentina
E-mail: gustavo@lifia.info.unlp.edu.ar

Library of Congress Control Number: 2010929193

CR Subject Classification (1998): H.3, H.4, I.2, C.2, H.5, J.1

LNCS Sublibrary: SL 3 – Information Systems and Application, incl. Internet/Web and HCI

ISSN	0302-9743
ISBN-10	3-642-13910-8 Springer Berlin Heidelberg New York
ISBN-13	978-3-642-13910-9 Springer Berlin Heidelberg New York

springer.com

© Springer-Verlag Berlin Heidelberg 2010
Printed in Germany

Typesetting: Camera-ready by author, data conversion by Scientific Publishing Services, Chennai, India
Printed on acid-free paper   06/3180

Preface

Web engineering is now a well-established and mature field of research with strong relationships with other disciplines such as software engineering, human–computer interaction, and artificial intelligence. Web engineering has also been recognized as a multidisciplinary field, which is growing fast together with the growth of the World Wide Web. This evolution is manifested in the richness of the Web Engineering Conferences which attract researchers, practitioners, educators, and students from different countries.

This volume contains the proceedings of the 10th International Conference on Web Engineering (ICWE 2010), which was held in Vienna, Austria, in July 2010. The ICWE conferences are among the most essential events of the Web engineering community. This fact is manifested both by the number of accomplished researchers that support the conference series with their work and contributions as well as by the continuing patronage of several international organizations dedicated to promoting research and scientific progress in the field of Web engineering.

ICWE 2010 followed conferences in San Sebastian, Spain; Yorktown Heights, NY, USA; Como, Italy; Palo Alto, CA, USA; Sydney, Australia; Munich, Germany; Oviedo, Spain; Santa Fe, Argentina; and Caceres, Spain.

This year's call for research papers attracted a total of 120 submissions from 39 countries spanning all continents of the world with a good coverage of all the different aspects of Web engineering. Topics addressed by the contributions included areas ranging from more traditional fields such as model-driven Web engineering, Web services, performance, search, Semantic Web, quality, and testing to novel domains such as the Web 2.0, rich Internet applications, and mashups. All submitted papers were reviewed in detail by at least three members of the Program Committee, which was composed of experts in the field of Web engineering from 23 countries. Based on their reviews, 26 submissions were accepted as full papers, giving an acceptance rate of 22%. Additionally, we had six papers in the Industry Track. The program was completed by 13 posters and demonstrations that were selected from 26 submissions and presented in dedicated sessions at the conference. Finally, the conference was also host to keynotes by Wendy Hall (University of Southampton, UK) Serge Abiteboul (INRIA, France), and Bredley P. Allen (Elsevier Labs, USA), as well as an outstanding collection of tutorials and workshops.

We would like to express our gratitude to all the institutions and sponsors that supported ICWE 2010, namely, the Vienna University of Technology, the Business Informatics Group, the Austrian Computer Society, the Austrian Federal Ministry of Science and Research, the Austrian Federal Ministry for Transport, Innovation and Technology, the Vienna Convention Bureau, SIEMENS AG, Google, and Austrian Airlines. The conference would not have been possible without the endorsement of the International World Wide Web Conference Committee (IW3C2), and the International Society for Web Engineering (ISWE). In this context, we would especially like to thank Bebo White and Martin Gaedke for their work as our liaisons to these two organizations. Thanks also to Geert-Jan Houben, who acted as liaison to the ICWE Steering Committee.

We are also indebted to the Chairs of the different tracks (Claudio Bartolini, Heiko Ludwig, Florian Daniel, Federico M. Facca, Jaime Gomez, Daniel Schwabe, Marco Brambilla, Sven Casteleyn, Cesare Pautasso, Takehiro Tokuda, Hamid Motahari, Birgit Pröll, Manuel Wimmer and Schahram Dustdar), to the members of the Program Committee, to the external reviewers, and to the local organizers. All of them helped with their enthusiastic work to make ICWE 2010 a reality and success. Finally, a special thanks to all the researchers and students who contributed with their work and participated in the conference, as well as to Easychair for the great support.

July 2010

Boualem Benatallah
Fabio Casati
Gerti Kappel
Gustavo Rossi

Conference Organization

General Chair

Gerti Kappel   Vienna University of Technology, Austria

Program Chairs

Boualem Benatallah   University of New South Wales, Australia,
     LIMOS, France
Fabio Casati   University of Trento, Italy
Gustavo Rossi   National University of La Plata, Argentina

Industrial Track Chairs

Claudio Bartolini   HP Labs, USA
Heiko Ludwig   IBM Research, USA

Workshop Chairs

Florian Daniel   University of Trento, Italy
Federico M. Facca   University of Innsbruck, Austria

Tutorial Chairs

Jaime Gomez   University of Alicante, Spain
Daniel Schwabe   PUC-RIO, Brazil

Demo and Poster Chairs

Marco Brambilla   Politecnico di Milano, Italy
Sven Casteleyn   Vrije University Brussels, Belgium

Doctoral Consortium Chairs

Cesare Pautasso   University of Lugano, Switzerland
Takehiro Tokuda   Tokyo Institute of Technology, Japan

Publicity Chairs

Hamid Motahari   HP Labs, USA
Birgit Pröll   University of Linz, Austria

Conference Steering Committee Liaison

Geert-Jan Houben   Delft University of Technology,
                        The Netherlands

ISWE Liaison

Martin Gaedke   Chemnitz University of Technology, Germany

IW3C2 Liaison

Bebo White   SLAC, USA

Local Community Liaison

Schahram Dustdar   Vienna University of Technology, Austria

Local Organization Chair

Manuel Wimmer   Vienna University of Technology, Austria

Program Committee

Silvia Abrahao	Florian Daniel	Jaime Gomez
Ignacio Aedo	Davide Di Ruscio	Michael Grossniklaus
Helen Ashman	Oscar Diaz	Kaj Granbak
Marcos Baez	Damiano Distante	Mohand-Said Hacid
Luciano Baresi	Peter Dolog	Simon Harper
Alistair Barros	Marlon Dumas	Birgit Hofreiter
Khalid Belhajjame	Schahram Dustdar	Geert-Jan Houben
Maria Bielikova	Toumani Farouk	Arun Iyengar
Davide Bolchini	Rosta Farzan	Nora Koch
Matthias Book	Howard Foster	Jim Laredo
Athman Bouguettaya	Flavius Frasincar	Frank Leymann
Chris Brooks	Piero Fraternali	Xuemin Lin
Jordi Cabot	Martin Gaedke	An Liu
Dan Chiorean	Irene Garrigos	David Lowe
Christine Collet	Dragan Gasevic	Maristella Matera
Sara Comai	Angela Goh	Hong Mei

Wolfgang Nejdl
Moira Norrie
Luis Olsina
Satoshi Oyama
Oscar Pastor Lopez
Cesare Pautasso
Vicente Pelechano
Michalis Petropoulos
Alfonso Pierantonio
Birgit Proell
I.V. Ramakrishnan
Werner Retschitzegger

Bernhard Rumpe
Fernando
    Sanchez-Figueroa
Daniel Schwabe
Michael Sheng
Robert Steele
Bernhard Thalheim
Massimo Tisi
Giovanni
    Toffetti Carughi
Takehiro Tokuda
Riccardo Torlone

Jean Vanderdonckt
Yannis Velegrakis
Eelco Visser
Petri Vuorimaa
Vincent Wade
Marco Winckler
Bin Xu
Yeliz Yesilada
Yanchun Zhang
Xiaofang Zhou

External Reviewers

Adrian Fernandez
Alessandro Bozzon
Alessio Gambi
Antonio Cicchetti
Armin Haller
Arne Haber
Asiful Islam
Axel Rauschmayer
Carlos Laufer
Christoph Dorn
Christoph Herrmann
Cristobal Arellano
David Schumm
Davide Di Ruscio
Dimka Karastoyanova
Dirk Reiss
Emmanuel Mulo
Fabrice Jouanot
Faisal Ahmed
Florian Skopik
Francisco Valverde
Ganna Monakova
Ge Li
Giacomo Inches
Giovanni Giachetti
Guangyan Huang
Huan Xia
Javier Espinosa
Jian Yu
Jiangang Ma

Jose Ignacio
    Panach Navarrete
Jozef Tvarozek
Juan Carlos Preciado
Junfeng Zhao
Jan Suchal
Kreshnik Musaraj
Lukasz Jusczyk
Marco Brambilla
Marino Linaje
Mario Luca Bernardi
Marian Aimko
Mark Stein
Markus Look
Martin Schindler
Martin Treiber
Markus Schindler
Michael Reiter
Michal Tvarozek
Michal Barla
Minghui Zhou
Mohamed Abdallah
Nathalie Aquino
Noha Ibrahim
Oliver Kopp
Pau Giner
Peep Kangas
Peter Vojtek
Philip Lew
Puay-Siew Tan

Sai Zeng
Sam Guinea
Santiago Melia
Sen Luo
Shiping Chen
Song Feng
Stefan Wild
Steve Strauch
Steven Volkel
Surya Nepal
Taid Holmes
Thomas Kurpick
Vinicius Segura
Wanita Sherchan
William Van Woensel
Xia Zhao
Xiao Zhang
Xin Wang
Xuan Zhou
Yanan Hao
Yanzhen Zou
Yixin Yan
Zaki Malik
Gaoping Zhu
Haichuan Shang
Weiren Yu
Wenjie Zhang
Zhitao Shen

Table of Contents

Development Process

Web 2.0

Linked Data

Performance and Security

Industry Papers

Demo and Poster Papers

Searching Repositories
of Web Application Models

Alessandro Bozzon, Marco Brambilla, and Piero Fraternali

Politecnico di Milano, Dipartimento di Elettronica e Informazione
P.za L. Da Vinci, 32. I-20133 Milano, Italy
{alessandro.bozzon,marco.brambilla,piero.fraternali}@polimi.it

Abstract. Project repositories are a central asset in software development, as they preserve the technical knowledge gathered in past development activities. However, locating relevant information in a vast project repository is problematic, because it requires manually tagging projects with accurate metadata, an activity which is time consuming and prone to errors and omissions. This paper investigates the use of classical Information Retrieval techniques for easing the discovery of useful information from past projects. Differently from approaches based on textual search over the source code of applications or on querying structured metadata, we propose to index and search the *models* of applications, which are available in companies applying Model-Driven Engineering practices. We contrast alternative index structures and result presentations, and evaluate a prototype implementation on real-world experimental data.

1 Introduction

Software repositories play a central role in the technical organization of a company, as they accumulate the knowledge and best practices evolved by skilled developers over years. Besides serving the current needs of project development, they have also an archival value that can be of extreme importance in fostering reuse and the sharing of high quality design patterns. With the spreading of open source software, project repositories have overcome the boundaries of individual organizations and have assumed a social role in the diffusion of coding and design solutions. They store billions of lines of code and are used daily by thousands of developers. State-of-the-practice project repositories mostly support source code or documentation search [4,10,14]. Several solutions are available, with different degrees of sophistication in the way in which queries are expressed, the match between the query and the indexed knowledge is determined, and results are presented. *Source code search engines* (e.g., Google code, Snipplr, Koders) are helpful if the abstraction level at which development occurs is the implementation code. However, searching project repositories at the source code level clashes with the goal of Model-Driven Engineering, which advocates the use of models as the principal artefact to express solutions and design patterns. Therefore, the question arises of what tools to use to leverage the knowledge implicitly

B. Benatallah et al. (Eds.): ICWE 2010, LNCS 6189, pp. 1–15, 2010.

stored in repositories of models, to make them play the same role in disseminating modeling best practices and foster design with reuse as code repositories had in fostering implementation-level best practices and code reuse. Approaches to model-driven repository search have been recently explored in the fields of business process discovery [3,5,12] and UML design [9,17].

The problem of searching model repositories can be viewed from several angles: the language for expressing the user's query (natural language, keywords, structured expressions, design patterns); the granularity at which result should be presented (full project, module or package, design diagram, design pattern, individual construct); the criteria to use in computing the match between the query and the model and for ranking results; the kind of metadata to collect and incorporate in the index (manually provided or automatically extracted).

The goal of this paper is to investigate solutions for making project repositories searchable *without requiring developers to annotate artifacts for the sole purpose of search*. The key idea is to exploit the structural knowledge embedded within application models, which can be expressed at a variable degree of abstraction (from Computation Independent, to Platform Independent, to Platform Specific). We concentrate on Platform Independent Models, because they describe the application, and not the problem, and are independent of the implementation technology. The contribution of the paper can be summarized as follows: 1) we introduce the notion of *model-driven project information retrieval system*, as an application of the information retrieval (IR) techniques to project repository search; 2) we identify the relevant design dimensions and respective options: project segmentation, index structure, query language and processing, and result presentation; 3) we implement an architecture for automatic model-driven project segmentation, indexing and search, which does not require the manual annotation of models; 4) we evaluate the approach using 48 industrial projects provided by a company, encoded with a Domain Specific Language.

The paper is organized as follows: Section 2 discusses the related work; Section 3 introduces the architecture of the model-driven project information retrieval system; Section 4 classifies the main design decisions; Section 5 illustrates the implementation experience; Section 6 presents the results of a preliminary performance assessment and user evaluation; finally, Section 7 draws the conclusions and discusses the ongoing and future work.

2   Related Work

Before illustrating the proposed solution, we overview the state-of-the-art in searchable project repositories, to better highlight the current limitations and original contribution of the work.

Component Search. Retrieval of software components, intended as mere code artifacts or as annotated pieces of software, is a well established discipline. The first proposals date back to the '90s. Agora [23] is a search engine based on JavaBeans and CORBA technologies that automatically generates and indexes a worldwide database of software artifacts, classified by component model. In the context of

SOA, Dustdar *et al.* [22] propose a method for the discovery of Web services based on a Vector Space Model characterization of their properties, indexed in a search engine. The work in [15] proposes a graph-representation model of a software component library, based on analyzing actual usage relations of the components and propagating a significance score through such links. The approach proposed in [7] combines formal and semi-formal specification to describe behaviour and structure of components, so as to make the retrieval process more precise.

Source Code Search. Several communities and on-line tools exist for sharing and retrieving code, e.g., *Google code, Snipplr, Koders,* and *Codase*[1].

In the simplest case, keyword queries are directly matched to the code and the results are the exact locations where the keyword(s) appear in the matched code snippets. Several enhancements are possible: 1) using more expressive query languages, e.g., regular expressions (in Google Codesearch) or wildcards (in Codase); restricting search to specific syntactical categories, like class names, method invocations, variable declarations, and so on (e.g., in Jexamples and Codase); restricting keyword search using a fixed set of metadata (e.g., programming language, license type, file and package names). Another dimension concerns how the relevance of the match is computed and presented to the user; the spectrum of solutions goes from the minimal approaches that simply return a list of hits without a meaningful ranking, to classical IR-style ranking based on term importance and frequency (e.g., TF/IDF), to composite scores taking into account both inherent project properties, e.g., number of matches in the source code, recency of the project, and social aspects, e.g., number of downloads, activity rates, and so on; for example, in SourceForge, one can rank results based on relevance of match, activity, date of registration, recency of last update, or on a combination calculated from such partial scores, and the system can account precisely for the rank value of each project over time.

Research works have applied IR techniques [10] and structural context techniques [14] for improving productivity and reuse of software. For example, the Sourcerer Project [4] provides an infrastructure for large-scale indexing and analysis of open source code that takes advantage of code structural information.

Model Search. Some approaches have addressed the problem of searching UML models. Early works exploited the XML format for indexing seamlessly UML models, text files, and other sources [11]. The work [13] stores UML artifacts in a central knowledge base, classifies them with WordNet terms and extracts relevant items exploiting WordNet classification and Case-Based Reasoning. The paper [17] proposes a retrieval framework allowing designers to retrieve information on UML models based on XMI representation through two query modalities: inclusion and similarity. Schemr [9] implements a novel search algorithm, based on a combination of text search and schema matching techniques, as well as a structurally-aware scoring methods, for retrieving database conceptual models with queries by example and keyword-based. Another branch of research applies

---

[1] Sites: `http://code.google.com`, `http://www.snipplr.com`, `http://www.koders.com`, `http://www.codase.com`

IR techniques to models and code together, for tracing the association between requirements, design artifacts, and code [24] [2].

Business Process Model Search. Several proposals have attempted to facilitate the discovery of business process models. Most of the approaches only apply graph-based comparison or XML-based querying on the business process specifications: Eyal *et al.* [5] proposed BP-QL, a visual query language for querying and discovering business processes modelled using BPEL. Lu and Sadiq [18] propose a way for comparing and retrieving business process variants. WISE [25] is a business process search engine that extracts workflow models based on keyword matching. These proposals offer a query mechanism based on the process model structure (i.e., the workflow topology) only. Other approaches adopt semantic-based reasoning and discovery: Goderis *et al.* [12] developed a framework for discovering workflows using similarity metrics that consider the activities composing the workflows and their relationships, implementing a ranking algorithm. [20] proposed a framework for flexible queries on BP models, for providing better results when too few processes are extracted. [3] proposes the BPMN-Q query language for visual semantic queries over BPMN models. Kiefer *et al.* [16] proposed the use of semantic business processes to enable the integration and inter-operability of business processes across organizational boundaries. They offer an imprecise query engine based on iSPARQL to perform the process retrieval task and to find inter-organizational matching at the boundaries between partners. Zhuge *et al.* [26] proposes an inexact matching approach based on SQL-like queries on ontology repositories. The focus is on reuse, based on a multi-valued process specialization relationship. The similarity of two workflow processes is determined by the matching degrees of their corresponding sub-processes or activities, exploiting ontological distances. The work [6] proposes a query by example approach that relies on ontological description of business processes, activities, and their relationships, which can be automatically built from the workflow models themselves.

Contribution. The approach described in this paper falls into the category of model-based search solutions, where it brings several innovations: (i) it automatically extracts the semantics from the searched conceptual models, without requiring manual metadata annotation; (ii) it supports alternative index structures and ranking functions, based on the language concepts; (iii) it is based on a model-independent framework, which can be customized to any DSL meta-model; (iv) it has been subjected to a preliminary evaluation on the relative performance and quality of alternative design dimensions.

3   IR Architecture Overview

Applying information retrieval techniques over model repositories requires an architecture for processing content, building up the required search indexes, matching the query to the indexed content, and presenting the results. Figure 1 shows the reference architecture adopted in this paper and the two main information flows: the *content processing flow* and the *query flow*.

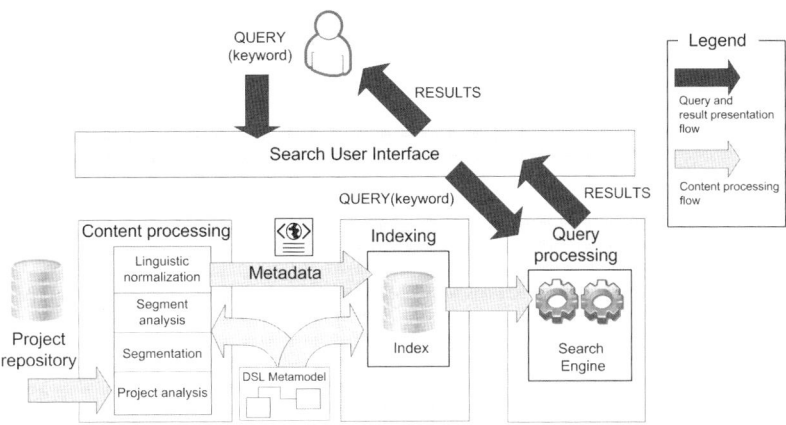

Fig. 1. Architecture of a content processing and search system

The *Content Processing Flow* extracts meaningful information from projects and uses it to create the search engine index. First, the *Content Processing* component analyzes each project by applying a sequence of steps: *project analysis* captures project-level, global metadata (e.g., title, contributors, date, and so on) useful for populating the search engine indexes; *segmentation* splits the project into smaller units better amenable to analysis, such as sub-projects or diagrams of different types; *segment analysis* mines from each segment the information used to build the index (e.g., model elements' names and types, model element relationships, designers's comments); *linguistic normalization* applies the text normalization operations typical of search engines (e.g., stop-word removal, stemming, etc. [19]) to optimize the retrieval performance. The information extracted from each project or segment thereof is physically represented as a *document*, which is fed to the *Indexing* component, for constructing the search engine indexes. Note that the metamodel of the DSL used to express the projects is used both in the Content Processing and in the Indexing components: in the former, it drives the model segmentation granularity and the information mining from the model elements; in the latter, it drives the definition of the index structure. For instance, the search engine index might be composed of several sub-indexes (called *fields*), each one devoted to a specific model element, so that a keyword query can selectively match specific model concepts.

The *query and result presentation flow* deals with the queries submitted by the user and with the production of the result set. Two main query modalities can be used: *Keyword Based*, shown in Figure 1, which simply looks for textual matches in the indexes, and *Content Based* (also known as Query by Example), not shown in in Figure 1, whereby a designer submits a model fragment as query, and the system extracts from it the relevant features (by applying the content processing flow) and matches them to the index using a given similarity criteria (e.g., text matching, semantic matching, graph matching).

4   Design Dimensions of Model-Driven Project Retrieval

The design space of a Model Driven Project IR System is characterized by multiple dimensions: the transformations applied to the models before indexing, the structure of the indexes, and the query and result presentation options.

Segmentation Granularity. An important design dimension is the **granularity** of indexable documents, which determines the atomic unit of retrieval for the user. An indexable document can correspond to:

- A whole design project: in this case, the result set of a query consists of a ranked list of projects.
- A subproject: the result set consists of ranked subprojects and each subproject should reference the project it belongs to.
- A project concept: each concept should reference its project and the concepts it relates to. The result set consists of ranked concepts, possibly of different types, from which other related concepts can be accessed.

Index structure. The structure of the index constructed from the models represents a crucial design dimension. An index structure may consists of one or more fields, and each field can be associated with an importance score (its weight). The division of the index into fields allows the matching procedure used in query processing to match in selected field, and the ranking algorithm to give different importance to matches based on the field where they occur.

The options that can be applied are the following:

- **Flat:** A simple list of terms is extracted from the models, without taking into account model concepts, relationships, and structure. The index structure is single-fielded, and stores undifferentiated bags of words. This option can be seen as a baseline, extracting the minimal amount of information from the models and disregarding any structure and semantics associated with the employed modeling language.
- **Weighted:** Terms are still extracted as flat lists, but model concepts are used in order to modify the weight of terms in the result ranking, so to give a significance boost to terms occurring in more important concepts. The index is single-fielded and stores weighted bags of words.
- **Multi-field:** Terms belonging to different model concepts are collected into separate index fields. The index is multi-fielded, and each field can be searched separately. This can be combined with the weighted approach, so as to produce a multi-field index containing weighted terms. The query language can express queries targeted to selected fields (e.g., to selected types of concepts, diagrams, etc).
- **Structured:** The model is translated into a representation that reflects the hierarchies and associations among concepts. The index model can be semi-structured (XML-based) or structured (e.g., the catalog of a relational database). Query processing can use a structured query language (e.g., SQL), coupled with functions for string matching into text data (e.g., indices for text objects).

Moving from flat to structured index structures augments the fidelity at which the model structure is reflected into the index structure, at the price of a more complex extraction and indexing phase and of a more articulated query language.

Query Language and Result Presentation. An IR system can offer different query and result visualization options. In the context of software model retrieval, the modalities that can be envisioned are:

- **Keyword-based search:** The user provides a set of keywords. The system returns results ranked according to their relevance to the input keywords.
- **Document-based search:** The user provides a document (e.g., a specification of a new project). The system analyzes the document, extracts the most significant words and submits them as a query. Results are returned as before.
- **Search by example:** The user provides a model as a query. The model is analyzed in the same way as the projects in the repository, which produces a document to be used as a query[2]. The match is done between the query and the project document and results are ranked by similarity.
- **Faceted search:** The user can explore the repository using *facets* (i.e., property-value pairs) extracted from the indexed documents, or he can pose a query and then refine its results by applying restrictions based on the facets present in the result set.
- **Snippet visualization:** Each item in the result set can be associated with an informative visualization, where the matching points are highlighted in graphical or textual form.

The abovementioned functionalities can compose a complex query process, in which the user applies an initial query and subsequently navigates and/or refines the results in an exploratory fashion.

Use Cases and Experiments. Although several combinations could be assembled from the above dimensions, we evaluate two representative configurations, compared to a baseline one. As reported in Section 6, the following scenarios have been tested: Experiment A (baseline): keyword search on whole projects; Experiment B: retrieval of subprojects and concepts with a flat index structure; Experiment C: retrieval of subprojects and concepts with a weighted index structure. Experiment B and C represent two alternative ways of structuring the index, both viable for responding to designer's query targeted at relevant subproject retrieval and design pattern reuse. Table 1 summarizes the design options and their coverage in the evaluated scenarios. In this work we focus on *flat* and *weighted* index structures. *Mulit-field* and *Structured* indexes will addressed in the future work.

---

[2] Here the term *document* means by extension any representation of the model useful for matching, which can be a bag of words, a feature vector, a graph, and so on.

Table 1. Summary of the design options and their relevance in the experiments

Option	Description	A	B	C
Segmentation Granularity				
Project	entire project	X		
Subproject	subproject		X	X
Single Concept	arbitrary model concepts		X	X
Index structure				
Flat	flat lists of words	X		
Weighted	words weighted by the model concepts they belong to			X
Multi-field	words belonging to each model concept in separate fields			
Structured	XML representation reflecting hierarchies and associations			
Query language and result presentation				
Keyword-based	query by keywords	X	X	X
Document-based	query through a document			
By example	query through a model (content-based)			
Faceted	query refined through specific dimensions	X	X	X
Snippets	visualization and exploration of result previews	X	X	X

5   Implementation Experience

To verify our approach we developed a prototype system that, given a meta-model and a repository of models conforming to such meta-model: 1) configures a general purpose search engine according to selected dimensions, 2) exploits metamodel-aware extraction rules to analyze models and populate the index with information extracted from them; 3) provides a visual interface to perform queries and inspect results.

The experiments adopt WebML as a Domain Specific Language [8]. The WebML metamodel [21] specifies the constructs for expressing of the data, business logics, hypertext interface, and presentation of a Web application. In WebML, content objects are modeled using Entity-Relationship or UML class diagrams. Upon the same content model, it is possible to define different *application models* (called *site views*), targeted to different user roles or access devices. Figure 2 (a) depicts an excerpt of the WebML meta-model describing the siteview construct. A site view is internally structured into *areas*, which in turn may contain *pages*. Pages comprise *content units*, i.e., components for publishing content, and units are connected to each other through *links*, which carry parameters and allow the user to navigate the hypertext. WebML also allows specifying *operations* implementing arbitrary business logic (e.g., to manipulate content instances), and *Web service* invocation and publishing. Finally, the language comprises a notion of *module*, which denotes a reusable model pattern. Figure 2 (b) presents a sample Web application that implements a product catalog, where users can see a list of product and select one for getting more details. Figure 2 (c) shows the corresponding XML project file encoding the model.

5.1   Content Processing

Content processing has been implemented according to the schema of Figure 1. The *segmentation* and *text-extraction* steps are implemented in a generic and model-independent way; they are configurable by means of model transformation

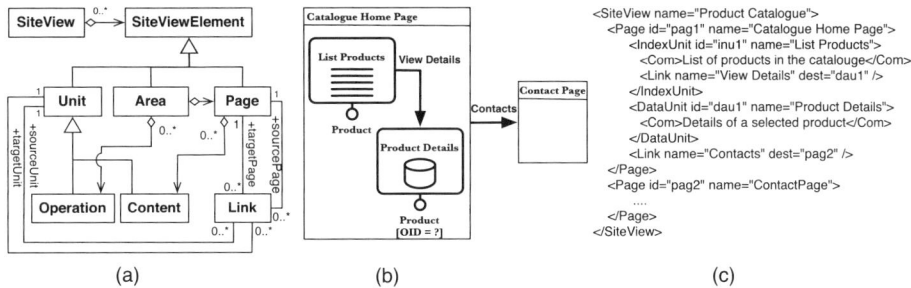

Fig. 2. a) Excerpt of the WebML metamodel. b) Example of WebML model. c) XML representation of the WebML model depicted in b).

rules encoded in XSLT, so to be adapted to the DSL of choice. These rules simply match each metamodel concept and decide what information to extract for populating the index. An auxiliary *Repository Analysis* component has been added to the architecture, which performs the offline analysis of the entire project collection to compute statistics for fine-tuning the retrieval and ranking performance: 1) a list of *Stop Domain Concepts*, i.e., words very common in the project repository (e.g., the name of meta-model concepts or terms that are part of the organization vocabulary and culture); 2) the *Weight* assigned to each model concept; presently, concepts weights are computed automatically based on the relative frequency of model concepts in the entire collection, but can be adjusted manually by the search engine administrator to promote or demote individual concepts.

Content processing has been implemented by extending the text processing and analysis components provided by Apache Lucene[3], an open-source search engine platform.

5.2   Index Structure and Ranking Function Design

Different index structures have been defined to cope with the evaluation scenarios: Experiment A and B have a flat index structure; Scenario C instead employs a weighed multi-field index structure, with the following fields: id | projectID | projectName | documentType | text, where the documentType field can have the values: DataModel, SiteView, Area, ServiceView, and Module, which represent the fundamental modularization constructs of WebML.

The relevance of the match is also computed differently in the scenarios. Experiment A uses the pure TF/IDF measure for textual match relevance [19], ignoring model structure; experiment B also uses traditional TF/IDF relevance, but due to segmentation the matching is performed separately on the different model concepts; finally, Experiment C exploits the model concepts also in the ranking function, which is defined as follows:

$$score(q, d) = \sum_{t \in q} \sqrt{tf(t, d)} \cdot idf(t)^2 \cdot mtw(m, t) \cdot dw(d) \tag{1}$$

---

[3] http://lucene.apache.org/java/docs/

where:

- $tf(t, d)$ is the *term frequency*, i.e., the number of times the term t appears in the document d;
- $idf(t)$ is the *inverse document frequency* of t, i.e., a value calculated as $1 + \log \frac{|D|}{freq(t,d)+1}$ that measures the informative potential carried by the term in the collection;
- $mtw(m, t)$ is the *Model Term Weight* of a term t, i.e., a metamodel-specific boosting value that depends on the concepts m containing the term t. For instance, in the running example, a designer can decide that the weight of terms t in a *module* should be double than all the other constructs. Then he will set $mtw(module, t)$ to 2.0 and mtw of all the others concepts to 1.0;
- $dw(d)$ is the *Document Weight*, i.e., a model-specific boosting value conveying the importance of a given project segment (i.e., project, sub-project, or concept, depending on the chosen segmentation policy) in the index.

The scoring function is implemented by extending the APIs of Lucene, which support both the Boolean and the Vector Space IR Models, thus enabling complex Boolean expressions on multiple fields.

5.3   Query and Result Presentation

Figure 3 shows the UI of the search system, implemented as a rich Web application based on the YUI Javascript library. The interface lets users express Boolean keyword queries, e.g., expressions like *"validate AND credit AND NOT card"*. Upon query submission, the interface provides a paginated list of matching items (A); each item is characterized by a a type (e.g., model elements like data model, area, module, etc.) and by the associated project metadata (e.g. project name, creation date, authors, etc.). The *Inspect Match* link opens a *Snippet* window (B) that shows all the query matches in the current project/subproject/construct, allowing users to determine which model fragments are useful to their information need. Users can also drill down in the result list by applying faceted navigation. The available facets are shown in the left-hand side of Figure 3. Each facet contains automatically extracted property values of the retrieved results. The selection of a facet value triggers a refinement of the current result set, which is restricted only to the entries associated to the facet value.

6   Evaluation

Evaluation has addressed the space and time efficiency of the system and the *designers' perception* of the results quality and of the usefulness of the proposed tool.

Experimental settings and dataset. A sample project repository has been provided by Web Models [1], the company that develops WebRatio, an MDD tool for WebML modeling and automatic generation of Web applications. The repository contains *48* large-size, real-word projects spanning several applications domains (e.g., trouble ticketing, human resource management, multimedia search engines, Web portals, etc.). Projects have been developed both in

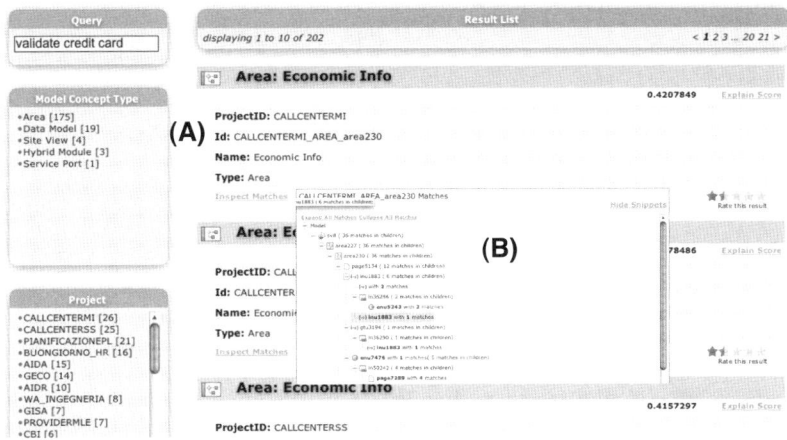

Fig. 3. Project repository search UI: result list(A) and result exploration(B)

Italian and English, and comprise about *250* different modeling concepts (resulting from WebML standard constructs and user-defined modeling elements), *3,800* data model entities (with about *35,000* attributes and *3,800* relationships), *138* site views with about *10,000* pages and *470,000* units, and *20* Web services. Each WebML project is encoded as an XML project file conforming with the WebML DTD, and the overall repository takes around 85MB of disk space. The experiments were conducted on a machine equipped with a 2GHz Dual-Core AMD Opteron and 2GB of RAM. We used Apache Solr 1.4 (`http://lucene.apache.org/solr/`) as Lucene-based search engine framework.

Performance Evaluation. Performance experiments measured index size and query response time, varying the number of indexed projects. Two scenarios were used: the baseline, in which whole projects are indexed as documents with no segmentation, and a segmentation scenario in which selected model concepts are indexed as separate documents. For both scenarios, three alternative configurations have been evaluated (keyword search, faceted search, and snippet browsing) so to exercise all *query language and result presentation* options reported in Table 1.

Figure 4 (a) shows the results of evaluating index size. For each of the six system configurations, all project in the repository are progressively indexed, disabling any compression mechanism in the search engine index. Size grows almost linearly with the number of projects in all configurations, thus showing good scalability. As expected, the basic *keyword search* scenario has the least space requirements in both the non-segmented and segmented configurations (about 10 times smaller than the repository size)[4]. The addition of *Faceted Search*, doubles the index size, because the original text of facet properties must

---
[4] Higher values at the beginning of the curves can be explained by low efficiency with small index sizes.

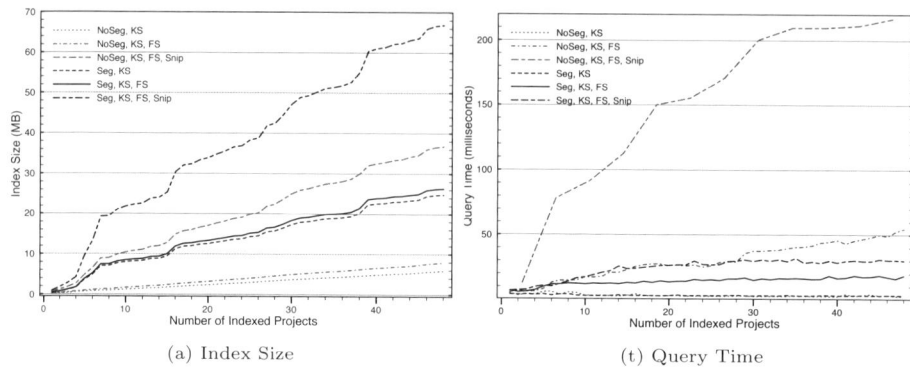

(a) Index Size   (t) Query Time

Fig. 4. Index size (a) and response time (b) of six system configurations (*NoSeg*: No Segmentation, *Seg*: Segmentation, *KS*: Keyword Search, *FS*: Faceted Search, *Snip*: Snippet)

be stored in the index to enable its visualization in the UI. *Snippet Visualization* is the most expensive option, especially when models are segmented and thus the number of indexed documents grows.

Figure 4 (b) shows the query response time with a varying number of indexed projects. We selected about 400 2-terms and 3-terms keyword queries, randomly generated starting from the most informative terms indexed within the repository. Each query has been executed 20 times and execution times have been averaged. Under every configuration, query time is abundantly sub-second, and curves indicate a sub-linear growth with the number of indexed projects, thus showing good scalability. Notice that the addition of *Faceted Search* and *Snippet Visualization* impacts on performances also for query latency time; differently from the previous evaluation, though, the most affected scenario is the one where no project segmentation is applied, thus introducing a penalty caused by the greater amount of information that needs to be retrieved for each query result item.

Preliminary User Evaluation. A preliminary user study has been conducted with 5 expert WebML designers to assess 1) the perceived retrieval quality of alternative configurations and 2) the usability and ease-of-use of the system.

The perceived retrieval quality has been tested with the three system configurations listed in Table 1, using a a set of ten queries manually crafted by the company managers responsible of the Web applications present in the repository. Designers could access the three system configurations and vote the appropriateness of each one of the top-10 items in the result set; they were asked to take into account both the element relevance with respect to the query and its rank in the result set. Votes ranged from 1 (highly inappropriate) to 5 (highly appropriate). Figure 5 shows the distribution of the 500 votes assigned to each system configuration. Experiment B and C got more votes in high (4-5) range

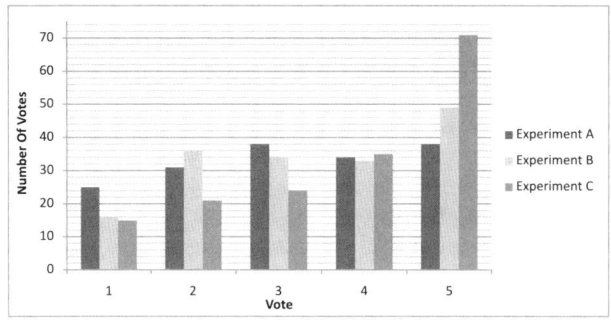

Fig. 5. Average vote distribution for the test queries

Table 2. User Evaluation: questionnaire items, average rates and variance

Item	Avg.	Var.
Features		
Keyword Search	3.6	0.24
Search Result Ranking	3.2	0.16
Faceted Search	3.8	0.16
Match Highlighting	3.6	0.24
Application		
Help reducing maintenance costs?	3.2	0.56
Help improving the quality of the delivered applications?	3.0	0.4
Help understanding the model assets in the company?	4.4	0.24
Help providing better estimates for future application costs?	2.8	0.56
Wrap-up		
Overall evaluation of the system	4.0	0.4
Would you use the system in your activities?	3.0	1.2

of the scale with respect to the baseline. Also, results were considered better in configuration C, which assigned weights to the terms based on the model concept they belonged to. The penalty of Configuration A can be explained by the lower number of the indexed documents (48), which may brings up in the top ten more irrelevant projects, having a relatively low score match with the query.

The study also included a questionnaire to collect the user opinion on various aspects of the system. Table 2 reports the questions and the collected results, in a 1 to 5 range. The results show that the system, although very prototypical, has been deemed quite useful for model maintenance and reuse, while its direct role in improving the quality of the produced applications is not perceivable. One could notice a certain distance between the overall judged quality and the likelihood of adoption, which is possibly due to three aspects: 1) such a tool would probably be more valuable in case of a very large corpus of projects; 2) UI usability is greatly hampered by the impossibility at present of displaying the matches in the WebRatio diagram editor; and 3) developers are not familiar with search over a model repository and would probably need some experience before accepting it in their everyday work life.

While our evaluation is insufficient to entail statistical relevance, we believe that the proposed user study suffices to support the motivations of our work and provides useful feedbacks to drive future research efforts.

7   Conclusions

In this paper we presented an approach and a system prototype for searches over model repositories. We analyzed the typical usage scenarios and requirements and we validated our claims with an implementation over a repository of Web application models specified in WebML. The experimental evidence show that the system scales well both in size and query response time; a preliminary user evaluation showed that the ranking of results based on a priori knowledge on the metamodel elements is gives better results with respect to the baseline solution of flat indexing of the text content of a project. Ongoing and future work includes the implementation of content-based search, the integration of the search interface in the diagram editing GUI of WebRatio, for visually highlighting the matches in the projects, the capture of the user feedback on the top-ranking results, to automatically learn how to fine-tune the model weights and improve precision and recall, a larger scale study on the scalability and effectiveness of the retrieval system, not limited to WebML models but also to UML artifacts, and the definition of benchmark criteria for model-driven repository search.

References

1. Acerbis, R., Bongio, A., Brambilla, M., Butti, S.: Webratio 5: An eclipse-based case tool for engineering web applications. In: Baresi, L., Fraternali, P., Houben, G.-J. (eds.) ICWE 2007. LNCS, vol. 4607, pp. 501–505. Springer, Heidelberg (2007)
2. Antoniol, G., Canfora, G., de Lucia, A., Casazza, G.: Information retrieval models for recovering traceability links between code and documentation. In: IEEE International Conference on Software Maintenance, p. 40 (2000)
3. Awad, A., Polyvyanyy, A., Weske, M.: Semantic querying of business process models. In: Enterprise Distributed Object Computing Conference (EDOC), pp. 85–94 (2008)
4. Bajracharya, S., Ossher, J., Lopes, C.: Sourcerer: An internet-scale software repository. In: ICSE Workshop on Search-Driven Development-Users, Infrastructure, Tools and Evaluation. SUITE '09, pp. 1–4 (May 2009)
5. Beeri, C., Eyal, A., Kamenkovich, S., Milo, T.: Querying business processes. In: VLDB, pp. 343–354. ACM, New York (2006)
6. Belhajjame, K., Brambilla, M.: Ontology-based description and discovery of business processes. In: Interval Mathematics. LNBIP, vol. 29. Springer, Heidelberg (2009)
7. Ben Khalifa, H., Khayati, O., Ghezala, H.: A behavioral and structural components retrieval technique for software reuse. In: Advanced Software Engineering and Its Applications. ASEA 2008, pp. 134–137 (December 2008)
8. Ceri, S., Fraternali, P., Bongio, A., Brambilla, M., Comai, S., Matera, M.: Designing Data-Intensive Web Applications. Morgan Kaufmann Publishers Inc., San Francisco (2002)

9. Chen, K., Madhavan, J., Halevy, A.: Exploring schema repositories with schemr. In: SIGMOD '09: Proc. of the 35th SIGMOD Int. Conf. on Management of data, New York, NY, USA, pp. 1095–1098. ACM, New York (2009)
10. Frakes, W.B., Nejmeh, B.A.: Software reuse through information retrieval. SIGIR Forum 21(1-2), 30–36 (1987)
11. Gibb, F., McCartan, C., O'Donnell, R., Sweeney, N., Leon, R.: The integration of information retrieval techniques within a software reuse environment. Journal of Information Science 26(4), 211–226 (2000)
12. Goderis, A., Li, P., Goble, C.A.: Workflow discovery: the problem, a case study from e-science and a graph-based solution. In: ICWS, pp. 312–319. IEEE Computer Society, Los Alamitos (2006)
13. Gomes, P., Pereira, F.C., Paiva, P., Seco, N., Carreiro, P., Ferreira, J.L., BentoI, C.: Using wordnet for case-based retrieval of uml models. AI Communications 17(1), 13–23 (2004)
14. Holmes, R., Murphy, G.C.: Using structural context to recommend source code examples. In: ICSE '05: Proceedings of the 27th international conference on Software engineering, pp. 117–125. ACM, New York (2005)
15. Inoue, K., Yokomori, R., Yamamoto, T., Matsushita, M., Kusumoto, S.: Ranking significance of software components based on use relations. IEEE Transactions on Software Engineering 31(3), 213–225 (2005)
16. Kiefer, C., Bernstein, A., Lee, H.J., Klein, M., Stocker, M.: Semantic process retrieval with iSPARQL. In: Franconi, E., Kifer, M., May, W. (eds.) ESWC 2007. LNCS, vol. 4519, pp. 609–623. Springer, Heidelberg (2007)
17. Llorens, J., Fuentes, J.M., Morato, J.: Uml retrieval and reuse using xmi. In: IASTED Software Engineering. Acta Press (2004)
18. Lu, R., Sadiq, S.: Managing process variants as an information resource. In: Dustdar, S., Fiadeiro, J.L., Sheth, A.P. (eds.) BPM 2006. LNCS, vol. 4102, pp. 426–431. Springer, Heidelberg (2006)
19. Manning, C.D., Raghavan, P., Schütze, H.: Introduction to Information Retrieval. Cambridge University Press, Cambridge (July 2008)
20. Markovic, I., Pereira, A.C., Stojanovic, N.: A framework for querying in business process modelling. In: Multikonferenz Wirtschaftsinformatik (February 2008)
21. Moreno, N., Fraternali, P., Vallecillo, A.: Webml modelling in uml. IET Software 1(3), 67–80 (2007)
22. Platzer, C., Dustdar, S.: A vector space search engine forweb services. In: ECOWS '05: Proceedings of the Third European Conference on Web Services, Washington, DC, USA, pp. 62–71. IEEE Computer Society, Los Alamitos (2005)
23. Seacord, R.C., Hissam, S.A., Wallnau, K.C.: Agora: A search engine for software components. IEEE Internet Computing 2(6), 62–70 (1998)
24. Settimi, R., Cleland-Huang, J., Ben Khadra, O., Mody, J., Lukasik, W., DePalma, C.: Supporting software evolution through dynamically retrieving traces to uml artifacts. In: Proceedings of 7th International Workshop on Principles of Software Evolution, pp. 49–54 (2004)
25. Shao, Q., Sun, P., Chen, Y.: Wise: A workflow information search engine. In: IEEE 25th International Conference on Data Engineering. ICDE '09, pp. 1491–1494 (2009)
26. Zhuge, H.: A process matching approach for flexible workflow process reuse. Information & Software Technology 44(8), 445–450 (2002)

Toward Approximate GML Retrieval Based on Structural and Semantic Characteristics

Joe Tekli[1], Richard Chbeir[1], Fernando Ferri[2], and Patrizia Grifoni[2]

[1] LE2I Laboratory UMR-CNRS, University of Bourgogne
21078 Dijon Cedex France
{joe.tekli,richard.chbeir}@u-bourgogne.fr
[2] IRPPS-CNR, via Nizza 128, 00198 Roma, Italy
{fernando.ferri,patrizia.grifoni}@irpps.cnr.it

Abstract. GML is emerging as the new standard for representing geographic information in GISs on the Web, allowing the encoding of structurally and semantically rich geographic data in self describing XML-based geographic entities. In this study, we address the problem of approximate querying and ranked results for GML data and provide a method for GML query evaluation. Our method consists of two main contributions. First, we propose a tree model for representing GML queries and data collections. Then, we introduce a GML retrieval method based on the concept of tree edit distance as an efficient means for comparing semi-structured data. Our approach allows the evaluation of both structural and semantic similarities in GML data, enabling the user to tune the querying process according to her needs. The user can also choose to perform either *template* querying, taking into account all elements in the query and data trees, or *minimal constraint* querying, considering only those elements required by the query (disregarding additional data elements), in the similarity evaluation process. An experimental prototype was implemented to test and validate our method. Results are promising.

Keywords: GML Search, Ranked Retrieval, Structural & Semantic Similarity, GIS.

1 Introduction

In recent times, the amount of spatial data, available in standalone as well as web-based Geographic Information Systems (GISs), is becoming huge and accessible to users who are generally non-experts. Most of the time, such users query data without a deep knowledge about the spatial domain they want to query, or they may not know how to formulate meaningful queries, resulting in a reduction of the quality of the query results. In order to overcome such limitations, the introduction of some query relaxation mechanisms, by which approximate and ranked answers are returned to the user, represents a possible solution. The need of answers that approximately match the query specified by the user requires the evaluation of similarity.

Another important new trend in GISs is the adoption of XML-based formats, particularly GML (Geography Mark-up Language) [18] as the main standard for

B. Benatallah et al. (Eds.): ICWE 2010, LNCS 6189, pp. 16–34, 2010.

exchanging geographic data and making them available on the Web. This language is based on W3C's XML (eXtensible Mark-up Language) encoding, as an efficient and widely accepted means for (semi-structured) data representation and exchange. In fact, a geographic entity in GML, consists of a hierarchically structured self-describing piece of geographic information, made of atomic and complex features (i.e., containing other features) as well as atomic attributes, thus incorporating structure and semantically rich data in one entity. Hence, the problem of evaluating GML similarity in order to perform approximate querying, can be reduced to that of performing XML-based search and retrieval, considering the nature and properties of geographic data and data requests.

A wide range of algorithms for comparing semi-structured data, e.g., XML-based documents, have been proposed in the literature. These vary w.r.t. the kinds of XML data they consider, as well as the kinds of applications they perform. On one hand, most of them make use of techniques for finding the edit distance between tree structures [3, 17, 26], XML documents being modeled as Ordered Labeled Trees (OLT). On the other hand, some works have focused on extending conventional information retrieval methods, e.g., [1, 6, 8], so as to provide efficient XML similarity assessment. In this study, we focus on the former group of methods, i.e., edit distance based approaches, since they target rigorously structured XML documents (i.e., documents made of strictly tagged information, which is the case of GML data, cf. Section 3) and are usually more fine-grained (exploited in XML structural querying [23], in comparison with *content-only* querying in conventional *IR* [22]). Note that information retrieval based methods target loosely structured XML data (i.e., including lots of free text) and are usually coarse-grained (useful for fast simple XML querying, e.g., keyword-based retrieval [8]).

Nonetheless, in addition to quantifying the structural similarities of GML features, semantic similarity evaluation is becoming increasingly relevant in geospatial data retrieval as it supports the identification of entities that are conceptually close, but not exactly identical. Identifying semantic similarity becomes crucial in settings such as (geospatial) heterogeneous databases, particularly on the Web where users have different backgrounds and no precise definitions about the matter of discourse [21]. Thus, finding semantically related GML modeled items, and given a set of items, effectively ranking them according to their semantic similarity (as with Web document retrieval [13]), would help improve GML search results.

In this study, we present the building blocks for a GML retrieval framework, evaluating both *structural* and *semantic* similarities in GML data, so as to produce approximate and ranked results. Our query formalism is based on approximate tree matching as a simple and efficient technique to query GML objects. It allows the formulation of *structure-and-content* queries with only partial knowledge of the data collection structure and semantics. In addition, our method allows both *template* and *minimum constraint* querying. According to the latter interpretation, the GML answer entity could contain additional elements w.r.t. those required by the query, such elements being disregarded in similarity evaluation. Yet, following the former strategy, all query and data elements are equally considered. The user can also tune the GML similarity evaluation process, by assigning more importance to either structural or semantic similarity, using an input structural/semantic parameter.

The remainder of the paper is organized as follows. Section 2 briefly reviews the state of the art in GML search methods and related XML similarity issues. Section 3 discusses the background and motivations of our study. In Section 4, we develop our GML approximate query evaluation approach. Section 5 presents our preliminary experimental tests. Section 6 concludes the paper and outlines future research directions.

2   State of the Art in GML and XML Retrieval, and Related Issues

Conventional geographic information science and retrieval have been concerned with managing and searching digital maps where geometry plays a major role (e.g., spatial browsing, exact querying based on geographic coordinates, …) [11]. Nonetheless, little support has been provided for managing geographic information based on text, in which references to locations are primarily by means of place names and textual descriptions, in addition to the geospatial data itself [10]. In this context, very few approaches have been proposed for GML-based geographic data search and retrieval in particular.

The few existing methods for managing and querying GML-based geographic information have tried to map GML data to classic spatial DBMS (e.g., Oracle Spatial, DB2 Spatial, PostGIS, …), e.g., [19, 27, 31]. This is connected with the genesis of GML, which was born as an interchange format for heterogeneous geographic database systems. Such methods usually underline the semi-automatic mapping of the GML application schema (describing the geographic data) to a bunch of object/relational schemas. They extend XML data storage in traditional DBMS to consider geospatial properties of GML (e.g., adding dedicated structures for storing geographic coordinates). Having mapped the GML data into object/relational DB structures, corresponding geographic data can be hence processed for classic DB querying. While such techniques might be efficient w.r.t. storage and indexing, they are limited to exact querying and retrieval functions, and do not allow approximate and ranked search results.

On the other hand, approximate querying methods developed for XML, e.g., [1, 6, 7, 14], cannot be straightforwardly applied to GML. First, such methods would have to consider the semantic richness of text-based geographic data in order to perform relevant GML querying. Second, most of these methods are based on underlying IR-concepts (most address the INEX evaluation campaigns) and target loosely structured XML (including lots of free text). They can be generally criticized for not sufficiently considering the structural properties of XML [20] (and consequently GML, which usually underlines rigorously structured data, due to the structured nature of geographic information).

Some methods have tackled the problem of searching rigorously structured XML data, by exploiting the concept of approximate tree matching [9, 23, 24]. In [23], the author propose an approach based on tree edit distance for evaluating the similarity between XML query and data trees. Similarity is evaluated in terms of the total cost needed to transform the query tree into one embedded in the data tree, and is used to rank the results. In a subsequent study [24], the authors propose to combine approximate tree embedding with TF-IDF ranking in evaluating query and data tree similarity. The classical notion of term (as a piece of text) is extended to structural term (as a sub-tree). Methods in [23, 24] only focus on the structural features of XML, disregarding semantic similarity. A method based on a similar approximate tree embedding technique is provided in [9] for querying MPEG-7 XML documents. Here, the authors introduce semantic similarity assessment between query and data tree labels, based on a dedicated knowledge base describing MPEG-7 concepts. Yet, the approach does not produce ranked results (it returns a Boolean value indicating whether the query tree is embedded or not in the data tree).

3 Background and Motivation

3.1 A Glimpse on GML

The Geography Markup Language (GML) [18] is an XML encoding for the transport and storage of geographic information, where real world entities can be represented as sets of GML features. Geometric features are those with properties that may be valued geometrically (e.g., types *Point*, *Line*, *Polygon*…, with geometric coordinates designating their positions, extents and coverage). Remaining non-geometric features provide textual descriptions of the geographic entity at hand. For instance, to model a *Hotel* in GML, one would define non-geometric features, such as *Name (Text)*, *Rank (Number)*, *Address (Text)*, … and geometric ones, e.g., *Location (Point)*, *Area (Polygon)*, …

Features/attributes and corresponding types, in a given GML modeled entity, are defined via the GML application schema to which the document, containing the GML entity model, conforms. The GML schema defines the features and attributes of the GML documents they describe, as well as their structural dispositions and the rules they adhere to in the documents. Similarly to schemas in traditional DBMS, GML schemas are valuable for the protection, indexing, retrieval and exchange of corresponding documents [18]. Figure 1 shows a sample GML document and part of its corresponding GML schema.

```
<?xml version="1.0">                                      <?xml version="1.0">
<xmlns:gml="http://www.opengis.net/gml" City.xsd ...>     <xs:schema xmlns:gml="http://www.opengis.net/gml" ...>
<City name= "Rome">                                        <element name="City" type=="CityType"/>
<ArtisticGuide>                                            <complexType name="CityType">
  <Monuments>                                              <sequence> <element name = "ArtisticGuide" type ="ArtisticGuideType"/>...
    <Cathedral name= "St Peter">                               </sequence>
      <Style>Renaissance</Style>                            <attribute name="Name" type="String">
      <Location>                                           </complexType>
        <Point>                                            <complexType name="ArtisticGuideType">
          <Coordinates>                                      <sequence> <element name = "Monuments" type=="MonumentsType"/>
            <Latitude>                                     </sequence>
              <Degrees>41</Degrees>                        </complexType>
              <Minutes>52</Minutes>                        <complexType name="MonumentsType">
            </Latitude>                                      <sequence> <element name="Cathedral" type="CathedralType"/> ...
              ...                                            </sequence>
          </Coordinates>                                   </complexType>
        </Point>                                           <complexType name="CathedralType">
      </Location>                                            <sequence>
      ...                                                     <element name="Style" type="String">
    </Cathedral>                                             <element name="Location" type="LocationType"> ...
  </Monuments>                                              </sequence>
</ArtisticGuide>                                           <attribute name="Name" type="string"/>
</City>                                                    </complexType> ...
```

Fig. 1. Sample GML document and part of the corresponding GML application schema

3.2 Querying GML Data

In order to allow efficient approximate and ranked GML querying on the Web, we underline the need for a technique to search GML data where users can express queries in the simplest form possible, taking into account the structured nature of GML, in a way that less control is given to the user and more of the logic is put in the ranking mechanism to best match the user's needs. In other words, we aim to simplify, as much as possible, the query model and predicates (developed in the following section) without however undermining query expressiveness. In this context, we distinguish between two different kinds of GML queries, i) *template* where the user specifies a sample snapshot of the GML data she is searching for (e.g., a piece of map, providing a somewhat complete description of the requested data), or ii) *minimal constraint* where the user only identifies the minimal requirements the data should meet in order to belong to the query answer set (e.g., providing a small description, or an approximate location to pinpoint a given geographic object). For instance:

- Q_1: *"Find Churches built prior to 1600".*
- Q_2: *"Find Cities containing gothic churches".*
- Q_3: *"Pinpoint Locations of churches in the city of Rome".*
- Q_4: *"Find all restaurants situated at latitude 41 degrees 55 minutes North, and longitude 12 degrees 28 minutes East".*

While queries Q_{1-3} are solely user-based, queries combining minimal constraint user preferences and geo-coordinates could be equally relevant. Consider for instance query Q_4 where the user searches, via her mobile GPS device, for restaurants in the vicinity of her current location. Such queries could also be viewed as of partial *Template* style, due to the presence of rather detailed geographic information provided by the GPS device.

Note that in this study, we do not aim to define a GML querying language (i.e., a formal syntax following which a GML query should be written), but rather the underlying querying framework. The user could formulate the query using plain text (guided by a dedicated GUI), or via some predefined syntax (e.g., a GML document fragment, or a pictorial representation converted into GML [5]) to be represented in our query model.

In addition, we emphasize on the need to consider the semantic meaning of GML entity descriptions and corresponding textual values, as a crucial requirement to perform approximate GML querying. For instance, a user searching for *cities* with *cathedrals* of *Gothic* style (query Q_1), would naturally expect to get as answers *cities*, *counties* or *towns* containing either *basilicas*, *cathedrals*, *churches* or *temples* which are of either *Gothic*, *Medieval* or *Pre-renaissance* styles, ranked following their degrees of semantic relevance to the original user request. The impact of semantic similarity on approximate GML querying is further discussed in the following section.

4   Proposal

4.1   GML Data and Query Models

As shown above, geographic entities in GML represent hierarchically structured (XML-based) information and can be modeled as Ordered Labeled Trees (OLTs)

[28]. In our study, each GML document is represented as an OLT with nodes corresponding to each subsumed feature and attribute. Feature nodes are labeled with corresponding element tag names. Feature values (contents) are mapped to leaf nodes (which parent nodes are those corresponding to their features' tag names) labeled with the respective values. To simplify our model, attributes are simply modeled as atomic features, corresponding nodes appearing as children of their encompassing feature nodes, sorted by attribute name, and appearing before all sub-element siblings.

Values could be of different types (*text*, *number*…), and user derived types could also be defined [18]. In the following, and for simplicity of presentation, we consider the basic *text*, *number* and *date* types in our discussion (from which derive most data-types, including geometric ones, e.g., *point*, *polygon*…). Note that our GML tree model itself, and the GML querying approach as a whole, are not bound to the types above, and could consider any other data-type, as we will show subsequently.

Definition 1 - GML Tree: Formally, we represent a GML document as a rooted ordered labeled tree $G = (N_G, E_G, L_G, T_G, g_G)$ where N_G is the set of nodes in G, $E_G \subseteq N_G \times N_G$ is the set of edges (feature/attribute containment relations), L_G is the set of labels corresponding to the nodes of G ($L_G = Fl_G \cup Fv_G \cup Al_G \cup Av_G$ such as Fl_G (Al_G) and Fv_G (Av_G) designate respectively the labels and values of the features (attributes) of G), T_G is the set of data-types associated to the feature and attribute nodes of G ($T_G = \{GeoEntity\} \cup FT \cup AT$, having $FT = AT = \{Text, Number, Date\}$), and g_G is a function $g_G : N_G \rightarrow L_G$, T_G that associates a label $l \in L_G$ and a data-type $t \in T_G$ to each node $n \in N_G$. We denote by $R(G)$ the root node of G, and by $G' \lhd G$ a sub-tree of G   •

Definition 2 - GML Tree Node: A node n of GML tree $G = (N_G, E_G, L_G, T_G, g_G)$ is represented by a doublet $n = (l, t)$ where $l \in L_G$ and $t \in T_G$ are respectively its label and node data-type. The constituents of node n are referred to as $n.l$ and $n.t$ respectively (Figure 2)   •

n.l	n.t

Fig. 2. Graphical representation of GML tree node n

Value data-types in our GML tree model are extracted from the corresponding GML schema. In other words, in GML tree construction time, the GML document and corresponding schema are assessed simultaneously so as to build the GML tree. Textual values are treated for stemming and stop word removal, and are mapped to leaf nodes of type *Text* in the GML tree. Numerical and date values are mapped to leaf nodes of types *Number* and *Date* respectively. As for the GML feature/attribute nodes themselves, they are assigned the data-type *GeoEntity*, their labels corresponding to the geographical entity names defined in the corresponding GML schema. To model the GML data repository, we connect all GML trees to a single root node, with a unique label (e.g., *'Root'*).

Consider for instance the GML data repository in Figure 3. It is made of two GML document trees describing *City* geographic entities (cf. extracts of GML document and schema in Figure 1). Geometric coordinates are depicted for the geographic entity

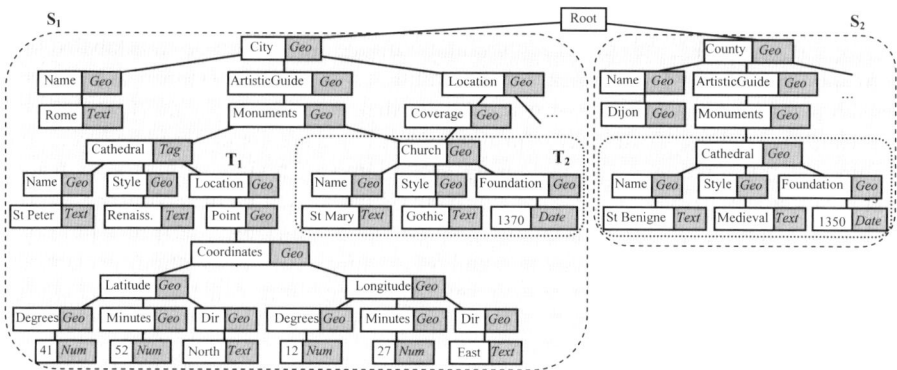

Fig. 3. Extract of a sample GML data repository (*Geo* stands for *GeoEntity*)

describing *St Peter cathedral* in *Rome* (latitude and longitude coordinates), and are omitted for remaining GML entities for clarity of presentation. Recall that most geographic data-types can be expressed in terms of basic types *Text*, *Number* and *Date*, which is the case of element *Point* (of derived GML *PointType*).

On the other hand, our definition of a GML query is simple and consists of a GML tree, similarly to GML documents, with special leaf nodes to represent query predicates. A query with an *Or* logical operator is decomposed into a *disjunctive normal form* [23], and is thus represented as a set of GML trees, corresponding to the set of conjunctive queries.

Definition 3 - GML Query: It is expressed as a GML tree, $Q = (N_Q, E_Q, L_Q, T_Q, g_Q, n_d)$ (cf. Definition 1) encompassing a *distinguished* node n_d underlining the matches in the data tree that are required as answers to the query (i.e., the query's return clause). The query's root node $R(Q)$ designates its search scope/context. Its set T_Q encompasses the *GeoEntity* type for distinguishing GML geographic entities, and predicate types P_t_i corresponding to every GML value data-type t_i considered in the GML data model (e.g., $T_Q = \{GeoEntity\} \cup \{P\_Text, P\_Number, P\_Date\}$)   ●

Definition 4 - GML Query Node: It is a GML tree node (cf. Definition 2) with additional properties to represent predicates. With $n.t = P\_t_i$ (predicate corresponding to GML data-type t_i), the node's label $n.l$ underlines a composite content made of the predicate operator $n.l.op$ and value $n.l.val$ (e.g., leaf node $Q_1[2]$ of query Q_1 in Figure 4 is of $Q_1[2].l.op = $ '<' and $Q_1[2].l.val = $ '*1600*', having $Q_1[2].t = P\_Date$, which underlines that the predicate value '*1600*' is of type *Date*)   ●

Note that each data-type has its own set of operators (e.g., $\{=, <, \leq, >, \geq\}^1$ for numbers and dates, and $\{=, like, ...\}$ for text). GML query trees, corresponding to the sample queries provided in Section 3.2, are depicted in Figure 4. Recall that query trees can be constructed via a dedicated GUI, which would suggest, on-the-fly, the list of possible query nodes following the context of the query at hand.

---

[1] The *difference* operator ($\neq$) is omitted due to its particular processing (to be addressed in an future study).

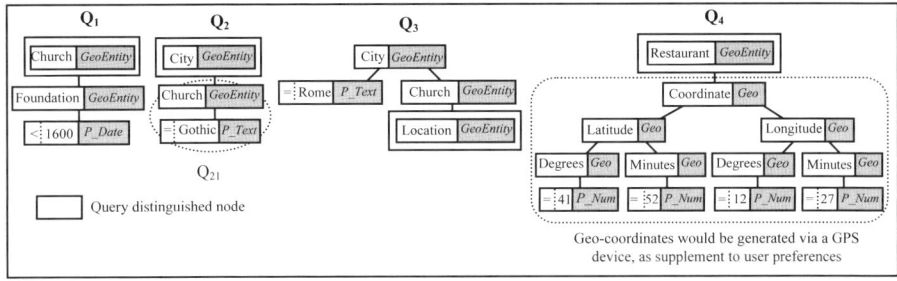

Fig. 4. Sample GML query trees

To the exception of the *containment* topological operator implicitly encoded in the GML hierarchy itself (cf. queries Q_2 and Q_3), we do not consider explicit spatial and temporal operators (e.g., *far*, *near*, *adjacent to*, …) in our current approach. These underline composite computational operations (e.g., a location point is *near* another location point if their distance, computed based on their coordinates, is below a certain threshold) and would induce more complex GML document and query graph structures instead of simple trees (introducing different kinds of cross links connecting GML entities). Recall that our current study sets the foundations toward approximate GML querying, to be consequently extended in addressing spatio-temporal relations.

Definition 5 – Predicate Satisfaction: Given a predicate GML query node q_i, and a data node s_j, s_j satisfies q_i ($s_j \models q_i$) if:

- The data node type corresponds to that of the query ($s_i.t \approx q_i.t$, i.e., $\forall\, t_r \in \{Text, Number, Date\}$, $q_i.t = P\_t_r \wedge s_j.t = t_r$),
- The data node label $s_j.l$ verifies the logical condition defined by $q_i.l$ ●

For instance, leaf node $T_2[6]$ of data tree T_2 in Figure 3, having $T_2[6].l = '1350'$ and $T_2[6].t = 'Date'$, satisfies predicate node $Q_1[2]$ of query Q_1 in Figure 4, with $Q_1[2].l = '<1600'$ and $Q_1[2].t = 'P\_Date'$.

Definition 6 - GML Query Scope: Given a GML query Q, the scope of Q is identified by its root node $R(Q)$, and corresponds to the GML data sub-trees, in the data repository, having identical or semantically similar enough root nodes as that of the query. ●

We assume that the user defines, with the query, the kind of GML data she is looking for, i.e. the scope/context of her query. If for instance the root of the query is labeled *Restaurant*, then GML data in the context of GML data entity *Restaurant*, or semantically similar entities such as *Pizzeria*, *Bar*, … would naturally interest the user.

Definition 7 - *Template* and *Minimal constraint* querying: A GML query Q could be either evaluated as a i) *template* of the GML data the user is searching for, ii) or could represent the *minimal constraints* the data should meet to belong to the query answer set. In the former case, all GML query and data nodes would be considered in query/data similarity evaluation. Following the latter strategy, only elements required by the query tree are taken into account in query/data similarity evaluation, additional elements in the data tree being disregarded in the evaluation process.   •

Note that geographic queries most likely follow the *minimal constraint* style, the user usually specifying her information needs in the simplest form possible (cf. queries Q_1, Q_2 and Q_3 in Figure 4). Nonetheless, *template* querying could be particularly useful in *search-by-document* and *search-by-image* systems for instance, where the query could be a whole geographic document or a piece of map the user is searching for in the geographic repository. A *template* style query could be any of the sub-trees S_1, S_2, T_1... in Figure 3.

4.2 GML Query Evaluation

The goal of this work is to develop a method for searching a GML data repository in order to identify portions of data that exactly or approximately match user requests. Having modeled both GML data and queries as trees, GML query evaluation can be reduced to the problem of searching the various data sub-trees, in the data repository, corresponding to the query's search scope (i.e., with matching root nodes), identifying those that share structural and semantic similarities with the query tree. The result of the query would be a set of data nodes matching the query's distinguished node, ranked by the similarity degree between the query tree and corresponding data candidate answer sub-tree. Thus, we propose a GML querying framework based on the concept of tree edit distance as a widely known and efficient means for comparing XML-based tree structures [3, 4, 17]. In addition to evaluating GML data structure, our method also integrates semantic similarity assessment [12, 13], so as to capture the semantic meaning of GML element labels/values.

   A simple motivating example, underlining the need to consider semantic similarity in GML querying, is that of evaluating query Q_2 of Figure 4, against the GML data repository G in Figure 3. Using structural-only similarity evaluation, one can realize that the only GML data tree to (actually) fulfill the data request of Q_2 (searching for *Cities* containing *Cathedrals* of *Gothic Style*) is S_1 (describing the *City* of *Rome*, containing data tree T_2 describing the *St. Mary Church* which is of *Gothic* style). That is due to the structural similarity between sub-tree Q_{21} and T_2. However, one can recognize that data tree S_1 (describing the *County* of *Dijon*) also fulfills the data request of query Q_2, since it contains tree T_3 (describing the *St. Benigne Cathedral* which is of *Medieval* style). This answer goes undetected using structure-only similarity evaluation, since the semantic similarity between *City/County*, *Church/Cathedral* and *Gothic/Medieval* are missed.

   In summary, our GML querying method consists of three main components: i) *CAT Identification* component for identifying GML data Candidate Answer Trees (following the query's scope), ii) *GML Tree Comparison* component for evaluating the structural and semantic similarity between the query tree and each of the candidate answer trees, iii) and the *Query Answer Identification* component for recognizing the GML data elements, in each data candidate answer tree, to be returned to the user (following the query's distinguished node). The overall system architecture is depicted in Figure 5.

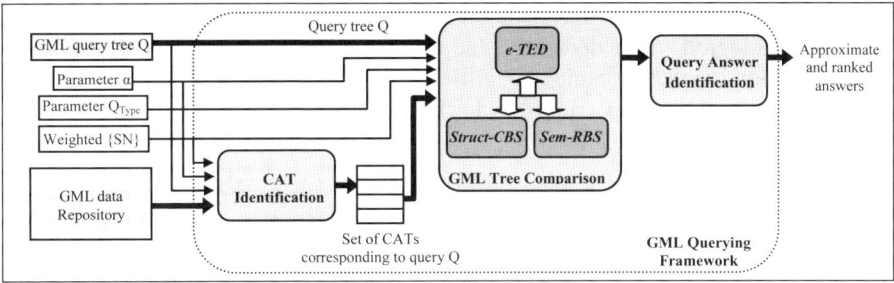

Fig. 5. Simplified activity diagram of our GML querying approach

4.2.1 GML Candidate Data Tree Identification

The first step in assessing a query is to identify its search scope. Following the traditional IR logic, whole physical files are considered as candidate answers. Nonetheless, GML documents differ in their granularity: some documents may contain information about *monuments*, others about *cities* containing hundreds of *monuments* (cf. Figure 3).

Obviously, it is not relevant to retrieve the entire *city* when the user is searching for certain *monuments*. Hence, the GML query search scope should be identified dynamically, w.r.t. the query at hand.

Following our GML data and query model, the query scope (cf. Definition 6) can be identified as the set of GML data sub-trees (which we identify as Candidate Answer Trees, *CATs*), in the data repository, having identical, or semantically similar enough, root nodes as that of the query (i.e., same/similar label, with the same datatype). Consider for instance query Q_1, searching for *churches* that have certain characteristics. When considering root node identity, query Q_1's *CATs* would be all data sub-trees having root node label *Church*, i.e., data tree T_2. When taking into account semantic similarity, Q_1's *CATs* would also encompass T_1 and T_3 of root nodes *Cathedral*. With queries Q_2 and Q_3, answer candidates would be data sub-tree S_1 (of root node *City*) when considering node identity, and would include S_2 (of root node *County*) when considering semantic similarity.

Definition 8. Candidate Answer Tree: Given a GML node similarity measure Sim_{GML}, reference semantic networks $\{SN\} = \{SN_{Geo}, SN_{Text}\}$ for evaluating the semantic similarity between GML *GeoEntity* and *Text* node labels, and a semantic similarity threshold α, the set of candidate answer trees Q_{CAT}, for a given query Q, in a GML data repository G, $Q_{CAT} = \{S \mid S \lhd G \wedge ((R(Q) = R(S)$ if $\alpha = 1) \vee Sim_{GML}(R(Q), R(S), \{SN\}) \geq \alpha$ otherwise$)\}$ ●

Our semantic similarity threshold also serves as a structural/semantic similarity parameter, underlying the extent of structural/semantic similarity considered while identifying candidate answers. It allows the user to assign more importance to the structural or semantic characteristics of GML data in answering the query at hand.

- For $\alpha = 1$, only *CATs* with root nodes identical to that of the query are the only ones considered. This corresponds to purely structural querying.

- For $0 < \alpha < 1$, *CAT*s with root nodes of semantic similarity higher than α are considered. As α decreases, the size of the answer set Q_{CAT} will increase, following the semantic similarities between query and *CAT* root nodes.
- For $\alpha = 0$, all data sub-trees in the GML data repository are considered as *CAT*s.

Parameter α is exploited throughout the querying framework to determine the amount of structural/semantic similarity considered in query/*CAT* comparison (cf. Section 4.2.2).

As for GML node similarity Sim_{GML}, it is evaluated w.r.t. the nodes' constituents, i.e. their labels and types and is developed subsequently.

4.2.2 GML Tree Comparison

Having identified the set of *CAT*s corresponding to the query at hand, the GML tree comparison component evaluates the structural and semantic similarity between the query tree and each of the *CAT*s, so as to provide the user with approximate and ranked results.

Our GML query/*CAT* tree comparison component combines and extends two recent approaches that target XML structure and semantic similarity respectively [25, 26]. It consists of four main modules for: i) identifying the *Structural Commonality Between two XML Sub-trees* (*Struct-CBS*) [26], ii) quantifying the *Semantic Resemblance Between two XML Sub-trees* (*Sem-RBS*) [25], and iii) computing *Tree Edit Distance* (*TED*). In short, the *TOC* algorithm makes use of *Struct-CBS* [26] and *Sem-RBS* [25] to structurally and semantically compare all sub-trees in the GML query tree and data tree (*CAT*) being compared. The produced sub-tree similarity results are consequently exploited as edit operations costs (node update, node insertion, tree insertion…) in an extension of Nierman and Jagadish [17]'s main edit distance algorithm. Here, *e-TED* identifies our extended edit distance algorithm (Figure 5).

Hence, the inputs to the GML tree comparison component are as follows:

- The GML query tree and data tree (*CAT*) to be compared,
- Parameter $\alpha \in [0, 1]$ enabling the user to assign more importance to the structural or semantic aspects of the GML query and data trees (*CAT*),
- Parameter Q_{Type} enabling the user to chose between *template* or *minimal constraint* querying.
- Reference semantic networks $\{SN\}=\{SN_{Geo}, SN_{Text}\}$ to be utilized for semantic similarity evaluation.

The GML tree comparison component outputs the similarity (edit distance) value between the pair of query tree and data tree (*CAT*) being compared, based on the sum of corresponding edit operations costs. Hereunder, we first i) develop the GML node semantic similarity measure Sim_{GML} exploited in computing edit operations costs, and then ii) show how the main tree edit distance algorithm *e-TED* exploits edit operations costs, and considers both template and minimal constraint querying in the GML query/data comparison process. Note that we skip the details concerning the inner-workings algorithms *Struct-CBS* and *Sem-RBS* mentioned above, since they have been thouroughly described in [25, 26].

4.2.2.1 GML Node Similarity Measure

As shown in Section 4.1, GML data (query) node labels either consist of geographic entity names, i.e., nodes of type *GeoEntity*, or geographic feature/attribute values (predicates), mainly *Text*, *Number* and *Date* (cf. Definitions 2 and 5). Obviously, it is meaningless to compare nodes encompassing different types of data (e.g., *GeoEntity* names with nodes bearing information of type *Date* or *Number*). Hence, we compute GML node similarity between corresponding node labels, given that the concerned nodes are of matching data-types, making use of similarity measures dedicated to the data-types at hand. We particularly focus on the semantic similarity Sim_{Sem} between nodes bearing conceptual information, i.e., nodes of types *GeoEntity* and *Text*, where information can be described via groups of concepts, organized in knowledge bases or semantic networks. Here, exsiting semantic similarity measures (e.g. Lin [12], Wu and Palmer [29]...) could be exploited, taking into account the concerned reference semantic network:

- We define SN_{Geo} as a semanitc network describing the semantic relations between the different geographical entities defined in the GML application schema (describing the data at hand), and exploit it in evaluating semantic similarity between *GeoEntity* node labels,
- We also exploit SN_{Text} as a more generic semantic network describing concepts found in everyday language (e.g., WordNet [15]), to compare GML element/attribute textual values.

As for *Number* and *Date* labels, they bear non-conceptual information, i.e., information that cannot be described with concepts, organized in knowledge bases. Various methods for comparing such non-conceptual information has been addressed in classic database systems [16], e.g.:

$$Sim_{Number}(n_1.l, n_2.l) = 1 - \frac{|n_1.l - n_2.l|}{|n_1.l| + |n_2.l|}$$

A similar (yet more intricate) variation could be exploited for comparing dates. Formally, given a GML query node qi and a data node sj, and considering the basic data-types mentioned above (GeoEntity, Text, Number and Date):

$$Sim_{GML}(q_i, s_j, \{SN\}) = \begin{cases} 1 & if & s_j \vDash q_i \\ Sim_{Sem}(q_i.l, s_j.l, SN_{Geo}) & else\ if & q_i.t = s_j.t = \text{'GeoEntity'} \\ Sim_{Sem}(q_i.l.val, s_j.l, SN_{Text}) & else\ if & q_i.t = P\_t_r \wedge s_j.t = \text{'Text'} \\ Sim_{Number}(q_i.l.val, s_j.l) & else\ if & q_i.t = P\_t_r \wedge s_j.t = \text{'Number'} \\ Sim_{Date}(q_i.l.val, s_j.l) & else\ if & q_i.t = P\_t_r \wedge s_j.t = \text{'Date'} \\ 0 & & otherwise \end{cases} \quad (1)$$

Recall that $s_i \vDash q_i$ underlines predicate satisfaction (Definition 5), i.e., query and data nodes are of corresponding types $q_i.t \approx s_j.t$, such as the data node satisfies the logical condition specified by the query. Similarity is obviously maximal (=1) when the data node satisfies the query's predicate node. If both query and data nodes underline the same data-type, similarity is evaluated following the corresponding similarity measure. Yet, if data-types are different, similarity is minimal (=0). Note that additional data-types could be considered in the same manner, by exploiting corresponding similarity measures.

4.2.2.2 Edit Operations Costs

Here, we provide the cost scheme for the *update* operation, as an example on how GML node similarity is exploited in computing edit operations costs. Remaing tree edit operations costs (i.e., node insertion/deletion, and tree insertion/deletion) follow similar costs schemes, integrating structural and semantic similarity scores accordingly. Given a GML query node $q \in Q$ (Definition 4) and GML data tree node $s \in S$ (Definition 2), the cost of the update operation $Upd(q, s)$ applied to q and resulting in GML query node q' such as *(s=q'* if *q.t='GeoEntity')* $\vee$ *(s ⊨ q'* otherwise)* (i.e., if q is of type predicate, P_t_i), would vary as:

$$Cost_{Upd}(q, s, \alpha, \{SN\}) = \begin{bmatrix} 1 - (1-\alpha) \times Sim_{GML}(q, s, \{SN\}) & if\ ((q \neq s) \wedge (s \not\models q)) \\ 0 & otherwise \end{bmatrix}$$

Parameter α is the structural/semantic parameter utilized in the *CAT Identification* component to assign more importance to either structural or semantic similarities:

- For $\alpha = 1$, only label equality/difference is considered in computing edit operations costs. Consequrently, *e-TED* will be considering the structural similarity between the query sub-tree Q_{Sb} (rooted at node q) and the CAT tree.
- For $\alpha = 0$, label semantic similarity is considered between corresponding GML node and *CAT* node labels. Hence, *e-TED* will evaluate the structural and semantic similarity between the sub-tree Q_{Sb} (rooted at q) and the CAT tree.

4.2.2.3 TED Algorithm Extended to Template/Minimal Constraint Comparisons

In short, *e-TED* starts by computing the cost of updating the root nodes of the trees being compared (Figure 6, line 4). Then, it computes the costs of deleting every first level sub-tree in the query tree (lines 5-10), and those of inserting every first level sub-tree in the *CAT* data tree (lines 10-16). Here, both structural and semantic similarity evaluation are considered when assigning edit operations costs (as briefly described above, [25, 26]).

On one hand, all (structurally and/or semantically weighted) operations are considered when performing *template* querying (i.e., all query/data tree elements are considered, which comes down to the classic *TED* formulation [17]). On the other hand, to allow *minimal constraint* querying, our *e-TED* disregards node and tree insertion operations in the computation process (Figure 6, lines 9 and 17). In other words, all additional elements in the *CAT* data tree will be disregarded in computing similarity, only considering those required by the query. Consequently, the algorithm recursively computes all combination of insertion, deletion and update operations to identify those yielding the minimum edit distance, i.e., the minimum cost edit script (lines 11-20). For instance, the result of comparing query Q_3 with data sub-tree S_1, following the minimal constraint strategy, is depicted in Figure 7. Here, only nodes required by the query are considered in the computation process, additional data nodes being disregarded (cf., edit distance mappings and mapping scores,

```
Algorithm e-TED()

Input: Query Tree Q and data tree S, parameter α for structural/semantic weighting, Q_Type
parameter, weighted semantic networks {SN}

Output: Edit distance between Q and S

Begin
M = Degree(Q)                        // The number of first level sub-trees in Q.        1
N = Degree(S)                        // The number of first level sub-trees in S.        2

Dist [][] = new [0...M][0...N]                                                           3
Dist[0][0] = Cost_Upd(R(Q), R(S), α, {SN})        //Update operation                     4

For (i = 1 ; i ≤ M ; i++)  { Dist[i][0] = Dist[i-1][0] + Cost_DelTree(Q_i) }             5

For (j = 1 ; j ≤ N ; j++)                                                                6
    {                                                                                    7
        If (Q_Type='Template') {Dist[0][j] = Dist[0][j-1] + Cost_InsTree(S_j) }          8
        Else {Dist[0][j] = Dist[0][j-1]}                    // Q_Type = 'Minimal Constraint'   9
    }                                                                                    10
For (i = 1 ; i ≤ M ; i++) {                                                              11
        For (j = 1 ; j ≤ N ; j++) {                                                      12
            Dist[i][j] = min{                                                            13
                Dist[i-1][j-1] + TED(Q_i, S_j),              //Dynamic programming       14
                Dist[i-1][j] + Cost_DelTree(Q_i),                                        15
                If (Q_Type='Template') { Dist[i][j-1] + Cost_InsTree(S_j) }              16
                           Else { Dist[i][j-1] }       // Q_Type='Minimal Constraint'    17
                }                                                                        18
            }                                                                            19
        }                                                                                20
Return  Dist[M][N]        // Sim =1 / (1 + Dist))                                        21
End
```

Fig. 6. Edit distance algorithm *TED*

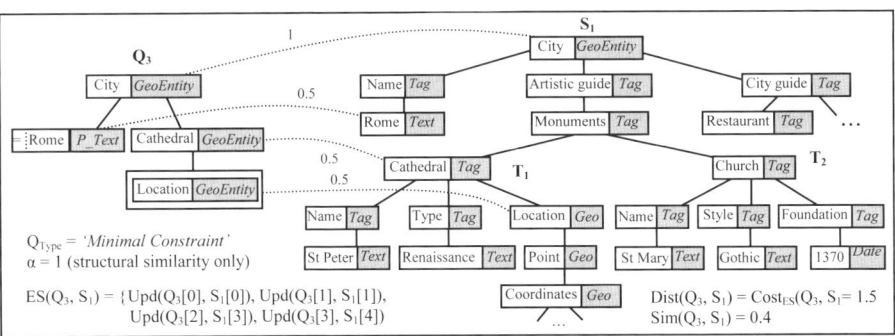

Fig. 7. GML query/CAT tree mappings

computational details being omitted due to space limitations). Note that with $Q_{Type}=$ *'Template'*, all additional nodes in S_1 (i.e., $S_1[6]$, $S_1[7]$, ...), would have to be considered in the similarity evaluation process, which would drastically decrease the similarity value.

Recall that similarity is computed based on the sum of the minimum cost edit operations corresponding to the query and *CAT* trees, i.e., inverse of edit distance (cf. Figure 6 line 21, and Figure 7 for computation example).

4.2.3 GML Query Answer Identification Component

The GML *Query Answer Identification* component underlines the elements in the data tree (*CAT*) which are to be returned to the user. These correspond to the nodes (along with their sub-trees), in the data tree, that match the GML query's *distinguished node* (cf. Definition 3). Such matches could be identified following a post-processing of the results (i.e., edit operations and mappings) produced by the *GML Tree Comparison component*.

In fact, one of the main advantages of using tree edit distance is that along the similarity (distance) value, a mapping between the nodes in the compared trees is provided in terms of the edit script, allowing the identification of correspondences between elements of the query tree and data tree (*CAT*) being compared. Consider for instance the edit distance mappings between GML query tree Q_3 and data CAT S_2, depicted in Figure 7. The number next to each mapping link designates its mapping score, which is inversely proportional to the cost of the corresponding edit operation. Consequently, mappings reveal the data node matching the distinguished query node, in our case $S_1[10]$=('*Location*', *GeoEntity*). Hence data node $S_1[10]$ is returned to the user, along with its sub-tree (i.e., the geographic coordinates of the *St Peter* cathedral in *Rome*).

In the case where multiple data nodes match the query's distinguished node, we simply identify those with the highest mapping scores, i.e., those corresponding to the most relevant mappings. Note that a dedicated threshold, specifying the minimum acceptable mapping score for a node to be considered as a relevant match to the query's distinguished node, can be considered. In addition, when the query's distinguished node is the same as its root node (e.g., queries Q_1 and Q_2), its matching node in the data *CAT* would be none other than the data tree's root node itself. Thus, the whole data tree would be returned to the user.

5 Experimental Evaluation and Validation Tests

We have implemented our GML query evaluation approach in the *XS3* prototype system[2].Hereunder, we provide preliminary *precision* and *recall* results w.r.t. a select collection of GML queries (including Q_1, Q_2, Q_3 and Q_4 of Figure 4) applied on a GML data repository constructed based on geographic data sampled from Wikipedia (Figure 3). The current data repository includes geographic information concerning 40 major historical and artistic monuments in the cities of *Rome, Dijon* and *Sao Paolo*. Ten queries were considered, distributed equally between *minimal constraint* (Q_{1-3} and Q_{5-6}) and partial *template* (Q_4 and Q_{7-10}) styles. Queries were first manually evaluated, identifying the set of relevant answers for each query, ranked following their order of relevance w.r.t. to the user (three different test subjects, one doctoral student and two post-doctoral researchers, were involved in the experiment). Manual answers were mapped to system generated ones so as to compute precision (*PR*), recall (*R*) and F-measure values (*F-value*) accordingly.

Results in Figure 8 depict overall *PR*, *R* and *F-value* results for each query. These underline our approach's applicability and potential in identifying relevant answers to

---

[2] Available online at http://www.u-bourgogne.fr/DbConf/XS3

simple GML queries. Note that in our evaluation, we adopted the *range query* formalism without however utilizing a predefined similarity threshold in identifying answers. We rather selected the whole set of ranked system generated answers (CATs) bound by the least similar relevant one, i.e., the last answer (CAT) to actually correspond to a user defined answer (which similarity value was considered as the *range query* threshold). This allowed us to verify the performance of our method in selecting relevant answers (achieving high *recall*, crucial for any method to be admissible in search applications [22]), and most importantly its effectiveness in filtering out non-relevant ones (*precision*).

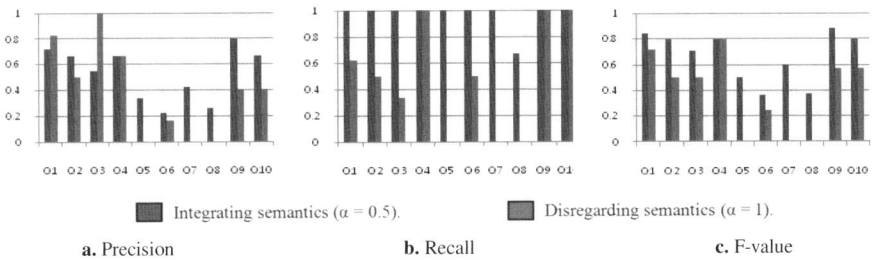

a. Precision   b. Recall   c. F-value

Fig. 8. Precision (*PR*), Recall (*R*) and F-measue (*F-Value*) results

High *recall* was particularly achieved when integrating semantic similarity evaluation, whereas quite a few relevant answers were missed when disregarding semantics (Figure 8.b). Semantic similarity evaluation also seemed crucial in amending *precision* (Figure 8.a). Queries Q_5, Q_7 and Q_8 are typical examples, where all relevant answers were missed by the system, when disregarding semantics (*PR=R=F-value*=0). However, the impact of semantic similarity evaluation seemed to decrease when searching geographic data based on their geometric attributes (e.g., coordinates) rather than textual descriptions, which was expected.

On the other hand, a major difference between the results achived with and without semantic similarity evaluation is *relevance ranking*. While single answers were usually obtained (for each query) when disregarding semantics and relying solely on GML data structure, the integration of semantic similarity resulted in the generation of a ranked set of answers, underlining their semantic similarities w.r.t. the geo-concepts in the query at hand. Ranking results are depicted in the *PR/R* graphs of Figure 9. Figures 9.a and 9.b show rather regular *PR/R* curves (*precision* decreasing gradually with the increase of *recall*), with queries Q_1 and Q_3 (Figure 9.a) clearly reflecting higher retrieval quality than their counterparts in Figure 9.b. Nontheless, some queries (e.g., Q_5, Q_7 and Q_9 of Figure 9.c) underlined relatively poor ranking capabilities, the system identifying and ranking non relevant answers (CATs) prior to relevant ones (*precision* starting at zero, and then increasing gradually w.r.t. *recall*, as relevant answers are added to the answer set). Further experiments are being conducted to analyze this effect, making use of dedicated relevance ranking metrics such as Kendall's tau and Spearman's footrule [2].

In addition, we have conducted timing experiments to verify the time complexity of the query evaluation process. Results show that the approach is linear in the size of

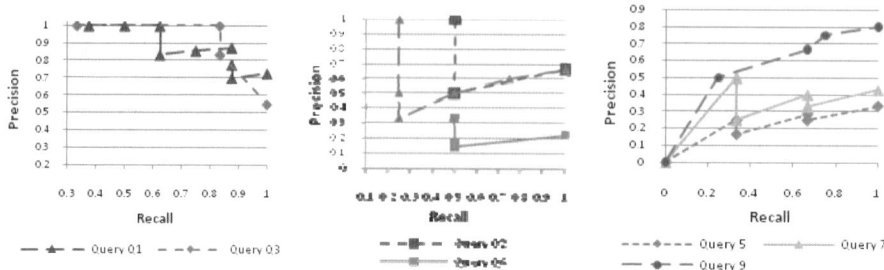

Fig. 9. *PR/R* graphs, obtained with semantic similarity evaluation

each of the query/data trees, as well as the size of the reference semantic network, when semantics comes to play, i.e., $O(|Q| \times |CAT| \times max(|SN_{Geo}|, |SN_{Text}|))$. Compexity graphs were omitted due to lack of space. Details concerning all experimental results are available online[3].

6   Conclusion

GML has been gaining growing attention as an effective means for geographic data representation and exchange in GISs on the Web. In this paper, we introduce the building blocks for an approximate GML retrieval method, considering both structural and semantic features of GML data, in the query evaluation process. Our query formalism is based on approximate tree matching as a simple and efficient technique to query GML. It allows the formulation of *structure-and-content* queries with only partial knowledge of the data collection structure and semantics, and enables both *template* and *minimum constraint* querying.

Preliminary experiments are promising, and underline the impact of semantic similarity on the query evaluation process. We are currently expanding our data testbed, in order to conduct more extensive experiments, also testing the ranking capabilities of the proposed methods using dedicated relevance ranking metrics such as Kendall's tau and Spearman's footrule [2]. We are also developing a web-based GUI to support the user in formulating queries, dynamically suggesting, following the corresponding input GML schema, the list of possible query nodes following the context of the query at hand. Considering spatio-temporal relations and predicates remains an obvious upcoming step. In this context, it might be interesting to extend our tree model to a more generic graph model, encompassing spatio-temporal links between geographic features, and thus try to adapt our tree edit distance algorithm accordingly.

References

[1] Amer-Yahia, S., Lakshmanan, L., Pandit, S.: FleXPath: Flexible Structure and Full-Text Querying for XML. In: Proc. of the ACM Inter. Conf. on Management of Data (SIGMOD), pp. 83–94 (2004)
[2] Bar-Ilan, J.: Comparing rankings of search results on the Web. Information Processsessing and Management (41), 1511–1519 (2005)

---

[3] At http://www.u-bourgogne.fr/DbConf/GMLSearch

[3] Chawathe, S.: Comparing Hierarchical Data in External Memory. In: Proceedings of VLDB, pp. 90–101 (1999)

[4] Dalamagas, T., et al.: A Methodology for Clustering XML Documents by Structure. Information Systems 31(3), 187–228 (2006)

[5] Ferri, F., Grifoni, P., Rafanelli, M.: The Management of Spatial and Temporal Constraints in GIS using Pictorial Interaction on the Web. In: Persson, A., Stirna, J. (eds.) CAiSE 2004. LNCS, vol. 3084, pp. 92–105. Springer, Heidelberg (2004)

[6] Fuhr, N., Großjohann, K.: XIRQL: A Query Language for Information Retrieval. In: Proc. of the ACM-SIGIR Conference, pp. 172–180 (2001)

[7] Grabs, T., Schek, H.-J.: Generating Vector Spaces On-the-fly for Flexible XML Retrieval. In: Proc. of ACM SIGIR Workshop on XML and Information Retrieval, pp. 4–13 (2002)

[8] Guo, L., et al.: XRANK: ranked keyword search over XML documents. ACM SIGMOD, 16–27 (2003)

[9] Hammiche, S., et al.: Semantic Retrieval of Multimedia Data. In: ACM MMDB Workshop, pp. 36–44 (2004)

[10] Jones, C., Purves, R.: Geographic Information Retrieval. J. of Geo. Info. Science 22(3), 219–228 (2008)

[11] Larson, R.: Geographic Information Retrieval and Spatial Browsing. In: GIS and Libraries: Patrons Maps and Spatial Information, pp. 81–124 (1996)

[12] Lin, D.: An Information-Theoretic Definition of Similarity. In: Proc. of the ICML Conference, pp. 296–304 (1998)

[13] Maguitman, A., et al.: Algorithmic Detection of Semantic Similarity. In: WWW Conference, pp. 107–116 (2005)

[14] Marian, A., et al.: Adaptive Processing of Top-k Queries in XML. In: ICDE Conference, pp. 162–173 (2005)

[15] Miller, G.: WordNet: An On-Line Lexical Database. International Journal of Lexicography 3(4) (1990)

[16] Motro, A.: Vague: A User Interface to Relational Databases that Permits Vague Queries. ACM Transactions on Office Information Systems 6(3), 187–214 (1988)

[17] Nierman, A., Jagadish, H.V.: Evaluating structural similarity in XML documents. In: Proc. of the ACM WebDB Workshop, pp. 61–66 (2002)

[18] Open Geospatial Consortium. Geography Mark-up Language, http://www.opengeospatial.org/standards/gml

[19] Paul, M., Gosh, S.K.: An Approach for Geospatial Data Management for Efficient Web Retrieval. In: Proc. of the 6th IEEE International Conference on Computer and Information Technology (2006)

[20] Pokorny, J., Rejlek, V.: Databases and Info. Systems, Frontiers in Artificial Intelligence and Applications. In: Barzdins, J., Caplinskas, A. (eds.) A Matrix Model for XML Data, pp. 53–64. IOS Press, Amsterdam (2005)

[21] Rodriguez, M.A., Egenhofer, M.J.: Comparing Geospatial Entity Classes: an Asymmetric and Content-Dependent Similarity Measure. Journal of Geographical Information Science 18(3), 229–256 (2004)

[22] Salton, G.: The SMART Retrieval System. Prentice Hall, New Jersey (1971)

[23] Schlieder, T.: Similarity Search in XML Data Using Cost-based Query Transformations. In: Proc. of the International ACM WebDB Workshop, pp. 19–24 (2001)

[24] Schlieder, T., Meuss, H.: Querying and Ranking XML Documents. Journal of the American Society for Information Science, Special Topic XML/IR 53(6), 489–503 (2002)

[25] Tekli, J., Chbeir, R., Yetongnon, K.: Extensible User-based Grammar Matching. In: ER Conf., pp. 294–314 (2009)

[26] Tekli, J., Chbeir, R., Yetongnon, K.: Efficient XML Structural Similarity Detection using Sub-tree Commonalities. In: Brazilian Symposium on Databases (SBBD) and SIGMOD DiSC, pp. 116–130 (2007)

[27] Torres, M., et al.: Retrieving Geospatial Information into a Web-Mapping Application using Geospatial Ontologies. In: Nguyen, N.T., Grzech, A., Howlett, R.J., Jain, L.C. (eds.) KES-AMSTA 2007. LNCS (LNAI), vol. 4496, pp. 267–277. Springer, Heidelberg (2007)

[28] World Wide Web Consortium. The Document Object Model (DOM) (May 2009), http://www.w3.org/DOM

[29] Wu, Z., Palmer, M.: Verb Semantics and Lexical Selection. In: Proc. of the 32nd Annual Meeting of the Associations of Computational Linguistics, pp. 133–138 (1994)

[30] Zhang, Z., Li, R., Cao, S., Zhu, Y.: Similarity Metric in XML Documents. In: Knowledge Management and Experience Management Workshop (2003)

[31] Zhu, F., Guan, J., Zhou, J., Zhou, S.: Storing and Querying GML in Object-Relational Databases. In: Proc. of the 14th Annual ACM Inter. Symp. on Advances in Geographic Information Systems, pp. 107–114 (2006)

Advancing Search Query Autocompletion Services with More and Better Suggestions

Dimitrios Kastrinakis[1] and Yannis Tzitzikas[1,2]

[1] Computer Science Department, University of Crete, Greece
[2] Institute of Computer Science, FORTH-ICS, Greece
{kastrin,tzitzik}@csd.uoc.gr

Abstract. Autocompletion services help users in formulating queries by exploiting past queries. In this paper we propose methods for improving such services; specifically methods for increasing the number and the quality of the suggested "completions". In particular, we propose a novel method for partitioning the internal data structure that keeps the suggestions, making autocompletion services more scalable and faster. In addition we introduce a ranking method which promotes a suggestion that can lead to many other suggestions. The experimental and empirical results are promising.

1 Introduction

Search query autocompletion is the process of computing in real time and suggesting to the user words or phrases which can complete the query that the user has already typed, on the basis of user queries which have been submitted in the past. This feature is very useful and popular in several domains and systems [6,15,16,5,1]. Initially it was used in the Amazon web site [15], providing suggestions for products related to the product a user is searching for. It has also been applied to provide *thesaurus based suggestions* [3]. Nowadays, web browsers (e.g. Mozilla Firefox) use this feature when the user is typing a URL in the address bar or completing forms, according to recent history. The same applies for the "Tab" key when using a command line interpreter (i.e. sh or bash shell in Unix). Autocompletion has been proposed also for assisting the formulation of *faceted-search queries* [5] and *underspecified SQL Queries* [14]. *Semantic autocompletion* [11] interfaces have recently become prevalent in semantic search and annotation applications. Other fields where autocompletion is involved include e-mail clients, source code editors, word processors, etc. Another notable application of autocompletion is in the field of *mashups* [1]. However, the most common and widely known use of autocompletion is the autosuggest feature used by Google, Yahoo!, Bing or Ask search engines. In this case, a user can see the most popular searches starting with the string currently being typed.

In the context of a WSE (Web Search Engine), the suggested completions are useful because in many cases the user does not know what words to use or how to describe his information need. Apart from that, this feature allows the user

B. Benatallah et al. (Eds.): ICWE 2010, LNCS 6189, pp. 35–49, 2010.

to find out, without any additional effort from his side, what is popular among the internet users given the current input string, as well as the number of hits that he will get if he submits each of the suggested queries. Our objective is to increase the number and the quality of suggested completions.

To increase the number of possible suggestions we need scalable data structures. Current techniques require loading in main memory a data structure based on the contents of the entire query log. However, for very large log files, this approach would occupy too much main memory space (or may not fit in main memory at all) and loading it would require much time. To tackle this problem, we propose a novel method for *partitioning* this data structure which allows loading only the fragment that is suitable for providing suggestions on the current input entered by a user. This approach is more scalable and it is faster since loading a fragment of the data structure requires less time.

Another key aspect for the success of autocompletion services is how the suggestions are ranked, since only a small number of the possible completions (usually less than 10) are prompted. In this paper we propose a ranking method which promotes those queries which are popular and are prefixes of other popular queries. We justify the advantages of this method by measurements over the query log of a real WSE.

In a nutshell, the key contribution of our work lies in: (a) showing how partitioning can enhance the scalability of autocompletion services and (b) proposing a new ranking method which promotes a query which is a prefix of other queries and if submitted, a relatively large number of results is retrieved. The rest of this paper is organized as follows. Section 2 discusses background. Section 3 elaborates on partitioning while Section 4 elaborates on ranking. Finally, Section 5 concludes and summarizes the advantages of the proposed methods.

2   Background and Related Work

Typically a WSE maintains a log file consisting of tuples of the form: $[ip\_addr, date, query, res\_num]$, where ip_addr is the IP address of the user who submitted the query, $date$ is the date when the query was submitted, $query$ is the query string submitted and res_num is the number of results this query yields. Of course, not every single query should be loaded from the log file. There are two main conditions that are usually adopted as filters for the queries (see Figure 1): (a) a query must not be more than D days old (e.g. a week) and (b) a query must not yield a null query result.

The queries (as well as their popularity scores) that satisfy the above conditions have to be inserted in a data structure with low complexity of retrieval, specifically $\mathcal{O}(n)$ for a string of n characters. The most appropriate data structure is a tree where every node contains a character. A query of n characters requires n nodes for insertion, one node in each level of the tree. This type of tree is called *trie*. An example of inserting queries in a trie is depicted at Figure 3a.

Let's now describe the on-line process. Consider a user who through a browser is typing a query in the query text box of a WSE which is equipped with an

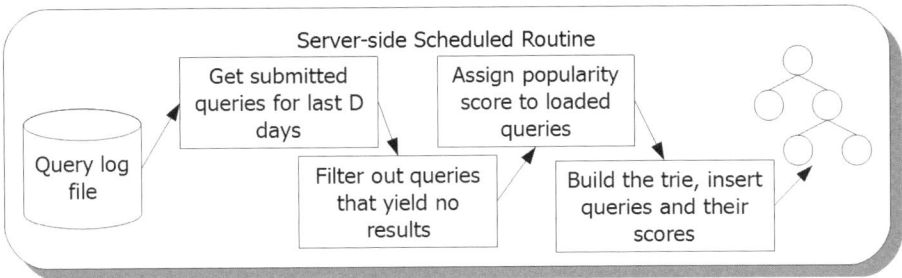

Fig. 1. Loading queries from the log file

autocompletion mechanism. Below we describe the chain of events that will occur. At first, the user types an input string *str* in the appropriate field (see Figure 2). The autocompletion client reads that string whenever a new keystroke occurs. Then, it is sent to the server that deploys the server side autocompletion component of the search engine, using asynchronous methods like *AJAX* [7]. Next, *str* is given as input for the trie, descending to the node containing its last character. A *Depth First Search* (see Figure 3b) is then applied below that node, collecting all past queries starting with *str*. The collected suggestions are ranked by their popularity score. Lastly, the server sends the *top-k* suggestions to the user.

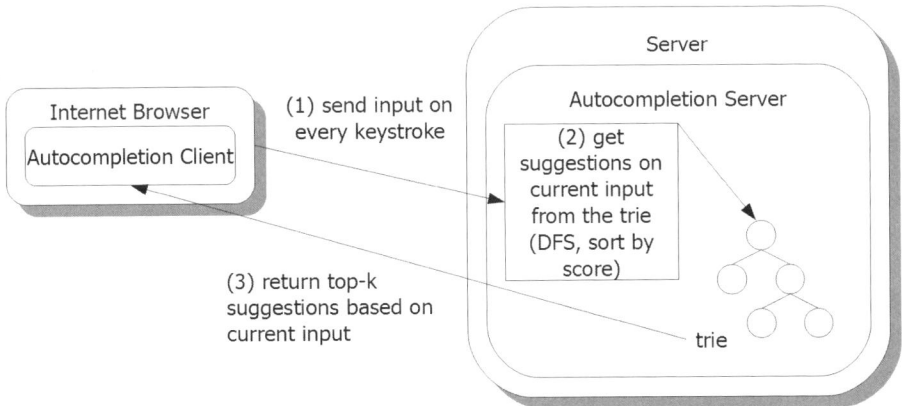

Fig. 2. Chain of events when auto completing a search query

3   Trie Partitioning

If the log file is very large, the trie containing the logged queries could be too big to fit in main memory. Even if the trie does fit in main memory, we may still want to reduce its main memory requirements in order to exploit the saved memory space for other reasons, e.g. for *result caching* [9] or *inverted list caching* [13].

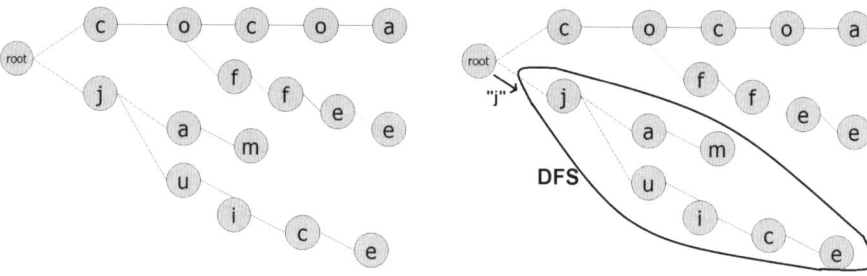

(a) Insertion of "cocoa", "coffee", "jam" (b) Retrieving suggestions "jam" and
and "juice" in a trie   "juice" from current input "j"

Fig. 3. An example of inserting and retrieving data from a trie for autocompletion
purposes

Except for space requirements, loading a huge trie can take a significant amount
of time. This is a major issue when dealing with autocompletion because a user
wants to see suggestions *during* typing. This means that the total time interval
between a keystroke and the visualization of suggestions must be short (less than
1 sec). Besides, we have to take into account that loading a trie from the disc
(if we have not already loaded it) is only one of the steps required (we have to
be aware of the delays during packet transfer, server load, etc.). Therefore we
should be able to keep the trie at a convenient size based on our space and time
constraints.

3.1   The Proposed Solution

Instead of keeping suggestions in one trie, we propose partitioning it into two or
more *subtries*. Eventually, the autocompletion mechanism will be using a forest
of tries, each time loading the proper one based on the user input. The rising
question is how should we partition the trie. At this point we should note that
tree partitioning methods [12] have been applied mainly to indexing structures
of database applications [2,10,8]. For instance, space-partitioning trees have been
implemented and realized inside PostgreSQL [8], resulting in performance gains.

Partitioning by Starting Characters: Here we propose a partitioning
method that is based on the starting characters (see Figure 4). Each subtrie
contains queries whose starting characters belong to a specified set of characters assigned to this trie. For example, if we assume that the query log contains
queries starting with latin characters only, then we can divide a trie into two
subtries: one containing all queries str where $str[0] \in \{a, b, \ldots, m\}$ and another
containing all strings str where $str[0] \in \{n, o, \ldots, z\}$. Let $A = \{a_1, \ldots, a_n\}$ be
the set of the first characters of queries loaded from the log file, so the maximum
number of tries is $|A|$. We can partition A to m ($m \leq |A|$) subsets $p_1, \ldots, p_m$,
where $p_i \cap p_j = \emptyset$, $i \neq j$. Each p_i is used for determining which strings are to be

inserted in a specific subtrie. Specifically, each p_i is associated with a trie that we will denote by tr_i. This association can be maintained by an index file that maps each partition p_i to a subtrie tr_i. Let $P = \{p_1, \ldots, p_m\}$ be the set that contains these subsets and let $T = \{tr_1, \ldots, tr_m\}$ be the set of all such tries. In section 3.2 we will propose a generalization of this method for the first k characters, which provides additional advantages. We should mention at this point that related works on trie partitioning, such as those proposed in [4], are not applicable for autocompletion because the length of the query string is progressively increasing (it is not fixed or a priori-known as it is assumed in [4]).

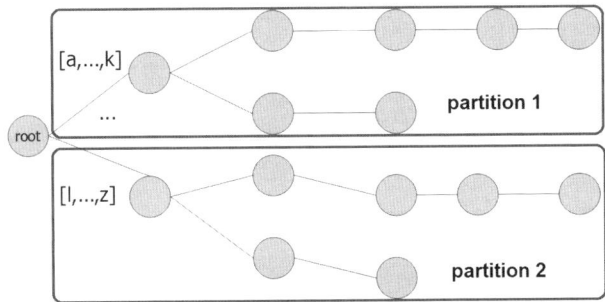

Fig. 4. Partitioning a trie based on the starting characters of inserted strings

3.2   Distributing the Starting Characters

The question that now arises is how we should form the partition P. We should not do this arbitrarily; imagine having 2 subtries and assigning: $\{a, b, \ldots, k\}$ to tr_1 and $\{l, m, \ldots, z\}$ to tr_2. Assume that the majority of queries in the log file start from the letters: 'm', 'n', 's', 't', 'o'. This means that tr_2 will be significantly larger than tr_1. This case is not better than keeping a single trie (because the size of tr_2 could be almost the size of the entire trie, and tr_2 could also be too big to host in main memory). In fact it is worse, because we may have to load the correct trie for a certain query, further delaying the appearance of suggestions to the user. Ultimately, a *distribution* of characters must be as smooth as possible, so that the tries eventually become of similar size.

Firstly, we analyze the query log file by counting the number of appearances of every starting character of each query. Specifically for every starting character c we compute its frequency $freq(c)$. Let $NTries$ be the number of subtries that we have decided to create, and $|Q|$ be the number of queries in the log file. To obtain a uniform distribution we would like $avg = \frac{|Q|}{NTries}$ queries to be inserted to each subtrie. One simple approach would be to assign n starting characters to a partition p_i until

$$\sum_{i=1}^{n} (freq(c_i)) \geq avg \, .$$

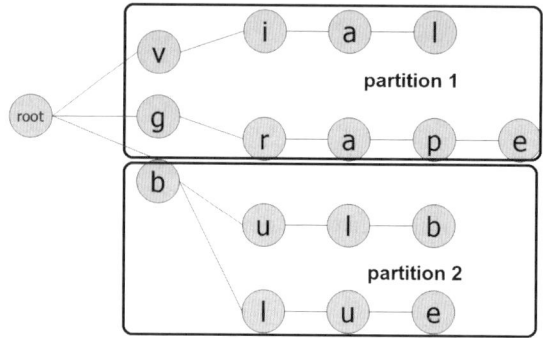

Fig. 5. Creating the partitions

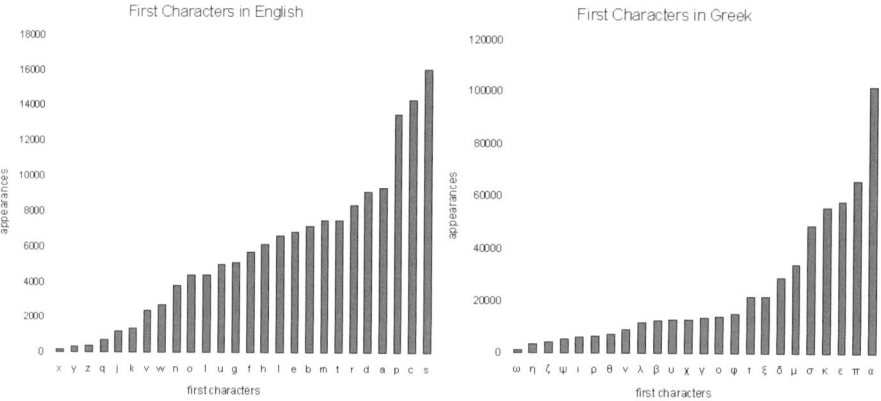

(a) Distribution of first letters in English   (b) Distribution of first letters in Greek

Fig. 6. Distributions of first letters in English and Greek

Figure 5 shows a simple example, where $Q = \{"blue", "bulb", "grape", "vial"\}$, $|Q| = 4$ and $NTries = 2$, so $avg = 4/2 = 2$. Partitions will be created as follows: $p_1 = \{b\}$ and $p_2 = \{g, v\}$. So $tr_1 = \{blue, bulb\}$ and $tr_2 = \{grape, vial\}$.

However the above method is effective if the number of queries that start from a certain character are less than avg. Moreover the distribution of the first letters greatly affect the uniformity of the subtrie-sizes that we can achieve. To clarify this aspect we used natural language dictionaries to count the distribution of the first characters. For instance, Fig. 6a shows the distribution of the first characters for the English language. The letter 's' is the first in 16,104 out of 150,843 words contained in the dictionary, therefore 10.68% of English words start with 's'. Figure 6b shows the distribution of Greek words. There are 102,201 words starting with α out of 574,737 words contained in the dictionary, therefore 17.78% of Greek words start with α. Fig. 7 shows the distribution of starting

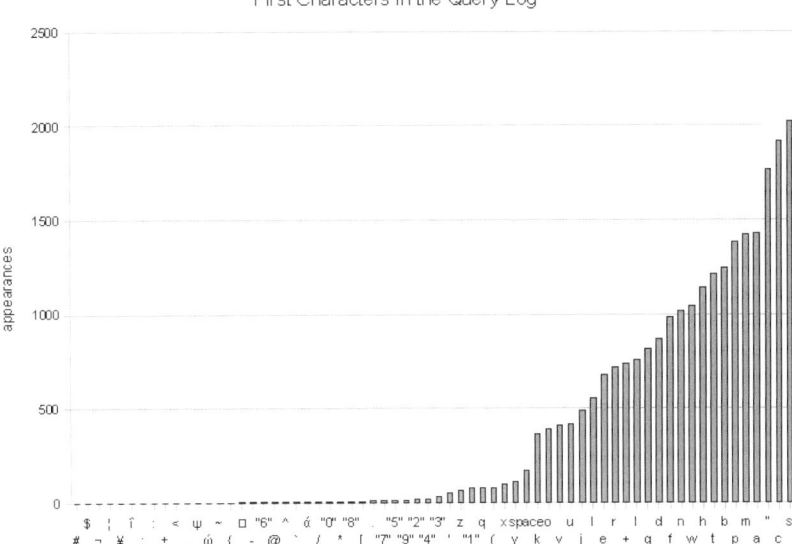

Fig. 7. Distributing queries by first characters, having used a query log from the Excite search engine

characters in queries stored in a log file[1]. Again, 's' is the most frequent first letter, appearing in 8.2% of all the queries.

The above facts motivate an alternative method for partitioning: instead of distributing queries by their first characters only, we distribute them based on their first k characters. This allows achieving smoothness even if we have a character c such that $frec(c) > avg$. For example, instead of creating a single partition for 's', multiple partitions of 's' are created based on the next letter (if $k = 2$). Since each subtrie now corresponds to a prefix of k characters, no suggestions are computed for the first $k - 1$ characters of the user input.

The exact algorithm for partitioning is shown in Fig. 8. At first it collects all possible prefixes of k characters from the query log and counts their frequency. Then it creates a partition and keeps populating it until the total frequency becomes greater than the desired partition capacity. After that, a new partition is created and so on. Notice that A is sorted lexicographically and each $p_i \in P$ is characterized by the *range* of prefixes that have been associated to it. For example, suppose that $k = 2$ and p_i has been associated to "sa", "se" and "so". In this case only "sa" and "so" are kept, meaning that the queries that will be assigned to subtrie tr_i are those whose prefix is in the range $["sa", \ldots, "so"]$. Also notice that the algorithm does not require building and keeping the entire trie in main memory. After collecting the k-prefixes it reads the queries from the

---

[1] The log contained queries submitted to the Excite (March 13, 1997) search engine. URL: www.excite.com

Alg. *Distribute k First Characters to Partitions*
Input: k, *Capacity // the desired partition capacity in queries,*
QL *// log file with all submitted queries.*
Output: $P = \{p_1, \ldots, p_m\}$.
(1) $A = \emptyset$;
(2) for each query $q \in QL$ do
(3) $a = q[1 \ldots k]$; *// a holds the first k chars of q*
(4)   if $a \in A$ then $a.frequency = a.frequency + 1$;
(5)   else
(6) $a.frequency = 1$;
(7) $A = A \cup \{a\}$;
(8)   end if;
(9) end for;
(10) sort(A); *// lexicographically*
(11) $i = j = 0$;
(12) while $i < |A|$
(13) $j++$;
(14)   while $p_j.size < Capacity$
(15) $i++$;
(16) $p_j = p_j \cup \{a_i\}$;
(17) $p_j.size = p_j.size + a_i.frequency$;
(18)   end while;
(19) $p_j.min = min_{lex}(p_j)$; *// the minimum lexicographically*
(20) $p_j.max = max_{lex}(p_j)$; *// the maximum lexicographically*
(21) $p_j.clear()$; *// we keep min and max only*
(22) end while;
(23) return P;

Fig. 8. Algorithm for distributing the k first characters of the logged queries to each p_i. Complexity: $\mathcal{O}(nk + mk \log m)$ where $n = |QL|$, $m = |A|$, and obviously $m \leq n$.

log file and distributes them to the appropriate partition. The above method of partitioning has to be repeated periodically.

3.3   Measurements

In this section we report experimental measurements[2]. We used the query log from the Excite search engine which contained $\sim 25,500$ distinct queries. For this data set we created various possible partitionings with $k = 1$ comprising 2, 3, 4, ..., 10 and 66 subtries (with $k = 1$ there are 66 distinct first characters in the log, i.e $|A| = 66$). Let LT be the time required to determine which trie to load, load that trie, and return the top 10 suggestions. For each case we computed the average LT of the generated tries as follows:

$$avg\_LT_{T_n} = \frac{\sum_{tr \in T_n} LT_{tr}(c)}{|T_n|} \quad ,$$

---
[2]   Programming Language: Java, Platform: CPU: Intel Core 2 Duo E6750 @ 2.66GHz, 4GB RAM, 2x 10,000RPM RAID0 disks, OS: Windows 7 x64.

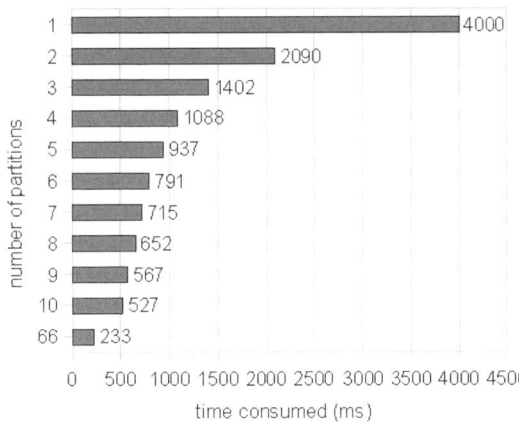

Fig. 9. Average LT for 1, 2, ..., 10 and 66 partitions and $k = 1$

where c is a single character given as input for the trie tr in order to get suggestions, $T_n = \{tr_1, \ldots, tr_n\}, n \in \{2, \ldots, 10, 66\}$. The measured times are shown in Fig. 9. Compared to non-partitioning, with 66 subtries and $k = 1$ we achieve a $Speedup = \frac{avg\_LT_{T_0}}{avg\_LT_{T_{66}}} = \frac{4000}{233} \simeq 17.1$ meaning that the partitioning is 17 times faster.

Fig. 10a shows how characters were distributed when partitioning was based on the first character only ($k = 1$), the first two characters ($k = 2$), and the first three characters ($k = 3$). The desired number of queries per partition was $avg = \frac{|Q|}{NTries} = \frac{25,500}{66} \simeq 386$, so the ideal distribution would have 386 queries per partition. Fig. 10b shows how "close" we are to the ideal distribution for $k = 1, \ldots, 5$, by plotting the standard deviation for each distribution, $(\sigma = \sqrt{\frac{1}{N} \sum_{i=1}^{N} (x_i - \mu)^2}$, where $N = NTries$, each x_i represents the number of queries in partition i, and μ is the mean value of these numbers). It is obvious from the plots, with $k \geq 2$, the distribution is dramatically smoother (closer to the ideal) than in the case of $k = 1$.

Synopsizing, we have seen that partitioning by the starting k-characters can lead to uniform in size subtries, which is crucial for respecting main memory constraints and reducing load times. The bigger k is, the smoother the distribution of queries to subtries becomes. The only drawback of a high k value is that we may not be able to compute suggestions for user inputs of size less than $k - 1$ characters. For example, suppose that $k = 2$ and assume that $P = \{[ab - az], [ba - bi], [bl - bz], ...\}$. If the user types "b" we cannot compute suggestions using one subtrie because b is distributed to the second and third trie, so we have to wait until the user types another character. Notice however, that if the user types "a" then we can compute suggestions since "a" is covered entirely by the first subtrie.

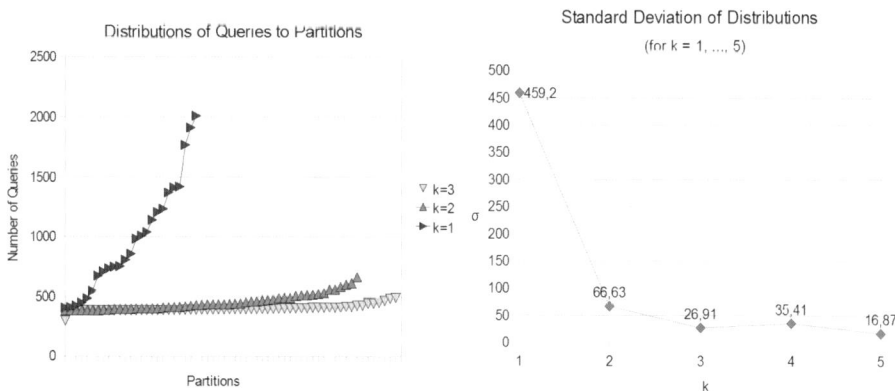

(a) Distributing queries to each partition for (b) Standard deviation of distributions for
$k = 1, 2, 3$ $k = 1, \ldots, 5$

Fig. 10. Distributing queries to partitions

4   Ranking Suggestions

Another key aspect for the success of an autocompletion service is how the
suggestions are ranked, since only a small number of the possible completions
are prompted. One might think of several criteria that could be used (in isolation
or in aggregation) for ranking suggestions, e.g. popularity, number of results, etc.
In this section we propose ranking methods that are based on both popularity
and number of results and also promote those queries which are popular and are
prefixes of other popular queries.

The computation of popularity is based on the contents of the query log file.
However note that just counting the submissions of a certain query from the
log file is not sufficient; imagine a user of a search engine entering a certain
query 1,000 times. For this reason we count only those submissions coming from
distinct sources only, specifically distinct IP addresses[3], therefore the query log
contains IP addresses.

Let $q \leadsto q'$ denote that q is a prefix of q'. Let q_u denote the query the user has
typed. We want to assign a score to each candidate completion q (where $q_u \leadsto q$).
First we introduce a metric that takes into account popularity and answer size:

$$PopSize(q) = \frac{freq(q)}{MAX\_freq} \cdot \frac{res\_num(q)}{MAX\_res\_num} \tag{1}$$

where

- $freq(q)$ is the number of distinct IP addresses that have submitted q,
- MAX_freq is the maximum frequency of the queries in the log,

---

[3] Distinct IP addresses do not necessarily imply distinct users. It is just a simple mean
of approaching the number of distinct users that entered a certain query.

- $res\_num(q)$ is the number of results the query q yields,
- MAX_res_num is the maximum number of results of the queries.

The above formula assumes that all queries are independent. However some queries are prefixes of other queries and this observation should be taken into account. For example, consider the queries $q_1 = "music"$ $q_2 = "music\ composers"$, $q_3 = "mammals"$, $q_4 = "mammals\ from\ Africa"$, and assume that all of them have the same popularity. Now suppose that we have to compute the top-2 suggestions. The queries q_1 and q_3 are good candidates since each is a prefix of another popular query (of q_2 and q_4 respectively).

Let $Reach(q)$ be the set of all queries that have q as a prefix, i.e. $Reach(q) = \{ q' \mid q \leadsto q' \}$. To exploit the above observation, here we propose another ranking formula, $PopSizeReach$ defined as:

$$PSR(q) = a * PopSize(q) + b * \frac{1}{|Reach(q)|} \sum_{q' \in Reach(q)} PopSize(q') , \quad (2)$$

where a and b are constants that range in $(0,1)$ and $a + b = 1$. The second part of the formula is the average $PopSize$ of the queries that have q as prefix.

An alternative approach is to increase the score gained for queries having large $Reach(q)$ by removing $\frac{1}{|Reach(q)|}$:

$$PSR_2(q) = a * PopSize(q) + b * \sum_{q' \in Reach(q)} PopSize(q') , \quad (3)$$

A probabilistic approach is also possible. Here we want to assign a score to each q (where $q_u \leadsto q$) that reflects the probability that the user will select q if he has typed q_u. The estimation of the probability is again based on the log file. Specifically, we define

$$Score(q) = \frac{DeepFreq(q)}{\sum_{q_u \leadsto q'} DeepFreq(q')} \quad (4)$$

where $DeepFreq(q) = freq(q) + \sum_{q \leadsto q'} freq(q')$.

Table 1 and 2 show the suggestions produced by the first two formulas when typing the query "books" and "news" respectively using the Excite query log. In this case, the computation of popularity ignored the number of results of each query because this information was not available in the query log. Because of this, (3) and (4) behave the same, producing identical suggestion rankings. Therefore only (4) is included in the tables. In Table 1, a notable change in ranking is "bookstore", which is 7th using (1) and first using (2). Similarly, in Table 2 "news" is 4th using (1), 3rd using (2) and 1st using (4). On the other hand, a long query has less probability of having many other queries that contain it as prefix. This fact is depicted in Table 2, where "newspaper vancouver washington" is 3rd using (1) but only 6th using (2).

However, the above examples are just indicative and they do now allow us to draw safe conclusions regarding these formulas. To evaluate a ranking method

Table 1. Actual suggestion rankings for query "books" using formulae (1), (2) and (4)

(1)	(2), $a = 0.5$, $b = 0.5$	(4)
bookstores and catalogue and 1-800	bookstore	books
books	bookstores and catalogue and 1-800	bookstore
books a million	books	bookstores and catalogue and 1-800
books and titles	books a million	books a million
books on tape	books and titles	books and titles
books: crime and punishment	books on tape	books on tape
bookstore	books: crime and punishment	books: crime and punishment

Table 2. Actual suggestion rankings for query "news" using (1), (2) and (4)

(1)	(2), $a = 0.5$, $b = 0.5$	(4)
newspaper clark county washington	newspaper clark county washington	news
newspapers	newspapers	newspaper
newspaper vancouver washington	news	newspaper clark county washington
news	newspaper	newspapers
news and server	newsgroups	newsgroups
news group	newspaper vancouver washington	newspaper vancouver washington
newsgroups	news and server	news and server
newspaper book reviews	news group	news group

through a user study is a laborious and expensive task and often yields results which are not repeatable. In general we can say that there is not yet any formal methodology and method for evaluating ranking methods for query completions. For instance [3] mentions a user study but does no report any concrete results, while the majority of works on autocompletion (e.g. [5]) focus on efficiency and the quality of suggestions is by no means evaluated. For this reason, we introduce a metric that can measure the *predictive power* of a ranking method. The key point is that it requires as input only a query log. The main idea is the following: we use a prefix of a submitted query and then measure the rank of the whole query in the list of completions suggested (and ranked) by the ranking formula under evaluation. It follows that a ranking formula f is better than a ranking formula f' if the ranks yielded by f are lower than those of f'.

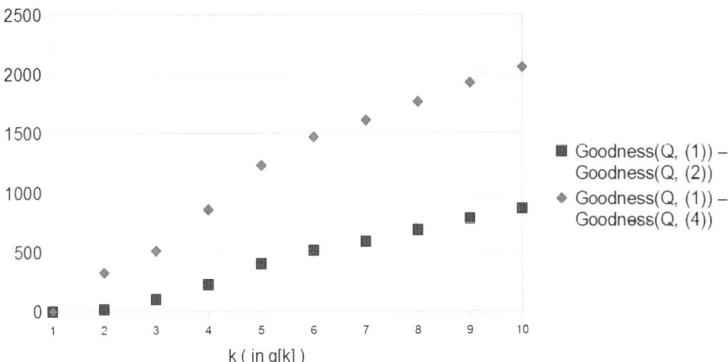

Fig. 11. Comparing the ranking methods using the difference of *Goodness* for each ranking method

Let $q[k]$ denote the k-char prefix of a query q of the log file. We shall use $Rank_{(1)}(q, q[k])$ to denote the position of q in the suggestions produced by formula (1) if we sort them in descending order. For example, if $q = "company"$, $q[2] = "co"$ and the suggestions produced by formula (1) are $\langle core,\ computer,\ company,\ car \rangle$ then $Rank_{(1)}("company", "co") = 3$. The less $Rank_f(q, q[k])$ is, the better formula f behaves assuming the given log file. To take into account all queries of a query log, we define the "goodness" of a scoring formula f over a query log Q as follows:

$$Goodness(Q, f) = \sum_{q \in Q} Rank_f(q, q[k])$$

As our objective is to comparatively evaluate scoring functions, it is not necessary to make any kind of normalization. A formula f is better than a formula $\acute{f}$ if $Goodness(Q, f) < Goodness(Q, \acute{f})$. Of course, one could generalize and consider all prefixes of q, not only one (i.e. the k-char prefix).

Returning to the problem at hand, we applied the above metric on formulas (1), (2), (3) and (4) for $k = 1, \ldots, 10$, using the Excite query log. Here, Q was accessed as a set that contained the distinct submitted queries, so duplicate queries were not included in ranking. Fig. 11 shows the plot of $Goodness(Q, (1)) - Goodness(Q, (2))$ and the plot of $Goodness(Q, (1)) - Goodness(Q, (4))$. Since the latter plot is higher than the first, it follows that (4) is better than (2), and this holds for all values of $k = 1, \ldots, 10$ (since both plots are positive, both (4) and (2) are better than (1)). The reason we did not include $Goodness(Q, (1)) - Goodness(Q, (3))$ in the plot was the unavailability of the number of results each query yields in the Excite log. Therefore (3) and (4) behave the same and produce the same *Goodness* results.

Another aspect of a ranking method is the amount of time required for computing the scores. Table 3 reports the times required for each ranking method using the Excite log (which contained $\sim 25,500$ distinct queries). Since we want to return suggestions to the end user in real time (otherwise the autocompletion

Table 3. Time consumed while computing scores for every ranking formula

$PopSize$ (1)	$PopSizeReach$ (2)	$PopSizeReach_2$(3)	$DeepFreq$ (4)
$\sim 0.35s$	$\sim 60s$	$\sim 60s$	$\sim 54s$

feature would be useless), we pre-compute the scores for every query stored in the log and this is done off-line by the autocompletion server as shown in Fig. 1. In this way the computation of query completions is almost instant.

5   Concluding Remarks

To make autocompletion services more scalable, we proposed a method for partitioning the trie of logged queries. This partitioning allows increasing the number of suggestions that can be hosted, and speeding up their computation at real time. As an example, consider an amount of main memory sufficient for hosting 25,500 different suggestions. By partitioning the trie with respect to the first character (i.e. $k = 1$), the same amount of memory can host 1,659,027 more suggestions, i.e. almost two orders of magnitude more, and the loading time is 17.1 times shorter. Since the first characters are not uniformly distributed in natural languages, we proposed a partitioning that is based on the first k characters which can be used for yielding uniform in size subtries.

Finally, we proposed a novel method for ranking suggestions, where the score of each suggestion depends on (a) its popularity (distinct submissions), (b) the number of results it yields if submitted, and (c) the suggestions that contain the current suggestion as prefix. To comparatively evaluate such ranking functions we introduced a metric measuring the predictive power of a ranking method and we identified the ranking method that prevails over a query log file of a real WSE. To the best of our knowledge no other work has elaborated on *index partitioning* or *structure-aware ranking*. We believe that these techniques can enhance autocompletion services in various applications.

References

1. Abiteboul, S., Greenshpan, O., Milo, T., Polyzotis, N.: Matchup: Autocompletion for mashups. In: IEEE International Conference on Data Engineering, Shanghai, China, pp. 1479–1482 (2009)
2. Aref, W.G., Ilyas, I.F.: Sp-gist: An extensible database index for supporting space partitioning trees. Journal of Intelligent Information Systems 17(2-3) (December 2001)
3. Arias, M., Cantera, J.M., Vegas, J.: Context based personalization for mobile web search. In: VLDB '08 (August 2008)
4. Baberwal, S., Choi, B.: Speeding up keyword search for search engines. In: 3rd IASTED International Conference on Communications, Internet, and Information Technology, St. Thomas, US Virgin Islands, pp. 255–260 (November 2004)

5. Bast, H., Weber, I.: When you 're lost for words: Faceted search with autocompletion. In: SIGIR '06 Workshop on Faceted Search, Seattle, Washington, USA (August 2006)
6. Bowman, D.E., Ortega, R.E., Hamrick, M.L., Spiegel, J.R., Kohn, T.R.: Refining search queries by the suggestion of correlated terms from prior searches. Patent Number: 6,006,225 (December 1999)
7. Draganova, C.: Asynchronous javascript technology and xml (ajax),
   http://www.myacrobatpdf.com/6319/asynchronous-javascript-technology-and-xml-ajax.html
8. Eltabakh, M.Y., Eltarras, R., Aref, W.G.: To trie or not to trie? realizing space-partitioning trees inside postgresql: Challenges, experiences and performance. In: Procs. of the 31st VLDB Conference, Trondheim, Norway (2005)
9. Fagni, T., Perego, R., Silvestri, F., Orlando, S.: Boosting the performance of web search engines Caching and prefetching query results by exploiting historical usage data. ACM Transactions on Information Systems (TOIS), 51–78 (2006)
10. Ghanem, T.M., Shah, R., Mokbel, M.F., Aref, W.G., Vitter, J.S.: Bulk operations for space-partitioning trees. In: 20th International Conference on Data Engineering (March 2004)
11. Hyvonen, E., Makela, E.: Semantic autocompletion. In: Mizoguchi, R., Shi, Z.-Z., Giunchiglia, F. (eds.) ASWC 2006. LNCS, vol. 4185, pp. 739–751. Springer, Heidelberg (2006)
12. Jacquet, P., Regnier, M.: Trie partitioning process: Limiting distributions, vol. 214. Springer, Heidelberg (1986)
13. Zhang, J., Long, X., Suel, T.: Performance of compressed inverted list caching in search engines. In: Proceedings of the 17th international conference on World Wide Web, Beijing, China (April 2008)
14. Mason, T., Lawrence, R.: Auto-completion of Underspecified SQL Queries. Springer, Heidelberg (2006)
15. Ortega, R.E., Avery, J.W., Frederick, R.: Search query autocompletion. Patent Number: US 6,564,213 B1 (May 2003)
16. Whitman, R.M., Scofield, C.L.: Search query refinement using related search phrases. Patent Number: US 6,772,150 B1 (August. 2004)

Designing Service Marts for Engineering Search Computing Applications

Alessandro Campi, Stefano Ceri, Andrea Maesani, and Stefania Ronchi

Politecnico di Milano, Italy
{campi,ceri,maesani,ronchi}@elet.polimi.it

Abstract. The use of patterns in data management is not new: in data warehousing, data marts are simple conceptual schemas with exactly one core entity, describing facts, surrounded by multiple entities, describing data analysis dimensions; data marts support special analysis operations, such as roll up, drill down, and cube. Similarly, Service Marts are simple schemas which match "Web objects" by hiding the underlying data source structures and presenting a simple interface, consisting of input, output, and rank attributes; attributes may have multiple values and be clustered within repeating groups. Service Marts support Search Computing operations, such as ranked access and service compositions. When objects are accessed through Service Marts, responses are ranked lists of objects, which are presented subdivided in chunks, so as to avoid receiving too many irrelevant objects – cutting results and showing only the best ones is typical of search services. This paper gives a formal definition of Service Marts and shows how Service Marts can be implemented and used for building Search Computing applications.

1 Introduction

Search Computing is a new paradigm for composing search services [7]. While state-of-art search systems answer generic or domain-specific queries, Search Computing enables answering questions via a constellation of cooperating search services, which are correlated by means of join operations. Search Computing aims at responding to queries over multiple semantic fields of interest; thus, Search Computing fills the gap between generalized search systems, which are unable to find information spanning multiple topics, and domain-specific search systems, which cannot go beyond their domain limits. Paradigmatic examples of Search Computing queries are: "Where can I attend an interesting scientific conference in my field and at the same time relax on a beautiful beach nearby?", "Where is the cinema closest to my hotel, offering a high rank action movie and a near-by pizzeria?", "Who is the best doctor who can cure insomnia in a nearby public hospital?", "Which are the highest risk factors associated with the most prevalent diseases among the young population?" These queries cannot be answered without capturing some of their semantics, which at minimum consists in understanding their underlying domains, in routing appropriate query subsets to each domain-expert search engine, and in combining answers from each engine to build a complete answer that is meaningful for the user.

B. Benatallah et al. (Eds.): ICWE 2010, LNCS 6189, pp. 50–65, 2010.

A prerequisite for setting such goal is the availability of a large number of valuable search services. We could just wait for SOA (Service Oriented Architecture) to become widespread. However, few software services are currently designed to support search, and moreover a huge number of valuable data sources (the so called "long tail" of Web information) are not provided with a service interface. In this paper, we therefore focus on the important issue of publishing service interfaces suitable for Search Computing so as to facilitate the widespread use of data sources on the Web and to simplify their integration in Search Computing applications. At the basis of our work, we observe the pervasive role of software services and SOA[1]. While the SOA principles are becoming widespread, however, we observe distinct standards, languages, and programming styles. Thus, the SOA vision is developed in a variety of directions, and we emphasize the relevance of supporting ordered queries upon data sources, by means of a new pattern.

In the framework of Search Computing, we define a **Service Mart** as the data abstraction for data source publication and composition. The goal of a Service Mart is to ease the publication of a special class of software services, called search services, whose responses are ranked lists of objects. Every Service Mart is mapped to one "Web object" available on Internet; therefore, we may have Service Marts for "hotels", "flights", "doctors", and so on. Thus, Service Marts are consistent with a view of the "internet of objects" which is gaining popularity as a new way to reinterpret concept organization in the Web and go beyond the unstructured organization of Web pages. Moreover, pairs of Service Marts are augmented with "connections", so as to support their linking; Service Marts and their connections constitute a resource network that can be used as a high-level interface for expressing search queries.

This paper is organized as follows. A general overview of the state of the art is presented in Section 2. Section 3 introduces Service Marts by illustrating their three levels of description (conceptual, logical, and physical). Section 4 illustrates the Search Computing framework so as to position Service Marts within the global reference architecture, and Section 5 illustrates mechanisms for service registration and adaptation. Finally, an example of usage of Service Marts is briefly described in Section 6, and Section 7 provides some conclusions.

2 Related Work

Current service description languages and protocols, such as WSDL and SOAP, provides limited support in mechanizing service recognition, combination, and automated negotiation. Several proposals, such as WSFL [19] or DAML-S [1], extend these languages and protocol in a semantic direction. OWL-S [20] is an attempt at formalizing the semantics of Web services using ontology technology. COSMO [22] provides concepts for reasoning about services, and for supporting operations, such as composition and discovery, which are performed on them at design and run-time. The framework facilitates the use of different service description

---

[1] Consider the prominent role given to Software and Services in Call 5 of the EU-funded Seventh Framework Programme (FP7), whose goal is to achieve an "Internet of the Future, where organizations and individuals can find software as services on the Internet, combine them, and easily adapt them to their specific context [10]".

languages tailored to different service aspects. Web Service Modeling Framework (WSMF) [13,14] is a fully-fledged modeling framework for describing the various aspects related to Web services using four main elements: ontologies providing the terminology, Web services providing access to resources, goals representing user desires, and mediators dealing with interoperability problems. Other ongoing efforts aim at creating a conceptual framework for service modeling; a prominent example is the W3C's Web Services Architecture [27]. In [9] authors propose a conceptual model that describes actors, activities and entities involved in a service-oriented scenario and the relationships between them. This work extends the W3C's Web Services Architecture, but the authors do not specify the semantics of the concepts described; their conceptual model is a glossary of terms.

Previous approaches to Web service combination is described by the SOA scientific community with two different flavors: orchestration and choreography. The distributed approach of choreographed services (e.g., using WS-CDL [25] or WSCI [24]) deals with service compatibility and compliance to predefined behavior descriptions but has limited relevance in the Search Computing context. The centralized approach of orchestrated services (e.g., using BPEL [21]) suits better the research problem addressed in this paper, as orchestrations are executable service compositions and services need not be aware of being the object of query execution. The research on composition and integration of data sources is very rich, some interesting results are in [2,11,12,26,28].

The work described in this paper is the result of a research stream starting with [23], in which the authors propose a Web service management system that enables querying multiple Web services in a transparent and integrated fashion and propose an algorithm for arranging a query's Web service calls into a pipelined execution plan that exploits parallelism among Web services. Subsequent work [5] established the theoretical framework for stating the problem of joining heterogeneous search engines; while [6] presented an overall framework for multi-domain queries on the Web. These works are the predecessors of the current research project named "Search Computing", whose preliminary results are described in [7].

Service Marts are specific "data patterns"; their regular organization helps structuring Search Computing applications. The search of patterns for easing data modeling for specific contexts is not new; the most well known data modeling pattern is the so called "data mart", used in data warehousing as conceptual schemas for driving data analysis. Data marts [3] are simple schemas having one core entity, describing facts, surrounded by multiple entities, describing the dimensions of data analysis. Such subschema allows a number of interesting operations for data selection and aggregation (e.g. data cubes, rollup, drilldown) whose semantic definition is much simplified by data characterization as either fact or dimension and by the regular structure of the schema. Analogously, a "Web mart" [8] is a pattern introduced in the Web design community to characterize the role played by data items in data-intensive Web applications. Web marts have a central entity, the core concept, which describes a collection of core objects, surrounded by other entities which are classified as "access entities", enabling selection of core objects through navigation, and "detail entities", describing core objects in greater detail. Thus it is possible to drive a design process that produces first-cut standard Web applications (e.g. sales, inventories, travels, and so on).

3 Service Marts

Service Marts are abstractions; publishing a Service Mart entails bridging an abstract description to several concrete implementations of services. Indeed, implementing a Service Mart may require the mapping to several data sources, each one configured either as Web services or as an API, or as a materialized data collection. Thus, the Service Mart concept offers an abstraction for giving a "regular" view of the world, together with a method and associated technology for building such a regular view out of concrete data sources. This section gives a top-down view of the definition of Service Marts, from the conceptual level through the logical to the physical level. It also describes composition patterns (at the conceptual and logical level) and introduce the service registration (at the physical level).

3.1 Conceptual Level

A Service Mart is an abstraction describing a class of Web objects. Thus, every Service Mart definition includes a name and a set of exposed attributes. Service Marts have atomic attributes and repeating groups consisting of a non-empty set of sub-attributes that collectively define a property. Atomic attributes are single-valued, while repeating groups are multi-valued. For example, a "Movie" Service Mart has single-value attributes ("Title", "Director", "Score, "Year", "Language") and repeating groups ("Genres", "Openings", "Actor"), each with sub-attributes. The schema of a repeating group is introduced by one level of parentheses:

Movie(Title, Director, Score, Year, Genres(Genre), Language,
Openings(Country, Date), Actors(Name))

Other Service Marts used in this paper describe cinemas and restaurants:

Cinema(Name, Address, City, Country, Phone, Movies(Title, StartTimes))
Restaurant(Name, Address, City, Country, Phone, Url, Rating, Category(Name))

Attributes and sub-attributes are typed and semantically tagged when they are defined. Repeating groups model many-valued properties (such as the "actors") within the object instances of the Service Mart (the "movie"). In this way, besides adding expressive power to Service Mart properties, they also model 1:M or M:N relationships, i.e. conceptual elements whose purpose is bridging real world objects. Concepts such as "acts-in" between "actor" and "movie" are modeled by repeating groups, by placing actors as a repeating group of movie or movies as a repeating group of actor (or both). This goal is consistent with keeping the Search Computing infrastructure as simple as possible, and also with keeping the connection between the two Service Marts as simple as possible. Of course, such a pattern introduces a limitation upon the ways of describing reality, which seems rather coercive if one considers the richness of data modeling choices offered by top-down design. But in our framework we do not use a top-down process; rather, we model data sources bottom-up, and then we look for their integration; moreover, most data sources have a simple schema, which can be well represented by a one-level nesting. Therefore, the expressive power of Service Marts seems to be appropriate for its purpose.

3.2 Logical Level

At the logical level each Service Mart is associated with one or more specific access patterns. An **access pattern** describe the way in which one can access the Service Mart. It is a specific signature of the Service Mart with the characterization of each attribute or sub-attribute as either input (I) or output (O), depending on the role that the attribute plays in the service call. In the context of logical databases, an assignment of labels I/O to the attributes of a predicate is called adornment, and thus access patterns can be considered as adorned Service Marts. Moreover, an output attribute is designed as ranked (R) if the service produces its results in an order which depends on the value of that attribute. To ease service composition, we assume that all ranked attributes return a normalized value within the interval $[0..1]^2$. For example, for the Service Mart "Movie" we can have the following access patterns:

$Movie_1(Title^O, Director^O, Score^R, Year^O, Genres.Genre^I, Language^O, Openings.Country^I, Openings.Date^I, Actor.Name^O)$

$Movie_2(Title^I, Director^O, Score^R, Year^O, Genres.Genre^O, Language^O, Openings.Country^I, Openings.Date^I, Actor.Name^O)$

In all cases, "Score" is an output attribute (ranging in $[0..1]$) used for ranking movies, which are presented in descending order of their score, i.e. with highest score movies first. The openings "Country" and "Date" are input parameters, which are used to extract movies shown in a specific country after a specific opening date (enabling the extraction of recent movies for that country). In the first access pattern, movies are retrieved by providing as input also one of their genres (thus modeling the request "search recent movies by genre"). In the second access pattern, movies are retrieved by providing as input also the title (thus modeling request "search recent movies by title"). Other access patterns could be used for accessing movies by providing the director or one actor. The choice of access patterns is a limitation on the way in which one can obtain data, typically imposed by existing service interfaces. Therefore, defining access patterns requires both a top-down process (from query requirements) and a bottom-up process (from service implementations). In general, this tension between top-down and bottom-up processes is typical of service design.

Sometimes access patterns have more attributes than the original Service Mart. Consider cinemas and restaurants: their address is a characteristic of the underlying object, but users searching for cinemas and addresses typically provide to the service and input address (e.g. their home or current location) and search by proximity. Thus, U versions of attributes "Address", "City" and "Country" denote the user's location, and T/R versions of the same attributes that represent the cinema/restaurant location;

$Cinema_1(Name^O, UAddress^I, UCity^I, UCountry^I, TAddress^O, TCity^O, TCountry^O, TPhone^O, Distance^R, Movie.Title^{O,} Movie.StartTimes^O)$

---

[2] We also consider the possibility of service interfaces providing two or more ranking attributes, in such case the service definition includes an aggregation function which indicates how to obtain a score in the $[0..1]$ interval as a function of the ranking attributes.

3.3 Connection Patterns

Connection patterns introduce a pair-wise coupling of Service Marts. Every pattern has a *conceptual name* and then a *logical specification*, consisting of a sequence of simple comparison predicates between pairs of attributes or sub-attributes of the two services, which are interpreted as a conjunctive Boolean expression, and therefore can be implemented by joining the results returned by calling service implementations. Connection patterns can be directed or undirected.

For example, Movies and Cinemas are connected via the undirected connection pattern "Shows", which uses a join on titles:

Shows(Movie,Cinema): [(Title=Title)]

The interpretation of joins within connection patterns is existential: if the movie's title is equal to the title of any movie shown in the cinema, then the predicate is satisfied, and the two instances of movie and cinema are composed to form an instance of the result; the two instances are composed without performing any selection on sub-attributes (in the example, if one title of cinema satisfies the join, then all movies shown at the cinema are selected). Using the existential interpretation of equality predicates in selection and joins involving sub-attributes as operands yields to a simple semantics and an efficient implementation of these operations.

Consider next a directed connection between cinemas and restaurants; a directed pattern can be used "from" the first "to" the second (the query search first for cinemas and next for nearby restaurants). The connections is specified as a conjunction of predicate expressions, relating the cinema address to the input location of a restaurant service, so that after determining a cinema (close to the user's address) the service will be invoked by using the cinema's location as input for the restaurant search:

DinnerPlace(Cinema, Restaurant): [(TAddress=UAddress),
 (TCity=UCity), (TCountry=UCountry)]

Logically, connection patterns are expressed among pairs of orderly type compatible attributes. A connection pattern must be *supported* by a pair of access patterns. All the attributes of both selected access patterns must have the same labels, either I or O, and they should not both be labeled I. If both the right and left operand have an O label, then the pattern is undirected, else if the left operand is labeled O and the right operand is labeled I then the pattern is directed from left to right.

Visually, Service Marts and logical connection patterns can be presented as **resource graphs**, where nodes represent marts and arcs represents logical connection patterns; directed connections include an edge. Thus, the Search Computing model of the Web presents a simplification of reality, seen through potentially very large resource graphs. Such representation enables the selection of interconnected concepts that support the creation and dynamic extension of multi-domain queries.

3.4 Physical Level

At the physical level of Service Marts we model **service interfaces**, where each service interface is mapped to a concrete data source. A service interface may not support some of the attributes of the Service Mart, e.g., because one source could miss a property; moreover, sources may be provided with type coercion facilities so as to adapt to a single type description. These provisions allow for a minimal amount of inconsistency between service interfaces and Service Marts.

A service interface is a unit of invocation and as such must be described not only by its conceptual schema or logical adornment, but also by its physical properties. There are a huge number of options for characterizing data-intensive services, both in terms of performance and quality. Service interfaces are described by four kinds of parameters:

- **Ranking descriptors** classify the service interface as a *search service* (i.e., one producing ranked result) or an *exact service* (i.e., services producing objects which are not ranked). Exact services are associated with a *selectivity*, which is a positive number expressing the average number of tuples produced by each call. When a search service is associated with an access pattern having one or more output attributes tagged as R, then the ranking is said explicit, else it is said opaque. Explicit ranking over a single attribute can be denoted as ascending or descending. Note that search services may not be present a result with ranked attributes; e.g., most commercial search engines can be characterized as Service Marts accepting input keywords and producing semi-structured output information which is mapped to a schematic representation, but they normally do not expose rankings.

- **Chunk descriptors** deal with output production by a service interface. The service is *chunked* when it can be repeatedly invoked and at each invocation a new set of objects is returned, typically in a fixed number, so as to enable the progressive retrieval of all the objects in the result; in such case, it exposes a *chunk size* (number of tuples in the chunk). Search Computing is focused on the efficient data-intensive computation and therefore most service interfaces are chunked. Of course, if the service is ranked, then the first chunk contains the objects with highest ranking, and subsequent chunks yields the next objects in the ranking; normally, with exact services a query should examine all chunks, while with search services a query can examine just the top chunks.

- **Cache descriptors** deal with repeated invocations of the service. A very efficient way to speed up service invocations consists in caching at the requester side the responses returned for given inputs, and then use such stored answers instead of invoking the service. But such policy is not acceptable with many services, e.g. those offering real-time answers. Hence, parameters indicate if a service interface is *cacheable* and in such case what is the cache *decay*, i.e. the elapsed time between two calls at the source that make the use of stored answers tolerable.

- **Cost descriptors** deal with associating each service call with a cost characterization; this in turn can be expressed as the *response time* (time required in order to complete a request-response cycle), and/or as the *monetary cost* associated with making a specific query (for those systems who charge their answers).

Every access pattern may have several service interfaces. For instance, the first access pattern of the Service Mart "movie" requires a physical service capable of filtering movies by time (e.g., whose opening date in US is recent enough) and genre (e.g.

action movies) and then extracting them ranked by their quality score. For this purpose we use the IMDB archive (http://www.imdb.com), which stores information about thousands of movies and enriches their description with a "score" attribute computed as the average of the scores assigned by worldwide user communities. We extract data by building an ad-hoc wrapper and using it to materialize all movie descriptions; this requires periodic downloads to maintain such data materialization up-to-date. Tools for data materialization and refreshments are described in Section 5. Similarly, for the Service Mart "cinema" we use "*Movie Showtimes - Google Search*" (http://www.google.com/movies), a service allowing the retrieval of all the cinemas nearby an input location that is expressed as a complete address (address, city, country) or as a city. The service returns results sorted by cinema distance from the input location, but it does not return the actual distance (therefore, ranking is opaque, and the implementation does not expose "Distance"). Finally, for the Service Mart "restaurant" we use the Yahoo Local source (http://local.yahoo.com/), a service that allow the users to find Businesses & Commercial Services (e.g. restaurants) that are in or nearby a requested address, city and state, or a specific zip code. These service interfaces support the connection patterns "shows" and "dinner place" described in the previous section.

4   The Search Computing Framework

Search Computing applications are built by exploiting a configurable software framework approach, illustrated in Figure 1. The *Service Mart Framework* provides the scaffolding for wrapping and registering data sources; the *Service Invocation Framework* masks the technical issues involved in the interaction with the registered Service Mart, e.g., the Web service protocol and data caching issues. Their features are discussed in the next section.

The *User Framework* provides functionality and storage for registering users, with different roles and capabilities, as discussed in Section 4.1. The *Query Framework* supports the management and storage of queries, which can be executed, saved, modified, and published for other users to see. The *Query Processing Framework* is the central component of the architecture, which provides service for executing multi-domain queries. The *Query Manager* takes care of splitting the query into sub-queries and binding them to the respective relevant data sources; the *Query Planner* produces an optimized query execution plan, which dictates the sequence of steps for executing the query. Finally, the *Execution Engine* actually executes the query plan, by submitting the service calls to designated services through the Service Invocation Framework, building the query results by combining the outputs produced by service calls, computing the global ranking of query results, and producing the query result outputs in an order that reflects their global relevance. These components are not the target of this paper, but are investigated in the Search Computing project. The very first results of this research stream are described in [5,6,7].

To obtain a specific Search Computing application, the general-purpose architecture of Figure 1 is customized with the help of tools targeted to programmers, expert users, and end users.

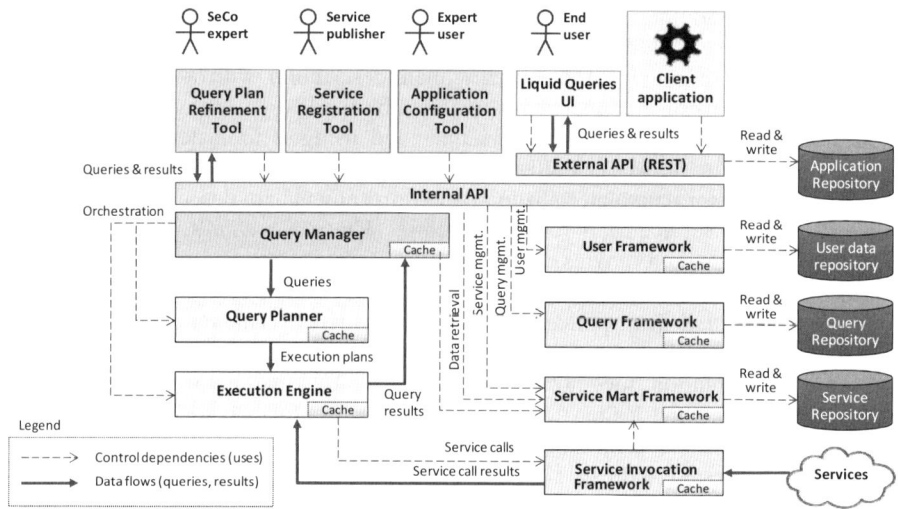

Fig. 1. Overview of the Search Computing framework

- **Service Publishers** register Service Mart definitions within the service repository, and declare the connection patterns usable to join them. The registration process is realized through a *Service Registration Tool* that: 1) helps the publisher in the specification of the SM, AP and SI attributes and parameters respectively and 2) it hides to the user the Internal API, that allow the communication between the services and the engine levels. The service publishers are in charge of implementing mediators, wrappers, or data materialization components, so as to make data sources compatible with the Service Mart standard interface and expected behavior.

- **Expert Users** configure Search Computing applications, by selecting the Service Marts of interest, by choosing a data source supporting the Service Mart, and by connecting them through connection patterns. They also configure the complexity of the user interface, in terms of controls and configurability choices to be left to the end user.

- **End Users** use Search Computing applications configured by expert users. They interact by submitting queries, inspecting results, and refining/evolving their information need according to an exploratory information seeking approach, which we call *Liquid Query* [4].

Search Computing aims at building two new communities of users: **Content providers**, who want to organize their content (now in the format of data collections, databases, Web pages) in order to make it available for search access by third parties, and **expert users**, who want to offer new services built by composing domain-specific content in order to go "beyond" general-purpose search engines such as *Google* and the other main players. The service registration framework aims at facilitating content providers in their task of publishing data sources.

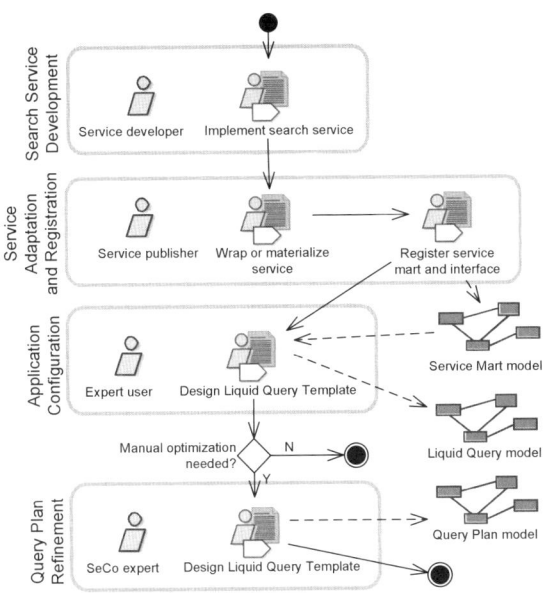

Fig. 2. Development process for SeCo applications (SPEM notation)

5 Web Service Registration and Adaptation

The trend towards supporting users in **publishing data sources on the Web** is a general one. Google, Yahoo and Microsoft are building environments and tools (Fusion Tables [15], Yahoo! BOSS [29]) for helping Web users to publish their data, with the goal of capturing the so-called "long tail" of data sources. In Search Computing, data sources should produce ranked output and data extraction should be performed incrementally, by "chunks"; users can suspend a search and then resume it, possibly guiding the way in which data sources should be inspected. We are building tools and/or providing best practices, applicable to data sources of various kinds, for enabling data providers to build "search" service adapters. We distinguish three different scenarios:

- Data can be queried by means of a Web service or combining the results of different Web services.
- Data are available on the Web but must be extracted from Web sites through wrappers.
- Data are not directly accessible and must first be materialized.

Results returned by a call to a service interface expose an **interchange format** written in JSON (JavaScript Object Notation) [17], a lightweight data-interchange format easy to read and write by humans and easy to parse and generate by machines. The format descends directly from the conceptual description of the Service Mart, therefore all instances of a Service Mart share the same interchange format, regardless of the service interface which produces them. Below is a JSON "movie" instance:

```
{"title": "Highlander",
 "director": "Russell Mulcahy",
 "score": "0.7",
 "year": "1986",
 "genres": [{"genre": "action"}],
 "openings": [{"country": "US", "date": "31-10-1986"}],
 "actors": [{"name": "Christopher Lambert"},
            {"name": "Sean Connery"}]}
```

5.1 Web Services

The typical service implementation is a real Web service registered in the platform. Web services return their output in arbitrary format, including but not limited to HTML, XML and JSON. Given that the Service Mart interchange format is a well-defined JSON structure, the service implementation developer must define a series of transformations on the results, and bundle them into a remote service implementation that hides the peculiar features of each remote source.

To tackle the need to combine the results of different Web services we built a Service Mart Framework containing some predefined software modules useful to manipulate data. The very first of them is the *invocation module* which invokes a service and returns a list of tuples; next, tuples are read by a *tuple reader* and possibly copied by a *tuple cloner*. Other modules perform projections, string replacements, computations of regular expressions, data conversions and splitting or concatenation of attributes. Once the services are transformed to return JSON, another step can be necessary in order to adapt the cardinality of the results returned in each service, which can be not appropriate (a search service could return too many results with each call, or even all the results together). In this case, a *chunker module* supports changing the chunk size: every call to the actual service is translated into the appropriate number of calls to the service implementation, which buffers results and produces chunks of the desired size.

5.2 Web Pages

The second types of sources we want to use are HTML pages. The Web is rich of good quality information stored in HTML Web sites. Wrappers are particular programs that can make available data published in the Web. In the context of Service Marts, wrappers can be used to capture data which is published by Web servers in HTML format, because in such case a data conversion is needed in order to support data source integration – data must be rearranged according to the Service Mart normalized schema. Another typical use of wrappers in Search Computing occurs when services respond with HTML documents which must be translated in the normal schema and encoded in JSON. For building wrappers, several systems are available; in particular we use Lixto [16]. By marking a region of an example Web document displayed on screen the user helps the tool to build a set of rules describing the structure of the pages of the Web site. These rules are used to generate a wrapper that "query" Web site in real time. Fig. 3 shows the relationships between data extracted on the Web and a tabular view on these data.

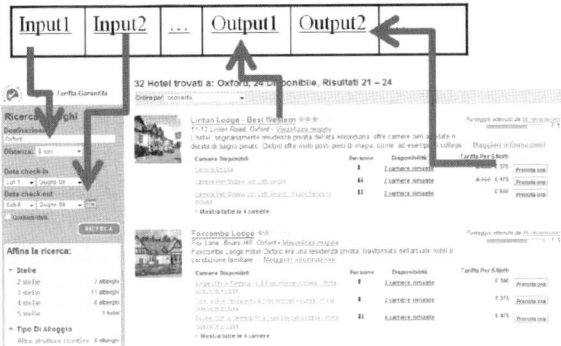

Fig. 3. Data extraction from query results published in HTML

5.3 Materialized Databases

Even if most service implementations require a *call to a remote service*, in some cases summarized and materialized data may need to be stored at the engine site. Data materialization is a general process, which can be applied to sources in order to transform their format, to eliminate redundancy, to improve their quality, and so on; data materialization moves data preparation from query execution to source registration time, together with a data materialization schedules setting the times when materialization should be repeated; therefore, data materialization is very useful for supporting efficient query execution. Intrinsic to the normalization process, however, is the capturing of a given snapshot of the data, which is not current; therefore the approach can be used only with data which rarely changes.

We developed a **materializer** specifically for use in Search Computing. The materializer is a software component whose objective is to read arbitrary data sources and organize data in a normalized format, suitable for data export according to a Service Mart definition. The materializer is organized with two logical layers: the data extraction layer operates directly upon the data sources, that can be of arbitrary formats (e.g., tables, XSL files, XML trees, and so on); its purpose is to transform the input data into relational tables of arbitrary format, called primary materialization; such tables are temporary, used only in the materializer, and invisible to the outside environment. A series of SQL procedures are applied to the primary materialization in order to produce a *normalized schema*. Such schema has maps every Service Mart to a primary table and every repeating group to an auxiliary table; the primary table has a system-generated unique identifier, while the auxiliary table has a composite identifier built with the primary table's identifier and a progressive number.

A materializer uses the modules described in Section 5.1 to combine results returned by different Web services and contains some new units that operate together with the unit previously defined. For example, Tuple writer unit writes data items as rows in a database table. Figure 4 shows an example of materialization process. When data materializations are stored according to the normalized schema, the service implementation is automatically built by using SQL queries whose code depends only on the service interface description. Note that data providers need not use the materializer, as long as they build tables according to the normalized schema.

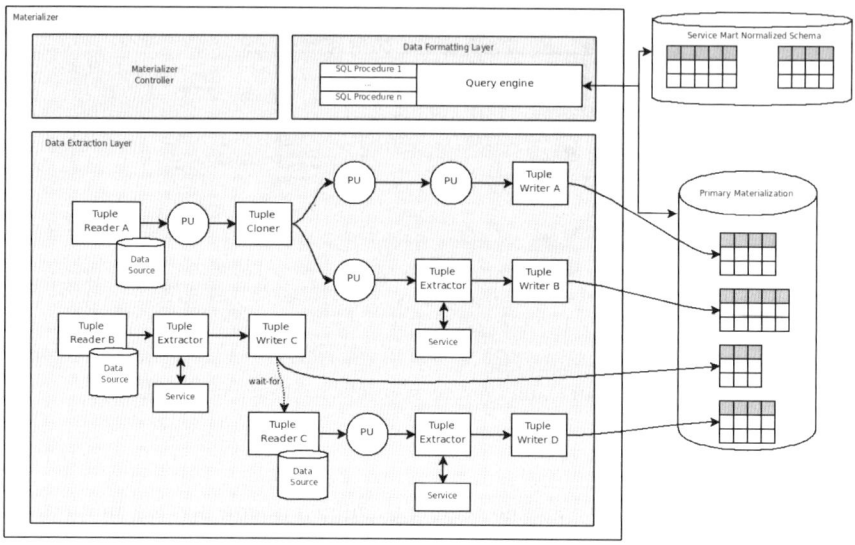

Fig. 4. Process description within the Materializer

Specifically, queries over stored tables perform selection based upon input and ranking using the ORDER BY clause. While selection, ranking and nesting are supported by standard SQL, chunking requires returning at each call the "top k" tuples; unfortunately, "top k" queries are not supported in standard SQL, but all commercial DBMS support them in a specific SQL dialect; some of them offer as well "interval" queries, enabling the extraction of the "next k" (defined as the tuples within the interval [k+1..2k+). MySQL offers "interval" queries through a LIMIT clause which returns at each query evaluation an ordered table with the tuples whose ranking falls between the first and second parameter. If we use such feature, a simple query pattern for extracting tuples from rank h to k (where k-h is the chunk size) is:

```
SELECT *
FROM Table
WHERE condition
ORDER BY rank DESC
LIMIT h,k
```

6 Applications and Use Cases

This section describes typical user interaction scenarios based on the running example presented in the previous sections. We describe a search interaction concerning close-by movies, cinemas, and restaurants, and the refinement and exploration of results through application of additional local filters. Suppose the user submit a query asking for a cinema with a high rated movie of a given kind next to a good vegetarian restaurant. Once the user submits the search parameters, the query is performed and results are calculated and displayed in the result table, as illustrated in Fig. 5. The result page is enriched with interaction options that the user can choose.

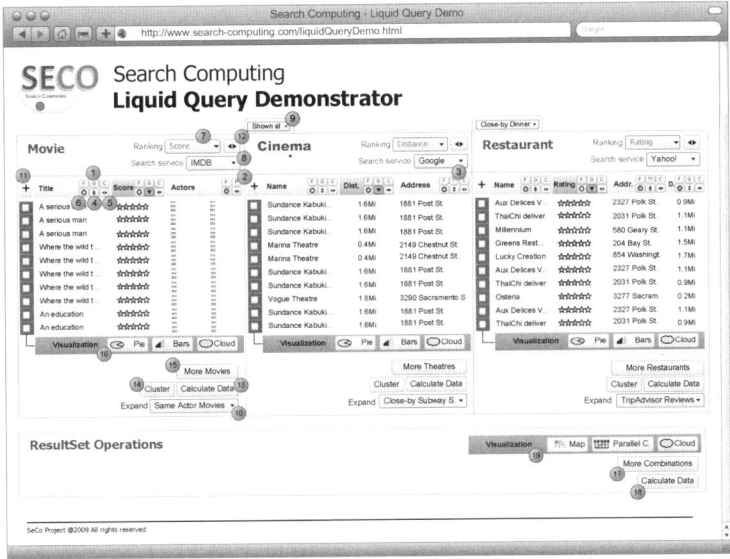

Fig. 5. Result interface

Once the results are shown, the user can interact with them through the available commands. Some operations (i.e., visualization options and expansion to new services) require the user to select a subset of result instances; selection is performed by means of checkboxes. When needed, a popup window asks for additional parameters or details on the operation to be performed.

Local filters on column values can be applied by clicking the "F" button on the column header of interest (e.g. the user may want to select only the restaurants having rating higher than three stars). Additional search dimensions (e.g. public transportation schedules, or other shows nearby for after dinner) can be added in order to augment a given solution set. Data summarization and visualization methods can also be used. The user interface and HCI aspects of Search Computing are further discussed in [4].

7 Conclusions

This paper has provided the definition of Service Marts as an interoperability concept for building Search Computing applications, with associated technologies for registering and adapting Web services. The Web world is described as a resource graph with Service Marts linked by and connection patterns, and then Service Marts are associated with service interfaces and implementations. Tool associated with Service Marts help the publication of arbitrary content (e.g., extracted from data sources or Web pages) in a standard format. Formats and extraction rules enable the execution of queries and the composition of query results.

Future work in this direction of research within the Search Computing project will address extending connection patterns so as to express semantically rich concepts in the context of search ("nearness" interpreted in the context of terminological, spatial, and temporal distance), so as to support richer corms of service compositions.

References

[1] Ankolenkar, A., et al.: DAML-S,
`http://www.daml.org/services/daml-s/2001/10/daml-s.html`

[2] Berners-Lee, T., Handler, J., Lassila, O.: The Semantic Web. Scientific American (May 2001)

[3] Bonifati, A., Cattaneo, F., Ceri, S., Fuggetta, A., Paraboschi, S.: Designing data marts for data warehouses. ACM Trans. Software Engineering Methodology 10(4), 452–483 (2001)

[4] Bozzon, A., Brambilla, M., Ceri, S., Fraternali, P.: Liquid query: multi-domain exploratory search on the Web. In: Proc. WWW-2010 Conference (2010, to appear)

[5] Braga, D., Campi, A., Ceri, S., Raffio, A.: Joining the results of heterogeneous search engines. Information Systems 33(7-8), 658–680 (2008)

[6] Braga, D., Ceri, S., Daniel, F., Martinenghi, D.: Optimization of multi-domain queries on the Web. In: Proc. VLDB, vol. 1(1), pp. 562–573 (August 2008)

[7] Ceri, S., Brambilla, M. (eds.): Search Computing - Challenges and Directions. LNCS, vol. 5950. Springer, Heidelberg (to appear, March 2010)

[8] Ceri, S., Matera, M., Rizzo, F., Demaldè, V.: Designing data-intensive Web applications for content accessibility using Web marts. Commun. ACM 50(4), 55–61 (2007)

[9] Colombo, M., Di Nitto, E., Di Penta, M., Distante, D., Zuccalà, M.: Speaking a common language: A conceptual model for describing service-oriented systems. In: Benatallah, B., Casati, F., Traverso, P. (eds.) ICSOC 2005. LNCS, vol. 3826, pp. 48–60. Springer, Heidelberg (2005)

[10] CORDIS (September 17, 2009), `http://cordis.europa.eu/fp7/ict/ssai/`

[11] De Witt, D.J., Ghandeharizadeh, S., Schneider, D.A., Bricker, A., Hsiao, H.-I., Rasmussen, R.: The Gamma Database Machine Project. IEEE TKDE 2(1), 44–62 (1990)

[12] Doan, A., Domingos, P., Halevy, A.Y.: Reconciling schemas of disparate data sources: A machine-learning approach. In: SIGMOD (2001)

[13] Fensel, D., Bussler, C.: The Web Service Modeling Framework WSMF. Electronic Commerce Research and Applications 1(2), 113–137

[14] Fensel, D., Musen, M.: Special Issue on Semantic Web Technology. IEEE Intelligent Systems (IEEE IS) 16 (2)

[15] Fusion Tables, `http://tables.googlelabs.com/`

[16] Gottlob, G., Koch, C., Baumgartner, R., Herzog, M., Flesca, S.: The Lixto data extraction project: back and forth between theory and practice. In: PODS '04 (2004)

[17] JSON, http://JSON.org/

[18] Levy, A.Y., Rajaraman, A., Ordille, J.J.: Querying Heterogeneous Information Sources Using Source Descriptions. In: VLDB (1996)

[19] Leymann, F.: WSFL,
`http://www-4.ibm.com/software/solutions/`
`Webservices/pdf/WSFL.pdf`

[20] Martin, D., Burstein, et al.: Bringing Semantics to Web Services with OWL-S. In: World Wide Web, vol. 10(3), pp. 243–277 (2007)

[21] OASIS. Web Services Business Process Execution Language. Technical report (2007),
`http://www.oasis-open.org/committees/wsbpel/`

[22] Quartel, D.S., Steen, M.W., Pokraev, S., Sinderen, M.J.: COSMO: A conceptual framework for service modeling and refinement. Information Systems Frontiers 9(2-3), 225–244 (2007)

[23] Srivastava, U., Munagala, K., Widom, J., Motwani, R.: Query optimization over Web services. In: VLDB '06, VLDB Endowment, pp. 355–366 (2006)

[24] W3C. Web Service Choreography Interface (WSCI) 1.0. W3C Note (August 2002)

[25] W3C. Web Services Choreography Description Language Version 1.0 (December 2004)

[26] Wang, J., Wen, J.R., Lochovsky, F., Ma, W.Y.: Instance-based schema matching for Web databases by domainspecific query probing. In: VLDB (2004)

[27] Web Services Architecture (2004), http://www.w3.org/TR/ws-arch/

[28] Wu, W., Yu, C., Doan, A., Meng, W.: An interactive clustering-based approach to integrating source query interfaces on the Deep Web. In: SIGMOD (2004)

[29] Yahoo! Search Boss, http://developer.yahoo.com/search/boss/

Engineering Autonomic Controllers for Virtualized Web Applications

Giovanni Toffetti[1], Alessio Gambi[1],
Mauro Pezzè[1,2], and Cesare Pautasso[1]

[1] University of Lugano
6904, Lugano, Switzerland
[2] University of Milano Bicocca
20126, Milan, Italy

Abstract. Modern Web applications are often hosted in a virtualized cloud computing infrastructure, and can dynamically scale in response to unpredictable changes in the workload to guarantee a given service level agreement. In this paper we propose to use Kriging surrogate models to approximate the performance profile of virtualized, multi-tier Web applications. The model is first built through a set of automated and controlled experiments at staging time, and can be later updated and refined by monitoring the Web application deployed in production. We claim that surrogate modeling makes a very good candidate for a model-driven approach to the engineering of an autonomic controller. Our experimental evaluation shows that the model predictions are faithful to the observed system's performance, they improve with an increasing amount of samples and they can be computed quickly. We also provide evidence that the model can be effectively used to synthetize an aggregated objective function, a critical component of the autonomic controller. The approach is evaluated in the context of a RESTful Web service composition case study deployed on the RESERVOIR cloud.

1 Introduction

More and more Web applications are hosted in Cloud computing environments to reduce their operational and maintenance costs. Cloud infrastructures build upon virtualization technology to simplify the deployment of Web applications and to enable application resources to be controlled dynamically [12]. Clouds offer the necessary flexibility to scale Web applications in order to support a variable number of clients during the runtime. These capabilities need to be balanced against increasing performance overhead and architecture complexity. Whereas the performance overhead may be acceptable for many applications [16], the problem of finding suitable deployment configurations of virtualized Web applications facing unpredictable client demand changes remains open.

In this paper we focus on multi-tier Web applications that are executed within virtualized infrastructures, and propose a method for automatically reconfiguring these applications in response to sudden and unpredictable changes in client workload that may derive for example from flash crowd or periodic peaks [2].

B. Benatallah et al. (Eds.): ICWE 2010, LNCS 6189, pp. 66–80, 2010.
© Springer-Verlag Berlin Heidelberg 2010

Introducing resource virtualization technology in infrastructures that host Web applications requires both to determine how many replicas of the Web application components to instantiate, and to address many details that include: how to assign each virtualized component to the physical resources, how to size the resources (CPU and memory) allocated to each virtual machine, how to bind service replicas with one another, and how to optimally distribute the client workload over a heterogeneous set of resources. Thus, the layer of abstraction introduced by virtualization in the Web application architecture augments the set of possible configuration decisions, and makes it very difficult to predict the effects of reconfiguration actions on the overall system performance.

To address these problems, we propose a model-driven approach to engineer an effective autonomic controller. Our models capture the relationship between many tunable configuration parameters and the expected performance of the Web application. In particular, we show how to apply multi-dimensional surrogate models [19] to study and predict the performance of virtualized Web applications. As more samples are observed, the prediction error of the models is reduced. We use the model to construct a utility function that can drive the controller self-configuration decisions. As opposed to other modeling approaches, the advantages of using surrogate models in this context are manifold. First, surrogate models are independent from the actual system complexity. As such, they can quickly predict the expected system behaviour [19]. They provide confidence measures that indicate the possible directions to follow when searching for optimal configurations. They effectively deal with highly-dimensional configuration spaces, and can be quickly updated at runtime. Thus, they support a continuous learning and prediction improving process. We use surrogate models to bridge the gap between measured non-functional system properties, like responsiveness, availability, and throughput, and the system configuration, to approximate the complex and unknown relation between them.

The rest of this paper is structured as follows. Section 2 defines the context and the architecture of the autonomic controller. Section 3 describes the case study use in this paper. Section 4 validates the approach experimentally by discussing the main research questions, the experimental methodology, the experimental results and their effective adoption for control. Section 5 outlines the related work. Finally, Section 6 summarizes the main contribution of this paper and delineates future research directions.

2   Architecture

Figure 1 shows the key elements of a virtualized Web application, and sketches a high-level representation of the controller architecture. The controlled system is composed of the virtualized Web application that runs on top of the physical hardware and network infrastructure. The system is designed with a public service interface and an internal management interface. The public service interface is used by the clients of the Web application, while the management interface enables both the monitoring of the system configuration (SC) and its performance (P) and the application of reconfiguration control actions (A).

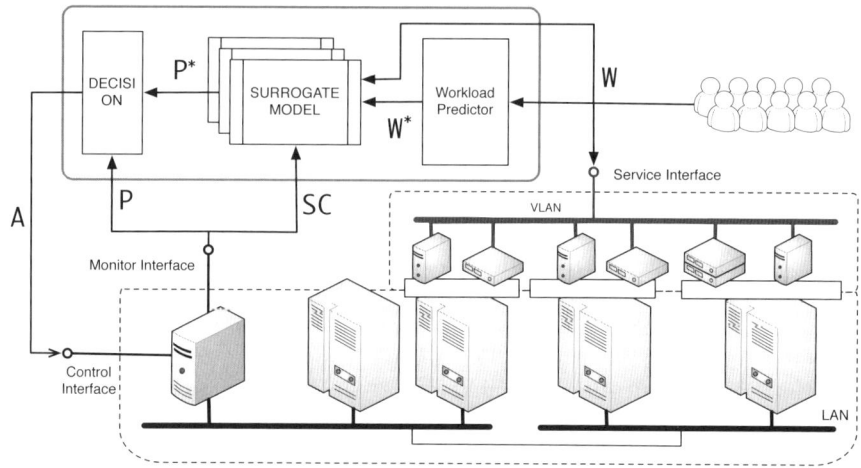

Fig. 1. Logical architecture of an autonomic, virtualized Web application

In production, the system is subject to a varying client workload (W) that can be characterized according to different dimensions (for instance, average inter-arrival times of request per request type, average request size, workload mix). We assume that the controller does not limit admission to the services and thus the workload cannot be altered by the controller. However, the controller could be fed both with information about the current workload (W) and the predicted future workload (W*). In control-theoretical terms, the workload is generally represented as a *disturbance*, that is a non-controllable system input [22]. The controller internal representation of the system is kept up-to-date by monitoring both layers of the controlled system. Monitoring data (the *system output*) coming from both the virtualized Web application components and the underlying physical infrastructure include the current system configuration, environmental information, as well as key performance indicators (KPIs) such as workload and performance measurements (for example, response times, throughputs and SLA violations). The KPIs are commonly used to express the goals (or service-level objectives, SLOs) of the controller. The controller aims to determine the optimal system configuration (*control input*) that meets the goals required to ensure that the performance of the Web application is acceptable.

Autonomic control approaches are characterized by the so-called MAPE-K closed-loop model, named after the basic activities that comprise the loop: Monitor, Analise, Plan and Execute with Knowledge [3,8]. Autonomic controllers are driven by control policies that express their main goals. For instance a control policy may give highest priority to preventing SLA violations, and, once SLAs are guaranteed, it may minimize the operational costs. Controllers *monitor* the systems key performance indicators that characterize the service quality. They *analyse* the collected data to diagnose for instance (potential) SLA violations. They *plan* a strategy to meet the control goals using *knowledge* about the expected system behaviour, usually expressed as models. They *execute* the control actions that implement the strategy. Finally, they evaluate the effects of the control actions updating their knowledge.

In a Cloud computing scenario, the possible control actions are amenable to virtual machine instantiation, de-instantiation, and placement, as well as setting the system to a specific point of its *configuration space*. In this paper, we focus on the critical *Knowledge* component of a controller. This component provides the configuration analyzer with the essential information to self-configure the virtualized Web application, as it contains the representation of the system used to find the appropriate configuration for a specific goal.

We capture the essential information of the knowledge component with surrogate models, that are mathematical approximations of unknown complex functions built from sampling [19]. Surrogate models provide both a predicted value and an accuracy measure of the actual function. They are widely used in engineering, when the sampling process is expensive, the exploration of the complete design space is not feasible, and an upper-bounded approximation is tolerable. For example, surrogate models are often used to reduce the highly expensive computer simulations or controlled experiments when exploring a vast design or configuration space.

Among many possible surrogate models, we use Kriging models [18] that interpolate the space with Gaussian processes to approximate the system behavior sampled through controlled experiments. Kriging models fit well our problem domain for several reasons. Modern Kriging provides an *exact* sample interpolation, i.e., the predicted outputs for the inputs that correspond to the samples match the measured outputs. Having a model that matches the samples exactly is important for the reliability of the predictions, and is often required in computer aided engineering. Kriging models cover the whole parameter space (the experimental area), thus they often produce better predictions than regression analysis [18]. Also, thanks to their excellent performance properties, Kriging models can be efficiently used to build controllers for run-time adaptation of the Web application configurations. As we show in this paper, Kriging models can be computed quickly, also in presence of many samples, and thus can deal with frequent updates.

3   Case Study

In our investigations we consider a composite RESTful Web service called Doodle Restaurant Map (DoReMap) [13]. DoReMap manages voting polls that are mashed up with a well-known map widget and facilitate agreements on nearby restaurants.

Users query the service for restaurants in the vicinity of a particular location. The service identifies a set of restaurants, and uses it to automatically create a voting poll so that users can cast their vote for a particular restaurant. To ease the choice, the voting poll is augmented with a map that shows the location of each restaurant.

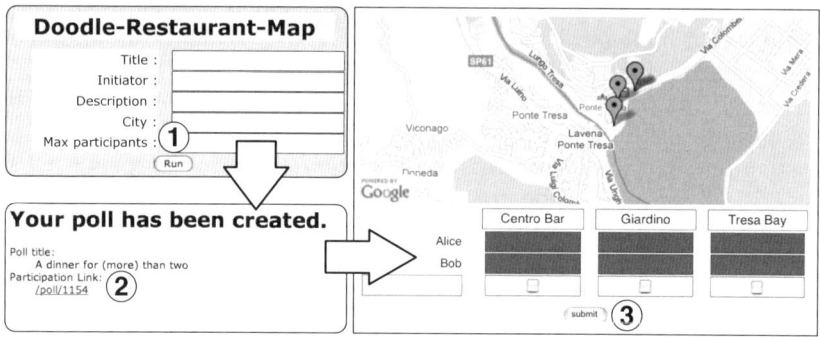

Fig. 2. The Doodle Restaurant Map (DoReMap) Web Application

Figure 2 visualizes the three main steps comprising the use of this service: (1) A user playing the role of *initiator* submits a poll creation form; (2) the system creates the poll and sends back to the initiator a *participation link*; (3) the initiator communicates the participation link to the other *participant* users that contact the poll, look at the restaurant locations on the map, and cast their votes. Once enough participants express their choices the poll is closed and no more votes can be submitted. The result of the poll might be inspected through the original participation link until the *initiator* deletes the poll from the system. We assume that DoReMap should comply with a simple SLA that specifies the maximum response time for each of the application requests, such as creation of the poll resource and vote.

3.1   Service Composition Model

The DoReMap service composes two atomic RESTful Web services: a restaurant search service, inspired by Yahoo! Local search engine API [1], used to query for restaurants near a given location, and a voting poll service inspired by the Doodle REST API[2], used to create, update and close polls.

The composition is modeled using the JOpera visual composition language [14], and is executed on the JOpera Engine version 2.4.9[3]. Figure 3 shows the control flow view of the JOpera composition model, limited to the creation of the poll and the handling of the client vote requests. The composition receives the input parameters from the initiator's form and then invokes the restaurant search service. Once the search is complete, the composition uses the data about the restaurants to create both the map that shows their location and the poll by invoking the voting poll service. When both steps have completed, the page hosting the poll mashup for the participants is generated and stored at a unique URI. This URI is then embedded as the participation link into the notification page that is sent back to the initiator.

---

[1] http://local.yahoo.com
[2] http://www.doodle.com
[3] http://www.jopera.org

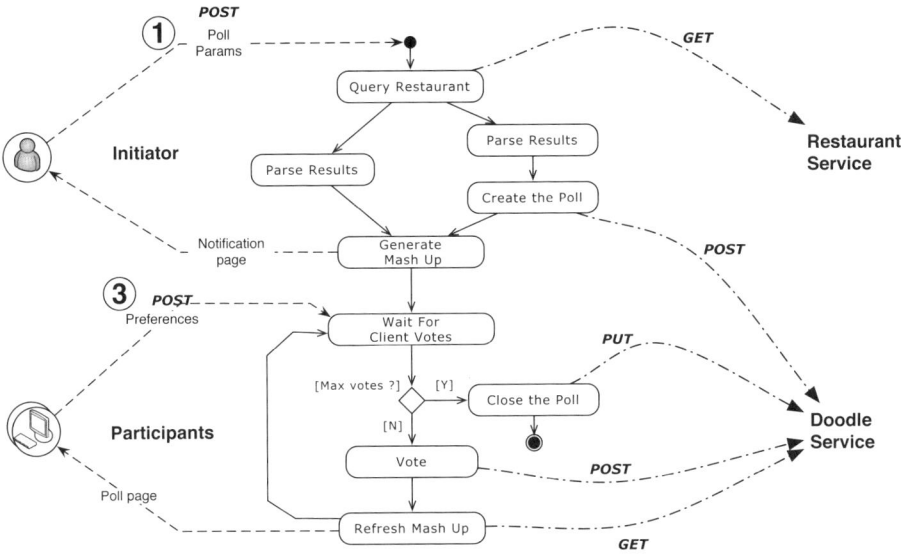

Fig. 3. Control flow model of the DoReMap service composition

Having completed the initialization stage, the composite service enters the main loop stage to collect the votes of the participants. At each vote, the composition checks the status of the poll service and updates it accordingly. When the status of the poll changes, the composition updates the poll page. The composite service continues executing until the number of votes reaches a threshold given by the initiator. At this point, the DoReMap service closes the poll, and keeps its final state published until explicitly deleted by a client.

3.2   System Architecture and Deployment

The architecture of DoReMap includes several components that are deployed inside four virtual servers interacting through a virtual network. As Figure 4 shows, the JOpera engine is deployed on its own virtual server (`JOperaAS`) to separate the logic implementing the composition from the component atomic services. Both sets of atomic services are designed as standard two-tier Web services and are composed of a REST front-end and a database back-end. They are deployed following different policies: the restaurant lookup service components are packaged into a single server, called `RestaurantAS`, because they are used only to read data, and because they should optimize the access to the data during the search; the voting services instead are deployed inside two separated servers, called respectively `DoodleAS` and `DoodleDB`, to separate the data access logic from the database tier.

Architecting the Web application as a loosely coupled composition of services deployed on separate virtual servers increases flexibility when it comes to deploying the composite service in the cloud. Each virtual server is packaged as a

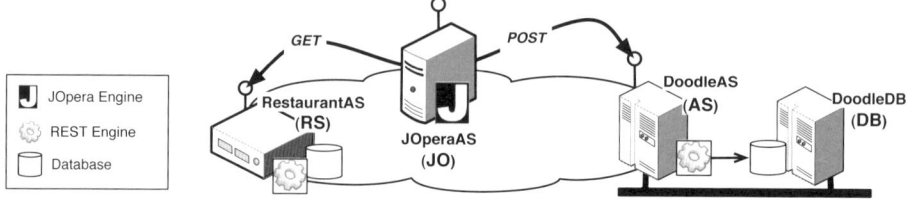

Fig. 4. Logical Architecture of the DoReMap Composite Service

disk image that can be seamlessly instantiated as virtual machine in the Cloud. To serve a demanding workload, multiple instances of critical services can be dynamically created without replicating the entire Web application [21]. For example, a growing number of concurrent clients might increase the number of requests at the composition layer. To prevent service saturation, new instances of JOpera components can be added to serve all requests without violating the SLA. Similarly, the system can respond to a changing workload mix by adding new `DoodleAS` and `RestaurantAS` server instances, thus scaling *horizontally*.

3.3   The RESERVOIR Cloud Computing Testbed

We executed the virtualized version of the DoReMap component services on an infrastructure developed within the FP7 RESERVOIR Project[4].

We executed the service on a partition of the RESERVOIR testbed cloud composed of six Blades IBM LS21 biprocessor dual-core Opteron 2218 at 2.6GHz with 8GB RAM DDR2 at 667MHz divided into two separated sites of three machines each, and linked by a dedicated high speed network. One of the machines was devoted to infrastructure services such as deployment and monitoring, while the remaining ones were used as raw resource pools, running the Web application virtual servers.

The RESERVOIR cloud infrastructure is designed to support run time deployment and live migration of virtual machines. This enables virtualized Web applications to be dynamically reconfigured and scaled to control the service behavior and performance by acting on the number (and the deployment) of virtual machines.

Combining this capability to quickly change the number of active virtual servers with the flexibility of a composite Web application, service providers can adapt the system to the actual load providing responsive services at reduced costs, as we show in the next Section.

4   Experimental Validation

In this section, we report the results of our experiments in building Kriging models to represent the behaviour of the DoReMap virtualized Web application. The

---

[4] http://www.reservoir-fp7.eu/

experiments aim to verify whether Kriging models can be used as an approximation of the behaviour of a realistic composed system, what kind of SLA-related metrics can be predicted accurately, and how they can be used to choose an optimal system configuration. In more details, the experiments address the following research questions:

Q1 How accurate is the prediction outside the training set?
Q2 How does the quality of the prediction increase with the number of samples?
Q3 How quickly can the model be computed/updated?
Q4 Can surrogate models be used to choose an optimal system configuration?

To answer **Q1** we first build surrogate models using a regular sample set in the feature space, and then we compare their predictions with respect to the system response measured at randomly chosen samples. We address **Q2** by measuring the prediction error of models generated with different sparse and small sample sets with respect to the model computed starting from the full sample set. We evaluate the cost of generating models (**Q3**) by benchmarking our algorithm with increasingly large samples. The speed of the fitting of Kriging models is a critical aspect to determine whether models can be kept up-to-date at run time and thus used to drive the adaptation decisions of an autonomic controller. We use the models to compute the objective function needed for the controller self-configuration functionality (**Q4**).

4.1   Experimental Setup

Our experiments aim to construct surrogate models that represent how different configurations (model input) impact on the system behaviour measured considering different KPIs (model output). As system configurations we consider the set of controllable system parameters (i.e., the number of VMs instantiated per each tier of the Web application) as well as the intensity of the workload (under our control at staging time, but not controllable in production). We measure the workload intensity in terms of number of clients that concurrently access the Web application.

The size of the system configuration space is limited by the available resources on which VMs can be allocated. In our case we deployed our experimental system on 20 physical CPU cores. We allocated the cores as follows: up to 8 cores to run the JOpera (JO) engines, up to 16 cores to run the Doodle Application Servers (AS), one core for the Restaurant Application Server (RS) and one core for the Database Server (DB). The allocation is constrained to at most 20 cores used simultaneously, and to each core dedicated to a single component to avoid contention that would complicate the interpretation of the results. The smallest working configuration thus requires each tier (JO+AS+RS+DB) to be instantiated, for a total of 5 cores, considering that each JOpera engine instance requires 2 cores. The largest configurations that we could test used 4 JOpera instances (consuming 8 cores) with 10 AS instances, and 16 AS instances with one instance of the JOpera engine. We used these configurations to determine the saturation point of the system (where the throughput stops increasing), and we observed that this occurs with about 100 concurrent clients. Thus, we need to explore 52 possible system configurations for each client workload (from 1 to 100 clients).

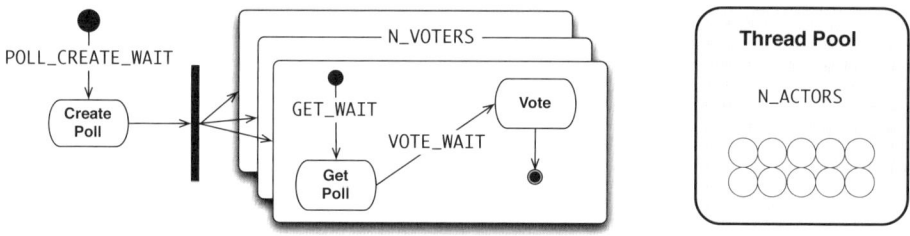

Fig. 5. Workload Model and Workload Generation Parameters

We sampled the parameter space through a batch of controlled experiments executed with Weevil [20]. To minimize undesired randomness, we repeated the experimental runs 5 times per sample, and we measured the output results for a given sample as the average over the run averages. To avoid measuring transient behaviours at component start-up or shut-down, each run lasted 5 minutes, and we discarded the first and last 10 seconds of observation. To stress the system, we used a synthetic client workload generated as a set of Poisson processes with different rates for each request type (POLL_CREATE_WAIT = 5 s, GET_WAIT = 2 s, VOTE_WAIT = 1 s, as shown in Figure 5). We controlled the workload intensity by selecting the number of concurrent client processes (N_ACTORS). Due to the specific nature of the Web application, in which the clients must follow a predefined navigation path by following hyperlinks (for instance get request only after poll creation with post, vote request after getting the available options), the effect of adding client processes to the workload is not reflected linearly in the measured system throughput. Rather, an undersized system configuration would result in a slower workload execution.

After collecting system response averages through reproduceable experimental scripts, we computed the Kriging model with the *octgpr*[5] Octave package. We set the parameters of the model training to high error tolerance to smoothly approximate the whole configuration space even with few samples, as shown in the results presented in the next section.

4.2   Results

We sampled and modeled the throughput (Figure 6.a) and the response time (Figure 6.b) that are the most critical KPIs. In both cases we show a projection of the 4-dimensional models by setting the workload intensity to 20, 40, and 60 concurrent clients. The x-y axes show the system configuration in terms of the number of VM instances running the JOpera engine and the Doodle Application

---

[5] http://octave.sourceforge.net/octgpr/index.html

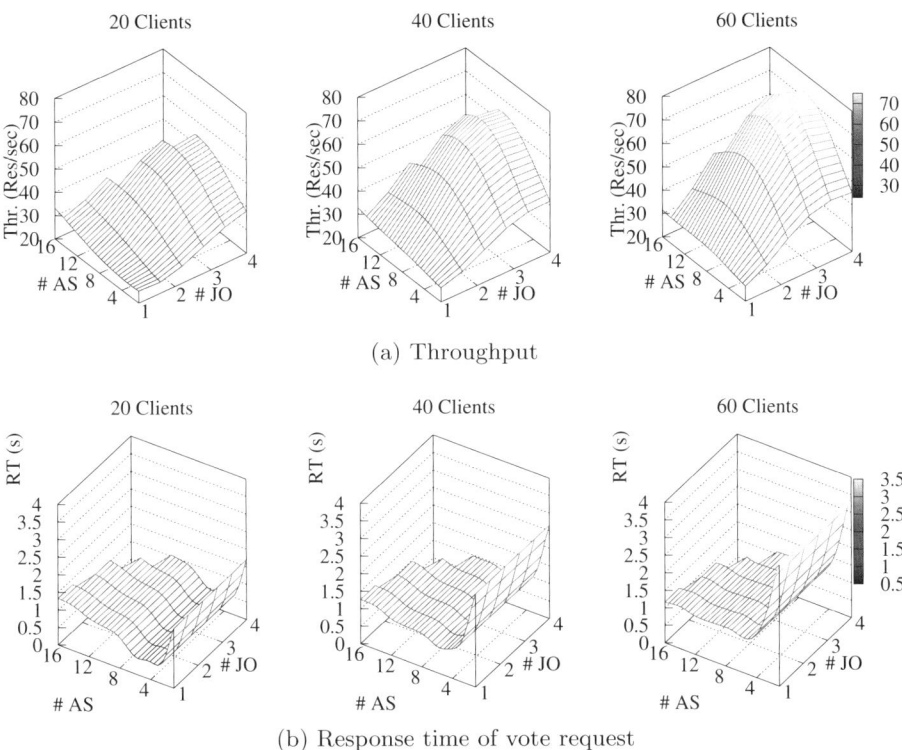

(a) Throughput

(b) Response time of vote request

Fig. 6. Surrogate model as a function of the number of JOpera VMs (# JO), Application Servers VMs (# AS), and Workload (# Clients) built from a regular 100-samples mesh

Server. The z axis shows the predicted throughput (in requests/second) or the response time (seconds).

The model reflects the scalablity of the system, as adding additional resources decreases the response time and increases the throughput. Also, the model predicts that for smaller workloads, the best performance (in terms of response time) is obtained with 7 replicas of the AS tier, while for a larger number of clients, the highest throughput is achieved with 4 JOpera engines and 10 instances of the application server.

Another feature of surrogate models concerns their ability to improve their predictions as more samples are fed into them. To study how the quality of the prediction increases with the number of samples, we built surrogate models using regular sampling patterns of 50, 75, 100 samples, and measured the prediction error with respect to a randomly generated validation set. Table 1 shows that the quality of the prediction increases with the coverage of the parameter space: the mean square error, the root mean square error and the average absolute error significantly decrease as the number of samples used to build the model

Table 1. System throughput: mean square error (MSE), root mean square error (RMSE), average absolute error (ME) and time needed to build the model (Time) with respect to sample sets of increasing size

Training set size	MSE	RMSE	ME	Time
50	117.47	10.838	7.4827	0.0241971 s
75	87.273	9.3420	6.1642	0.048146 s
100	71.402	8.4500	5.2788	0.087447 s

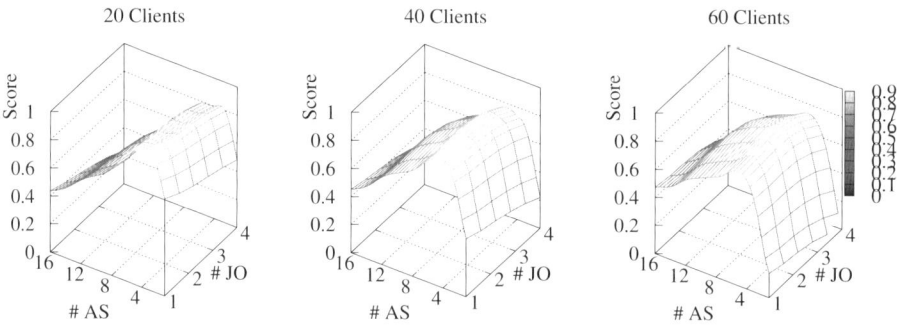

Fig. 7. Aggregated Objective Function (AOF)

increases. The last column of the table shows that the model can be computed in a small amount of time even when using the full set of samples. Thus, these models can be used within the closed control loop envisaged in the architecture of our controller, since they do not introduce a significant delay compared, for example, to the time required to start or shut down a new VM instance.

4.3   Objective Function

Finding a suitable configuration for a system using a set of models like the ones presented in the previous section becomes a multi-objective optimization problem. Several techniques are available to solve the problem, for instance, by ranking objectives (minimize SLA violations, then maximize throughput, and minimize operating costs), or with a single aggregated objective function (AOF), or with Pareto optimization methods.

To show that the surrogate models can be used as suitable input to analyze and optimize configurations, we define an objective function that aggregates them. In this proof of concept, we aggregate the response times for vote requests (R), system throughput (T), and VM operating costs (C) per hour[6] in the following form:

$$AOF(R, T, C) = \alpha * (2s - R) + \beta * T - \gamma * C \tag{1}$$

---

[6] As cost indication, we used Amazon EC2 Ireland prices mapping AS, DB, and RS to *Small* machines at \$0.095 per hour and JO to *Large* at \$0.38 per hour.

The AOF translates our service level objectives by giving a negative score for response times above 2 seconds. The second term gives a positive score to high throughputs. This is compensated by subtracting the operating costs of allocating more resources to the system. Parameters α, β, γ let the service designer provide a relative weight for each criterion.

Figure 7 shows the normalized AOF for varying client workloads using values $[\alpha = 10, \beta = 200, \gamma = 5000]$. We observe that the optimal configuration predicted by aggregating the surrogated models into this objective function varies with the number of expected clients respectively to 1 JO and 3 AS for 20 concurrent clients; 2 JO and 5 AS for 40 clients; 3 JO and 6 AS for 60 clients. This results indicates that our approach to modeling the system configuration and its performance can lead to a useful objective function that can be embedded into an autonomic controller.

Additional criteria built from surrogate models can be considered in the AOF (for instance response times for other requests, predicted percentage of SLA violations), as well as other more business-related metrics (for example the difference between the revenue in terms of successfully served requests versus the cost of violations).

5   Related Work

The study and design of controllers for virtualized Web applications is a lively research area. We can give a rough classification of different approaches according to the type of the control technology they adopt (e.g., rule based, control theory) and the knowledge representation that they use (i.e., white-box vs. black-box).

Rule based approaches do not have an explicit representation of the system: domain experts embed their knowledge in event-condition-action rules (ECA) that are evaluated at run time to trigger system adaptation. These controllers have limited capabilities as their effectiveness is bound to the domain experts' ability to define rules, and they do not have built-in learning mechanisms [9].

Control theoretic approaches apply classic control theory and describe system behaviour by means of first principle models or transfer functions. These approaches rely on mathematically-sound control techniques: well known results guarantee the stability of the control under the strong hypothesis of linear system behavior. However, for real systems, model identification (e.g., estimating the transfer function) becomes a difficult and time consuming activity [22]. Basic control theory approaches do not have learning capabilities. Still, more advanced controllers, such as self-tuning regulators (STR), can adapt to the actual system behaviour using different techniques for on-line model parameters estimation [10].

White box feedback loop approaches for controlling virtualized Web applications leverage knowledge of system internals to construct analytical representations in form of Queue Network models: simple *product form* versions of QN that can be analytically solved on-line [1]. Very specialized versions of QNs have been proposed for particular domains, such as multi-tier virtualized Web applications [4]. Other forms, such as Layered Queue Networks (LQNs), can express more details on system resource contentions and end-to-end behavior [15,5]. Queue networks give reliable system performance predictions and do not require any training of the model. However each change of system configuration requires the entire computation of the model and potentially the re-estimation of all its parameters [17,7].

Black box approaches trade model identification and parameter estimation with feature identification and model training, either in an experimental setting at staging time or through continuous learning while the system is in production. For example, artificial neural networks (ANN) are defined in terms of number of neurons and layers and must be extensively trained before deployment. Once deployed they may need to be retrained if the quality of their predictions decreases [11]. In the cited work, the authors use a ANN to predict if a composite service violates its SLA while continuously retraining the network through monitoring data. A different kind of black box model is employed in [6]. These authors exploits Bayesian Networks (BNs) to predict SLA violations caused by performance problems in a three-tiered application. In the approach, BNs are periodically updated from monitoring data, and the controller can query the model to obtain a probabilistic measure of SLA violation in the near future given the actual working conditions. In general, approaches based on ANNs and BNs are more demanding in terms of samples and training time required to build a reliable model than Kriging [18], hence we deem our solution more appropriate for autonomic controllers for virtualized Web applications where the parameter configuration space is very large and continuous learning is required.

6   Conclusion and Future Work

In this paper we propose to apply Kriging surrogate models to approximate the behavior of a virtualized Web application. This helps systems to automatically and dynamically control how the application is deployed on a cloud infrastructure based on its incoming client workload. We discussed how the main features of Kriging models closely match the requirements of such controllers by providing complete, precise, and quickly update-able representations of the complex multidimensional functions tying system configurations with different performance metrics. We presented our experience in applying our approach to a case study application deployed on a research cloud testbed showing the viability of the approach.

Our current research work concentrates on completing the development and study of a fully functional controller. As a first step, we plan to automate the experiments by actively using the surrogate model error prediction to drive the sampling phase. The second step will be catering for runtime monitoring of the relevant system KPIs and updating the surrogate model accordingly. The final step will be to define suitable optimization policies to make reconfiguration decisions based on aggregated objective functions such as the one presented in this paper. In the long term, we plan to automatically leverage additional knowledge about the system. For instance, we used the model of the service composition to identify an effective configuration space sampling strategy. That same knowledge can be combined at runtime with surrogate model prediction adopting different strategies to provide the autonomic controller with an improved representation of the system internals.

Acknowledgments

We wish to thank Antonio Carzaniga for the insigtful comments and discussions on the paper. This work is partially supported by the European Community under the IST programme of the 7th FP for RTD - project RESERVOIR contract IST-215605, by the S-Cube NoE and by the Swiss National Science Foundation SOSOA project (SINERGIA grant nr. CRSI22_127386).

References

1. Abrahao, B.D., Almeida, V., Almeida, J.M., Zhang, A., Beyer, D., Safai, F.: Self-adaptive SLA-driven capacity management for internet services. In: Proc. of IFIP/IEEE International Symposium on Integrated Network Management, pp. 557–568 (2006)
2. Almeida, V.A., Menascé, D.A.: Capacity planning: An essential tool for managing web services. IT Professional 4, 33–38 (2002)
3. Brun, Y., Serugendo, G.D.M., Gacek, C., Giese, H., Kienle, H.M., Litoiu, M., Müller, H.A., Pezzè, M., Shaw, M.: Engineering self-adaptive systems through feedback loops. In: Cheng, B.H.C., de Lemos, R., Giese, H., Inverardi, P., Magee, J. (eds.) Software Engineering for Self-Adaptive Systems. LNCS, vol. 5525, pp. 48–70. Springer, Heidelberg (2009)
4. Cunha, I., Almeida, J.M., Almeida, V., Santos, M.: Self-adaptive capacity management for multi-tier virtualized environments. In: Proc. of IFIP/IEEE International Symposium on Integrated Network Management, pp. 129–138 (2007)
5. D'Ambrogio, A., Bocciarelli, P.: A model-driven approach to describe and predict the performance of composite services. In: Proc. of the 6th International Workshop on Software and Performance, pp. 78–89 (2007)
6. Duan, S., Babu, S.: Proactive identification of performance problems. In: Proc. of ACM SIGMOD international conference on Management of data, pp. 766–768 (2006)
7. Ghezzi, C., Tamburrelli, G.: Predicting performance properties for open systems with KAMI. In: Proc. of the International Conference on the Quality of Software Architectures, pp. 70–85 (2009)
8. IBM. An Architectural Blueprint for Autonomic Computing. Technical report, IBM (2003)
9. Jung, G., Joshi, K., Hiltunen, M., Schlichting, R., Pu, C.: Generating adaptation policies for multi-tier applications in consolidated server environments. In: Proc. of International Conference on Autonomic Computing, pp. 23–32 (2008)
10. Karlsson, M., Covell, M.: Dynamic black-box performance model estimation for self-tuning regulators. In: Proc. of the International Conference on Autonomic Computing, pp. 172–182 (2005)

11. Leitner, P., Wetzstein, B., Rosenberg, F., Michlmayr, A., Dustdar, S., Leymann, F.: Runtime prediction of service level agreement violations for composite services. In: Proc. of the Workshop on Non-Functional Properties and SLA Management in Service-Oriented Computing (2009)
12. Lenk, A., Klems, M., Nimis, J., Tai, S., Sandholm, T.: What's inside the cloud? an architectural map of the cloud landscape. In: Proc. of the Workshop on Software Engineering Challenges of Cloud Computing, pp. 23–31 (2009)
13. Pautasso, C.: Composing RESTful services with JOpera. In: Bergel, A., Fabry, J. (eds.) Software Composition. LNCS, vol. 5634, pp. 142–159. Springer, Heidelberg (2009)
14. Pautasso, C., Alonso, G.: The jopera visual composition language. Journal of Visual Languages and Computing 16, 119–152 (2005)
15. Rolia, J., Casale, G., Krishnamurthy, D., Dawson, S., Kraft, S.: Predictive modelling of SAP ERP applications: Challenges and solutions. In: Proc. of the International Workshop on Run-time mOdels for Self-managing Systems and Applications, pp. 2–10 (2009)
16. Sotomayor, B., Keahey, K., Foster, I.: Overhead matters: A model for virtual resource management. In: Proc. of International Workshop on Virtualization Technology in Distributed Computing, pp. 35–42 (2006)
17. Urgaonkar, B., Pacifici, G., Shenoy, P., Spreitzer, M., Tantawi, A.: Analytic modeling of multitier internet applications. ACM Transactions on the Web 1(1), 2–37 (2007)
18. van Beers, W., Kleijnen, J.: Kriging interpolation in simulation: a survey. In: Proc. of Conference on Winter Simulation, pp. 113–121 (2004)
19. Wang, G.G., Shan, S.: Review of metamodeling techniques in support of engineering design optimization. Mechanical Design 129(4), 370–380 (2007)
20. Wang, Y., Rutherford, M.J., Carzaniga, A., Wolf, A.L.: Automating experimentation on distributed testbeds. In: Proc. of International Conference on Automated Software Engineering, pp. 164–173 (2005)
21. Wei, Z., Dejun, J., Pierre, G., Chi, C.-H., van Steen, M.: Service-oriented data denormalization for scalable web applications. In: Proc. of the International Conference on World Wide Web, pp. 267–276 (2008)
22. Zhu, X., Uysal, M., Wang, Z., Singhal, S., Merchant, A., Padala, P., Shin, K.: What does control theory bring to systems research? SIGOPS Oper. Syst. Rev. 43(1), 62–69 (2009)

AWAIT: Efficient Overload Management for Busy Multi-tier Web Services under Bursty Workloads[*]

Lei Lu[1], Ludmila Cherkasova[2], Vittoria de Nitto Personè[3],
Ningfang Mi[4], and Evgenia Smirni[1]

[1] College of William and Mary, Williamsburg, VA 23187, USA
{llu,esmirni}@cs.wm.edu
[2] Hewlett-Packard Laboratories, Palo Alto, CA 94304, USA
lucy.cherkasova@hp.com
[3] Universitá degli Studi di Roma "Tor Vergata", Rome, Italy
denitto@info.uniroma2.it
[4] Electrical and Computer Engineering, Northeastern University, Boston, MA
ningfang@ece.neu.edu

Abstract. The problem of service differentiation and admission control in web services that utilize a multi-tier architecture is more challenging than in a single-tiered one, especially in the presence of bursty conditions, i.e., when arrivals of user web sessions to the system are characterized by temporal surges in their arrival intensities and demands. We demonstrate that classic techniques for a session based admission control that are triggered by threshold violations are ineffective under bursty workload conditions, as user-perceived performance metrics rapidly and dramatically deteriorate, inadvertently leading the system to reject requests from already accepted user sessions, resulting in business loss. Here, as a solution for service differentiation of accepted user sessions we promote a methodology that is based on blocking, i.e., when the system operates in overload, requests from accepted sessions are not rejected but are instead stored in a blocking queue that effectively acts as a waiting room. The requests in the blocking queue implicitly become of higher priority and are served immediately after load subsides. Residence in the blocking queue comes with a performance cost as blocking time adds to the perceived end-to-end user response time. We present a novel autonomic session based admission control policy, called *AWAIT*, that adaptively adjusts the capacity of the blocking queue as a function of workload burstiness in order to meet predefined user service level objectives while keeping the portion of aborted accepted sessions to a minimum. Detailed simulations illustrate the effectiveness of *AWAIT* under different workload burstiness profiles and therefore strongly argue for its effectiveness.

1 Introduction

One of the most challenging problems for public Internet and e-commerce sites is the delivery of performance targets to users given the unpredictability of Web accesses. As Internet services become indispensable both for businesses and personal productivity,

* This work was partially supported by the National Science Foundation under grants CNS-0720699, CCF-0811417, and CCF-0937925.

B. Benatallah et al. (Eds.): ICWE 2010, LNCS 6189, pp. 81–97, 2010.

the efficient management of Internet services under periods where the system is over-loaded or simply highly variable, is of critical importance. There is a host of solutions to maintain user-perceived performance levels in the form of service-level objectives (SLOs) that focus mainly on admission control and/or techniques for service differen-tiation that are threshold based [13,6,7,19] but their effectiveness can be compromised if the workload is *bursty*, i.e., it is characterized by sudden temporal "surges" in the intensity of user arrivals [23] and user demands [22]. While capacity planning of sys-tems under bursty workload conditions has been recently demonstrated as critical for business success [22,23], the problem of efficient admission control and service differ-entiation under temporal workload bursts remains largely unexplored.

To get the intuition why threshold, usage based techniques may not be effective if the system is subject to bursty conditions, let us consider a system that provides web ser-vices and which is built according to the widely used multi-tiered paradigm. Typically, access to a web service occurs in the form of a *session* consisting of many individual requests. For a customer trying to place an order, or a retailer trying to make a sale, the real measure of a web server performance is its ability to process the *entire* sequence of requests needed to complete a transaction. Session-based admission control (SBAC) has been proposed as a solution to the above problem [13] and its gist can be summarized as follows: the system accepts a new session only when the system has enough capacity to process all future requests related to the session, i.e., the system can guarantee that the session completes successfully. If the system is functioning near its capacity, a new session will be rejected (or redirected to another server if one is available).

The original session-based admission control (SBAC) [13] is proposed for a single-tier web server, and its implementation is usage-based. SBAC accepts a new session only when the server CPU utilization is below a certain threshold. However, burstiness in the user arrival flows results in sudden, nearly simultaneous arrivals of requests in the system. The experiments presented in [23] show that under bursty arrivals SBAC is ineffective in maintaining a low ratio of aborted sessions due to a slow reaction to bursts.

Conventional wisdom suggests that the original session-based admission control can be extended for a multi-tiered system in a straightforward way: it should simply be employed at the bottleneck tier. Yet, if burstiness exists in the flows of a multi-tiered system (irrespective of its source, in the arrivals or service) then burstiness triggers the phenomenon of persistent bottleneck switch, i.e., the bottleneck continuously shifts to another tier [22], making control at the bottleneck tier an elusive task.

In this paper, we depart from threshold usage-based policies, and instead we dynam-ically control the number and the type of user requests admitted for processing into the multi-tier system. When the system enters the overload state, we advocate request buffer-ing from the already accepted sessions in a so-called "blocking" queue, that effectively acts as a waiting room. This blocking queue differentiates among the requests of already accepted sessions to those of new sessions, and implicitly gives them higher priority. To this end, we borrow ideas from the theory of queueing networks with blocking [5,25].

Blocking of accepted sessions during workload surges may be very effective in dif-ferentiating accepted sessions from new sessions, but the performance of accepted ses-sions is still directly bounded by the time the requests spent in the blocking queue. That is, if the time spent is so long that results in SLO violations, it is desirable to limit the capacity of the blocking queue in order to bound the user end-to-end time. We perform a sensitivity study to explore the different fixed blocking queue limits under a variety of burstiness profiles and conclude that the effectiveness of blocking is strongly related to the workload burstiness. To address this issue, we propose a parameter-free, auto-nomic session-based admission control policy called *AWAIT* that adjusts the blocking queue capacity in response to workload burstiness. We perform detailed simulations using the parameterized TPC-W benchmark with extended functionality for generating bursty session arrivals [23] to demonstrate the effectiveness and robustness of the new strategy. *AWAIT* supports a simple and inexpensive implementation. It does not require significant changes or modifications to the existing Internet infrastructure, and at the same time, it significantly improves the performance of overloaded multi-tier web sites.

This paper is organized as follows. Section 2 presents results that motivate this work. Section 3 presents the admission control algorithm and illustrates its robustness un-der different burstiness profiles by showing that it consistently meets the sought after performance goals while optimizing its performance targets. Section 4 positions this contribution within the context of related work. Section 5 summarizes the paper.

2   Capacity Planning and Admission Control

In this section, we present a short case study that illustrates how burstiness may im-pact in an unexpected way the performance of admission control. The basic model of an e-commerce site that we use in this paper is based on the TPC-W benchmark that is implemented as a typical multi-tier application which consists of a web server, an appli-cation server, and a back-end database. The web server and the application server reside usually within the same physical server, which is called front server. After a new session connection is generated, client requests circulate among the front and database server before they are sent back to the client. After a request is sent back, the client spends an average think time E[Z] before sending the following request. A session completes after the client has generated a series of requests.

Overload management is a critical business requirement for today's Internet services. A common approach to handle overload is to apply specific resource limits that typi-cally bound the number of simultaneous socket connections or threads. For example, in traditional web servers that employ thread-per-connection implementation, the server configuration specifies the number of processes (and connections) that are allocated for admitting the user requests. As an example, in the Apache web server [4], when all the server threads are busy, the system stops accepting new connections. The same princi-ple applies for providing the basic overload protection in multi-tier applications. The system administrators may set limits on the number of simultaneous client sessions (we call them *active requests*) in the system. Limiting the active requests is critical for qual-ity of service: setting this limit too low results in achieving a good response time but at a price of lower system throughput (and a high number of dropped user sessions). Setting this limit too high may lead to a better throughput and reduced drop rates at a price of a much higher response time.

Capacity planning is routinely used to determine the base number of active requests in order to strike a balance among the expected customer response times and dropped

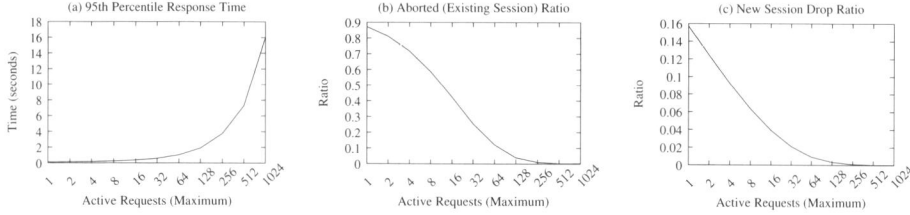

Fig. 1. Capacity Planning study for SBAC under exponential (i.e., not bursty) new session arrivals. Performance measures are presented as a function of the maximum number of active requests in the system.

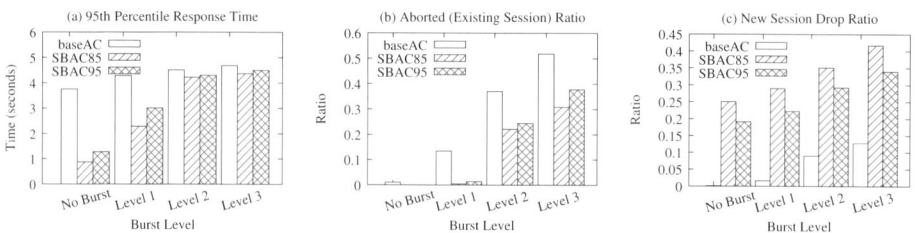

Fig. 2. Three different burstiness profiles. The capacity planning results and SLO targets are now violated. It appears that a queue size of 256 (i.e., maximum active requests for the *baseAC* configuration) is not sufficient to meet SLO requirements.

sessions. Figure 1 illustrates the results of a capacity planning study that models a typical TPC-W 2-tier implementation (i.e., a front server and a database server). The TPC-W defines 14 transactions, each of which can be generally classified as "browsing" or "ordering". Here, we assume that we use the ordering mix, that consists of 50% browsing and 50% ordering transactions. Request service times in the front and database servers are derived using the models presented in [22] that have been shown to capture very accurately the performance and behavior of multi-tier applications. Consistent with the specifications of the TPC-W benchmark, the average user think time is equal to 7 seconds, exponentially distributed. Inter-arrival times of new sessions are assumed to be exponentially distributed, i.e., there is no burstiness in the arrival stream of new sessions.

Each session consists of a sequence of requests (i.e., essentially visit "rounds" to the front and database server that define the a session length) that is uniformly distributed with parameters 5 and 35, that is with expected mean equal to 20.

It is a typical situation when after a certain waiting time an impatient client might "click again" and reissue the original request. Client request timeouts and retries can be added to our model to reflect a more complex and realistic scenario but according to a sensitivity analysis in [13] this will just decrease the useful system throughput (due to the processing overhead of these additional requests) but does not fundamentally change the results of our study. In this paper, we use a simplified model without request timeouts and retries in order to focus on the effects of burstiness.

Figure 1(a) illustrates the 95th percentile of user end-to-end response time as a function of the predefined value of the maximum active requests in the system. Figures 1(b) and 1(c) present the aborted rate of existing sessions and the drop rate of new sessions, respectively, as a function of the allowed active requests. The figure suggests that given this TPC-W mix, one may use 256 as the recommended limit on active requests, since this value strikes a good balance among all desired measures. In all experiments in the remaining of this paper we set the limit on active requests equal to 256.

Session-based admission control (SBAC) [13] is a very effective policy for web servers and is based on monitoring the CPU utilization of the web server. SBAC accepts a new session only when the system utilization is below a certain threshold, to guarantee a successful session completion. If the observed utilization is above a specified threshold, then for the next time interval, the admission controller rejects all new sessions and only serves requests from already admitted sessions. Once the observed utilization drops below the given threshold, the admission controller changes its policy for the next time interval and begins admitting and processing new sessions again. A web server employs a configurable size "listen queue" for buffering the incoming requests. If requests from sessions that are already accepted arrive when the queue is full, then they are aborted. The *useful throughput* of the system is measured as a function of the number of completed sessions. Aborted requests of already accepted sessions are highly undesirable because they compromise the server's ability to process all requests needed to complete a transaction and result in wasted system resources.

We have implemented the SBAC mechanism in a simulation model of a client-server system that is built according to the TPC-W specifications. The SBAC mechanism uses a front server utilization threshold for admitting new sessions.[1] Figure 2 illustrates the ineffectiveness of the threshold-based techniques in presence of bursty arrivals. We compare the results of two different admission control strategies. A first strategy (called *baseAC*) employs a traditional overload control based on admitting a fixed, predefined number of active requests for processing. Here, we set *ActiveRequests* = 256 as suggested by capacity planning (see Figure 1). The second strategy is SBAC where the front server utilization threshold is set to 85% and 95% respectively. The three burstiness profiles that we used here are further discussed and described in Section 3.2.

Figure 2(a) illustrates the 95th percentile of user response time. SBAC is effective in maintaining good response times under bursty arrivals but at the expense of a relatively high ratio of aborted sessions as well as a high ratio of rejected new sessions, see Figure2(b) and 2(c). The *baseAC* strategy does not differentiate between the requests from new and existing sessions and this leads to a very high ratio of aborted sessions. While both of these threshold-based strategies might be a reasonable choice under non-bursty traffic, they clearly exhibit their deficiencies under bursty traffic conditions. This simple experiment shows that the admission control mechanism has to take traffic burstiness into account and adapt the system configuration and/or thresholds in order

---

[1] For the TPC-W testbed used in our experiments of the ordering mix, SBAC uses the CPU utilization of the front server because the front server is the system bottleneck for this particular mix. In general, admission should be based on the utilization of the bottleneck resource, e.g., if the DB tier is a bottleneck then its CPU utilization should be used for SBAC decisions.

to effectively deal with bursty traffic conditions. In the next section, we present a new algorithm that effectively deals with the above problem.

3   *AWAIT* Algorithm

In this section, we describe *AWAIT*, a novel session-based admission control algorithm that aims to provide an additional support for dealing with bursty session arrivals. *AWAIT* has two different mechanisms to regulate request acceptance for processing. The first mechanism uses a counter of *ActiveRequests* that is defined according to capacity planning for achieving a given SLO for response time. Until this counter reaches its maximum any incoming request is accepted, this request may represent a new session or it may belong to an already accepted session. The second mechanism uses a special queue, called *blocking queue*, which is created to store the requests from already accepted sessions after the number of *ActiveRequests* reaches its maximum capacity. Via this mechanism, the *AWAIT* controller rejects new session requests if *ActiveRequests* reached its capacity but the system still admits requests from earlier accepted sessions. When the blocking queue becomes full, then incoming requests from accepted sessions are unavoidably aborted. This is undesirable because it leads to business loss.

The capacity of the blocking queue is a critical parameter for the performance of the accepted sessions since the time spent there contributes to the user end-to-end time, thus may violate the target SLOs. The larger the capacity of the blocking queue, the longer the contribution of the time waiting there to the user end-to-end time. Similarly, the larger the capacity of the blocking queue, the smaller the expected aborted ratio of accepted requests. Striking a good balance between these two conflicting measures is the purpose of *AWAIT*.

To ease the presentation of *AWAIT*, we first present a static version that considers a fixed blocking queue size. In the adaptive version of *AWAIT*, the size of this blocking queue is autonomically adjusted according to the burstiness of the workload while ensuring that the response time SLOs are met.

3.1   Static *AWAIT*

To formally describe the *AWAIT* algorithm, we introduce the following notions:

- *New session request* – a request that is generated by a new client (i.e., it is a first request in a new session);
- *Accepted session request* – a request that is issued by a client within an already accepted session;
- *ActiveRequests* – a counter that reflects the number of accepted requests which are currently in processing by the system. These active requests could be either of new sessions or of already accepted sessions. The maximum value for this counter is set to a value defined by capacity planning (see Section 2). Let us denote this value as A;
- *BlockedRequests* – a counter that reflects the number of blocked requests which are received from the clients of already accepted sessions and which are stored in the *BlockingQueue*. Note this difference: the blocking queue stores requests from

already accepted sessions only. Let B denote the maximum value of this counter that also defines the capacity of this queue;
- *AdmitNew* – a boolean variable that defines whether a new session can be accepted by the system. If $AdmitNew = 1$ then a new session can be accepted by the system. If $AdmitNew = 0$ then all the new sessions are rejected by the system;

Now, we describe the iteration steps of the algorithm. Let a new request *req* arrive for processing. The system can be in one of the following states.

- $AdmitNew = 1$ and $ActiveRequests < A$.
  This state corresponds to normal system processing when there is enough system capacity for processing requests from new sessions as well as requests from already accepted sessions. Therefore, independent on the request type *req* is accepted for processing and the counter *ActiveRequests* increases by one. When this counter reaches its maximum value A, then $AdmitNew = 0$, and this corresponds to a new system state when any requests from new sessions are rejected.
- $AdmitNew = 0$ and $BlockedRequests < B$.
  In this state the incoming requests are treated differently depending on their type. If the incoming request is from a new session then it is rejected. If it is part to an already accepted session, then it is stored in the $BlockingQueue$ and the queue's counter is updated.
- $AdmitNew = 0$ and $BlockedRequests = B$.
  This state reflects to the situation when $BlockedRequests$ has reached its maximum value B. Any incoming request, independent on its type, is rejected. If the request comes from an already accepted session, then its entire session is *aborted*.

Now, we describe how the system counters $ActiveRequests$ and $BlockedRequests$ are updated when a processed request leaves the system, i.e., the reply is sent to the client. The system can be in one of the following states (similar to the states described above).

- If $ActiveRequests < A$,
  then $ActiveRequests \leftarrow ActiveRequests - 1$.
- If $AdmitNew = 0$, $ActiveRequests = A$, and $BlockedRequests = 0$,
  then $ActiveRequests \leftarrow ActiveRequests - 1$ and $AdmitNew = 1$, i.e., the admission control status changes and the system again starts accepting both types of requests: from new sessions and already accepted sessions.
- If $AdmitNew = 0$, $ActiveRequests = A$, and $0 < BlockedRequests \leq B$,
  then one of the blocked requests is accepted for processing in the system and only the counter $BlockedRequests$ is updated: $BlockedRequests \leftarrow Blocked Requests - 1$.

We call this version of algorithm the *conservative AWAIT*. Under this algorithm the differentiation of requests from new and accepted sessions starts when the counter $ActiveRequests$ reaches its maximum value A. Then new sessions are rejected and requests from accepted sessions have extra buffering facility in the blocking queue. Once the $ActiveRequests$ counter gets below A, then the admission restriction is lifted and new session requests are again accepted.

We also introduce a different version of the algorithm, called *aggressive AWAIT*, which at a first glance is only slightly different from the *conservative AWAIT* above. However, the performance evaluation of these two versions shows a surprising difference in behavior and in the numbers of aborted and rejected sessions. As we see later, the *aggressive AWAIT* decreases forcefully the number of aborted sessions while supporting the same useful system throughput as the conservative *AWAIT*.

For the *aggressive AWAIT* strategy we introduce the additional variable $Overload$:

– *Overload* is a boolean variable that defines whether the system is under severe overload. Typically, $Overload = 0$ while the system can process all the requests from the already accepted sessions. $Overload = 1$ when system observes an aborted request from the accepted session. This may happen when $ActiveRequests = A$ and $BlockedRequests = B$, and the incoming request is from an accepted session. The aborted session triggers an "emergency situation" that is treated aggressively. New session requests are not accepted during overload until all the queues in the system are flushed. This helps in providing a prolonged preferential treatment of requests from the accepted sessions to rapidly overcome the overload state.

When the overload condition is triggered, i.e., $Overload = 1$, there are slightly different rules for updating the system state when a processed request leaves the system:

– If $AdmitNew = 0$, $Overload = 1$, $ActiveRequests = A$, and $Blocked Requests = 0$,
  then $ActiveRequests \leftarrow ActiveRequests - 1$, but the system is considered to be still under severe overload and its admission control status does not change: the system still rejects requests from new sessions and only processes requests from the already accepted sessions.
– If $Overload = 1$ and $ActiveRequests = 0$, then the operation of the system goes back to normal: $Overload = 0$ and $AdmitNew = 1$.

The pseudo-code shown in Figure 3 summarizes both versions of the *AWAIT* algorithm: conservative and aggressive. To unify the description, in the conservative version of the algorithm the state of variable $Overload$ does not change, i.e., $Overload = 0$.

In sum, the rationale for the *conservative* versus the *aggressive* version of the algorithm is the following. If the system operates under a burst, then queues tend to build up fast. An accepted session that is aborted signals the system about insufficient resource capacity for processing requests from already accepted sessions. To mitigate the performance effects of this, it is more effective to completely dedicate system resources for processing only the accepted session requests by flushing the system queues at the expense of a higher ratio of rejected new sessions. This strategy benefits accepted sessions by implicitly giving them priority and "reserving" the system for exclusive processing of accepted session requests (until overload subsides). In the following subsection, we present experimental evidence that shows the relative performance of the *conservative* versus the *aggressive* version of the algorithm.

3.2   Performance Evaluation of *AWAIT*

We evaluate the performance of *AWAIT* via trace driven simulation. Because our purpose is to evaluate the different proposed algorithms under varying burstiness levels,

```
For every request req that arrives for processing
  if (AdmitNew AND ActiveRequests < A)
    accept req
    ActiveRequests = ActiveRequests + 1
    if (ActiveRequests == A)
      AdmitNew = 0
  else if (!AdmitNew AND BlockedRequests < B)
    if (type(req) == NewSession)
      reject req
    if (type(req) == AcceptedSession)
      accept req into BlockingQueue
      BlockedRequests = BlockedRequests +1
  else if (!AdmitNew AND BlockedRequests == B)
    reject req      //Reject all requests
    if (type(req)==AcceptedSession)  //Accepted session is aborted
      Overload=1    //Aggressive version: trigger overload state

For every request req that leaves the system
  if (ActiveRequests < A)
    ActiveRequests = ActiveRequests -1
  else if (ActiveRequests==A AND 0<BlockedRequests≤B
    move one request from blocking queue to queue
    BlockedRequests = BlockedRequests -1
  else if (ActiveRequests == A AND BlockedRequests == 0)
    ActiveRequests = ActiveRequests -1
    if (!Overload)
      AdmitNew = 1
  if (ActiveRequests==0 ) //Aggressive version: queues flushed
    Overload = 0     //Restore overload state to normal
    AdmitNew = 1     //Start admitting new sessions
```

Fig. 3. *AWAIT*: Admission control algorithm, aggressive version. The *conservative AWAIT* is obtained by removing the statements labeled **Aggressive version**.

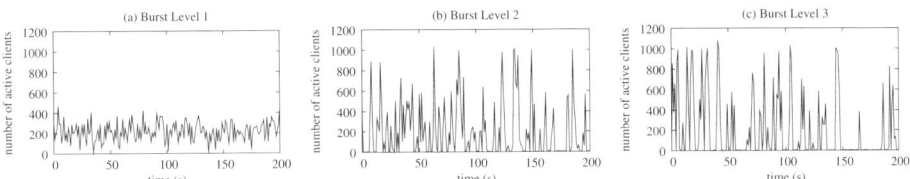

Fig. 4. The burstiness profiles of the three arrival MAPs

we conducted experiments assuming that arrivals of new sessions are bursty. We use a Markovian Arrival Process (MAP) to generate three arrival processes with the same mean and variance but with distinctive burstiness profiles. For details on the generation of the three MAP processes as well as on their effectiveness in mimicking bursty arrivals such those reported in the 1998 World Cup web server we direct the reader to [23]. The burstiness profiles (i.e., the number of arrivals as a function of time) for the three MAPs that we use for the arrival process are illustrated in Figure 4.

The service processes at the front server and the database server are also modeled via MAPs (see [22]) that accurate capture the service demands of TPC-W's ordering mix.[2] Each session consists of a sequence of requests that defines a *session length*. MAPs have

---

[2] Experiments with TPC-W's ordering and browsing mixes were also conducted. Results are qualitatively the same as with the ordering mix and are not reported here due to lack of space.

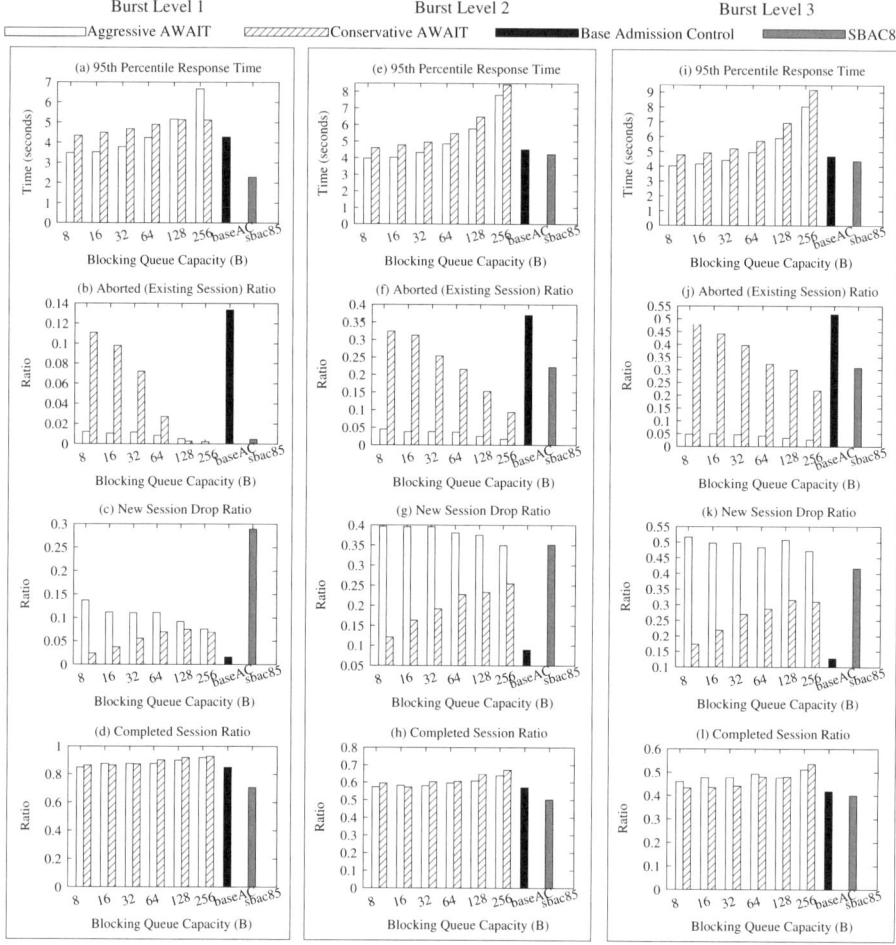

Fig. 5. *AWAIT* with fixed size of the blocking queue. The graphs illustrate performance values for the aggressive and conservative versions (see white and shaded bars, respectively) for various fixed sizes of the blocking queue B. In all experiments, the limit of accepted requests A is set to 256, based on capacity planning.

been shown to be surprisingly compact yet very effective models of the service process in multi-tiered systems, modeling implicitly conditions such as caching or database locks (see [22]). Session lengths are uniformly distributed between parameters 5 and 35, that is with expected mean equal to 20.

Figure 5 illustrates the performance of the two versions of *AWAIT* as a function of the capacity of the blocking queue B. For reference, we also report on the performance of the system with simple admission control based on the number of *ActiveRequests* only (labeled: "baseAC") as well as the performance of SBAC with CPU utilization threshold set to 85%. Note that for *all* experiments, we set the *ActiveRequests* counter to 256, as suggested by the capacity planning study of Section 2.

The figure presents results for the three burstiness profiles in the arrivals of new sessions. First, one can easily see that the effect of the degree of burstiness in the arrivals dramatically impacts the user perceived performance, see the 95th percentiles of user response times for the various policies, first row of graphs. Looking just at the percentiles, it is clear that the addition of the blocking queue deteriorates the user end-to-end times but the real benefit of blocking can be seen in the decrease of the aborted session ratio, see the second row of graphs, as well as in the decrease of new session drop ratio, see the third row of graphs. The useful throughput of the system (measured in successfully completed sessions) is shown in the last row of graph that demonstrate the improved metric for both versions of *AWAIT* strategy compared to SBAC and *baseAC*.

Under low burstiness conditions, see first column of graphs, it is apparent that SBAC remains a good choice, at the expense of a very high percentage (nearly as high as 30%) of new session rejections. The aggressive and conservative versions of *AWAIT* result in longer response times but in significantly lower drop ratios, see Figures 5(b) and 5(c). With higher burstiness levels, the aggressive version results in better response time percentiles, see Figures 5(e) and 5(i).

The effectiveness of the aggressive version to keep the aborted session ratio low is apparent across all burstiness levels, see Figures 5(b), 5(f), and 5(j) (second row of graphs). These figures show that the aggressive version very effectively differentiates between existing and new sessions, and treats existing sessions preferentially.

Naturally, because of the limited system capacity, if the number of accepted sessions that are aborted is low, then the ratio of rejected new sessions is bound to increase. This effect is shown for the aggressive policy in the third row of graphs in Figure 5, but this is unavoidable since our purpose is to bias the system for processing the requests of already accepted sessions against admitting new sessions, especially under periods of bursty traffic.

However, intuitively, there is an additional concern on the effectiveness of the aggressive *AWAIT* strategy compared to its conservative version: "flushing" the system queues might result in a less efficient resource usage and potentially may lead to a lower useful throughput. The last row of graphs in Figure 5 answers this question. It shows that the useful throughput of the system measured in successfully completed sessions is very similar for both conservative and aggressive versions of *AWAIT* and also significantly higher than under earlier proposed SBAC strategy or the simple *baseAC* policy.

In summary, the Figures 5 shows that the aggressive *AWAIT* minimizes the number of aborted sessions while meeting service SLOs. Yet, its performance is sensitive to the capacity of the blocking queue B. In the following section we present an adaptive algorithm that changes the blocking queue capacity as a function of the workload burstiness in order to adaptively meet SLO targets.

3.3   Adaptive *AWAIT* Strategy

Here, we show how we can adjust on-the-fly the size of the blocking queue B in order to achieve a certain predefined SLO. Larger blocking queues result in longer user response times but have less aborted sessions.

To dynamically adjust the blocking queue size, we use historical information of the achieved 95th percentiles of all requests served by the system (irrespective of the blocking queue capacity used – this value should reflect the target system SLO as the size of the blocking queue is transparent to the user) but also response time percentiles that correspond to *every* other blocking queue capacity B used since the inception of the system. We use this information to decide whether the current blocking queue capacity is sufficient or not. Changing the blocking queue capacity B throughout the lifetime of the system is critical as during workload surges smaller B's result in better performance rather than large B's.[3] To make readily available the values of the 95th percentiles of the user response times, we maintain for each blocking capacity B a corresponding his-togram of the user response times for that B. Therefore, for each completed request, two response time histograms are updated: the histogram of all requests in the system (irre-spective of the blocking capacity B) and the histogram that corresponds to the current block capacity B used.

We decide whether to change the capacity of the blocking queue for every group of $K = 10,000$ requests served.[4] The adaptive algorithm then compares the achieved response time percentiles of *all* jobs in the system and the response times percentiles of the current configuration B with the target SLOs. If both percentiles are less than the SLO and there are aborted sessions, then it is clear that we can reduce the aborted ratio because there is room to increase B (since response times percentiles do not violate the SLO). If both percentiles are greater than the SLO, then the blocking queue should be reduced in an effort to meet the SLO target. If none of the above two conditions are met, we opt to leave the blocking queue capacity in its current level, otherwise the system may suffer from thrashing. For example, if the response time percentile of all requests is violated, but the percentile of the current B is not, the algorithm still stays with the current blocking queue size B, since the system is on a positive state and its accumulated statistics eventually will correct the percentile of all requests.

The steps of increase/decrease of the blocking queue capacity can be arbitrary. In the experiments presented in this section, the capacity of the blocking queue B can have sizes as small as 1 and as large as 120. The increase/decrease step is equal to 5 for values of B less than 10 and equal to 20 for values of B greater than 20. We stress that other step values could also work, their selection may affect though how quickly the algorithm converges to a desirable B range. Figure 6 summarizes the algorithm.

The effectiveness of the adaptive *AWAIT* strategy is illustrated in Figure 7. Here, we experimented with the three different burst levels but also using different target SLOs. The figure illustrates how the blocking queue size changes as a function of the number of requests that are processed by the system for the various experiments. In each graph we also report on the achieved 95th percentile of the round-trip time, as well as on the aborted and new session drop ratios. The figure shows that the adaptive *AWAIT* is remarkably robust: it reaches the target SLOs exceptionally well for all cases,

---

[3] This may initially seem counter-intuitive as workload surges would result in large numbers of requests that are simultaneously in the system. However, in order to maintain the target SLOs during a surge it is necessary to limit the blocking queue capacity, otherwise the time spent there dominates user response times and SLOs are violated.

[4] We selected $K = 10,000$ to be able to collect meaningful statistics for a group of requests. K should be large enough for accumulating meaningful statistics, but different values, e.g., $K = 5,000$ or $K = 15,000$ will work too.

```
For every aborted session
    AbortedSessions++
For every finished request
    counter++
    update total_RT_histogram (all requests, irrespective of B)
    update current_B_RT_ histogram (with current blocking queue B)
    if (counter == K)
        if (total_RT_percentile < SLO AND current_B_RT_percentile < SLO
        AND AbortedSessions > 0)
            increase current blocking capacity B   // Reduce aborted ratio
        if (total_RT_percentile > SLO AND current_B_RT_percentile > SLO)
            reduce current blocking capacity B   // Meet SLO target
        counter = 0
        AbortedSessions = 0
```

Fig. 6. Policy for adapting the blocking queue size B in the enhanced, adaptive *AWAIT* strategy

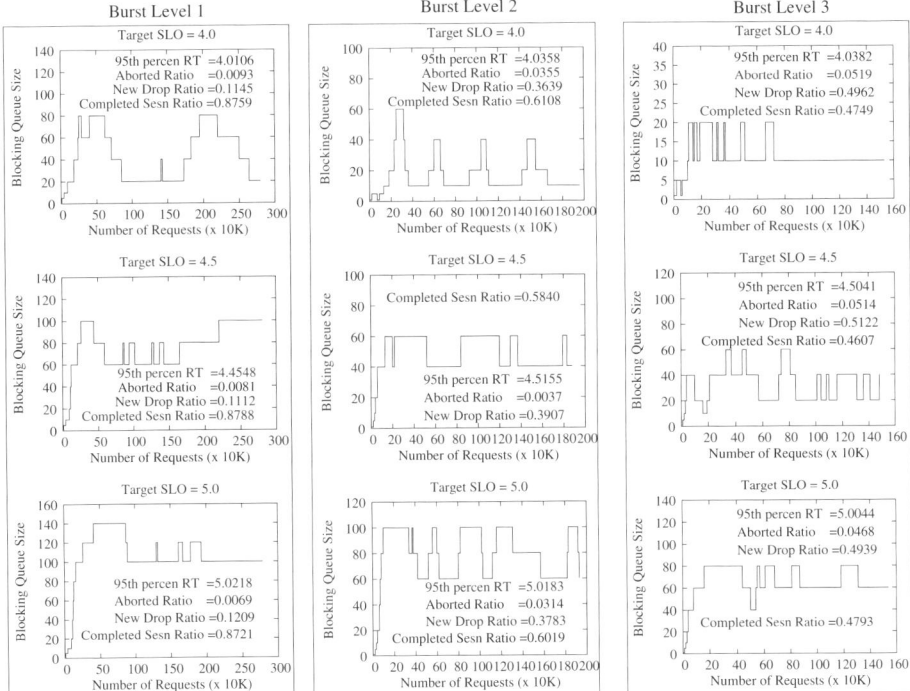

Fig. 7. Adaptive *AWAIT*: illustration of how the capacity of the blocking queue B changes as a function of the workload

while maintaining very low aborted rates. For each burst level, as the target SLO increases, the algorithm effectively increases the blocking queue capacity while reducing the aborted ratio. If we maintain the same SLO but change the burstiness of arrivals, the algorithm decreases the capacity of the blocking queue B. In all experiments, requests from existing sessions are preferentially treated as low aborted ratios across all experiments are reported, and the ratio of successfully completed sessions is higher under the adaptive *AWAIT* policy compared to the aggressive static *AWAIT* strategy introduced in Section 3.1. These results demonstrate the effectiveness and robustness of the proposed autonomic mechanism of the aggressive *AWAIT* policy.

Note that the target SLO can be achieved with a fixed blocking size queue, but the size of the blocking queue needs to differ depending on the degree of burstiness (e.g., $SLO = 4$ sec can be achieved with a blocking queue size set to 8 for the burst levels 2 and 3, but if the system operates under burst level 1, then the blocking queue size should be set to 32, see Figure 5). Note that any fixed configuration does not adapt to a changing traffic pattern. The proposed adaptive strategy is specially designed to "auto-tune" the blocking queue size for achieving and supporting a given SLO.

4   Related Work

There has been a lot of research in the areas of overload control, service differentiation, request scheduling, and request distribution for Web servers and web server clusters. Due to space limitations, we provide a very brief overview here.

The use of admission control for an overload management has been proposed and explored in several systems. Iyer et al. [18] employ a simple admission control mechanism based on bounding the length of the Web server listen queue. The authors try minimizing the work spent on a request which is eventually not serviced due to overload. They analyze different queue management approaches and use multiple thresholds, though they do not specify how these thresholds should be set to meet a given performance target. Cherkasova and Phaal [13] introduce session-based admission control, driven by a CPU utilization threshold, which performs an admission decision based on user sessions rather than individual requests, and during the overload rejects new sessions while serving requests from already accepted sessions. Carlstrom and Rom [9] proposed a performance model for scheduling client requests and session-level admission control using generalized processor scheduling discipline. To improve the efficiency of session-based admission-control mechanisms and reduce its overhead, Voigt et al. [28,29] present several kernel-level mechanisms for overload protection and service differentiation. The earlier works consider a single tier web server, and the proposed techniques do not directly provide a solution for a multi-tier system.

Many of the proposed techniques are based on fixed policies, such as bounding the maximum request rate of requests to some constant value. For example, PACERS [11] limits the number of admitted requests based on estimated web server capacity. The authors use a very simple simulated service where request processing time is linear function of the requested Web page size. Similar ideas (and similar problems with fixed threshold settings) are pursued in [8]. Web2K presents a mechanism prioritizing requests into two classes: premium and basic. Connection requests are forwarded into two different request queues, and admission control is performed using two metrics: the accept queue length and measurement-based predictions of arrival and service rates from that class. Bartolini et al., in their recent work [6,7], introduce a quite elaborate session admission algorithm, called AACA, that self-configures a dynamic constraint on the rate of incoming new sessions to satisfy guarantees of the Service Level Agreements (SLA). However, the rate limitation for the next iteration interval is based on a relatively straightforward prediction of the session arrival rate from the previous interval measurements.

Many earlier papers combine differentiated service with admission control [3,15,19,20,28]. Kanodia and Knightly [19] develop an admission control and service differentiation mechanism which is based on a general framework of request and service envelopes. Such envelopes statistically describe the server's request load and service capacity as a function of interval length.The proposed mechanism integrates latency targets with admission control and improves the percentage of requests that meet their QoS delay requirements. The approach is evaluated via a trace-driven simulation.

A number of systems have explored a controlled content adaptation [1,10,17] for scaling web site performance, i.e., degrading the quality of static Web content by reducing the resolution and the number of images delivered to clients.

Several research papers have examined how control theory can be applied in the context of Web servers [2,21,24]. Lu et al. [21] present a control-theoretic approach to provide guaranteed relative delays between different service classes. Main challenge in such works is that good models of system behavior are difficult to derive. Web applications are subject to widely varying traffic patterns and resource demands. The considered papers make use of linear models, which may be inaccurate in describing systems with bursty loads and resource requirements.

Many earlier papers study request and connection scheduling for improving Web server performance [12,14,16]. While shortest job first scheduling for static content Web sites can improve performance of a web server, it can not prevent it from overload though. Elnikety et al. [16] present an elegant solution for admission control and request scheduling for multi-tier e-commerce sites,. Their method is based on measuring the execution costs of requests online, distinguishing different request types, and performing both overload protection and preferential scheduling using a straightforward control mechanism. They implement their admission control using proxy, called Gatekeeper, with standard software components on the Linux operating system. There were a few other works close to Gatekeeper in spirit, SEDA [31] is a prime example of such work. In SEDA, applications consist of a network of event-driven stages connected by explicit queues. SEDA makes use of a set of dynamic resource controllers by preventing resources from being over-committed when demand exceeds service capacity. It keeps stages within their operating regime despite large fluctuations in load and allows services to be well-conditioned to load, i.e., preventing their performance degradation under severe overload. The authors describe several control mechanisms for automatic tuning and load conditioning, including thread pool sizing, event batching, and adaptive load shedding.

5   Conclusions

We presented an autonomic policy for service differentiation and admission control during overload for multi-tiered system management that offer web services. We focused on the pitfalls of existing policies under bursty conditions and remedy the problem by proposing the concept of a blocking queue where requests from already accepted sessions are stored if the system operates in overload. This blocking queue benefits performance by minimizing the dropped requests of already accepted sessions but also contributes to the end-to-end user perceived system response time. We proposed a novel autonomic algorithm, called *AWAIT*, that can limit the increase of the end-to-end response times within predefined SLO targets while dynamically adjusting the capacity of the blocking queue to the workload burstiness. Detailed simulations with the widely used TPC-W e-commerce benchmark under a variety of workload burstiness levels support the effectiveness and robustness of *AWAIT*.

The current algorithm adapts the blocking queue capacity to shield the offered web service from bursty arrivals, to provide service differentiation, and to prevent the system from overload. It complements the basic overload mechanism that sets a limit on the number of active client requests that are simultaneously processed by the system. Currently, this limit is defined by capacity planning. In our future work, we plan to automate the capacity planning step as well, i.e., to adjusts the value of this basic parameter on-the-fly when the workload profile experiences significant changes. We are also working on theoretically determining the ideal blocking queue capacity given a level of workload burstiness.

References

1. Abdelzaher, T., Bhatti, N.: Web content adaptation to improve server overload behavior. Computer Networks 31(11-16) (1999)
2. Abdelzaher, T., Shin, K.G., Bhatti, N.: Performance guarantees for Web server end-systems: A control-theoretical approach. IEEE Transactions on Parallel and Distributed Systems 13(1) (2002)
3. Almeida, J., Dabu, M., Manikutty, A., Cao, P.: Providing differentiated levels of service in Web content hosting. In: Workshop on Internet Server Performance, Madison, WI (June 1998)
4. Apache Software Foundation. The Apache Web server, http://www.apache.org
5. Balsamo, S., de Nitto Personè, V., Onvural, R.: Analysis of Queueing Networks with Blocking. Kluwer Academic Publishers, Dordrecht (2001)
6. Bartolini, N., Bongiovanni, G., Silvestri, S.: An autonomic admission control policy for distributed web systems. In: Proc. of the Intl. Symp. on Modeling, Analysis, and Simulation of Computer and Telecommunication Systems, MASCOTS '07 (2007)
7. Bartolini, N., Bongiovanni, G., Silvestri, S.: Self-* overload control for distributed web systems. In: Proc. of the Intl. Workshop on Quality of Service, IWQoS (2008)
8. Bhoj, P., Ramanathan, S., Singhal, S.: Web2K: Bringing QoS to Web servers. Technical Report HPL-2000-61, HP Labs (May 2000)
9. Carlstrom, J., Rom, R.: Application aware admission control and scheduling in web servers. In: Proc. of INFOCOM (2002)
10. Chandra, S., Ellis, C., Vahdat, A.: Differentiated multimedia Web services using quality aware transcoding. In: Proc. of INFOCOM (2000)
11. Chen, X., Mohapatra, P., Chen, H.: An admission control scheme for predictable server response time for Web accesses. In: Proc. of the 10th World Wide Web Conference (WWW), Hong Kong (2001)
12. Cherkasova, L.: Scheduling strategy to improve response time for Web applications. In: Bubak, M., Hertzberger, B., Sloot, P.M.A. (eds.) HPCN-Europe 1998. LNCS, vol. 1401. Springer, Heidelberg (1998)
13. Cherkasova, L., Phaal, P.: Session-based admission control: A mechanism for peak load management of commercial Web sites. IEEE Transactions on Computers 51(6) (June 2002)

14. Crovella, M., Frangioso, R., Harchol-Balter, M.: Connection scheduling in Web servers. In: Proc. of the USENIX Symposium on Internet Technologies and Systems, USITS (1999)
15. Eggert, L., Heidemann, J.: Application-level differentiated services for Web servers. World-Wide Web Journal 2(3) (August 1999)
16. Elnikety, S., Nahum, E., Tracey, J., Zwaenepoel, W.: A method for transparent admission control and request scheduling in e-commerce web sites. In: Proc. of the World Wide Web Conference, WWW (2004)
17. Fox, A., Gribble, S.D., Chawathe, Y., Brewer, E.A., Gauthier, P.: Cluster-based scalable network services. In: Proc. of the 16th ACM Symposium on Operating Systems Principles, SOSP (1997)
18. Iyer, R., Tewari, V., Kant, K.: Overload control mechanisms for Web servers. In: Workshop on Performance and QoS of Next Generation Networks, Nagoya, Japan (November 2000)
19. Kanodia, V., Knightly, E.W.: Ensuring latency targets in multiclass Web servers. IEEE Transactions on Parallel and Distributed Systems 13(10) (October 2002)
20. Li, K., Jamin, S.: A measurement-based admission-controlled Web server. In: Proc of INFOCOM (2000)
21. Lu, C., Abdelzaher, T.F., Stankovic, J.A., Son, S.H.: A feedback control approach for guaranteeing relative delays in Web servers. In: IEEE Real-Time Technology and Applications Symposium (2001)
22. Mi, N., Casale, G., Cherkasova, L., Smirni, E.: Burstiness in multi-tier applications: Symptoms, causes, and new models. In: Issarny, V., Schantz, R. (eds.) Middleware 2008. LNCS, vol. 5346, pp. 265–286. Springer, Heidelberg (2008)
23. Mi, N., Casale, G., Cherkasova, L., Smirni, E.: Injecting realistic burstiness to a traditional client-server benchmark. In: Proc. of the 6th Intl. Conference on Autonomic Computing, ICAC (2009)
24. Parekh, S., Gandhi, N., Hellerstein, J.L., Tilbury, D., Jayram, T., Bigus, J.: Using control theory to achieve service level objectives in performance management. In: Proc. of the IFIP/IEEE International Symposium on Integrated Network Management (2001)
25. Perros, H.G.: Queueing networks with blocking. Oxford University Press, Oxford (1994)
26. Shen, K., Tang, H., Yang, T., Chu, L.: Integrated resource management for cluster-based Internet services. In: Proc.of Operating Systems Design and Implementation, OSDI (2002)
27. TPC-W Benchmark, http://www.tpc.org
28. Voigt, T., Tewari, R., Freimuth, D., Mehra, A.: Kernel mechanisms for service differentiation in overloaded Web servers. In: Proc. of the USENIX Annual Technical Conference (2001)
29. Voigt, T.: Overload Behaviour and Protection of Event-driven Web Servers. In: Proc. of International Workshop on Web Engineering (2002)
30. Welsh, M., Culler, D.: Adaptive Overload Control for Busy Internet Servers. In: Proc. of the USENIX Symposium on Internet Technologies and Systems (USITS) (2003)
31. Welsh, M., Culler, D., Brewer, E.: SEDA: An architecture for well-conditioned, scalable Internet services. In: Proc. of the 18th Symposium on Operating Systems Principles, SOSP (2001)

Normative Management of Web Service Level Agreements

Caroline Herssens[1], Stéphane Faulkner[2], and Ivan J. Jureta[2]

[1] PRECISE, LSM, Université catholique de Louvain, Belgium
[2] PRECISE, LSM, University of Namur, Belgium
caroline.herssens@uclouvain.be,
{stephane.faulkner,ivan.jureta}@fundp.ac.be

Abstract. Service Level Agreements (SLAs) are used in Service-Oriented Computing to define the obligations of the parties involved in a transaction. SLAs define these obligations, including for instance the expected service levels to be delivered by the provider, and the payment expected from the client. The obligations of the parties must be made explicit prior to the transaction, and a mechanism should be available to control the interaction, in order to ensure that the obligations are met. We outline a norm-oriented multiagent system (NoMAS) architecture that is combined with the service-oriented architecture in order to support the definition, management, and control of SLAs between the service clients and service providers.

Keywords: SLA, management, mutual obligations, supervision, norm oriented multi-agent systems.

1 Introduction

We focus in this paper on the critical task of ensuring that the contractual obligations of the parties – the service providers and the service clients – involved in a transaction are respected by these parties within a service-oriented system. Their obligations are typically outlined in a service-level agreement (SLA). An SLA is a contract between the said parties, who specify the quality-of-service (QoS) levels that should be met [17]. QoS is a combination of several quality properties, e.g., availability, reliability, cost, response time [21]. A provider can propose the same service at different quality QoS levels. When a service client requests the execution of a given functionality, it advertises its QoS expectations. The service selected for the service execution will be the one that best satisfy client expectations about QoS properties. Prior to the transaction, the client and the provider enter into a contract by signing an SLA, and thereby specify quality levels to be observed during the service execution [17]. SLAs are used in the QoS management context in order to know what clients requirements to meet, how to manage clients expectations, how to regulate resources and to control costs [26].

The use of SLAs in managing the transaction between a provider and a client requires appropriate conceptual foundations and associated computational mechanisms. SLAs require an architecture if they are to enable the interactions

B. Benatallah et al. (Eds.): ICWE 2010, LNCS 6189, pp. 98–113, 2010.

between stakeholders. This architecture must support a specification language used by the stakeholders to communicate their expectations and capabilities. Similarly, a specification language to define the elements of the SLA is needed. Beside the architecture, the SLA management needs an incentive mechanism. To stimulate the correct behavior of stakeholders, these must have mutual obligations. E.g., the provider has the obligation to meet a given QoS level and the client has the obligation to pay according to that quality level. However, the client pays for the service after its execution by the provider. It follows that the client's payment can be adapted to the QoS level delivered by the provider. Finally, stakeholders can behave in an opportunistic manner, i.e., the client can underevaluate the QoS level perceived and the provider can exaggerate the QoS level offered. To prevent such situations, the architecture must introduce a third-party controller to monitor the SLA execution.

The architecture must support the adaptation of SLA during the service execution. This architecture needs to be responsive and flexible. Norm-oriented Multi-Agent Systems (NoMAS) provide these characteristics. Normative agents refer to agents conforming to norms. The core idea of this paper is to adapt the analogy of norms and agents to the issue of SLA and stakeholders. An SLA will be described as a set of norms to be fulfilled by the different agents of the system. The stakeholders of the service execution will be represented by normative agents complying to norms that restrict their behavior. The architecture supports a language enabling the communication between stakeholders. This language allows to express norms to be followed by the normative agents of the MAS. The elements constitutive of the SLA are defined by obligations norms regulating the stakeholders.

Contributions. We propose an architecture based on normative agents in order to: (i) enable the communication between stakeholders involved in the SLA with a common language; (ii) define SLAs that meet provider capabilities and client requirements; (iii) manage the service execution and check the conformance of the quality level expected and observed; (iv) ensure the execution of the mutual obligations according to the SLA contract. We propose to achieve the SLA compliance through two particular mechanisms: *mutual obligations*, which motivate the fulfillment of respective obligations of the involved stakeholders; and a *supervised interaction* with a third-party controller, which monitors and evaluates the SLA execution and penalizes the agents that does not fulfill their obligations.

Organization. Section 2 presents the conceptual foundations of the approach. Section 3 outlines the management architecture and the management of SLA. Section 4 proposes an evaluation of the proposed approach. Section 5 summarizes the related work and the Section 6 concludes this paper.

2   Case Study and Conceptual Foundations

This Section covers the conceptual foundations of our SLA management approach. We first describe the case study used to illustrate the approach proposed in this paper. We briefly introduce the Service Level Agreement concept. We also outline the mutual obligations and the supervised interaction used throughout the approach.

2.1   Case Study

We refer in this paper to a case study coming from the European Space Agency (ESA) program on Earth observation. This program allows researchers to access and use the infrastructure operated and the data collected by the agency[1]. The data and infrastructure of the ESA are accessed through web services. In order to facilitate the discussion and delimit our example, we focus on one part of the overall system. The MERIS/MGVI service is a service able to use the MERIS instrument data provided by the Envisat satellite of the ESA to compute the vegetation indexes for a given period of time and region of the world. A vegetation index measures the amount of vegetation on the Earth's surface. The data on the vegetation index can be obtained for any time range and it is possible to delimit the region of the world that is of interest. This service is subject to one particular QoS characteristic: the latency is initially situated between 4 and 6 hours by day of the selected period. E.g.: if the time range selected is from October 24th 2009 to October 26th 2009, the execution time needed to compute the vegetation index is set between 12 to 18 hours. The length of the selected period impacts then strongly the time needed to fulfill the request. The SLA specification between stakeholders of this service must clearly constrain the execution time prior to all remaining QoS properties. The different concepts presented in this paper will be illustrated with the MERIS/MGVI service and, specially about the execution time of this service.

2.2   Service Level Agreement

A Service Level Agreement is a contract between the service provider and the service client specifying mutually agreed obligations of the provision of a service [6,30]. The SLA concerns the non-functional properties of the service [17], i.e., quality properties. When clients can choose among a set of functionally equivalent web services, Quality of Service (QoS) considerations become the key criteria for service selection. As a consequence, SLA about nonfunctional properties must be defined and managed between service clients and providers [17].

The specification of QoS obligations of a SLA starts from a set of Service Level Objective (SLOs) [26]. A Service Level Objective is a guarantee of a particular state of the SLA parameters in a given time period [13]. All quality properties advertised by the provider are associated to an SLO as illustrated in Example 1. Each SLO has a functional part that refers to the QoS concerned and a guarantee part (italicized in Example 1) applied on the functional part. With SLOs, the SLA covers all quality properties defined in the QoS request of the service client.

---

[1] http://gpod.eo.esa.int

Example 1. The provider of the MERIS/MGVI service shall execute the service *within 5 hours by day of the selected period.*

Example 1 is the SLO stating the maximum execution time of the agreement defined between the provider and the client. As referred in Section 2, the execution time of the MERIS/MGVI service is very important and needs to be clearly defined in the SLA. The SLA definition is communicated between the different stakeholders of the service execution. To assure the interoperability of SLA definitions, their specifications need to be written in a language common to providers and clients. The Web Service Level Agreement (WSLA) language [13] is one of the main standard for specifying SLAs. The Example 2 illustrates how the SLO agreement of the Example 1 is specified with WSLA.

Example 2.

```
<ServiceLevelObjective name=''exectime''>
  <Obliged>provider</Obliged>
  <Validity>
    <Start>2009-10-25T08:00:00.000-05:00</Start>
    <End>2009-10-30T08:00:00.000-05:00</End>
  </Validity>
  <Expression>
    <Predicate xsi:type=''wsla:Less''>
      <SLAParameter>ExecutionTime</SLAParameter>
      <Value>ExecutionTimeThreshold</Value>
    </Predicate>
  </Expression>
  <EvaluationEvent>NewValue</EvaluationEvent>
<ServiceLevelObjective>
```

The `ExecutionTimeThreshold` used in Example 1 is a constant that assigns a name to a simple value that can be referred in other definitions [13]. This threshold corresponds to the maximal execution time expected by the client, i.e., 5 hours by day of the selected period to compute.

2.3   Mutual Obligations

Delivering the service at the quality level specified in the SLA is an obligation for the service provider. However, the service client has an obligation to provide all the information needed for the service execution (i.e.: inputs needed for web service execution), but also to pay for the service execution. Interactions between the provider and the client involve mutual obligations [20]. Such bilateral obligations motivate the SLA conformance. Indeed, breaches to some obligations of one party can compromise the fulfillment of obligations of the other party. Both parts have interest in achieving their obligations to meet the contract. Goodin [10] outlines the possible structures of mutual obligations. The SLA of web services can be defined as mutually conditional obligations. With mutual conditional obligations, each party is obliged to discharge his obligations if and only if the other party discharges his obligations. E.g., if the service provided does not meet the contracted execution time, the client has not to pay the amount initially set. The SLA defines mutual obligations compelling the respective behavior of stakeholders.

The service execution is made of bilateral obligations, i.e., unilateral obligations from the provider about the service level execution and unilateral obligations of the client about payment or rating [16,22]. We consider in the remainder of this paper that the client obligation is only about payment. However, other contractual obligations can be used as feedback rating as requests frequency.

The execution of obligations occurs sequentially: obligations of one of the stakeholders are executed before the obligations of the other. E.g., the provider's obligations are executed before the client's and the level of payment can consequently be adapted to the degree, to which the provider conforms to the obligations. Adaptations of the client obligations according to the observed quality level must be specified in the initial SLA. In the classification of mutual obligations [10], SLA contracts are diachronic mutual obligations, because one party is supposed to discharge its obligations before the other party does the same. The consequence is that initial contract must specifies the expected penalties if the defined quality level is not met [16]. The SLA contract implies that the penalty is initially accepted by the provider. Clearly, the efficiency of the relationship existing between the client and the provider improves if specifications of penalties for cases of contract breaches are present.

2.4   Supervised Interaction

Stakeholders of the service execution must achieve their respective obligations to conform to the initial SLA. If they are not supervised, they can adopt an opportunistic behavior, i.e., not fulfill their obligations or fulfilling them at a level lower than expected. E.g., the service client can reduce the payment even if the quality level provided meets is expectation. To prevent such situations, we propose to monitor the service execution with a third-party. This third-party will act as a controlling authority and allows to ensure the correct execution of the SLA. It is a witness of the service execution and stimulates the conformance to SLA for both involved parts. The third-party allows deterrence-based trust, i.e., you trust the other party because there is a very strict rule normative or legal system of rules, and the agent is punished for any violation of rules [7]. The third-party is the controller that controls the compliance of both parts to rules defined in the SLA, it measures the efficiency of the stakeholders transactions and computes their respective reputations [23]. An analogy of the third-party controller is the ebay online auction website[2]. The evaluation system of ebay prevents the opportunistic behavior of the stakeholders of the transaction.

This authority has an additional role in managing the SLA. Namely, it is in charge of collecting and computing metrics. It collects and stores metrics defined in the SLA and computes them to compare observed and expected results. Such metrics are used to establish the trust value of stakeholders involved in transactions. The measurement of quality values is allowed by existing metrics

---

[2] http://www.ebay.com

such as those discussed in [8]. If an SLA is breached, the third-party controller sends notifications to the involved stakeholders. The third-party is independent of the parties involved in the actual transaction, given that its aim is to prevent opportunistic behavior.

3   The Architecture and the Process for SLA Management

To solve the issues of SLA definition and its management during the service execution, we propose to use a normative MAS. Our proposed system will allow the SLA management and stimulates the SLA compliance through the respect of mutual obligations and the supervision of an authority. We first introduce our agent architecture that monitors the SLA through the service stakeholders in Subsection 3.1. We then explain how SLAs are managed with this architecture through the definition of norms associated to the stakeholders in Subsection 3.2.

3.1   SLA Management Architecture

We chose to use a normative multi-agent system to monitor the execution of SLA. A normative multiagent system (MAS) involves normative mechanisms, which allow agents to adopt norms and specify how agents can modify these norms [4]. Norms can increase the efficiency of agent reasoning while their explicit representation supports reasoning about a wide range of behaviour types in a single framework [9]. Agent norms describe the obligations, permissions and prohibitions of a norm addressee to pursue certain activities, either to achieve a state of affairs or to perform an action [18]. The behaviour of an agent is monitored by its norms defining its permissions and obligations. Such a normative system allows deterrence based trust, the agent is punished for any violation of rules of the normative system [7].

The stakeholders of the service execution and the third-party controller are managed by normative agents. Norms condition the behavior of agents, the SLA is defined by obligations and prohibitions restraining the set of possible actions. We manage SLAs with normative agents coordinated within a suitable architecture. Three kinds of normative agents step in this architecture: the provider agents, the client agents, and the cluster agents. These are illustrated in Figure 1.

A cluster agent (A_{Clus}) is dedicated to each existing cluster of web services. A cluster of web services gathers functionally equivalent web services by providing several web services inside a unique wrapper. This wrapper is used by service clients as a standard web service. Services in a same cluster can be offered by different providers. The cluster selects the service that best satisfies the QoS expectations of the client in the cluster with an appropriate selection method [12]. This method relies on QoS advertisements of the provider and QoS expectations of the service client. These advertisements and expectations can have be made with WSLA [13] or another common appropriate language.

A service provider agent (A_P) is dedicated to each existing provider in the service cluster. A provider can offer several services in the cluster, i.e., the same

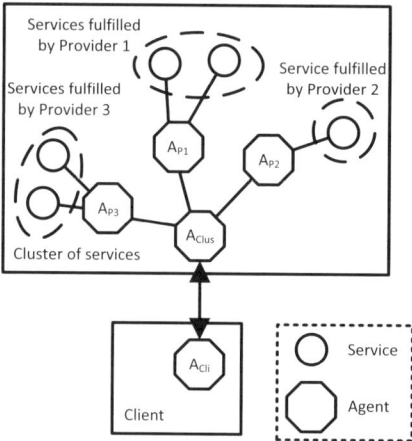

Fig. 1. SLA management architecture

functionality at different quality levels. This is illustrated in Figure 1 with the provider 1 and 3 each offering services with the same functionality and different quality characteristics. Providers can also offer services into different clusters, as they provide different functionalities. Provider agents advertise their QoS possibilities to the cluster agent.

A client agent (A_{Cli}) is assigned to each client requesting a service. The service request includes particular QoS expectations of the service client about the service execution.

Once the SLA has been negotiated [30] between stakeholders of the service execution it can be defined with an appropriate language. The cluster agent is responsible of the stakeholders conformance to the SLA defined. The cluster agent is also in charge of the collection and evaluation of metrics of the service execution. It collects information about QoS observed at each service transaction. It is able to compute statistical data on the service and able to determine if the service execution met the defined SLA. It is the third-party controller of the service execution, it controls the SLA compliance of the provider and the client agent. Agents of the SLA architecture are normative agents, conforming to norms derived from the SLA initially defined between the client and the provider. This architecture can be easily supported by existing normative MAS frameworks [9,18]. In the remainder, we focus on how normative agents can support the management of SLA. The normative MAS infrastructure and communication is out of scope of this paper.

In the context of our case study, an agent is dedicated to each provider able to propose a service functionally equivalent to the MERIS/MGVI service introduced in the case study. A client agent is dedicated to the requester of the MERIS/MGVI service. A cluster agent is responsible for the supervision of interactions occurring between the client agent and the providers offering the functionality.

3.2   SLA Management Process

The management of SLAs with the proposed architecture covers several steps:
(1) the definition of the SLA between the client and the provider; (2) the control
of provider obligations, i.e., the service execution; (3) the penalties to apply to
the provider if the SLA is not met, and; (4) the control of the client obligations,
i.e., payment or evaluation. To fulfill these steps, the agents of the architecture
introduced in Subsection 3.1 will take on different roles.

Step 1: Definition of the SLA. The SLA is defined between the service client
and the service provider from WSLA specification advertised by the provider.
The WSLA specification is extended to include the mutual obligations of stakeholders. The defined SLA must cover expectations about the quality level of the
service execution but also the penalties associated to breaches of SLA. These
penalties are defined according to the importance of quality properties involved
and according to the importance of breaches. Moreover, initial SLA specifications
can also be enriched with complex rules, dependent rules or normative rules [24].
Such extensions allow the definition of enriched contracts, e.g., graduated rules
are rules sets which specify graduated range for certain parameters so that it can
be evaluated whether the measured values exceed, meet or fall below the defined
service levels. To define these extended SLAs and include mutual obligations of
service providers and clients, usual languages as WSLA [17] or WSOL [28] need
to be enriched. To this aim, we choose to express the different SLOs of the initial
SLA with obligation norms associated to involved agents of the architecture to
benefit from the information added by more complex rules. To express SLO with
normative obligations, we refer to the work of Kollingbaum [18,19] about supervised interaction. Each SLO of the SLA contracted between the provider and
the client is expressed with the NoA language [18] interpretable by all agents of
the architecture. Moreover, complex conditions and penalties associated to SLO
failures are also expressed with this language in further steps. The Example 3
illustrates the conversion of the SLO specified with WSLA in Example 2 into an
obligation norm of the service provider agent specified with the NoA language.

Example 3.

```
obligation(
ServiceProvider,
achieve ServiceExecutionUnderTimeThreshold (ServiceProvider, Service,
  ExecutionTimeThreshold),
ServiceExecuted (ServiceProvider, Service) and
  ExecutionTime (Service) <= ExecutionTimeThreshold
ServiceExecutionUnderTimeThreshold (ServiceProvider, Service,
  ExecutionTimeThreshold))
```

This obligation states that the provider must achieve the execution of the MERIS/
MGVI service under the time limit (`ServiceExecutionUnderTimeThreshold`)
specified in the initial WSLA specification (`ExecutionTimeThreshold`). The

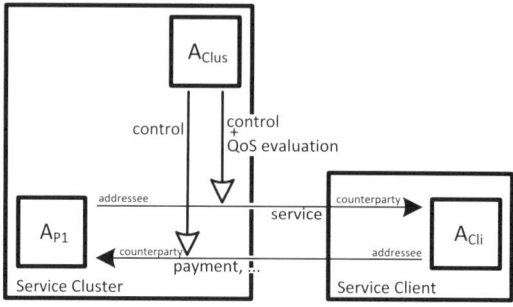

Fig. 2. Roles fulfilled by normative agents

normative agents of the architecture monitor the execution of services through their norms. However, these agents can make a choice whether to obey the norms in specific cases. If the service provider is not able to achieve all SLOs of its contract, it can violate some of them to assure the fulfillment of remaining norms. This situation arises due to unexpected events (i.e.: additional requests, hardware failures) or because the provider amplified its capabilities to be selected. Among all SLOs defined with obligations norms between the client and the provider, some can be met and some can not.

Step 2: Control of provider obligations. Control is enabled through mechanisms of normative agents. Each agent of the architecture fulfills one or several roles in the contract management. The SLA contract is then monitored through these different roles: the addressee commits an obligation defined in the contract; the counter-party is the recipient of the obligation fulfilled, and; the authority is a witness of the contract. The authority is in charge of the correct execution of the contract and imposes sanctions in case of defective behavior of the addressee. The different roles of client, provider and cluster agents are illustrated in Figure 2. The provider illustrated in the service cluster of this example is the provider 1 among those proposed in Figure 1.

The interactions between the service provider and the service client (i.e., the service execution and its payment) are restrained by obligation norms associated to these roles. The SLOs specifying the expected QoS level of the MERIS/MGVI service appear as norms. The provider agent is the addressee in these norms, while the client is the counter-party in the transaction. To control the achievement of this contract, the cluster agent acts as an authority. As stated in Subsection 3.1, the cluster agent is responsible for collecting and computing the metrics in order to control the SLA execution. It is then able to determine if the service provider meets the SLOs defined through obligation norms. The cluster agent will impose sanctions when the quality level provided does not meet the level contracted in the SLA. Such sanctions appear as penalties applied to the provider. As stated before, these penalties are part of the initial SLA. In the

MERIS/MGVI instance, the decreasing of payment is proportional to the reduction of quality level provided. Sanctions are expressed by obligations norms to be followed by the cluster agent. When the provider chooses or is forced to breach a norm specifying one of its SLO, the cluster agent captures it and activates a specific penalty. There can be several norms specifying different penalties corresponding to the spreading of the breach. The Example 4 illustrates a specification of one such penalty.

Example 4.

```
sanction(
ServiceCluster,
perform EvaluationTime (ServiceProvider, Service, ExecutionTimeThreshold,
  ExecutionTimeThreshold 2, AmountPenalty),
ServiceExecuted (ServiceProvider, Service) and
  ExecutionTime (Service) > ExecutionTimeThreshold and
  ExecutionTime (Service) <= ExecutionTimeThreshold2
TimePenalty(ServiceProvider, AmountPenalty))
```

When one of the SLOs of the MERIS/MGVI service is not met, a sanction is applied by the cluster agent according to the importance of the breach. As stated in Section 2, the execution time is a critical issue for the MERIS/MGVI service, sanctions to apply must penalize all provider weaknesses about delays. The Example 4 illustrates one sanction: if the execution time observed is above the SLA time limit (`ExecutionTimeThreshold`) but is under the second time limit of the breach scale (`ExecutionTimeThreshold2`), the decreasing of payment (`AmountPenalty`) applied is proportional to the observed level on the breach scale. The cluster agent independently estimates the equality of the quality level provided and the amount to pay. Moreover, according to characteristics of mutual obligations in SLAs, the user must discharge its obligations only if the provider has discharged its owns obligations.

Step 3: Penalties to apply. When the cluster agent observes that the SLA is not fulfilled by the provider agent, it notifies the service client through the application of a sanction. The client agent will then reflect this sanction on its own behavior. The mutual obligations of SLAs are diachronic; the client obligations are adapted to the provider fulfillment of its owns obligations. In the Example 5, the `TimePenalty` is a constant defining the payment reduction of the MERIS/MGVI service initiated by the sanction of the Example 4. There can be several payment reduction to apply, corresponding to different level of breach or to different QoS properties involved in the SLA. According to the importance of the breach, the client agent follows the norm defining the corresponding penalty. The obligation of the Example 5 is the client obligation to pay for the MERIS/MGVI service execution, i.e., the contractual obligation of the client. However, the initial payment amount (`Amount`) is reduced by the penalty (`AmountPenalty`) induced by the time sanction illustrated in Example 4. The payment of the service is an obligation norm in which the client agent is the addressee and the provider agent is the counter-party as illustrated in Figure 2.

Example 5.

```
obligation(
ServiceClient,
achieve ServicePayment (ServiceClient, ServiceProvider,
  Amount - AmountPenalty),
ServiceExecutionUnderExecutionTimeThreshold (ServiceProvider, Service,
  ExecutionTimeThreshold)) and
  TimePenalty(ServiceProvider, AmountPenalty)
achieve ServicePayment (ServiceClient, ServiceProvider,
  Amount - AmountPenalty))
```

Step 4: Control of client obligations. The third-party controller checks the execution of the unilateral obligations of the service provider as detailed in Step 2. Similarly, the third-party controller must verify the obligations of the service client, the client can be subject to different categories of obligations i.e.: its payment to the provider after the service execution. To control the right execution of the payment obligation illustrated in Example 5, the cluster agent acts as the third-party controller. It is the authority of the payment transaction as illustrated in Figure 2. It must check that the right amount has been deposited to the provider. If the client fails to pay or deposits a bad amount, the cluster agent must apply a penalty. The Example 6 illustrates the sanction applied by the cluster agent to the client of the MERIS/MGVI service when the payment obligation is not met. With such sanctions, the cluster agent avoids the non payment of the service client. Indeed, if the payment is not made or if it is insufficient, the client is labeled as a bad payer (`PaymentPenalty(ServiceClient)`) by the third-party controller. The cluster agent can then reject future requests of bad payers on its cluster.

Example 6.

```
sanction(
ServiceCluster,
perform CheckPayment ServiceClient, ServiceProvider, Amount, AmountPenalty),
not ServicePayment (ServiceClient, ServiceProvider, Amount - AmountPenalty)
PaymentPenalty(ServiceClient))
```

4   Evaluation

Supervised interaction and mutual obligations ensure that delivery of services is better managed. We conduct some experiments in order to evaluate the effect of these mechanisms. These experiments simulate services transactions between users and providers and measure their utility with and without the utilization of such mechanisms. The utility denotes the abstract quality whereby an object serves our purposes, and becomes entitled to rank as a commodity [15]. We suppose here that the utility increasing is constant for each new transaction initiated. Each transaction initiated by a client involves a cost decreasing of its cumulated utility while each successful transaction increases its cumulated utility. The ratio over the increasing induced by the success of the transaction and the decreasing due to the cost of the transaction must be positive. E.g.: in our simulations, the increasing of utility is set to 1 while the service execution succeeds and the utility decreasing of the service payment is 0.8. The net utility of a service transaction is then 0.2. We generate 200000 transactions from 10000 different providers to 100 different users. Each service is executed 20 times by each service client. To simulate the opportunistic behavior of providers, we define 30% of opportunistic providers that do not fulfill their transactions 70% of time. Without mutual obligations and supervised interaction, the decreasing of client utility involved by the service payment occurs even while services executions fail.

To simulate the supervised interaction effect, we introduce a simple trust model. The trustworthiness of each provider is collected by the third-party controller. The third-party controller monitors all services executions and dismisses providers that fail 10 services executions previously supervised. While services executions occur without supervised interaction, the clients collect themselves information about past executions and dismiss providers that failed 3 of their own previous transactions.

To simulate the interest of mutual obligations, we introduce a variable payment model. The service client can reduce the initial payment while the provider obligations are not met. The utility decreasing of the service client can be less important when the service execution fails. E.g.: in our simulations, the utility decreasing involved by the payment is 0.8 when the service execution succeeds and is reduced to 0.2 while the service execution fails. Without mutual obligations, the payment has to be done and the decreasing of the client utility is fixed to 0.8.

However, the third-party controller offering such monitoring mechanisms has to be payed. We designed two different scenarios to simulate the payment of the third-party controller: a variable and a fixed remuneration. The variable remuneration implies a decreasing of the client utility at each service execution. This variable cost must be proportional to benefit of a transaction. E.g.: if the net utility before the remuneration of the third-party controller is 0.2, the third-party fee of each transaction can be 0.02. The fixed cost allows clients to benefit from third-party mechanisms after a single payment. It implies an important decreasing of the client utility. E.g.: in our simulations, we set the initial utility of the client to -1000 while the third-party controller relies on a fixed cost.

To evaluate benefits from supervised interaction and mutual obligations, we observe the mean cumulated utility of users during 200000 services executions. To highlight the benefits of third-party mechanisms, we design 7 models: (1) services executions without supervised interaction (s.i.) and without mutual obligations (m.o.); (2) services executions without s.i. and with m.o. at a variable cost; (3) services executions without s.i. and with m.o. at a fixed cost; (4) services executions with s.i. and without m.o. at a variable cost; (5) services executions with s.i. and without m.o. at a fixed cost; (6) services executions with s.i. and m.o. at a fixed cost, and; (7) services executions with s.i. and m.o. at a variable cost. We then measure the difference between the optimal cumulated utility and the cumulated utility of our different models (i.e., the optimal client utility is get while the service client never pays for the services executions that fail).

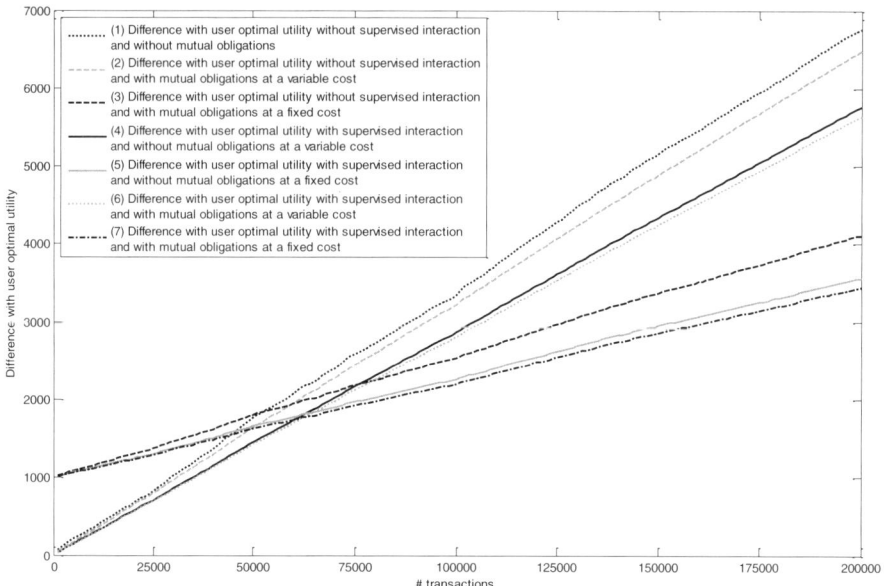

Fig. 3. Simulation Results

The results of our experiments are highlighted in Figure 3. The model nearest to the optimal client utility is the model (7) that provides both supervised interaction and mutual obligations with an initial fixed cost. However, this model becomes the best only when the initial cost is balanced by its profitability (after approximatively 64500 services executions) while at the beginning the most profitable model is the model (6) that provides both supervised interaction and mutual obligations at a variable cost. The profitability of each model is dependent from the third-party controller payment scenario but the utilization of third-party mechanisms always improve the client utility. The less profitable model is the (1) that provides neither supervised interaction nor mutual obligations. Models offering only mutual obligations ((2) and (3)) improve lightly the client utility while models providing only supervised interaction ((4) and (5)) ameliorate strongly the client utility. The combination of both mechanisms (models (6) and (7)) outperforms other models and highlight the interest of normative agents to control SLA of stakeholders transactions. The experiments conducted here to evaluate the client utility can be transposed to the provider utility. We can also simulate opportunistic client that do not fulfill their obligations and evaluate the mean utility of providers.

5   Related Work

QoS properties and SLA management need appropriate architectures to be handled during the service execution. Campbell et al. [5] propose the Quality of

Service Architecture (QoS-A) incorporating the notion of flow, service contract and flow management through QoS properties. Barbosa et al. outline in [3] different architectural configurations to enable the auditing of SLA and to evaluate their efficiencies. The WSLA framework [17] introduces a runtime architecture comprising several SLA monitoring services. Some services may be outsourced to third parties to increase the objectivity in the evaluation of the services. The QoS Mission-Action-Resource (Q-MAR) model [14] and the Grid Quality of Service Management (G-QoSM) framework [2] also propose to distribute the SLA monitoring to the different components of the system. Paschke et al. [24] introduce a Rule-Based Service Level Management (RBSLM) architecture in which SLAs are represented with declarative rules and managed through logical concepts and rule languages. Although all these architectures allow one to observe when a contract is violated, most of them do not prevent such violations and do not clearly define corrective actions to take. In our proposal, the third-party monitors stakeholders behaviors and the mutual obligations of the stakeholders define penalties to apply while the contract is not fulfilled. The BREIN project[3] offers an architecture enabling the management of SLAs through their whole life-cycle [1]. The SLA management is enabled by taking into account the policies of the parties and their respective business goals. The BREIN SLA management offers preventive monitoring to react to upcoming violations and a prioritization of SLAs. Our normative management of SLAs adapt contracts at runtime in response to unexpected violations in order to maximize stakeholders satisfaction.

Agents systems are well fitted to monitor activities requiring negotiation between stakeholders as SLA management or e-commerce mediation [11,27]. Other existing SLA architectures relies on multi agent systems [29]. Yan et al. [30] introduces a MAS architecture supporting the negotiation of services involved in a composition. In comparison with existing MAS architectures, our proposal is supported by normative agents. Normative agents allow to constrain the stakeholders behavior with norms defining the SLA to be achieved. They are particularly relevant to the SLA management issue. Normative agents are also used by Pitt et al. [25]. They propose a framework for QoS management which combines events, metrics and parameters with organizational intelligence offered by norm-governed multi-agent systems. Although their proposal monitors QoS information, they did not tackle the SLA conformance issue. One of the strongest point of our work is that the agreements between clients and providers are defined and monitored through norms associated to roles of agents and not to agents or components of the architecture. These roles allow the architecture to offer more flexibility, e.g., the provider can be easily substituted when unexpected failures occurs.

6   Conclusions and Future Work

We propose in this paper an architecture enabling the management of SLA. This architecture relies on a MAS and supports a normative definition of SLA.

---

[3] http://www.eu-brein.com

The MAS enables the communication between stakeholders involved in the SLA. Each party of the SLA is defined with an obligation norm that constrains stakeholders behaviors. The architecture checks the conformance of the stakeholders to the SLA. To stimulate the proper execution of the SLA, its execution is driven by mutual obligations and supervised by a third-party controller. The architecture benefits from the potential autonomy assured by normative agents. The normative architecture enables the interactions between provider and client and also the evaluation of the quality level of such interactions.

Future work will concern the implementation and the extension of the architecture to support actions to take while an SLA is breached. Rather than penalize the provider or the client, the architecture will propose corrective actions. To ensure the SLA conformance, the architecture will take advantage of the multiple services providing the same functionality inside the cluster.

References

1. Final brein architecture d4.1.3 v2 - wp 4.1 architectural design. Technical report, BREIN project (2009)
2. Al-Ali, R.J., Rana, O.F., Walker, D.W., Jha, S., Sohail, S.: G- qosm: Grid service discovery using qos properties. J. of Computing and Informatics 21(4), 363–382 (2002)
3. Barbosa, A.C., Sauvé, J., Cirne, W., Carelli, M.: Evaluating architectures for independently auditing service level agreements. Future Gener. Comput. Syst. 22(7), 721–731 (2006)
4. Boella, G., Torre, L., Verhagen, H.: Introduction to normative multiagent systems. Comput. Math. Organ. Theory 12(2-3), 71–79 (2006)
5. Campbell, A., Coulson, G., Hutchison, D.: A quality of service architecture. SIGCOMM Comput. Commun. Rev. 24(2), 6–27 (1994)
6. Cappiello, C., Comuzzi, M., Plebani, P.: On automated generation of web service level agreements. In: Krogstie, J., Opdahl, A.L., Sindre, G. (eds.) CAiSE 2007 and WES 2007. LNCS, vol. 4495, pp. 264–278. Springer, Heidelberg (2007)
7. Castelfranchi, C., Tan, Y.-H. (eds.): Trust and deception in virtual societies. Kluwer Academic Publishers, Norwell (2001)
8. Cherkasova, L., Fu, Y., Tang, W., Vahdat, A.: Measuring and characterizing end-to-end internet service performance. ACM Trans. Internet Technol. 3(4), 347–391 (2003)
9. Dignum, F., Morley, D.: Towards socially sophisticated bdi agents. In: Proc. of ICMAS '00, p. 111. IEEE Computer Society, Los Alamitos (2000)
10. Goodin, R.: Structures of mutual obligations. J. of Soc. Pol. 31(4), 579–596 (2002)
11. He, M., Jennings, N.R., Leung, H.-F.: On agent-mediated electronic commerce. IEEE Trans. on Know. and D. Eng. 15(4), 985–1003 (2003)
12. Herssens, C., Jureta, I., Faulkner, S.: Capturing and using qos relationships to improve service selection. In: Bellahsène, Z., Léonard, M. (eds.) CAiSE 2008. LNCS, vol. 5074, pp. 312–327. Springer, Heidelberg (2008)
13. International Business Machines (IBM). Web service level agreement (wsla) language specification (2003)
14. In, H.P., Kim, C., Yau, S.S.: Q-mar: An adaptive qos management model for situation-aware middleware. In: Yang, L.T., Guo, M., Gao, G.R., Jha, N.K. (eds.) EUC 2004. LNCS, vol. 3207, pp. 972–981. Springer, Heidelberg (2004)

15. Jevons, W.S.: Theory of Utility. In: The Theory of Political Economy (1965)
16. Kaminski, H., Perry, M.: Employing Intelligent Agents to Automate SLA Creation. In: Emerging Web Services Technology, pp. 33–46. Springer, Heidelberg (2007)
17. Keller, A., Ludwig, H.: The wsla framework: Specifying and monitoring service level agreements for web services. J. Netw. Syst. Manage. 11(1), 57–81 (2003)
18. Kollingbaum, M.: Norm-governed Practical Reasoning Agents. PhD thesis (2005)
19. Kollingbaum, M.J., Norman, T.J.: Supervised interaction - a form of contract management to create trust between agents. In: Falcone, R., Barber, S.K., Korba, L., Singh, M.P. (eds.) AAMAS 2002. LNCS (LNAI), vol. 2631, pp. 108–122. Springer, Heidelberg (2003)
20. Lockemann, P., Nimis, J., Braubach, L., Pokahr, A., Lamersdorf, W.: Architectural Design. In: Multiagent Engineering, pp. 405–429. Springer, Heidelberg (2006)
21. Menascé, D.A.: Qos issues in web services. IEEE Intern. Comp. 6(6), 72–75 (2002)
22. Morgan, G., Parkin, S., Molina-Jimenez, C., Skene, J.: Monitoring Middleware for Service Level Agreements in Heterogeneous Environments. In: Challenges of Expanding Internet: E-Commerce, E-Business, and E-Government, pp. 79–93. Springer, Heidelberg (2005)
23. Mui, L., Mohtashemi, M., Halberstadt, A.: Notions of reputation in multi-agents systems: a review. In: AAMAS '02, pp. 280–287 (2002)
24. Paschke, A., Dietrich, J., Kuhla, K.: A logic based sla management framework. In: SWPC '05: Semantic Web Policy Workshop at ISWC '05 (2005)
25. Pitt, J., Venkataram, P., Mamdani, A.: Qos management in manets using norm-governed agent societies. In: Dikenelli, O., Gleizes, M.-P., Ricci, A. (eds.) ESAW 2005. LNCS (LNAI), vol. 3963, pp. 221–240. Springer, Heidelberg (2006)
26. Sahai, A., Machiraju, V., Sayal, M., Jin, L.J., Casati, F.: Automated sla monitoring for web services. In: IEEE/IFIP DSOM, pp. 28–41 (2002)
27. Sierra, C., Dignum, F.: Agent-mediated electronic commerce: Scientific and technological roadmap. In: Sierra, C., Dignum, F.P.M. (eds.) AgentLink 2000. LNCS (LNAI), vol. 1991, pp. 1–18. Springer, Heidelberg (2001)
28. Tosic, V., Patel, K., Pagurek, B.: Wsol - web service offerings language. In: Bussler, C.J., McIlraith, S.A., Orlowska, M.E., Pernici, B., Yang, J. (eds.) CAiSE 2002 and WES 2002. LNCS, vol. 2512, pp. 57–67. Springer, Heidelberg (2002)
29. Trzec, K., Huljenic, D.: Intelligent agents for qos management. In: AAMAS '02, pp. 1405–1412 (2002)
30. Yan, J., Kowalczyk, R., Lin, J., Chhetri, M.B., Goh, S.K., Zhang, J.: Autonomous service level agreement negotiation for service composition provision. Future Gener. Comput. Syst. 23(6), 748–759 (2007)

Combining Schema and Level-Based Matching
for Web Service Discovery

Alsayed Algergawy[1], Richi Nayak[2], Norbert Siegmund[3],
Veit Köppen[3], and Gunter Saake[3]

[1] Department of Computer Science, University of Leipzig, Germany
[2] Queensland University of Technology, 2434 Brisbane, Australia
[3] School of Computer Science, University of Magdeburg, Germany
algergawy@informatik.uni-leipzig.de, r.nayak@qut.edu.au,
{n.siegmund,veit.koeppen,saake}@iti.cs.uni-magdeburg.de

Abstract. Due to the availability of huge number of Web services (WSs),
finding an appropriate WS according to the requirement of a service consumer is still a challenge. In this paper, we present a new and flexible
approach, called SeqDisc, that assesses the similarity between WSs. In
particular, the approach exploits the Prüfer encoding method to represent WSs as sequences capturing both semantic and structure information of service descriptions. Based on the sequence representation, we
develop an efficient sequence-based schema matching approach to measure the similarity between WSs. A set of experiments is conducted on
real data sets, and the results confirm the performance of the proposed
solution.

Keywords: Web service, WS discovery, WSDL, Schema matching.

1 Introduction

Web Services (WSs) have emerged as a popular paradigm for distributed
computing, and sparked a new round of interest from research and industrial
communities. By adopting service oriented architectures (SOA) using WS technologies, enterprises can flexibly solve enterprise-wide and cross-enterprise integration challenges [8]. These advantages of WSs can also be used within a
network of embedded systems which access and retrieve information from each
other (e.g., a logistic hub consisting of sensors, PDAs, etc. [18]). Web services can
then be used as an abstract interface for the devices to overcome communication
and data integration problems resulting from the heterogeneity of the devices.
Therefore, WSs can also be used to achieve the interoperability of a complex
and heterogeneous system.

The research community has identified two major areas of interest: *Web service discovery* and *Web service composition* [15]. In this paper, we present the
issue of locating WSs efficiently. As the number of WSs increases, the problem
of locating desired service(s) from a large pool of WSs becomes a challenging research problem [20,11,13,7,4]. In addition, if WSs are generated on demand (e.g.,

B. Benatallah et al. (Eds.): ICWE 2010, LNCS 6189, pp. 114–128, 2010.

to fulfill a certain task for a defined time, see [14,18]), it is difficult to discover the most suitable services. Several solutions have proposed, however, most of them suffer from the following two main disadvantages. Firstly, A large number of these solutions are syntactic-based. These methods use simple keyword search on Web service descriptions, and traditional attribute-based matchmaking algorithms to locate Web services according to a service request. However, these mechanisms are insufficient in the Web service discovery context since since they do not capture the underlying semantic of Web services and/or they partially satisfy the need of user search. This is due to the fact that keywords are often described by a natural language. As a result, the number of retrieved services with respect to the keywords are huge and/or the retrieved services might be irrelevant to the need of their consumers [15]. More recently, this issue sparked a new research into the Semantic Web where some research uses ontology to annotate the elements in Web services [5,16]. Nevertheless, integrating different ontologies may be difficult while the creation and maintenance of ontologies may involve a huge amount of human effort. To address the second aspect, clustering algorithms are used for discovering WSs. However, they are based on keyword search [11,16,15].

Secondly, most of the existing approaches are not scale well to large-scale and to large numbers of services, service publishers, and service requesters. This is due to the fact that they mostly follow a centralized registry approach. In such an approach, there is a registry that works as a store of WS advertisements and as the location where service publication and discovery takes place. The scalability issue of centralized approaches is usually addressed with the help of replication (e.g., UDDI). However, replicated Registries have high operational and maintenance cost. Furthermore, they are not transparent due to the fact that updates occur only periodically.

We see Web service discovery as a matching process, where available services' capabilities satisfy a service requester's requirement. There are two main aspects that should be considered during solving the matching process: the *quality* of the discovered service and the *efficiency* especially in large-scale environments. To obtain a better quality, not only is the textual description of Web services sufficient, but also the underlying structures and semantics should be exploited. Also to get a better performance, an efficient methodology should be advised.

In this paper, we propose a flexible and efficient approach, called *SeqDisc*, for assessing the similarity of Web services, which can be used to support locating WSs. We first represent WS document specifications described in WSDL as rooted, labeled trees, called *service trees*. By investigating service trees, we observe that each tree can be divided into two parts (subtrees), namely the *concrete* and *abstract* parts. We discover that the concrete parts from different WSDL documents have the same hierarchal structure, but may have different names. Therefore, we develop a *level-based matching* approach, which computes the name similarity between concrete elements at the same level. However, the abstract parts of the WSDL documents have differences in structure and semantics. To efficiently access the abstract elements, we represent them using the Prüfer encoding method [17], and then apply our sequence-based schema matching approach to the sequence representation. A set of experiments is conducted in order to validate our proposed approach employing real data sets. The experimental results showed that the approach performs well.

2 Preliminaries

A Web service is a software component identified by an URI, which can be accessed via the Internet through its exposed interface. Three fundamental layers are required to provide or use WSs [6]. First, WSs must be network-accessible to be invoked, HTTP is the de-facto standard network protocol for Internet available WSs. However, other network protocols can be used to enable the use of Web services in other kinds of networks (e.g., sensor networks). Second, WSs use XML-based messaging for exchanging information, and SOAP[1] is the chosen protocol. Finally, it is through a service description that all the specification for invoking a WS are made available; WSDL[2] is the de-facto standard for XML-based service description.

2.1 Web Service Modeling

In this paper, we represent a WSDL specification as a rooted labeled tree, called *service tree, ST*, defined as follows:

Definition 1. *A service tree, (ST), is a 3-tuple element; $ST = (N, E, Lab)$, where: $N = \{n_{root}, n_2, ..., n_n\}$ is the set of nodes representing WSDL document elements, where n_{root} is the root node of the tree; $E = \{(n_i, n_j)|n_i, n_j \in N\}$ is the set of edges representing the parent-child relationship between WSDL document elements; and Lab is a set of labels associating to WSDL document elements describing the properties of them.*

Examining the hierarchical structure of the WSDL document, we found that a *service* consists of a set of *ports*, each containing only one *binding*. A binding contains only one *portType*. Each portType consists of a set of *operations*, each containing an *input message* and a set of *output messages*. A message includes a set of *parts*, where each part describes the logical content of the message. All WSDL document elements (except part elements) have two main properties: the *type* property to indicate the type of the element (*port, binding, operation,...*) and the *name* property to distinguish between similar type elements.

From the hierarchal structure of a service tree, we divide its elements into a *concrete* part and *abstract* part. The intuition for this classification is that service trees representing different web services have the same structure from the root node to the part node, while the structure of the remaining depends on the content of operation messages. The following are definitions for concrete and abstract parts of a service tree.

---

[1] http://www.w3.org/TR/soap/
[2] http://www.w3.org/TR/wsdl20/

```
<?xml version="1.0"?>
<definitions name="WSDL1">
  <types>
    <schema ...">
      <complexType name="POType">
        <all>
          <element name="id" type="string"/>
          <element name="name" type="string"/>
          <element name="items">
            <complexType>
              <all>
                <element name="quantity" type="int"/>
                <element name="product" type="string"/>
              </all>
            </complexType>
          </element>
        </all>
      </complexType>
    </schema>
  </types>
  <message name="getDataRequest">
    <part name="id"type="xsd1:string"/>
  </message>
  <message name="getDataResponse">
    <part name="data" element="POType"/>
  </message>
  <portType name="Data_PortType">
    <operation name="getData">
      <input message="getDataRequest"/>
      <output message="getDataResponse"/>
    </operation>
  </portType>
  <binding name="getDataSoapBinding"
           type="tns:Data_PortType">

  </binding>
  <service name="getDataService">
    <port name="getDataPort"
          binding="getDataSoapBinding">
      ....
    </port>
  </service>
</definitions>
```

```
<?xml version="1.0"?>
<definitions name="WSDL2">
  <types>
    <schema ...">
      <complexType name="MyProduct">
        <all>
          <element name="id" type="int"/>
          <element name="name" type="string"/>
          <element name="price" type="double"/>
          <element name="part">
            <complexType>
              <all>
                <element name="name" type="string"/>
              </all>
            </complexType>
          </element>
        </all>
      </complexType>
    </schema>
  </types>
  <message name="getProductRequest">
    <part name="id"type="int"/>
  </message>
  <message name="getProductResponse">
    <part name="data" element="MyProduct"/>
  </message>
  <portType name="Product_PortType">
    <operation name="getProduct">
      <input message="getProductRequest"/>
      <output message="getProductResponse"/>
    </operation>
  </portType>
  <binding name="getProductSoapBinding"
           type="tns:Product_PortType">

  </binding>
  <service name="getProductService">
    <port name="getProductPort"
          binding="getProductSoapBinding">

    </port>
  </service>
</definitions>
```

(a) WSDL1: *getData* Web Service.   (b) WSDL2: *getProduct* Web service.

Fig. 1. Two WSDL specifications

Definition 2. *A concrete part of a service tree (ST) is the subtree (ST_C) extending from the root node to the portType element, such that $ST_C = \{N_C, E_C, Lab_C\} \subset ST$, $N_C = \{n_{root}, n_{port1}, n_{binding1}, n_{portType1}, ..., n_{portType_l}\} \subset N$, where l is the number of concrete elements in the service tree.*

Definition 3. *An abstract part of a service tree (ST) is the set of subtrees rooted at operation elements, such that $ST_A = \{ST_{A_1}, ST_{A_2}, ..., ST_{A_k}\}$, where k is the number of operations in the service tree.*

A service tree comprises a concrete part and an abstract part, i.e., $ST = ST_C \cup ST_A$. To assess the similarity between two WSs, we consequently compare their concrete and abstract parts. The problem of measuring similarity between Web services is converted into the problem of tree matching- comparing their concrete and abstract parts.

Let us now introduce an example of assessing the similarity between two WSs, which is taken from [20]. As shown in Fig. 1, we have two WSs described by two WSDL documents $WSDL1$ and $WSDL2$, respectively. $WSDL1$ contains one operation , *getData*, that takes a string as input and returns a complex data type named POType, which is a product order. The second document contains one operation, *getProduct*, that takes an integer as input and returns the complex data type MyProduct as output.

3 The SeqDisc Approach

The proposed SeqDisc approach is based on the exploitation of the structure and semantic information from WSDL documents. The objective is to develop a flexible and efficient approach that measures the similarity between WSs. The measured similarity is used as a guide in locating the desired service. To realize this goal, we first analyze WSDL documents and represent them as service trees using Java APIs for WSDL (JWSDL) and a SAX parser for the contents of the XML schema (the types element). Each service tree is examined to extract its concrete and its abstract parts. The concrete parts from different service trees have the same hierarchal structure. Hence, the similarity between concrete parts of two web services is computed using only concrete part element names by comparing elements with the same level. We call this type of matching level-based matching. Abstract parts from different service trees have different structures based on the message contents. Therefore, we propose a sequence-based abstract matching approach to measure the similarity between them. By the two mechanisms we gain a high flexibility in determining the similarity between WSs. In Section 6, we show two possibilities to compute the similarity. The first is to exploit abstract parts (operations), while the second is to use both the abstract and concrete parts. Furthermore, the proposed approach scales well. As it will be shown, the level-based matching algorithm has a linear time complexity as a function of the number of concrete elements, while the sequence-based matching algorithm benefits from the sequence representation to reduce time complexity.

4   Level-Based Matching

Once obtaining the concrete parts of service trees, $ST_{C1} \subset ST1$ and $ST_{C2} \subset ST2$, we apply our level-based matching algorithm that linguistically compares nodes at the same level, as shown in Fig. 2(a). The level-based approach considers only semantic information of concrete elements. It measures the elements (tag names) similarity by comparing each pair of elements at the same level based on their names.

Algorithm 1 accepts the concrete parts of the service tress, ST_{C1}, ST_{C2}, and computes the name similarity between the elements of the concrete parts. It starts by initializing the matrices, wherein the name similarities are kept. We have three levels for each service tree, $line\,2$. When the loop index equals 1, $i = 1$, the algorithm deals with the port nodes, when $i = 2$ it deals with the binding nodes, and with the portType nodes when $i = 3$. To compute the similarity between elements at the same level, the algorithm uses two inner loops, $lines\,3\,\&\,5$. It first extracts the name of the node j at the level i, $line\,4$, and the name of the node k at the same level, $line\,6$. Then, the algorithm uses a name similarity function to compute the name similarity between the names of the nodes, $line\,7$. Finally, depending on the level, it stores the name similarity matrix into the corresponding element matrix.

To compute the name similarity between two element names represented as strings, we first break each string into a set of tokens T_1 and T_2 using a customizable tokenizer using punctuation, upper case, special symbols, and digits, e.g. getDataService $\rightarrow$ {get, Data, Service}. We then determine the name similarity between the two sets of name tokens T_1 and T_2 as the the average best similarity of each token with a token in the other set. It is computed as follow:

$$Nsim(T_1, T_2) = \frac{\sum_{t_1 \in T_1}[\max_{t_2 \in T_2} sim(t_1, t_2)] + \sum_{t_2 \in T_2}[\max_{t_1 \in T_1} sim(t_2, t_1)]}{|T1| + |T2|} \quad (1)$$

Algorithm 1. Level-based matching algorithm

Require: Two concrete parts, ST_{C1} & ST_{C2}
Ensure: $3Name$ $similarity$ $matrices$, $NSimM$
1:   $PortSimM[][] \Leftarrow 0$, $BindSimM[][] \Leftarrow 0$, $PTypeSimM[][] \Leftarrow 0$;
2:   **for** $i = 1$ to 3 **do**
3:      **for** $j = 1$ to l **do**
4:        $name_1 \Leftarrow getName(ST_{C1}(i,j))$;
5:        **for** $k = 1$ to l' **do**
6:          $name_2 \Leftarrow getName(ST_{C2}(i,k))$;
7:          $NSimM[i][j] \Leftarrow NSim(name_1, name_2)$;
8:        **end for**
9:      **end for**
10:     **if** $i = 1$ **then**
11:        $PortSimM \Leftarrow NSimM$;
12:     **else if** $i = 2$ **then**
13:        $BindSimM \Leftarrow NSimM$;
14:     **else**
15:        $PTypeSimM \Leftarrow NSimM$;
16:     **end if**
17: **end for**

To measure the string similarity between a pair of tokens, $sim(t_1, t_2)$, we use two string similarity measures, namely the edit distance and trigrams [10]. The name similarity between two nodes is computed as the combination (weighted sum) of the two similarity values. The output of this stage is 3 ($l \times l'$) name similarity matrices, $NSimM$, where l is the number of concrete part elements of ST_{C1} and l' is the number of concrete part elements of ST_{C2} per level (knowing that the number of *ports*, the number of *bindings*, and the number of *protType* are equal). In the running example, see Fig. 2(a), $l = 1$ and $l' = 1$.

Algorithm Complexity. The algorithm runs three times, one for every level. Through each run, it compares l elements of ST_{C1} with l' elements of the second concrete part. This leads to a time complexity of $O(l \times l')$, taking into account that the number of elements in each level is very small.

5   Abstract Matching

In contrast to concrete parts, the abstract parts from different service trees have different structures. Therefore, to compute the similarity between them, we should capture both semantic and structural information of the abstract parts of the service trees. To realize this goal, we propose a sequence-based matching approach to achieve this goal. The approach consists of two stages: *Prüfer Sequence Construction* and *Similarity computation*.[3] The Pre-processing phase is considered with the representation of each abstract item (subtree) as a sequence representation using the Prüfer encoding method. The sequences

---

[3] For more details about our sequence-based schema matching approach, see [3].

should capture both semantic and structure information of the service tree. The similarity computation phase aims to assess the similarity between abstract parts of different service trees exploiting both information to construct an operation similarity matrix.

The outline of the algorithm implementing the proposed schema matching approach is shown in *Algorithm* 2. The algorithm accepts two sets of abstract parts of the service trees input, $ST_{A1} = \{ST_{A11}, ST_{A12}, ..., ST_{A1k}\}$ and $ST_{A2} = \{ST_{A21}, ST_{A22}, ..., ST_{A2k'}\}$, where each item in the sets represents an operation in the service tree. k and k' are the number of operations in the two abstract parts, respectively. We first analyze each operation (abstract item) and represent it as a Consolidated Prüfer Sequence (CPS) using the Prüfer encoding method. Then, the algorithm proceeds to compare all CPS pairs to assess the similarity between every operation pair using our developed sequence matching algorithms. The returned similarity value is stored in its corresponding position in the operation similarity matrix, $OpSimM$.

Prüfer Sequence Construction. This aims to represent every item in the abstract part set (operation) as a sequence representation using the Prüfer encoding method. The semantic and structural information of service tree operations are captured in Label Prüfer Sequences (LPSs) and Number Prüfer Sequences (NPSs), respectively. The two sequences form what is called a Consolidated Prüfer Sequences ($CPS = (NPS, LPS)$) [19]. They are constructed by doing a *post-order traversal* that tags each node in the service tree operation with a unique traversal number, as shown in Fig. 2(b) for $ST1$. NPS is then constructed iteratively by removing the node with the smallest traversal number and appending its parent node number to the already structured partial sequence. LPS is constructed similarly but by taking the node labels of deleted nodes instead of their parent node numbers.

Example 1. Consider the abstract parts of the two service trees $ST1$ & $ST2$ shown in Fig. 2(b). $CPS(ST_{A1}) = (NPS, LPS)$, where $NPS(ST_{A1}) = (2\ 10\ 8\ 8\ 7\ 7\ 8\ 9\ 10\ -)$ and $LPS(ST_{A1}).name = (id, getDataReequest, id, name, quantity, product, item, POType, getDataResponse, getData)$.

This sequence representation of service trees makes the proposed framework able to cope with the two mentioned aspects in Section 1. From the quality

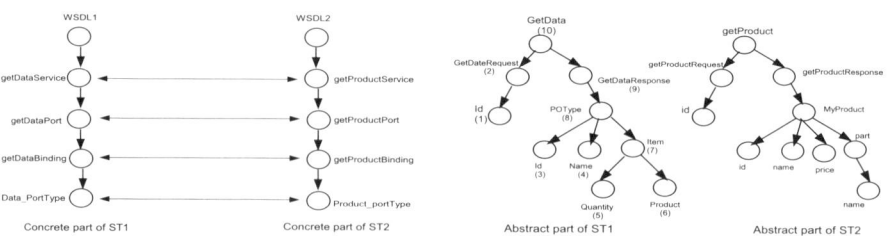

(a) Concrete parts of WSDL1 & WSDL2. (b) Abstract parts of $WSDL1$ & $WSDL2$.

Fig. 2. Concrete & abstract parts of $WSDL1$ & $WSDL2$

Algorithm 2. Schema matching algorithm

Require: Two abstract parts, ST_{A1} & ST_{A2}
       $ST_{A1} = \{ST_{A11}, ST_{A12}, ..., ST_{A1k}\}$
       $ST_{A2} = \{ST_{A21}, ST_{A22}, ..., ST_{A2k'}\}$
Ensure: *Operation similarity matrix, OpSimM*
 1: $OpSimM[][] \Leftarrow 0$;
 2: **for** $i = 1$ to k **do**
 3:   $CPS_1[i] \Leftarrow buildCPS(ST_{A1i})$
 4: **end for**
 5: **for** $j = 1$ to k' **do**
 6:   $CPS_2[j] \Leftarrow buildCPS(ST_{A2j})$
 7: **end for**
 8: **for** $i = 1$ to k **do**
 9:   **for** $j = 1$ to k' **do**
10:     $OpSimM[i][j] \Leftarrow computeSim(CPS_1[i], CPS_2[j])$;
11:   **end for**
12: **end for**
13: *return OpSimM*;

---

point of view, CPS captures both semantic information in LPSs and structure information in NPSs, which increases quality of Web service discovery. From performance point of view, CPS provides several structural properties, which enable dealing with service trees in an efficient manner.

Similarity Computation. This stage aims to compute the similarity between abstract parts (operations) from different service trees. This task can be stated as follows: Consider we have two Web service document specifications $WSDL1$ and $WSDL2$, each contains a set of operations represented as the abstract part of the corresponding service tree. $ST_{A1} = \{ST_{A11}, ST_{A12}, ..., ST_{Ak}\}$ represents the operation set belonging to $WSDL1$, while $ST_{A2} = \{ST_{A21}, ST_{A22}, ..., ST_{A2k'}\}$ is the operation set of $WSDL2$. The task at hand is to construct a $k \times k'$ operation similarity matrix, $OpSimM$, where k is the number of operations in $WSDL1$ and k' is the number of operations in $WSDL2$. Each entry in the matrix, $OpSimM[i][j]$, represents the similarity between operation ST_{A1i} from the first set and operation ST_{A2j} from the second. The similarity computation algorithm operates on the sequence representations of service trees, see *Algorithm, line* 6, and consists of three steps.

1. *Linguistic matcher.* First, a linguistic similarity algorithm is used to compute a degree of linguistic similarity between the elements of service tree operation pairs exploiting their semantic information represented in LPSs. The output of this step are $k \times k'$ linguistic similarity matrices, $LSimM$. Equation 2 gives the entries of a matrix, where $Nsim(T_i, T_j)$ is computed using the same procedure in Eq. 1, $DataType$ is a similarity function to compute the type/data type similarity between nodes, and $combine_l$ is an aggregation

function that combines the name and data type similarities.

$$LSimM[i,j] = combine_l(Nsim(T_i, T_j), DataType(n_i, n_j)) \qquad (2)$$

2. *Structural matcher.* Once a degree of linguistic similarity is computed, we use the *structural algorithm* to compute the structural similarity between abstract part elements. This matcher is based on the node context, which is reflected by its ancestors and its descendants. The descendants of an element include both its immediate children and the leaves of the subtrees rooted at the element. The immediate children reflect its basic structure, while the leaves reflect the element's content. We consider three kinds of node contexts depending on its position in the service tree: *child, leaf,* and *ancestor* context. The context of a node is the combination of its ancestor, its child, and its leaf context. Two nodes will be structurally similar if they have similar contexts. To measure the structural similarity between two nodes, we compute the similarity of their child, leaf, and ancestor contexts utilizing the structural properties carried by sequence representations of service trees as follows:

   - **Child Context Similarity**; The child context of a node is the set of its immediate children. This set can be easily extracted from the CPS representation of operations considering each entry in CPS represents an edge from the parent node NPS to its immediate child node LPS. To compute the child context similarity between two nodes $n_i \in CPS1$ and $n_j \in CPS2$, we first extract the child context set for each node, then we get the linguistic similarity between each pair of children in the two sets. We select the matching pairs with maximum similarity values, and finally we take the average of best similarity values.
   - **Leaf Context Similarity**; The leaf context of a node is the set of leaf nodes of subtrees rooted at the node. This set can be efficiently extracted from CPS representation. To determine the leaf context similarity between two nodes $n_i \in CPS1$ and $n_j \in CPS2$, we extract the leaf context set for each node, then we determine the gap between the node and its leaf context set as a vector, and finally we use the cosine measure between the two vectors.
   - **Ancestor Context Similarity**; The ancestor context of a node is the path extending from the root node to the node. To measure the ancestor context similarity between two nodes $n_i \in CPS1$ and $n_j \in CPS2$, first we extract each ancestor context from CPS representation, say path P_i for n_i and P_j for n_j, then we compare the two paths. To compare between paths, we use the scores established in [9].

The output of this step are $k \times k'$ structural similarity matrices, $SSimM$. Equation 3 gives entries of a matrix, where *child, leaf,* and *ancestor* are similarity functions that compute the child, leaf, and ancestor context similarity between nodes respectively, and $combine_s$ is an aggregation function combining these similarities.

$$SSimM[i,j] = combine_s(child(n_i, n_j), leaf(n_i, n_j), ancestor(n_i, n_j)) \qquad (3)$$

3. After computing both linguistic and structural similarities between Web service tree operations, we combine them. The output of this phase are $k \times k'$ total similarity matrices, $TSimM$. Equation 4 gives the entries of a matrix, where *combine* is an aggregation function combining these similarities.

$$TSimM[i, j] = combine(LSimM[i, j], SSimM[i, j]) \qquad (4)$$

Operation Similarity Matrix. We use $k \times k'$ total similarity matrices to construct the Web service operation similarity matrix, $OpSimM$. We compute the total similarity between every operation pairs by ranking element similarities in their total similarity matrix per element, selecting the best one, and averaging these selected similarities. Each computed value represents an entry in the matrix, $OpSimM[i, j]$, which represents the similarity between operation op_{1i} from the first set and operation op_{2j} from the second set.

Example 2. Applying the sequence-based matching approach to abstract parts illustrated in Fig. 2(b), we get $OpSim(getData, getProduct) = 0.75$.

Algorithm Complexity. The worst case time complexity of the schema matching algorithm can be expressed as a function of the number of nodes in each operation, the number of operation in each WS, and the number of WSs. Let n be the average operation size, k be the average operation number, and S be the number of input WSs. Following the same process in [3], it can be proven that the overall worst-case time complexity of the schema matching algorithm between two WSs is $O(n^2 k^2)$.

Matching Refinement. For every WS pairs we have two sets of matrices: three $NSimM$ matrices that store the similarity between concrete elements, and one $OpSimM$ that stores the similarity between two WS operations. This provides the SeqDisc approach more flexibility in assessing the similarity between services. As a consequence, we have two different possibilities to get the similarity.

Using only abstract parts; Given, the operation similarity matrix, $OpSimM$, that stores the similarity between operations of two WSs, how to obtain the similarity between them. We can simply get the similarity between the two services by averaging the similarity values in the matrix. However, this method produce smaller values, which do not represent the actual similarity among services. And due to uncertainty inherent in the matching process, the best matching can actually be an unsuccessful choice [12]. To overcome these shortcomings, similarity values are ranked up to top-2 ranking for each operation. Then, the average value is computed for these candidates.

Using both abstract and concrete parts; The second possibility to assess the similarity between WSs is to exploit both abstract and concrete parts. For any operation pair, $op_{1i} \in WSDL1$ and $op_{2j} \in WSDL2$, whose similarity is greater than a predefined threshold (i.e. $OpSimM[i, j] > th$), we increase the similarity of their corresponding parents (portType, binding, and port, respectively).

6   Experimental Evaluation

In order to evaluate the degree to which the SeqDisc approach can distinguish between WSs, we need to obtain families of related specifications. We found such a collection published by XMethods[4] and QWS data set [1]. We selected 78 WSDL documents from six different categories. Table 1 shows these categories and the number of Web services inside each one. Using the "analyze WSDL" method

Table 1. Data set specification

Category	No. of WSs	NO. of operations	Size (KB)
Address	13	50	360
Currency	11	88	190
DNA	16	48	150
Email	10	50	205
Stock quite	14	130	375
Weather	13	110	266

provided by XMethods, we identify the number of operations in each WS, and get the total number of operations inside each category, as shown in the table. All the experiments below share the same design: each service of the collection was used as the basis for the desired service; this desired service was then matched against the complete set to identify the best target service(s).

6.1   Experimental Results

We use precision, recall, and F-measure to evaluate the effectiveness of the SeqDisc framework. We have two possibilities to assess Web discovery process by finding the similarity between Web services depending on the exploited information of WSDL specifications.

Assessing the WS similarity using only abstract parts (operations).
In the first set of experiments, we match abstract parts of each service tree from each category against the abstract parts of all other service trees from all categories. Then, we select a set of candidate services, such that the similarity between individual candidate services and the desired one is greater than a predefined threshold. Precision and recall are then calculated for each service within a category. These calculated values are averaged to determine the average precision and recall for each category. Precision, recall and F-measure are calculated for all categories and illustrated in Fig.3(a). There are several interesting findings, which are evident in this figure. First, the SeqDisc approach has the ability to discover all WSs from a set of relevant services. As can be seen, across different six categories, the approach has a recall rate of 100% without missing any candidate service. This ability reflects the strong behavior of the approach of exploiting both semantic and structural information of WSDL specifications in an effective way. Second, Fig. 3 also shows that the ability of the approach to provide relevant WSs from a set of retrieved services is reasonable. The precision of the approach across six categories ranges between 64% and 86%. This means that while the approach does not miss any candidate service, however, it produces false match candidates. This is due to the WS assessment approach is based on lightweight semantic information and does not use any external dictionary or ontology. Finally, based on precision and recall, our framework is almost accurate with F-measure ranging from 78% to 93%.

---

[4] http://www.xmethods.net

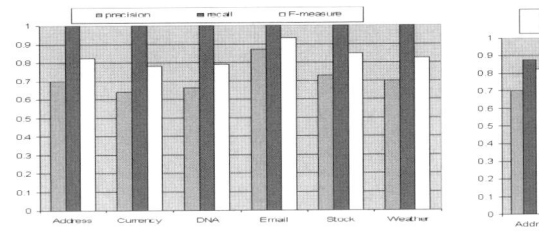

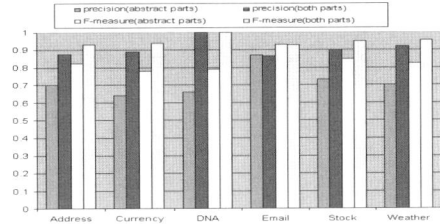

(a) Quality measures (abstract parts only).   (b) Quality measures (both parts).

Fig. 3. Effectiveness evaluation of SeqDisc

Assessing the WS similarity using abstract and concrete parts. In this set of experiments, we matched the whole parts (both abstract and concrete) of each service tree against all other service trees from all categories. Then, we selected a set of candidate services, such that the similarity between individual candidate services and the desired one is greater than a predefined threshold. Precision and recall are then calculated for each service within a category. These calculated values are averaged to determine the average precision and recall for each category. Precision and F-measure are calculated for all categories and illustrated in Fig. 3(b). We also compared them against the results of the first possibility. The results are reported in Fig. 3(b). The figure represents a number of appealing findings. (1) The recall of the approach remains at the unit level, i.e. no missing candidate services. (2) Exploiting more information about WSDL documents improves the approach precision, i.e. the number of false retrieved candidate services decreases across six different categories. The figure shows that the precision of the approach exploiting both concrete and abstract parts of service trees ranges between 86% in the Email category and 100% in the DNA category. (3) The first two findings lead to the quality of the approach is almost accurate with F-measure ranging between 90% and 100%.

Effect of Individual Matchers. We also performed another set of experiments to study the effect of individual matchers (linguistic and structure) on the effectiveness of WS similarity. To this end, we used data sets from the Address, Currency, DNA, and Weather domains. We consider the linguistic matcher utilizing either abstract parts or concrete and abstract parts. Figure 4 shows matching quality for these scenarios.

The results illustrated in Fig. 4 show several interesting findings. (1) Recall of the SeqDisc approach has a value of 1 across the four domains either exploiting only abstract parts or exploiting both parts, as shown in Fig. 4(a,b). This means that the approach is able to discover the desired service even if the linguistic matcher is only used. (2) However, precision of the approach decreases across the tested domains (except only for the DNA domain using the abstract parts). For example, in the Address domain, precision decreases from 88% to 70% utilizing both parts, and it reduces from 92% to 60% utilizing both parts in the Weather domain. This results in low F-measure values compared with the results shown in Fig. 3. (3) Exploiting both abstract and concrete parts outperforms exploiting

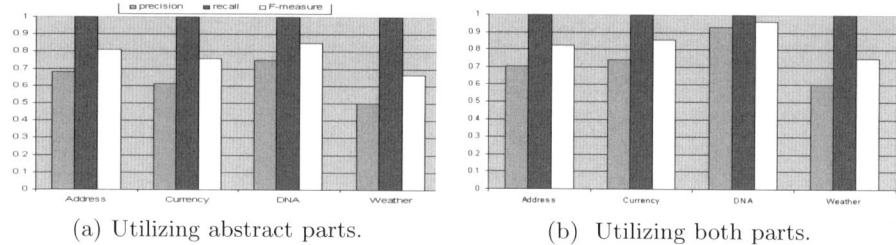

(a) Utilizing abstract parts.   (b) Utilizing both parts.

Fig. 4. Effectiveness evaluation of SeqDisc

only the abstract parts. This can be investigated by comparing results shown in Fig. 4(a) to results in Fig. 4(b).

In summary, using only the linguistic matcher is not sufficient to assess the similarity between WSs. Hence, it is desirable to consider other matchers. As the results in Fig. 3 indicate that the SeqDisc approach employing the structure matcher is sufficient to assess the similarity achieving F-measure between 90% and 100%.

Performance Comparison. Besides studying the performance of the SeqDisc approach, we also compared it with the discovery approach proposed in [7], called *KerDisc*[5]. To assess the similarity between the a service consumer request (user query) and the available WSs, the KerDisc approach first extracts the content from the WSDL documents followed by stop-word removal & stemming [7]. The constructed support-based semantic kernel in the training phase is then used to find the similarity between WSDL documents and a query when the query is provided. The topics of WSDL documents which are most related to the query topics are considered to be the most relevant. Based on the similarity computed using the support-based semantic kernel, the WSDLs are ranked and a list of appropriate Web services is returned to the service consumer.

Both SeqDisc and KerDisc have been validated using the data sets illustrated in Table 1. The quality measures have been evaluated and results are reported in Fig. 5. The figure shows that, in general, SeqDisc is more effective than KerDisc. It achieves higher F-measure than the other approach across five domains. It is worth noting that the KerDisc approach indicates low quality across the Address and Email domains. This is due to the two domains have common content, which produces many false positive candidates. The large number of false candidates declines the approach precision. Compared to the results of SeqDisc using only the linguistic matcher shown in Fig. 4(b), our approach outperforms across the Address and DNA domains, while the KerDisc approach is better across the other domains. This reveals two interesting findings: (1) KerDisc can effectively locate the desired service among heterogeneous WSs, while it fails to discover the desired service among a set of homogeneous services. In contrast, our approach could effectively locate the desired service among either a set of homogeneous or a set of heterogenous services. (2) SeqDisc clarifies the importance of exploiting the structure matcher.

---

[5] We give the approach this name for easier reference.

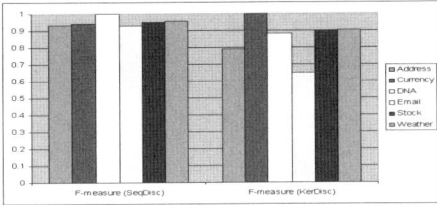

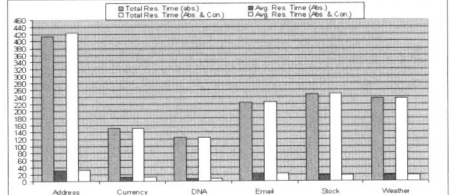

Fig. 5. Effectiveness comparison

Fig. 6. Response time response

Efficiency Evaluation. From the response time point of view, Fig. 6 gives the response time that is required to complete the task at hand, including both pre-processing and similarity measuring phases. The reported time is computed as a total time and an average time. The total time is the time needed to locate desired Web services belonging to a certain category, while the average time is the time required to discover a Web service of the category. The figure also shows that the framework needs 124 seconds in order to identify all desired Web services in the DNA category, and it requires 7 seconds to discover one service in the category, while it needs 3.7 minutes to locate all services in the Email category. We also considered the response time and compared it to the response time of the first set (i.e, using only the abstract parts). The results are calculated and listed in Fig. 6. The figure shows that the response time required to locate the desired Web service using both abstract and concrete parts equals to the response time when only using abstract parts, or needs a few milliseconds more.

7   Conclusions

We introduced a new and flexible approach to assess the similarity between WSs, which can be used to support a more automated WS discovery framework. The approach makes use of the whole WSDL document specification and distinguishes between the concrete and abstract parts. The concrete parts from different Web services have the same hierarchal structure, hence we devised a level-based matching approach. The abstract parts have different structures, therefore, we developed a sequence-based schema matching approach to compute the similarity between them. We have conducted a set of experiments to validate our approach. Our experimental results have shown that our method is accurate and scale-well. However, we are still a long way from automatic service discovery. In our ongoing work, we plan to complete the service discovery framework exploiting more WSDL features, such as text values associated to each element through documentation.

Acknowledgements. This work is an extended version of the paper presented in [2]. The work of A. Algergawy is supported by the BMBF, grant 03FO2152. While, the work of N. Siegmund and V. Köppen is also funded by the BMBF, project 01IM08003C.

References

1. Al-Masri, E., Mahmoud, Q.H.: Qos-based discovery and ranking of web services. In: ICCCN 2007, pp. 529–534 (2007)
2. Algergawy, A., Schallehn, E., Saake, G.: Efficiently locating web services using a sequence-based schema matching approach. In: 11th ICEIS(1) (2009)
3. Algergawy, A., Schallehn, E., Saake, G.: Improving XML schema matching performance using prüfer sequences. DKE 68(8), 728–747 (2009)
4. Anadiotis, G., Kotoulas, S., Lausen, H., Siebes, R.: Massively scalableweb service discovery. In: AINA '09 (2009)
5. Atkinson, C., Bostan, P., Hummel, O., Stoll, D.: A practical approach to web service discovery and retrieval. In: ICWS 2007, pp. 241–248 (2007)
6. Avila-rosas, A., Moreau, L., Dialani, V., Miles, S., Liu, X.: Agents for the grid: A comparison with web services (part ii: Service discovery). In: AAMAS '02, pp. 52–56 (2002)
7. Bose, A., Nayak, R., Bruza, P.: Improving web service discovery by using semantic models. In: Bailey, J., Maier, D., Schewe, K.-D., Thalheim, B., Wang, X.S. (eds.) WISE 2008. LNCS, vol. 5175, pp. 366–380. Springer, Heidelberg (2008)
8. Cabral, L., Domingue, J., Motta, E., Payne, T.R., Hakimpour, F.: Approaches to semantic web services: an overview and comparisons. In: Bussler, C.J., Davies, J., Fensel, D., Studer, R. (eds.) ESWS 2004. LNCS, vol. 3053, pp. 225–239. Springer, Heidelberg (2004)
9. Carmel, D., Efraty, N., Landau, G.M., Maarek, Y.S., Mass, Y.: An extension of the vector space model for querying XML documents via XML fragments. SIGIR Forum 36(2) (2002)
10. Cohen, W., Ravikumar, P., Fienberg, S.: A comparison of string distance metrics for name-matching tasks. In: IIWeb, pp. 73–78 (2003)
11. Dong, X., Halevy, A., Madhavan, J., Nemes, E., Zhang, J.: Similarity search for web services. In: VLDB 2004, pp. 372–383 (2004)
12. Gal, A.: Managing uncertainty in schema matching with top-k schema mappings. Journal on Data Semantics 6, 90–114 (2006)
13. Hao, Y., Zhang, Y.: Web services discovery based on schema matching. In: ACSC 2007, pp. 107–113 (2007)
14. Köppen, V., Siegmund, N., Soffner, M., Saake, G.: An architecture for interoperability of embedded systems and virtual reality. IETE Tech. Rev. 26(5) (2009)
15. Ma, J., Zhang, Y., He, J.: Efficiently finding web services using a clustering semantic approach. In: CSSSIA 2008, p. 5 (2008)
16. Nayak, R., Lee, B.: Web service discovery with additional semantics and clustering. In: WI 2007, pp. 555–558 (2007)
17. Prufer, H.: Neuer beweis eines satzes uber permutationen. Archiv fur Mathematik und Physik 27, 142–144 (1918)
18. Siegmund, N., Pukall, M., Soffner, M., Köppen, V., Saake, G.: Using software product lines for runtime interoperability. In: RAM-SE, pp. 1–7 (2009)
19. Tatikonda, S., Parthasarathy, S., Goyder, M.: LCS-TRIM: Dynamic programming meets XML indexing and querying. In: VLDB '07, pp. 63–74 (2007)
20. Wang, Y., Stroulia, E.: Flexible interface matching for web-service discovery. In: WISE 2003, pp. 147–156 (2003)

Web Messaging for Open and Scalable Distributed Sensing Applications

Vlad Trifa[1,2,*], Dominique Guinard[1,2], Vlatko Davidovski[1],
Andreas Kamilaris[3], and Ivan Delchev[2]

[1] Institute for Pervasive Computing, ETH Zurich, Switzerland
[2] SAP Research, Zurich, Switzerland
[3] University of Cyprus, Nicosia, Cyprus
vlad.trifa@ieee.org

Abstract. Future Web applications will increasingly require real-time data from the physical world collected by a myriad of sensors and actuators. Currently, integration of such devices require customized solutions due to the lack of widely adopted protocols for devices. Because the Web architecture offers a high degree of interoperability and a low entry barrier, we propose to leverage the Web to build hybrid applications that combine the physical world with Web content. Our work builds upon recent developments in Web push techniques and extends them for embedded devices with a RESTful messaging system. Our results illustrate that fully Web-based distributed sensing applications are not only feasible - but actually desirable - because Web standards offer an ideal compromise between performance and functionality.

1 Introduction

In the last decade computers have been silently pervading every corner of our lives. Among them, networks of tiny sensors that gather data about the real world – called *Wireless Sensor Networks* (WSN) – have been increasingly used in various disciplines ranging from civil engineering to biology. Because of their deeply embedded nature, WSN have the potential to revolutionize our understanding of natural and physical systems. As these systems will mature, the data they collect will increasingly need to be available on the Web in real time. A scenario for such applications is a heterogeneous city-wide Web API as shown in Fig. 1, to share real time information about the status of a city with different consumers. This data could be private and secure information (alarms, fire alerts, household energy consumption), but also public information (air and noise pollution levels, traffic status, etc).

Programming wireless sensor networks is challenging because devices have limited energy and computational resources that must be carefully managed. In addition, WSNs are highly dynamic and transient: connectivity is often unpredictable, devices disappear and reappear, new ones are added to extend the

---

* Corresponding author.

B. Benatallah et al. (Eds.): ICWE 2010, LNCS 6189, pp. 129–143, 2010.

network or to replace failed ones. In such conditions, a high-level *data-centric* approach in which information is delivered based on content rather than explicit addressing of individual nodes can simplify the development of applications that integrate data from these devices. Although consumer electronics with Internet access are becoming increasingly popular [1], a common ground that enables seamless integration of devices with applications is still lacking. Indeed, the myriad of existing protocols and standards for networked devices turns each network into isolated islands that hardly interact with each other.

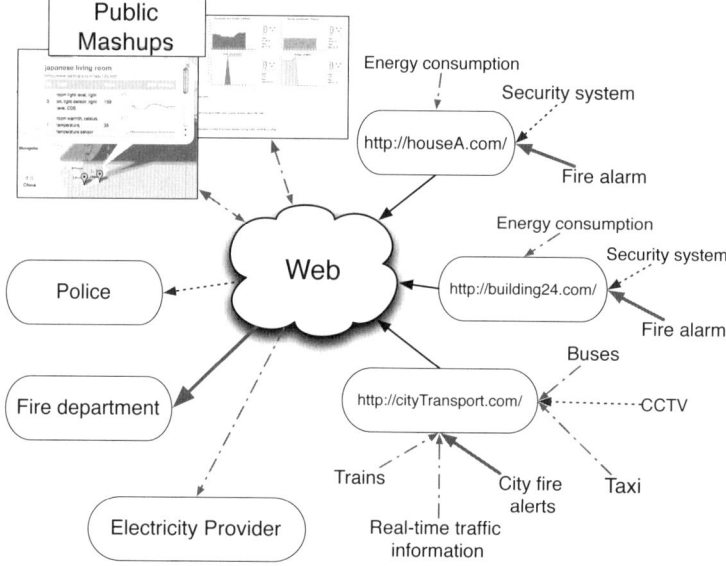

Fig. 1. Future distributed sensing applications will require a scalable infrastructure where new devices and services can be added and used with minimal effort

Because the Web is widely used, open, and easy to integrate, the Web of Things vision proposes to reuse the wide-spread Web infrastructure and standards (HTTP/XML/RSS) to connect embedded devices [2]. In this article, we extend the Web of Things paradigms to support more scalable and interoperable WSN applications by leveraging the event-driven nature of publish-subscribe systems.

A priori, HTTP seems not suited for building WSN applications because of its request/response nature. However, the recent success of Web push tools and techniques have enabled the development of event-driven applications directly over the Web. Our results show that RESTful messaging for embedded devices is a viable and scalable approach for building more open and programmable distributed sensing applications. To our knowledge, our work is the first to show the feasibility of a fully Web-based distributed sensing application that combines the recent advances in Web technologies to design an HTTP-based event-driven programming model for sensor networks.

After introducing the related work in Section 2, we survey Web messaging techniques and analyze their suitability for embedded devices in Section 3. Based on these findings, we designed RMS (Restful Messaging System), a scalable HTTP-based messaging system for distributed sensing applications which is described in Section 4. In Section 5 we evaluate the performance of RMS both with a simulation and with real devices. Finally, in Section 6 we discuss our results and their applicability for future distributed sensing, and Section 7 concludes this article.

2 Background

The Web of Things is the intersection of two fields that have been rarely associated in the past – networked embedded sensors and Web engineering. In this section we provide the required background and related work required to understand our contributions.

2.1 Networked Sensors and Actuators

Most sensor network applications share a common goal: gather, process, and store data collected by physically distributed sensors. To simplify the development and deployment of such applications, early approaches have explored the use of declarative data-centric models and query languages that consider sensor networks as a single logical unit, among which TinyDB [3]. However, such approaches are limited to heterogeneous systems and are not suited to the loosely coupled nature of the Web of Things, where new devices are added/removed at runtime.

Integration of physical things with the Web has already been proposed almost a decade ago [4,5]. However, these early projects focused mainly on embedding *information* about physical objects onto embedded Web servers and linking the objects with their virtual counterparts on the Web, which is fundamentally different from *integrating* devices into the Web as we propose here. Not only devices can be interacted with and controlled through an open Web API, but their status and functions can be searched, browsed, and used just like any other Web content. More recent projects [6] have investigated how to access embedded devices using the REST paradigm (see Section 2.2), but they mainly focused on isolated experiments and didn't address more scalable and heterogeneous systems.

The term *Sensor Web* refers to a global network of Web-connected sensors, and several projects have been proposed to build such a worldwide Sensor Web, as for example IRISnet [7], Senseweb [8], or Pachube[1]. In these projects, a unique endpoint is used to register and store data collected by many physical sensors. Such a central point of failure is against the distributed nature of the Web. Additionally, direct interaction among devices is not allowed as commands have to pass through the server. Stream Feeds [9] have proposed to extend the Web

---

[1] http://pachube.com

feed model (RSS/ATOM) to accommodate the large size and real-time nature of sensor data streams, however beyond a promising idea, a more substantial description and evaluation with actual devices is lacking and the project seems discontinued.

In this paper, we extend the simplistic stream feed model and propose a general-purpose, distributed, and flexible publish/subscribe system that is particularly suited for the requirements of the Web of Things. The novelty of our approach lies in the fact that RMS enables to collect and use data streams from heterogeneous sensors directly over the Web. This significantly lowers the access barrier for Web developers that can now rapidly integrate real time data in their Web applications using the tools they already know – which until now required advanced knowledge in embedded systems.

2.2   RESTful Web Services

HTTP is a rather simple protocol that follows the *Representational State Transfer (REST)* architectural style [10]. REST defines a few design constraints for building Internet-scale applications that are more flexible and simpler to use. Every component of an application is an URI-addressable resource whose representation is manipulated with a uniform and fixed interface, defined by the four main HTTP verbs (GET, POST, PUT, and DELETE). Interactions are stateless, thus servers do not keep the state of applications, which improves significantly the scalability of the system. HTML is the primary resource representation which is universally understood and is a lightweight language providing hypermedia features, while XML and JSON are the preferred representation for machine readable data.

Existing middleware for sensor networks have predominantly focused on data collection and processing, therefore ad-hoc interaction with devices is difficult, if not impossible. In a RESTful application, there is no need for any special interface as the application fully blends into the Web. Interacting with it does not use any special API beyond a HTTP client to access resources and manipulate them. Because of the low access barrier, more and more Web application have been switching from SOAP to REST as basis for their API, and this observation serves as motivation to apply the same paradigm to interact with embedded devices, as we suggested in this article.

2.3   Web-Based Sensor Networks

The core idea of the Web of Things is to enable Web-based interactions with embedded devices. This requires their functionality to be accessible through URIs that can be manipulated using HTTP. For example, one could send commands to actuators (*turn on LEDs*), retrieve sensor data (*get temperature sensor reading*), or change application state (*change the sampling frequency*) directly through a RESTful API [6]. As described in [2], two solutions are possible for *Web-enabling* sensor networks: directly on devices or through a proxy.

Device-level enablement. In this case, each device runs an actual Web server that processes and serves HTTP requests directly. Using HTTP for embedded devices has been often criticized because of the high memory footprint of HTTP servers and because of the verbosity of the HTTP protocol. However, recent work has shown that embedded Web servers can run on resource-constrained devices [11], requiring as little as 8 kB of memory [12]. Additionally, as the software and operating system for embedded devices are usually event-driven, their underlying architecture lends itself well for the construction of efficient event-driven HTTP servers [12]. Such event-driven Web applications are the most desirable approach for energy-constrained devices that sleep most of the time.

Proxy-level enablement. When device-level support for HTTP is not possible or desirable, Web integration can take place on a smart proxy (referred to as *gateways* hereafter) which hides the actual communication protocol used by devices behind a uniform Web interface, as proposed in [13]. Although gateways are not required for devices that support HTTP directly, they can nevertheless augment the functionality of single devices and improve the overall performance of sensing networks. Because gateways are much less constrained than sensor nodes, they can serve as distributed master nodes that can manage sensor networks. In addition, gateways can be delegated tasks that would be too expensive in terms of CPU and energy to be run on resource-constrained embedded devices. For example, caching data from the sensors for concurrent read accesses, buffer incoming request for devices, perform aggregation of data and local mashups, or manage security policies and access control to devices, could be all taken care of by gateways.

3   Web-Based Messaging

As the size of distributed sensing applications increase, so does the necessity to integrate disparate hardware and software platforms. Interfacing different messaging protocols is a complex procedure, and bridging different middlewares is prone to severely hinder the performance and scalability of such a future distributed systems. *Publish-subscribe systems* (pub/sub) are commonly used in distributed computing and large enterprise applications, because they allow decoupling data producers and consumers. Loose coupling allows more flexible and scalable systems where new entities can be easily added or removed. Highly scalable and efficient messaging protocols with various features have been proposed, however none of them directly integrates with the Web, as an additional protocol must be implemented on top of HTTP. XMPP is an XML-based messaging protocol widely used for chat servers. It is based on a decentralized network of servers that provide a multi-hop routing delivering messages from one client to the other. Because it is based on XML, it remains quite heavy for lightweight messaging with resource-constrained devices.

Only recently sensor networks have began to explore pub/sub messaging for building complex and interoperable applications that scale, as pub/sub shields applications from the underlying complexities of the WSN and provides a simple

– yet powerful – interface to interact with a WSN. MQTT-S [14] is a messaging protocol designed for tiny devices. TinyDDS [15] is another publish/subscribe middleware that enables interoperability between WSNs.

Web Syndication. A rudimentary form of Web messaging is *content syndication*. Feed formats such as RSS, or the more well-defined *Atom*, have become popular formats for exchanging machine-readable data over the Web. Atom offers a convenient metaphor to deal with time-ordered collections of entities, therefore would be particularly suited for storage, query, and retrieval of stored sensor data. However, Atom is limited by the pull-based mode, which cannot meet the requirements of event-driven applications. A push-based pub/sub protocol for the Web would simplify significantly the integration between WSNs and applications, however, such a messaging protocol that fully integrates with the Web has not yet been proposed.

Comet. *Comet* (also called HTTP streaming or server push) has become an increasingly popular technique to implement server side eventing for Web applications that circumvent the limitations of the traditional HTTP polling. Comet enables a Web server to push data back to the browser without the client requesting it explicitly by keeping the TCP/IP connection open after an initial response has been sent to the client. Since Web browsers (and HTTP) were not designed to support server-initiated notifications, Comet is a hack implemented through several specification loopholes. Comet servers and clients frequently use named channels or topics which are useful when different objects want to send data to a number of other interested parties. The main advantage of Comet is that standard Web clients (in particular browsers) can receive notifications pushed from servers in near real-time, even when behind firewalls or NAT.

Web Hooks. Web hooks are another solution for HTTP eventing that enable users to receive events and data in real time from applications through user-defined callbacks over HTTP. Once an event occurs, the application will POST data to the callback URLs specified by the users at subscription time. This pattern has been used by the PayPal service which allows you to specify a URL (on your own online commerce site) that will be triggered by PayPal once a payment has been accepted. Web hooks enables Web applications to synchronize data with other applications, but also to process, filter, or aggregate data from different sources and to notify people via email, IRC, Jabber, or Twitter. However, clients that want to receive notifications must also run a Web server where notifications will be posted. Web hooks are an elegant, clean, and RESTful solution for bi-directional Web eventing, unfortunately cannot be used when clients are behind NAT or a firewall, as they do not have a public network address.

The growing need for push-based communication on the Web is further supported by the introduction of *Web Sockets* and server-sent events in the HTML 5 specification, and also by the browser-side Web server embedded in the Opera Unite browser. An interesting recent project is RestMS[2], which offers a

---
[2] http://www.restms.org

RESTful interface to the Advanced Message Queuing Protocol (AMQP) and defines the behavior of a set of feed, join, and pipe types that provide an AMQP-interoperable messaging model. PubSubHubbub[3] (PuSH) is a lightweight and open server-to-server publish/subscribe protocol based on Web hooks as an extension to Atom and RSS. Servers can get near-instant notifications when a topic (feed URL) they're interested in is updated. However, these solutions were not designed for embedded devices as they are rather verbose. Based on these observations, we have designed RMS, a fully Web-based publish/subscribe mechanism designed to meet the requirements of distributed Web-based sensing applications, as will be described in the next section.

4   RMS: RESTful Messaging for Devices

Nowadays, integration of different WSNs is done in a fairly rigid manner: devices are tightly coupled to custom bridges that connect them to the outside world because of the lack of a common, widely adopted application protocol for sensor networks. In contrast, a uniform Web-based messaging would allow devices, gateways, brokers, and applications to transparently interact with each other in an ad-hoc manner. Devices can directly exchange information transparently with each other and with other Web resources thanks to the loose coupling of REST. Furthermore, as gateways are optional, they can easily be bypassed in case they fail, which increases the overall robustness of the whole system.

Two main classes of WSN applications exist: event-driven (where notifications are sent sporadically when an event occurs) and stream-based (sensor data is collected periodically and sent to a sink to be processed and/or stored). To support such interaction models, we developed the RESTful Messaging System (RMS), which is a lightweight pub/sub messaging suited for devices. In essence, our system is similar and directly mappable to RestMS and PubSubHubbub (PuSH). However, as we target embedded devices, we tried to keep it as simple as possible. Rather than creating a custom protocol on top of HTTP (such as XMPP or PuSH), we implement the core functionality of a pub/sub system solely using RESTful design patterns. More elaborate pub/sub protocols such as XMPP have a higher barrier of adoption and are somewhat complex for embedded devices. Also, just like SOAP-based Web services, packets are opaque therefore cannot be interpreted and acted upon by 3rd party proxies.

The gateway offers a RESTful API to use the eventing system and provides the following resources to manage interactions:

– /rms/channels every sub-resource represents a hierarchical channel where entities can post data to. For example, /rms/channels/ethz/ifw/floor/d/ 49.2 identify the channel related to the office No. 49.2 of the D floor, in our building (called ifw) at our school (ETH Zurich).
– /rms/subscriptions contains each subscription of entities to individual channels.

---

[3] http://pubsubhubbub.googlecode.com

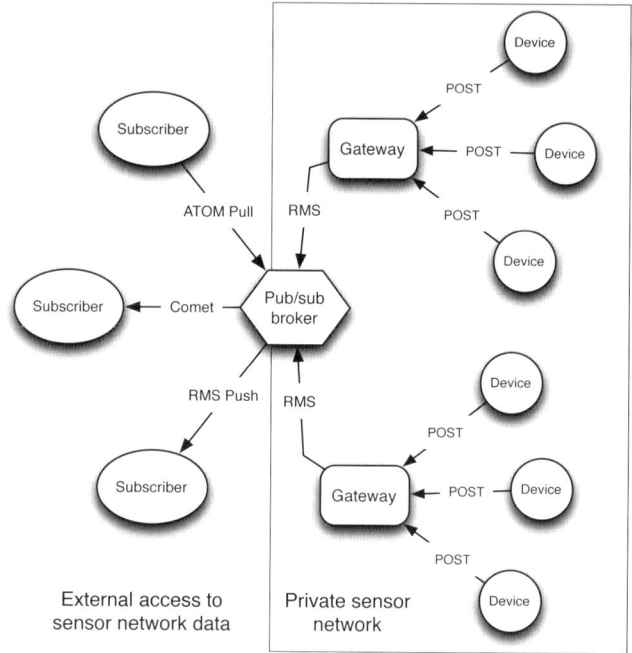

Fig. 2. The general model of Web-based messaging. Devices can POST their data on the gateway, using standard HTTP POST, AtomPub, or other proprietary protocols. Gateways will then forward it to another gateway or to a powerful broker using RMS. External users can fetch the data using the most appropriate method for their needs.

- /rms/channels/*/publishers contains all entities that are publishing data on the channel *.
- /rms/channels/*/subscribers contains all entities that are subscribed to events on the channel *.

A subscriber that wants to receive notifications about a channel, creates a new subscription by POSTing the following HTTP request to the gateway:

```
POST /rms/ethz/ifw/floor/d/49.2/subscriptions
Host: gateway_ip
Content-type: application/x-www-form-urlencoded
Content: cb-url=http://sub_ip/callback_url
```

Each new message posted on this channel will be POSTed back to all subscribers using the Web hook pattern to the URL they specified with the cb-url parameter. Any entity can publish data to the gateway by POSTing the following request on the gateway:

```
POST /rms/channels/ethz/ifw/floor/d/49.2
Host: gateway_ip
Content-type: application/x-www-form-urlencoded
Content: pub-url=http://device1&temperature=21
```

This device has never been registered with the gateway, so it includes its URL in the posted parameters, providing the gateway with a possibility to automatically scan it for semantic description of its capabilities and get required information. The message is posted directly to the channel `/ethz/ifw/d/49.2` without any need to previously create it. Other parameters are treated as tags (in this case `temperature`). In addition to specifying the channel or topic, publishers can also annotate messages with free text tags. Similarly, subscribers can easily filter out messages that contain or not specific tags.

Comet implementations such as CometD[4] treat topics as paths to support hierarchical relationships between topics. This is quite useful for example for permission models or aggregated data, where a subscriber to a parent topic can receive all messages from child topics. Paths can directly map with URI (e.g., `ethz/ifw/floor_d/49.2`), which can be directly integrated with the Web. Our gateway is implemented in OSGi[5] and supports Web hooks and Comet-based eventing. Comet data is accessible by replacing `rms/` by `cometd/` in the URL above. Therefore, users can can see in real time messages on a particular channel by pointing their Web browser to following URL:
`http://gateway_ip/cometd/channels/{channel_path}`

Because our gateway uses the internal eventing capabilities offered by OSGi as abstract model for notifications, additional notification protocols can be easily added (SMS, Twitter, e-mail, etc).

5 Evaluation

To evaluate the performance of RESTful messaging under extreme load conditions, we have conducted an experiment using simulated devices and subscribers connected through an RMS broker running a desktop computer (1.1 GHz, 2GB Ram, Gigabit Ethernet and Gentoo Linux). The HTTP clients (subscribers) were simulated on another machine (2x2.13 GHz, 8GB Ram, Gigabit Ethernet, Gentoo Linux). In the second part of this section, we describe a second experiment done to measure the performance of RMS in real-world conditions through an actual sensing system deployed using real sensor nodes.

Synthetic Load Simulation. First, we simulate many devices attached to the gateway, each one generating an event at a random interval between 1 and 5 seconds. Three runs have been performed with respectively 50 devices, 100 devices, and 200 devices attached. The time required to receive, process, and deliver all the events to one client has been measured, and results are shown in Figure 3.

Second, we evaluate the scalability of RMS with respect to concurrent subscribers for the same event triggered by a device. A test client started an event sink to receive events on respectively 50, 100, and 200 different ports, and for each port an event subscription was posted to the gateway. The gateway generated

---

[4] `http://cometd.org/`
[5] `http://www.osgi.org`

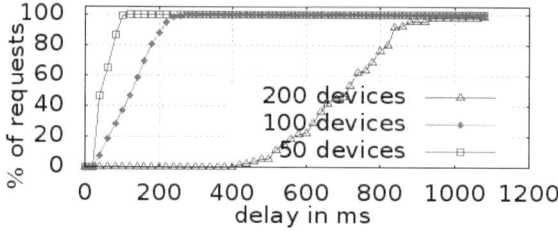

Fig. 3. *Many devices:* Response time to deliver events to a client with 50 devices (box), 100 devices (circle), and 200 devices (triangle) attached

artificial events (containing the generation time) that were delivered to all the subscribed ports. The test client measured the arrival time and from that computed the delay for each arriving event, and results are shown in Figure 4.

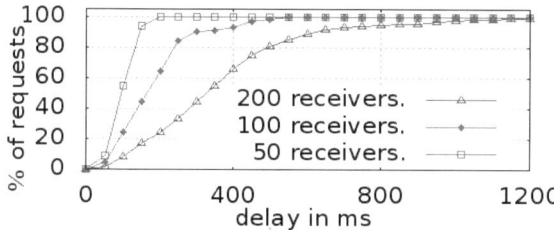

Fig. 4. *Many receivers:* Event delivery times measured for 50 subscribers (box), 100 subscribers (circle), 200 subscribers (triangle)

Third, we want to evaluate the performance of the classic request/response pattern under heavy load where many devices are attached to a gateway and a varying number of clients access simultaneously devices through HTTP. The test client started several concurrent threads that accessed the gateway and its devices randomly to simulate real clients accessing the gateway. In the first case, 4000 devices, three test runs with 100 clients, 50 clients and 25 clients, and results are shown in Table 1. In the second test 1000 virtual devices are attached to the gateway and three test runs have been performed with respectively 375 clients, 750 clients, and 1500 clients, and results are shown in Tables 2.

For a *moderate* number of devices (50 and 100) the gateway is able to dispatch RMS messages efficiently, with respectively 52 ms and 125 ms average delivery time. Doubling the number of devices results in a approximate doubling of the latency (2.4x). However, with 200 devices the performance drops significantly (5.59x slower than with 100). The third test shows that gateways can handle over 750 concurrent read requests per second from 1000 devices and 80% of these requests will be answered within 224 ms.

Table 1. Response time in milliseconds [ms] to deliver a request from a gateway with 4000 simulated devices attached to resp. 25 clients, 50 clients, and 100 clients

	25 clients	50 clients	100 clients
Min	3	3	3
Max	326	253	1248
RT 50%	16	40	56
RT 80%	44	111	172
Mean	63	67	136

Table 2. Response time in milliseconds [ms] to deliver a request from a gateway with 1000 simulated devices attached and resp. 375 clients, 750 clients, and 1500 clients

	375 clients	750 clients	1500 clients
Min	2	2	2
Max	9250	21054	21261
RT 50%	29	48	124
RT 80%	135	224	3059
Mean	614	860	1686

Real deployment. In this second experiment, we test the performance of RMS to transmit data from a real sensor network (devices used were TMotes running TinyOS), where the Web-enablement is done at the gateway level. Devices broadcast an event every second that are received at a time t_1 by a base station (A) attached to a WSN gateway running an RMS broker. Each event from the sensor network is then posted using RMS (Web hooks) to subscribers (in this case a sink laptop on the same LAN). Each event from the WSN is also caught using a spy base station (B) which starts a timer (also at time t_1). The timer is stopped when the corresponding notification is received from the WSN gateway though RMS at a time t_2. The time difference $t_2 - t_1$ corresponds to the time required to create and dispatch the RMS message from the WSN gateway and receive it on the sink.

Similarly to our results in the first experiment with simulated devices, a fully HTTP-based messaging system can transmit data from real sensor nodes with reasonable delivery times, as shown in Fig. 6. In all the cases, events needed less than 60 ms to be pushed from the publisher to the subscriber. This latency can be considered tolerable even for emergency and time-critical deployments. Obviously, the delay would be significantly larger for subscribers not on the same LAN as the RMS broker, or if the sensor network used a multi-hop topology.

WSN deployments for environmental monitoring rarely have more than 50 devices per gateway and sampling rate is rarely higher than one sample per second. Based on our results, we conclude that a fully HTTP-based solution is largely sufficient for collecting data in typical sensor network scenarios even when sub-second latency with hundreds of concurrent requests is needed.

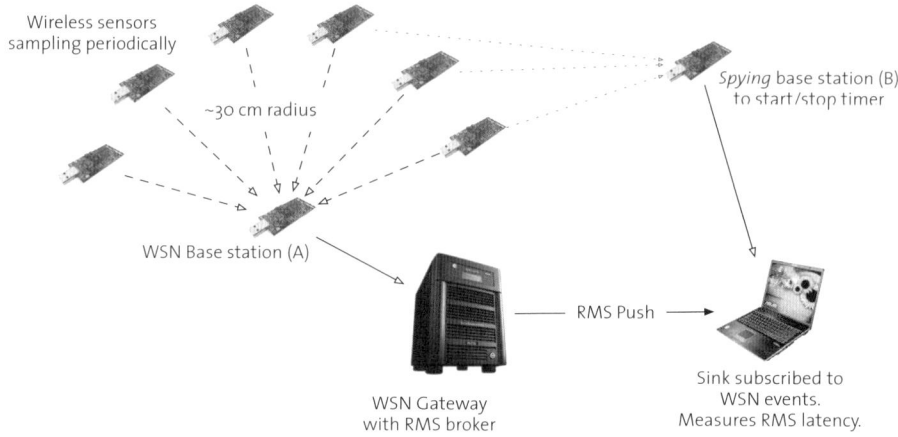

Fig. 5. Experimental setup to measure the latency of RMS. TMotes broadcast periodically numbered events caught by base station (A). The sink subscribes to all WSN events and each time it catches an event from the WSN (through its spy base station B) it starts a timer that will be stopped when it receives a RMS notification about the same event from the WSN Gateway.

6   Discussion

Nowadays, Web applications routinely integrate data from multiple sources such as RSS/Atom feeds, blogs, maps, etc. As the number of networked sensing devices will increase, so will the incentives to share and integrate the data they produce with Web applications. So far, only few projects have explored the Web of Things beyond linking physical objects with Web pages. Different middlewares have been proposed for the integration of heterogeneous sensors with applications, however they would introduce a strong coupling between components that is inappropriate with the ad-hoc nature of dynamic sensor networks, as all devices in the global network should settle on a single protocol, which is unlikely to happen.

The work presented here differs from middleware-based approaches in that we leverage *only* the ubiquitous modern Web architecture as abstraction layer for the peculiarities of various hardware and software platforms available for sensor networks. Enabling RESTful access with embedded devices significantly lower the access barrier to consume real-time data from the physical world. Simplified access to WSN data over the Web fosters the development of physical Web applications, as devices would have a Web API just like other Web resources. Programming with them could be done using highly popular and relatively simple languages such as JavaScript, DHTML, PHP, or even simple and visual mashup editors such as Yahoo Pipes. Web integration would be maximized as devices could be searched for, browsed, linked to, and used just like other Web content.

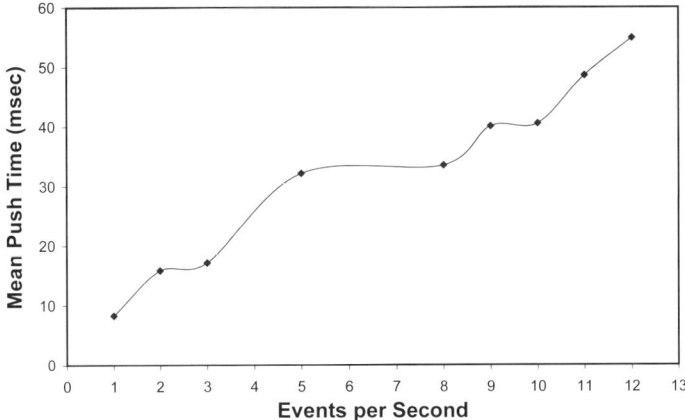

Fig. 6. Average delivery time for events generated on sensor nodes using a 1-hop sub-scriptions. Each devices generates one message per second, and increasing the number of devices also increases messages per second (here we use between 1 and 12 devices sending concurrent messages). Variability (not shown here) is mainly due to the physical nature of radio communication, which is hard to control.

Publishers and subscribers are loosely coupled to each other through the use of RMS. Unlike traditional pub/sub systems, a fully Web-based pub/sub system further decouples participants because only an HTTP client is needed and no additional protocol need to be implemented. Web standards maximize interoperability and loose coupling between components, which are desirable properties for scalable distributed applications. Along with the event-driven interaction model of RMS, these properties match well the dynamic nature of WSNs.

Future sensing applications will require enabling easy and timely access to physical data, and the solution we propose lowers the barriers to access WSN data while being fully integrated with the Web. As more embedded devices will be present in our daily environments, one could install gateways on home (or enterprise) routers where the number of attached devices is moderate (e.g. a WiFi, mobile phone, network attached storage...). For these appliances the performance of the system should therefore be sufficient.

In contrast to other optimized messaging systems, RMS suffers from the overhead of the HTTP protocol. Although our results are encouraging, optimizations and enhancements of the RMS broker implementation could increase the performance and throughput of the system. As RMS mainly consists of an HTTP server and client, it scales with the hardware. That is, running the broker on a more powerful machine would allow to attach more devices and serve more clients simultaneously.

7   Conclusion

As more embedded devices will be connected to the Internet, efficient solutions will be needed to collect, process, and store the data they will generate. In this article, we have described how to reuse the ubiquitous Web standards to build a scalable infrastructure for connecting embedded devices, the Web of Things. Our solution not only supports the request-response model of HTTP, but also leverages the recent development in the real-time Web to offer an efficient Web-based publish/subscribe system.

Our main contribution is to provide quantitative results to support the idea that using HTTP as application protocol for distributed sensing applications is not only a feasible, but also a desirable solution for integrating physical devices with applications. This is especially true when integration primes over raw performance in terms of latency and throughput, as the advantages brought by using Web technologies at the device-level outweighs the loss in performance. Our results show that HTTP-based messaging can support hundreds of concurrent users accessing hundreds of devices simultaneously with a sub-second latency, which is sufficient for most monitoring applications that use sensor networks. We have pointed out how the loss in performance in comparison to traditional messaging systems could be compensated by using gateways to improve the performance, scalability, and functionality of the applications running within sensor networks.

References

1. Dunkels, A., Vasseur, J.: IP for Smart Objects Alliance. Internet Protocol for Smart Objects (IPSO) Alliance White paper (September 2008)
2. Guinard, D., Trifa, V., Wilde, E.: Architecting a mashable open world wide web of things. Technical Report 663, Institute for Pervasive Computing, ETH Zurich (February 2010)
3. Madden, S.R., Franklin, M.J., Hellerstein, J.M., Hong, W.: TinyDB: an acquisitional query processing system for sensor networks. ACM Trans. Database Syst. 30(1), 122–173 (2005)
4. Ljungstrand, P., Redström, J., Holmquist, L.E.: WebStickers: using physical tokens to access, manage and share bookmarks to the Web. In: DARE '00: Proceedings of DARE 2000 on Designing Augmented Reality Environments, New York, NY, USA, pp. 23–31. ACM, New York (2000)
5. Kindberg, T., Barton, J., Morgan, J., Becker, G., Caswell, D., Debaty, P., Gopal, G., Frid, M., Krishnan, V., Morris, H., Schettino, J., Serra, B., Spasojevic, M.: People, places, things: web presence for the real world. Mob. Netw. Appl. 7(5), 365–376 (2002)
6. Luckenbach, T., Gober, P., Arbanowski, S., Kotsopoulos, A., Kim, K.: TinyREST - a protocol for integrating sensor networks into the internet. In: Proc. of REALWSN (2005)
7. Gibbons, P., Karp, B., Ke, Y., Nath, S., Seshan, S.: IrisNet: an architecture for a worldwide sensor Web. IEEE Pervasive Computing 2(4), 22–33 (2003)
8. Kansal, A., Nath, S., Liu, J., Zhao, F.: SenseWeb: an infrastructure for shared sensing. IEEE Multimedia 14(4), 8–13 (2007)
9. Dickerson, R., Lu, J., Lu, J., Whitehouse, K.: Stream Feeds: an Abstraction for the World Wide Sensor Web. In: Proceeding of the 1st Internet of Things Conference (IOT), Zurich, Switzerland (2008)

10. Fielding, R.T.: Architectural Styles and the Design of Network-based Software Architectures. PhD thesis, University of California, Irvine (2000)
11. Yazar, D., Dunkels, A.: Efficient Application Integration in IP-Based Sensor Network. In: Proc. of the First ACM Workshop on Embedded Sensing Systems for Energy-Efficiency in Buildings (BuildSys), at SenSys '09 (2009)
12. Duquennoy, S., Grimaud, G., Vandewalle, J.J.: Consistency and scalability in event notification for embedded web applications. In: 11th IEEE International Symposium on Web Systems Evolution (WSE'09), Edmonton, Canada, Edmonton, Canada (September 2009)
13. Trifa, V., Wieland, S., Guinard, D., Bohnert, T.M.: Design and implementation of a gateway for web-based interaction and management of embedded devices. In: Proceedings of the 2nd International Workshop on Sensor Network Engineering (IWSNE'09), Marina del Rey, CA, USA (June 2009)
14. Hunkeler, U., Truong, H.L., Stanford-Clark, A.: MQTT-S - A publish/subscribe protocol for Wireless Sensor Networks. In: Proceedings of the Third International Conference on COMmunication System softWAre and MiddlewaRE (COMSWARE 2008), Bangalore, India, pp. 791–798 (January 2008)
15. Boonma, P., Suzuki, J.: Middleware support for pluggable non-functional properties in wireless sensor networks. In: SERVICES '08: Proceedings of the 2008 IEEE Congress on Services - Part I, Washington, DC, USA, pp. 360–367. IEEE Computer Society, Los Alamitos (2008)

On Actors and the REST

Janne Kuuskeri and Tuomas Turto

Department of Software Systems
Tampere University of Technology
PL 553, 33101 Tampere, Finland
{janne.kuuskeri,tuomas.turto}@tut.fi

Abstract. The prevalence of RESTful services requires that we pay closer attention to how the principles that underlay REST are realized in actual services being implemented. This is especially crucial as REST is being applied to problem domains that require complex operations such as transactions. In this paper we investigate the relationship between RESTful web services and the actor model of computation. We suggest that by formulating RESTful services as a network of actors we can achieve deeper understanding what it means for a service to be RESTful.

1 Introduction

In his thesis Fielding [9] discusses how the Web's architecture as a distributed hypermedia system evolved in its early stages. In particular, the thesis describes how the Representational State Transfer, or REST, architectural style was used to guide the development. Given that REST pinpoints the architectural constraints that have made the Web a success, it has been argued that also the web services should be architected in a similar fashion [16]. These so-called RESTful services would then embrace the way applications on the Web are intended to function.

Often, however, the well-intended discussion on the principles of RESTful web services degenerates into heated debate about the merits of REST compared to the big web services implemented using the WS-* stack [20]. Although the architectural choice between a RESTful approach and the WS-* stack does require significant consideration and there are arguably better problem domains for each of them [14], this sort of comparative argumentation does not deepen our knowledge of RESTful services as much as it could.

Although the key principles of REST, such as the use of resources and the hypertext as an engine of application state in the sense of [9], are widely agreed upon, there seems to be conceptual confusion regarding how they do manifest and how they should manifest themselves in actual RESTful services. Discussion about the fundamentals of REST and a deeper understanding of these key issues is urgently required as RESTful services embed more complex behavior such as transactions [15]. Furthermore, as REST itself is being extended [7], we need solid understanding on how to apply the ideas behind REST.

In this paper we investigate the principles of REST in the framework of the actor model of computation [3]. The actor model has previously been used to

B. Benatallah et al. (Eds.): ICWE 2010, LNCS 6189, pp. 144–157, 2010.

reason about distributed systems in general [1] and as the basis for Internet-wide middleware development [2]. Also, from a more pragmatic point of view, programming languages based on the actor principles, such as Erlang [4], have received acclaim for their ability to implement RESTful services in a natural style. We conjecture that it is indeed the actor model that makes them especially suitable for implementing RESTful services.

The main contribution of this paper is an explicit investigation into the relationship between the actor computation model and the principles of REST. We show how a restricted actor model can be used to understand the principles underlying the RESTful paradigm and especially how an actor system embodies the idea of hypertext as an engine of application state. Moreover, we suggest a notation for describing RESTful systems.

The rest of the paper is structured as follows. In Section 2 we review the key ideas behind RESTful services and introduce the actor model of computation. Next, in Section 3, the underlying principles of REST and the actor model are related to each other. In Section 4 we introduce a notation for expressing REST-ful services in a restricted form of actors and apply the result to an example. Section 5 discusses the pros and cons of the suggested approach and Section 6 provides a review of related work. Section 7 concludes the paper with some final remarks.

2 Background

In this section we introduce REST and the actor model of computation separately in order to provide the necessary background for the rest of the article.

2.1 RESTful Architectural Style

Although REST is a general architectural style for building network-based software, in this article we consider RESTful interfaces only in the context of web services. In this resource oriented architecture, resources are exposed by the servers and consumed by the clients using HTTP methods [8]. A resource is accessed via a URL and its state is transferred using its representation. A key characteristic of a RESTful interface is the clear division of application state between the client and the server. In the following we inspect these essential features of REST in more detail.

Resources and URLs. In a RESTful interface, everything is a resource. Conversely, everything that can be represented by a URL can be a resource. A resource can be static (e.g. /blog/2010-01-03) or its representation can change over time (e.g. /blog/latest). Moreover, the resource can also represent a list of things (e.g. /blog/2010/). Although every resource must be addressable by a URL, the REST itself does not mandate any scheme for constructing URLs. Human readable URLs are preferred, as they should suggest how to use the interface, but not required by REST. However, the URL should not contain any information about any operation that might be applied to the resource (e.g. /add/ or /remove/).

Connectedness. A property that is closely related to addressability is connectedness. For a client to be able to consume a RESTful service, it needs to know the addresses of all the necessary resources. These resource identifiers can be pre-configured into the client but this is not desirable because it does not enforce connectedness. Instead, the server should guide the client by letting it know about all meaningful resources and then the client can make the decision which path to choose. The server can do this be sending hypermedia links to other resources it exposes. This pattern is also known as Hypermedia as the Engine of Application State.

Methods of REST. For RESTful services, the most commonly used HTTP methods are POST, GET, PUT and DELETE. These methods define the basic CRUD[1] operations for resources. These methods can be categorized in accordance to how they operate on resources. The GET method is said to be *safe* because it has read only semantics on the resource, implying that it is meant only for information and data retrieval without any side effects. With PUT and DELETE, GET is also *idempotent*, since it does not matter whether the operation is applied once or several times on a resource. The end result is always the same. For instance, it does not make a difference if a resource /blog/2010-02-09 is deleted once or twice; it will not exist afterwards.

Statelessness. Honoring the principles of HTTP, REST is strictly stateless. Any application state is stored only on the client. The server, on the other hand, only stores the resources and their states. This also means that each request to a resource must be self-containing; i.e. each request must contain all the information needed to carry out the request. So, the server does not need to – nor it should – know anything about any previous or future request made by the client. Each request should happen in complete isolation from any other request.

Uniform Interface. A very unique characteristic of REST is the uniform interface that it imposes on the services that employ it. Where RPC style services expose bespoke objects and methods for clients, RESTful interfaces expose resources with a fixed set of HTTP methods and return values for each resource. Most importantly this is a profound change in the way the interface for a web service is designed. Everything needs to be designed in terms of resources as opposed to functionality.

2.2 The Actor Model of Computation

Hewitt et al. [11] originally suggested the actor model of computation in the context of enhancing programming languages for artificial intelligence. Our exposition follows [3] that describes actors as a model for concurrent computation in distributed systems.

---

[1] Create, Read, Update, Delete.

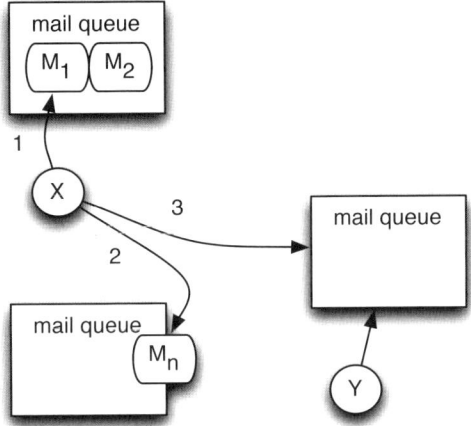

Fig. 1. Sending messages and creating actors

Actors. In the actor model of computation a system is composed of a multitude of computational agents. Each of these computational agents is an actor. An actor is an active self-contained entity that is identified by its address. Actors encapsulate all the necessary information that is required, as specified by the actor's behavior, for it to function as a part of the system. Actors do not share any state and communicate only by passing messages.

For each address there exists a corresponding conceptual mailbox that queues the incoming messages. In the actor model message passing guarantees that messages sent to an actor are eventually delivered, but it does not guarantee the time it takes to deliver a message, nor does it specify the order in which sent messages arrive at the target actor's mail queue. Furthermore, it is not possible to send a message to an arbitrary actor without first knowing its address.

Operational behavior. The driving force in an actor system is the process of sending and receiving messages. In fact, it is the only way the computation in an actor system makes progress. Each time an actor receives a message, it

- must decide on its replacement behavior.
- can create new actors.
- can send messages to other actors.

These three actions can occur concurrently. Once the replacement behavior has been decided another instance of an actor machine is created to represent the actor. This new actor machine is then able to process the next message from the actor's mail queue.

This process is shown in Figures 1 and 2^2. In Figure 1 an actor with two messages in its message queue is shown. The actor is represented by the actor machine X that processes the first message in the mail queue, M_1 (1). As a part

---

[2] Figures adapted and extended from Figure 3.2 in [3].

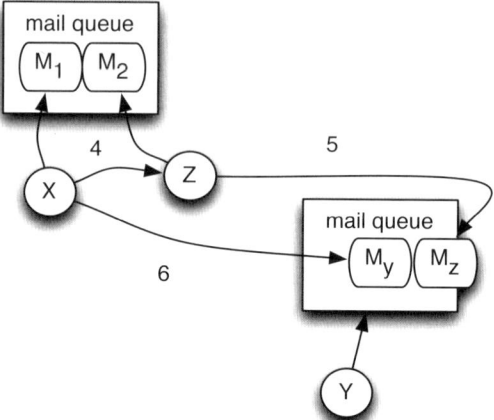

Fig. 2. Concurrent actor machines

of its behavior it can send messages to other actors (2) and create new actors (3). Once the actor machine X has done enough work to decide its replacement behavior with respect to the message it received and its environment, another actor machine Z that corresponds to this replacement behavior is created (4). This is depicted in Figure 2. It is important to note that once the replacement behavior has been decided, both actor machines run concurrently. That is, the new actor machine Z can create new actors or send messages to existing actors (5), while X is free to do the same (6).

3   Relating Actors and REST

In this section we relate the essential concepts of the actor model to those of REST and investigate the relationship. Although it might seem, as actors represent a model of computation and REST is an architectural style, that the connection between them is far-fetched, this is not the case. Let us consider the following rule of thumb given in the literature. Hewitt [10] defines the actor programming methodology to consist of

1. Deciding on the natural kinds of actors (objects) to have in the system to be constructed.
2. Deciding for each kind of actor what kind of messages it should receive.
3. Deciding for each kind of actor what it should do when it receives each kind of message.

On the other hand, adapting from Richardson and Ruby [16], for RESTful web services we get the following

1. Figure out the data set and split it into resources.
2. Name the resources with URLs and expose a subset of the uniform interface.

3. Design the representation(s) for the resources.
4. Integrate the resources with one another using hypermedia links.

Both approaches consider self-standing individual entities as the basic elements: actors in the actor model of computation and resources in REST. These elements are then identified by unique addresses, and they communicate by passing messages. In this section we first elaborate the relationship and then present an example where a RESTful service is represented in terms of the actor model.

3.1   Comparison

In Section 2 we reviewed the essential concepts in RESTful interfaces and in the actor model of computation. Table 1 shows the suggested correspondence. In the following, we investigate each pair individually.

Actor/Resource. Both actors and resources are meant to denote an entity that is self-contained and isolated. Moreover, both actors and resources are the basic components fundamental to the respective approach. In addition, the external interface of the constructed system is defined in terms of actors or resources. In RESTful services the resources available define the vocabulary for the web service, and in an actor system the *external actors* – those initially visible outside the system – determine the entities to which messages can be sent.

Mailbox/URL. In the actor model, the actors are identified by their mailbox addresses. Similarly, a RESTful design uses URLs to denote the resources the service exposes. In both scenarios the mailbox address or the URL is the only way for an external client to interact with the system. Moreover, both approaches allow entities that forward the received messages to other addresses. This corresponds to the use of aliases (e.g. /blog/latest/).

Acquaintance/Hypertext Link. In the actor model a node can only communicate with another node if it has its address. When the system is initialized, some set of *external* addresses is typically declared. In this way the actors whose addresses are revealed comprise the external interface of the system. When outside actors communicate with the external interface, they may receive addresses of actors that are not part of the external interface. Thus the topology of actor connections is not static but grows when the system is used.

Table 1. Counterparts in Actors and REST

Actor model	REST
Actor	Resource
Mailbox name	URL
Acquaintance	Hypertext link
Message	HTTP request
Behavior	Resource state change
Customer	Client

In the context of RESTful services, this behavior of actors giving out addresses of other actors corresponds to returning URLs in the resource's representation. This is especially important as now the evolving graph of actor's acquaintances makes the concept of hypermedia as the engine of application state explicit.

Messages/HTTP Request and Response. The actor model is based on asynchronous message passing with guaranteed delivery. REST, on the other hand, is built on top of HTTP and thus uses the traditional request/response paradigm. This might seem like a mismatch but, if necessary, the synchronous behavior of HTTP can be emulated by actors.

With respect to messages and their processing, the actor model is a lot more general. In the actor model of computation the messages can in principle be of any kind. HTTP, on the other hand, specifies pre-defined set of methods and return values. From the correspondence between the actor model and REST, this means that while using actors for analyzing RESTful services, we must limit ourselves to those of HTTP.

Behavior/Resource State Change. In the actor model, the functionality of an actor is defined by its behavior. This behavior changes over time as the actor processes messages and new replacement behaviors are decided. In REST, a resource has state. Also this state changes over time as requests are processed. For example, if a resource receives a PUT message it has to update its internal state to reflect the new representation of the data it received.

Customer/Client. The term customer refers to the sender of the message in the actor model. This way it becomes an acquaintance of the actor that receives the message. In REST the customer refers to the client of the request, which is most often the browser. In the actor model a "return value" of a behavior simply means sending a message to the customer. This is analogous to RESTful approach, where a response message always follows a request message. The only difference is that in actor model the response message is not mandatory.

3.2   Example

We now take a simple blogging service as an example and apply the ideas presented above. To keep the example simple, we omit issues such as user accounts, authentication, blog comments, and so forth. Thus the system under investigation consists solely of blog posts and their relationships.

We can consider the blogging service in two distinct ways. First, we can consider it to be a RESTful web service. On the other hand, we can start to analyze the service as a network of actors that process the messages of HTTP. To properly illustrate the relationship, we show both sides.

In REST, and with actors, we must first decide on the resources and then move on to the behavior. Since our service is simple, we only have a few resources. The root resource is the container for all blog entries. We give the name /blog for this resource. From the actor point of view, this means that the actor responsible

Table 2. Supported messages and their corresponding behaviors

	/blog	/blog/2010-01-10
POST	Create actor /blog/2010-01-10 with default behavior and send 201 to customer	No effect, send 405 to customer
GET	Send 200 with addresses of the available blog entries (actors) to the customer	Send 200 with the contents of the blog entry to customer
PUT	No effect, send 405 to customer	Change the behavior of the actor to reflect the updated content. Send 200 to customer.
DELETE	No effect, send 405 to customer	In future forward all messages to actor representing status 404. Send 200 to customer

for all posts should have a mailing address /blog. When new entries are added, they become new resources having names (addresses) like /blog/2010-01-10 and /blog/2010-01-15. In the actor model this means that new actors are created whose addresses correspond to the URLs.

Because RESTful interfaces always implement a subset of the uniform interface instead of defining methods on the interface, we specify the supported set of the HTTP methods. This is traditionally done using a table that specifies the resource, method and the intended effect. For our example, such table is shown in the Table 2. Note that this time we formulate the intended effect in the language of actors. Numerical return codes used in the table are from the HTTP specification.

4   Applying Actors to a RESTful Interface

In order to better examine the relationship between the actor model and REST we need a notation for defining actor based RESTful web services. Using the notation we are then able to apply it to some use cases to gain better understanding on how the relationship would work in practice.

4.1   The Notation

Notations for actor systems are usually full fledged programming languages such as PLASMA [10] and Act2 [17]. However, also simpler alternatives exist. The minimalistic example in Listing 1.1 shortly demonstrates the notation used by Agha in [3]. This is not a programming language but a notation used to illustrate actor behaviors. The example defines a piggy bank actor that can be used to deposit money but one is able withdraw money only by breaking it. When the actor receives a deposit message, it replaces its behavior with the new balance. Conversely, when the actor receives a break message, it sends the balance to customer and is replaced by a behavior that ignores all messages (sink).

```
piggy_bank with acquaintance balance
  if message is deposit
    become new piggy_bank with balance + amount
  if message is break ∧ balance ≠ 0
    send amount to customer
    become sink
```

Listing 1.1. Piggy Bank Actor

```
piggy_bank[balance] =
  POST[amount] →
    send 200 to customer
    become (new piggy_bank(balance + amount))
  GET →
    send 405 to customer
  PUT →
    send 405 to customer
  DELETE →
    send balance to customer
    become 404
```

Listing 1.2. RESTful Piggy Bank

Obviously, the example only covers a small subset of all the features in the actor model, but it does portray a notation that can be used to describe an actor and its behavior. In the listing the actor has the balance as its only acquaintance as indicated by the acquaintance keyword. The balance becomes actor's acquaintance when the actor's behavior is created. The replacement behavior is defined using the become keyword.

From Agha's notation we derive our own RESTful actor notation. This notation is presented in Listing 1.2, which demonstrates a similar piggy bank service as the previous example. The difference is that this time the actor can be thought of as a RESTful web service. Our notation resembles Agha's notation but also shows clear connection to REST via its HTTP keywords. In the listing, piggy_bank defines the behavior of the actor. It has one acquaintance as denoted by the variable balance within brackets. The actor implements handlers for all four messages but it only supports the POST and DELETE messages. When a POST message is received, the actor sends 200 OK to customer, whereas when a DELETE message is received, the balance is sent to customer after which the resource becomes unavailable. Other messages are responded with 405 Method Not Allowed. The keyword become is omitted in cases where the replacement behavior is the same as the one being executed.

4.2   Example

By now we have established the analogy between the actor model and REST. We have also defined the notation for applying the actor model to a RESTful interface. Hence, we have the elements to take the blog example presented in Section 3.2 and demonstrate how to depict it using our notation.

```
blog[list] =
  GET →
    send 200[self/latest] to customer
  PUT →
    send 405 to customer
  POST[c] →
    new blog_entry(call POST[c] to list) @ self/current-date
    send 201[self/current-date] to customer
  DELETE →
    send 405 to customer

blog_entry[item] =
  GET →
    send 200[call GET to item] to customer
  PUT[c] →
    send PUT[c] to item
    send 200 to customer
  POST →
    send 405 to customer
  DELETE →
    send DELETE to item
    send 200 to customer
    become 404

# create the blog and bind it to an address
new blog(new list) @ /blog
```

Listing 1.3. Blog example

The Listing 1.3 defines two actor behaviors: `blog` and `blog_entry`. The `blog` defines the top level actor for all `blog_entry` actors. For the sake of brevity we have left out code listings for `list` and `list_item` actor behaviors, which are acquaintances of `blog` and `blog_entry` respectively. The list actor is responsible for storing the blog content. There are a couple of new notations used in Listing 1.3:

- `self` refers to the address of the actor instance itself.
- Symbol @ makes actor external by binding it to the given address.
- `call` is similar to `send` but synchronous. That is, the actor blocks the execution until it has received the response from the actor it `calls`. The *call expression* of the actor model is defined in [3].

Next, we examine the most interesting parts of the Listing 1.3.

`blog.GET`: The `self/latest` is the address of the latest blog entry in the blog. Interestingly, messages sent to this address end up in the mailbox of different actors over time. Note that this behavior differs from the one presented in Table 2.

`blog.POST`: The `call POST[c] to list` sends POST message to the list and synchronously waits for the list item that is returned by the list. The list item is then given as acquaintance to the new blog_entry actor. The new blog entry is bound to blog's own address appended with current date. The current-date is expected to be a primitive returning current date.

blog_entry.GET: The actor retrieves the contents of the blog entry synchronously from the list_item actor. Next, it sends 200 with the content to the customer.

blog_entry.DELETE: The actor deletes the contents of the blog entry by sending DELETE to its list item. Furthermore, it creates a replacement behavior 404 for itself.

The last line in the listing creates the blog and binds it to address /blog. Also note that the created list actor is not explicitly bound to any address. This means that it is not an external actor but visible only for the blog actor.

5 Discussion

The previous sections have motivated the relationship between the actor model of computation and RESTful services. Moreover, we have investigated the relationship in more detail and provided an example. In this section we discuss the pros and cons of the suggested approach to understand RESTful services as a network of actors.

5.1 Resources and Communication

The presented correspondence between REST and actor model of computation puts emphasis on resources and actors, message passing in the form of a subset of the uniform interface, and revealing the hypertext as an engine of application state via returned actor addresses. However, it is important to note what is abstracted away: selection of representation, cookies, and HTTP protocol headers.

At this abstraction level the emphasis is on the resources and actors themselves. Therefore it helps the designer to see the system being built as a network of entities with their internal state and behavior. As it also abstracts away implementation details such as databases, we must model the information storing using actors. This makes the concurrent nature of web services more visible as opposed to hiding it by delegating the concurrency problems to the database.

The use of actors also provides us with a built-in mechanism for inter-resource communication. So far, this has been a property of the framework actually used to implement a service. Also in these cases often the database has been used as an arbitrer to store the shared information. The actor view of RESTful services, on the other hand, does not make a distinction between local and external actors. Hence, external web services can be thought of as being actors too and they can be accessed using the same message passing paradigm as local actors.

5.2 Naming

The naming of resources poses a problem for our approach. In the original actor model the names of individual actors are opaque. That is, there is no external representation that can be resolved by some mechanism to an actor. Naturally, when the addresses of actors are represented by URLs, there is an implicit assumption that a suitable URL corresponds to an actor. Indeed, the presented relationship builds on this assumption.

The opaqueness of actor naming also means that in our notation we have to provide means by which an actor is bound to a URL. The universal actor model [19] has investigated the naming of actors, but their model is not directly suitable for us, as it relies on an own naming scheme. The problem is made more difficult by the fact that when names are identified by URLs, there is nothing that stops the client from guessing arbitrary URLs and seeing whether they resolve to actual actors.

The most difficult problem related to naming occurs in conjunction with the PUT method. From a RESTful point of view, when a new resource is created using PUT, the distinction between a PUT and a POST is that a POST is targeted towards an existing resource whereas with PUT the client is in charge of naming the new resource. When considering the service as a network of actors, the POST case is easy: there exists a corresponding actor that in due course creates more actors if necessary. With PUT, however, there is necessarily no existing actor with the given URL and the actor model of computation does not allow an infinite number of actors.

5.3   Notation

In order to discuss RESTful systems as a network of actors, we have presented an informal notation. Although in the actor model there are no restrictions on the messages actors may send and receive, we must limit our notation to include only the ones supported by HTTP – the HTTP methods and status codes. In addition, the actor model imposes no predefined semantics for actor behavior. In contrast, RESTful actors must adhere to semantics of HTTP methods when responding to messages. This means, for example, that an actor is not allowed to create new actors when receiving a GET message.

As mentioned, we abstract away issues such as the content type of the representation. In addition, the notation is meant to be suggestive and it is not specified formally. Nevertheless, the notation puts emphasis on the inherent concurrency available when implementing RESTful services. The notation shows when it is possible to decide on the new behavior and to create a new actor machine respectively. Most important benefit of the notation is that it provides a tool for the developer to discuss a service especially from the point of view of REST, without implementation details.

6   Related Work

Actors have previously been used to analyze distribution in web applications [6]. Closer to our approach, however, is the actor based research done on middleware systems. Especially the research on Worldwide Computing Middleware [2] and the related research on universal actor model [18] is related to our work.

Although the Worldwide Computing Middleware (WCM) has identified many of the same connections between HTTP and actors (see the Table 1.1 of [2] and our Table 1) as we have, there are crucial differences in the overall approach. The WCM has a more generic approach and considers issues such as mobility that are outside the scope of this paper. Furthermore, the system they envisage would work outside the Web proper, although utilizing many of the familiar web concepts. Our approach, on the other hand, focuses strictly on RESTful services.

Research related to understanding RESTful interfaces has also been done in the context of modeling [12] [13]. However, the emphasis is placed on the process of developing a model that has the required RESTful properties. Our approach, on the other hand, investigates REST in a context of a computation model. This model of message passing has been previously suggested for web services [5], but not specifically, as far as we know, in the context of REST.

7   Conclusions

In recent years, REST has gained popularity as an architectural style of choice for big and complex web services. Unfortunately many of these services fall short of truly embracing the REST principles. To alleviate this problem and to gain better understanding of RESTful services in general, we need solid foundation as to how RESTful web services should be designed and modeled.

In this paper we have established the relationship between the actor model of computation and REST. We have presented the analogy in detail and identified the mismatching features between the two approaches. In order to apply the actor model to a RESTful web service, we have defined a notation for it. Using this notation we have designed a simple blogging service to illustrate the usefulness of the approach.

Building on the work reported in this paper, our next objective is to formalize the notation and build an environment where we are able to visualize the network of actors created by their behaviors. We hope that this visualization will help us to better understand complex RESTful web services.

References

1. Agha, G., Thati, P., Ziaei, R.: Actors: A Model For Reasong About Open Distributed Systems. Formal Methods for Distributed Processing - An Object Oriented Approach, ch. 8. Cambridge University Press, Cambridge (2001)
2. Agha, G., Varela, C.A.: Worldwide computing middleware. In: Singh, M. (ed.) Practical Handbook on Internet Computing. CRC Press, Boca Raton (2004) (invited book chapter)
3. Agha, G.A.: Actors: A model of concurrent computation in distributed systems. Technical Report 844, MIT Artifical Intelligence Laboratory (June 1985)
4. Armstrong, J.: Programming Erlang: Software for a Concurrent World. The Pragmatic Bookshelf (June 2007)
5. Böhm, A., Kanne, C.-C.: Processes Are Data: A Programming Model for Distributed Applications. In: Vossen, G., Long, D.D.E., Yu, J.X. (eds.) WISE 2009. LNCS, vol. 5802, pp. 53–56. Springer, Heidelberg (2009)

6. Chang, P.H., Agha, G.: Supporting reconfigurable object distribution for customized web applications. In: The 22nd Annual ACM Symposium on Applied Computing, SAC (2007)
7. Erenkrantz, J.R., Gorlick, M., Suryanarayana, G., Taylor, R.N.: From representations to computations: the evolution of web architectures. In: ESEC-FSE '07: Proceedings of the the 6th Joint Meeting of the European Software Engineering Conference and the ACM SIGSOFT Symposium on The foundations of Software Engineering, pp. 255–264. ACM, New York (2007)
8. Fielding, R., Gettys, J., Mogul, J., Frystyk, H., Masinter, L., Leach, P., Berners-Lee, T.: Hypertext Transfer Protocol – HTTP/1.1. RFC 2616 (Draft Standard) (June 1999), updated by RFC 2817 http://www.ietf.org/rfc/rfc2616.txt
9. Fielding, R.T.: Architectural Styles and the Design of Network-based Software Architectures. Ph.D. thesis, University of California, Irvine (2000)
10. Hewitt, C.: Viewing control structures as patterns of passing messages. A.I. Memo 410, MIT Artifical Intelligence Laboratory (December 1976)
11. Hewitt, C., Bishop, P., Steiger, R.: A universal modular actor formalism for artificial intelligence. In: IJCAI'73: Proceedings of the 3rd International Joint Conference on Artificial Intelligence, pp. 235–245. Morgan Kaufmann Publishers Inc, San Francisco (1973)
12. Laitkorpi, M., Koskinen, J., Systa, T.: A uml-based approach for abstracting application interfaces to rest-like services. In: WCRE '06: Proceedings of the 13th Working Conference on Reverse Engineering, pp. 134–146. IEEE Computer Society Press, Washington (2006)
13. Laitkorpi, M., Selonen, P., Systa, T.: Towards a model-driven process for designing restful web services. In: ICWS '09: Proceedings of the 2009 IEEE International Conference on Web Services, pp. 173–180. IEEE Computer Society Press, Washington (2009)
14. Pautasso, C., Zimmermann, O., Leymann, F.: Restful web services vs. "big" web services: making the right architectural decision. In: WWW '08: Proceeding of the 17th International Conference on World Wide Web, pp. 805–814. ACM, New York (2008)
15. Razavi, A., Marinos, A., Moschoyiannis, S., Krause, P.: RESTful Transactions Supported by the Isolation Theorems. In: Gaedke, M., Grissnikalus, M., Diaz, O. (eds.) ICWE 2009. LNCS, vol. 5648, pp. 394–409. Springer, Heidelberg (2009)
16. Richardson, L., Ruby, S.: RESTful Web Services. O'Reilly, Sebastopol (2007)
17. Theriault, D.G.: Issues in the design and implementation of act2. Technical Report 728, MIT Artifical Intelligence Laboratory (June 1983)
18. Varela, C.: Worldwide Computing with Universal Actors: Linguistic Abstractions for Naming, Migration, and Coordination. Ph.D. thesis, U. of Illinois at Urbana-Champaign (2001)
19. Varela, C.A., Agha, G.: Programming dynamically reconfigurable open systems with SALSA. In: ACM SIG-PLAN Notices. OOPSLA'2001 Intriguing Technology Track Proceedings, vol. 36(12), pp. 20–34 (December 2001)
20. Weerawarana, S., Curbera, F., Leymann, F., Storey, T., Ferguson, D.F.: Web Services Platform Architecture: SOAP, WSDL, WS-Policy, WS-Addressing, WS-BPEL, WS-Reliable Messaging and More. Prentice Hall PTR, Upper Saddle River (2005)

Multi-level Tests for Model Driven Web Applications

Piero Fraternali[1] and Massimo Tisi[2]

[1] Politecnico di Milano, Dipartimento di Elettronica e Informazione
Milano, Italy
piero.fraternali@polimi.it
[2] AtlanMod, INRIA & Ecole des Mines de Nantes
Nantes, France
massimo.tisi@inria.fr

Abstract. Model Driven Engineering (MDE) advocates the use of models and transformations to support all the tasks of software development, from analysis to testing and maintenance. Modern MDE methodologies employ multiple models, to represent the different perspectives of the system at a progressive level of abstraction. In these situations, MDE frameworks need to work on a set of interdependent models and tranformations, which may evolve over time. This paper presents a model transformation framework capable of aligning two streams of transformations: the forward engineering stream that goes from the Computation Independent Model to the running code, and the testing stream that goes from the Computation Independent Test specification to an executable test script. The "vertical" transformations composing the two streams are kept aligned, by means of "horizontal" mappings that can be applied after a change in the modeling framework (e.g., an update in the PIM-to-code transformation due to a change in the target deployment technology). The proposed framework has been implemented and is under evaluation in a real-world MDE tool.

1 Introduction

In Model Driven Engineering (MDE), models incorporate knowledge about the application at hand, at a specific level of abstraction. An MDE environment usually comprises several models, connected by semantic relationships. The knowledge embodied in more abstract models is primarily used for forward engineering, that is, the progressive refinement towards models that are more concrete, and eventually towards the final implementation code. For instance, a well-known way of structuring the refinement process is provided by the Model Driven Architecture (MDA)[21] scheme that distinguishes three main levels of abstraction: Computation Independent Models (CIM), Platform Independent Models (PIM) and Platform Specific Models (PSM). The translation from one level to the following can be performed manually or, in generative software engineering, it can be driven by automatic transformations between models.

B. Benatallah et al. (Eds.): ICWE 2010, LNCS 6189, pp. 158–172, 2010.

Models have a range of application that goes beyond code generation [23]. In particular, several works use MDE as a support to testing [22,5,20,6]. In these works we can identify a common approach, consisting in producing a set of test cases from the analysis of a CIM or PIM and in executing them on the running software. When the process is automated, model transformations are used to build the testing artifacts. In these approaches the main challenge is in producing tests that have the highest chance of revealing errors.

In this paper we focus on the problem of defining, managing and executing test cases for applications modeled at multiple levels of abstraction, in automated MDE environments. We ignore the issue of generating the right test cases (for that topic, we refer the reader to the aforementioned papers), and concentrate on the problem of aligning the CIM-PIM-PSM transformation stream of the code generator to the parallel CIT-PIT-PST[1] stream used to produce and maintain test cases. In multi-level environments in which a certain number of CIMs, PIMs and PSMs have a parallel lifecycle, this problem is rather complex. For example, if one of the forward engineering transformations is updated, it is not obvious how to modify the "parallel" test transformations.

We introduce a model-transformation framework for test cases, and a prototype implementation of this framework that relies on concrete modeling languages: the CIM level consists of BPMN models [27], which express the multi-actor processes served by the application; at the PIM level, we use a specific Web application DSL, WebML [10], which expresses the data, business logic and front-end interface of the Web/SOA application that supports the business processes specified at the CIM level. The CIM to PIM to PSM transformation is provided by a commercial tool suite, called WebRatio [3]. The paper concentrates on the chain of transformations for producing tests and its contribution can be summarized as follows:

- suitable metamodels for representing test cases for Web applications at different levels of abstraction (CIM/BPMN and PIM/WebML);
- a mechanism for supporting automatic alignment of the Platform Independent Test specifications after the manual refinement of a partial CIM to PIM transformation;
- a mechanism to co-evolve the PIM to PSM transformation and the parallel PIT to PST transformation, which ensures that tests are automatically regenerated after the regeneration of the application code for a different platform.

The rest of the paper is organized as follows: Section 2 presents the case study used throughout the paper; Section 3 overviews the framework and describes the testing metamodels; Sections 4 illustrates how to keep test representations synchronized, when models and transformations evolve; Section 5 compares our contribution to the related work; Section 6 draws the conclusions.

---

[1] Computation Independent Test (CIT), Platform Independent Test (PIT), Platform Specific Test (PST).

2   Case Study

As a case study, we consider a simple application that manages the creation of an expense report by an employee and all the following approval steps.

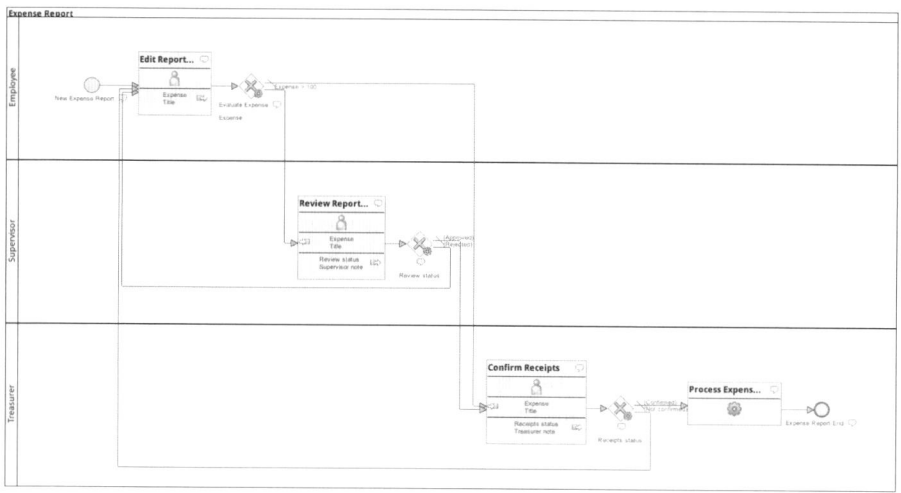

Fig. 1. BPMN model of the Product Catalog Application

The case study is first modeled at the CIM level by the BPMN model shown in Figure 1. The model has three lanes, representing the actors that take part to the process, i.e. employee, supervisor and treasurer. The application process starts with the *Edit Report* activity that allows the employee to insert the title and the total amount of the expense. The values are stored in the *Title* and *Expense* parameters and evaluated by a condition in the following gateway. If the expense exceeds 100$ then the process flow goes to the supervisor, while a smaller expense is directly managed by the treasurer. In the first case the supervisor checks the report parameters and sets the *Review status* parameter to "Approved" or "Rejected" (*Review Report* activity). If the value is "Approved" then the flow goes to the treasurer, otherwise the rejection is sent back to the employee. Finally the treasurer has to set the "Receipts status" parameter in the *Confirm Receipt* activity, and explain in the "Treasurer notes" parameter the reasons of this choice. If the value of "Receipts status" is "Confirmed" then the expense report is directly sent to the company account system by the *Process Expense* activity.

The model in Figure 1 is created using the BPMN modeling tool in the WebRatio toolsuite [3]. The toolsuite can automatically translate this process model into a Web application model, represented in the WebML language. The generated WebML application is specified on top of a data model by means of one or more *site views*, comprising *pages*, possibly clustered into *areas*, and containing various kinds of data publishing components (*content units* in the WebML jargon) connected by *links*. The WebRatio generator from BPMN to WebML creates:

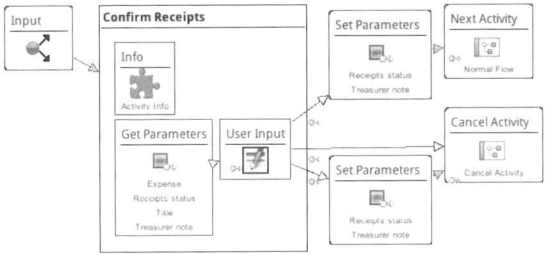

Fig. 2. Generated WebML hypertext for the Confirm Receipt activity

– a generic data model for the process execution (with entities like *User*, *ActivityInstance*, *Parameter*),
– two standard site views for authentication and process administration,
– one site view for each lane, for the orchestration of the process,
– one module, i.e. a composable partial site-view, for each activity.

Figure 2 shows for example the WebML translation of the *Confirm Receipt* activity. The *Input* unit represents the entry point of the module. The units have outgoing links, which enable navigation and parameter passing. For example, *Input* activates the *Confirm Receipts* page. The page displays the name of the current activity (by the *Info* unit) and retrieves from the application context the needed parameters by the *Get Parameters* unit. The retrieved parameters are *Title* and *Expense*, needed to evaluate the expense report, but the unit also looks for pre-existing values of *Receipt Status* and *Treasurer notes*, that could have been saved by the treasurer in a previous access to this activity. The link outgoing from *Get Parameters* communicates these values to the *User Input* unit that denotes a data entry form. The parameter values are used to pre-fill four corresponding input fields in the form. The user can edit these values and select one of the three outgoing navigable links. He can 1) store the new values of the parameters and pass to the *Next Activity*, giving the control to the corresponding module or 2) cancel the activity without touching the current value of the parameters or 3) save a temporary value for the *Receipt status* and *Treasurer notes* before cancelling the activity.

The WebML model enriches the BPMN process scheme with operational details. For example, the parameter saving functionality is not explicitly defined by the BPMN model but added automatically to the WebML design by the WebRatio generator. Furthermore, the application developer can manually modify the generated WebML model to add collateral functions not described at the CIM level. For example, it could be useful to give to the treasurer the possibility to review the history of past expense reports before taking a decision. To model this functionality the designer edits the WebML model to obtain the diagram in Figure 3. In the final model the treasurer can navigate a new link that takes him to the Expense Log page, showing a table of all the registered expenses (by the

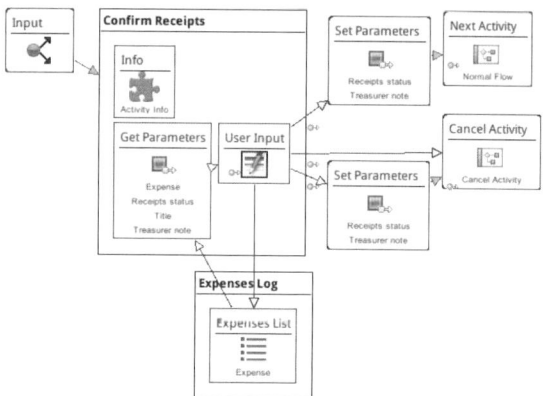

Fig. 3. Edited WebML hypertext for the Confirm Receipt activity

Expenses List index unit). From this page the user has to return to the Confirm Receipts page, to take a definite decision.

The final WebML PIMs can be automatically translated into a running application, by means of the WebRatio code generator. The generator produces all the implementation artifacts for the Java2EE deployment platform, exploiting the popular MVC2 Struts presentation framework and the Hibernate persistence layer. In particular, the View components can utilize any rendition platform (e.g., HTML, FLASH, Ajax), because the code generator is designed to be extensible: the generative rules producing the components of the View adopt a template-based style and thus can incorporate examples of layout for the various WebML elements (pages and content units) coded in arbitrary rendition languages. The user provided templates (like the main code generator) are written in the Groovy language, which allows a Java-like syntax encapsulated into scriptlets, to create template-based transformations.

Once the generated application has been deployed, the application models can be exploited to generate sets of testing sessions, to optimize some testing accuracy metric, e.g., by using techniques like the ones in [9]. For instance, the testing policy could require all the paths of the BPMN model to be exercised by at least one test. A testing session generated at the CIM level is expressed using the concepts that appear in the BPMN model. In the subsequent sections, we will use the following example:

1. an employee starts the process instance
2. the employee creates the report named "Car Rental" for 50$
3. a treasurer receives a report named "Car Rental"
4. the treasurer accepts the receipt
5. the expense report is sent to the company account system

From this high-level test we want to generate the correspondent platform-independent and executable versions.

3   Model-Driven Test Representation

Figure 4 shows an overview of the models involved in our framework. For each one of the MDA abstraction levels, we define both a metamodel of the Web application and a metamodel of the test case:

– at the CIM level, the modeling language is BPMN and the Computation Independent Tests (CITs) are modeled in our BPMN-Test metamodel;
– at the PIM level, the modeling language is WebML and the Platform Independent Tests (PITs) are modeled in our WebML-Test metamodel;
– at the PSM/code level, Platform Specific Tests (PSTs) are Web navigations represented as scripts of a Web testing suite (e.g. the Canoo WebTest system[2]).

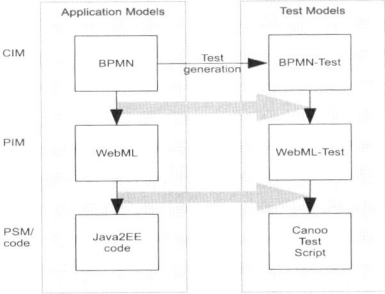

Fig. 4. Overview of the transformation framework

In the design of the test case metamodels we seek maximum simplicity and extensibility. The metamodels are based on a common core. They comprise a container class *TestSuite* that can be decorated with information about the application configuration. *TestSuite* contains multiple *Tests* composed by ordered sets of *Steps*. Each *Step* specifies the identifier of an application session, e.g. useful for distinguishing actions performed by concurrent users of the system. *Step* is specialized in two abstract classes that have to be subclassed for each concrete test case metamodel: an *ActionStep* activates some elements of the application model, referenced by an identifier, while an *AssertionStep* represents the evaluation of a predicate over the application state. Given an application domain, new *ActionStep* or *AssertionStep* subclasses can always be defined for domain-specific tests. Excerpts from the BPMN-Test and WebML-Test metamodels are shown in Figures 5 and 6. It is easy to identify in the two metamodels a reference to the specific concepts of the respective application models. A BPMN-Test model allows one to initiate a process instance, follow its links and insert values in the process instance variables. The only assertion provided checks the value of a process instance variable, but new assertions can be introduced by defining new subclasses. Finally a *Not* assertion allows one to negate the truth value

---

[2] http://webtest.canoo.com

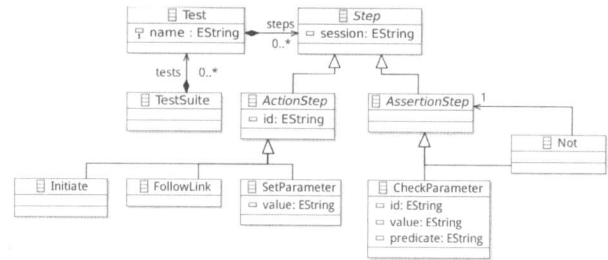

Fig. 5. BPMNTest Metamodel

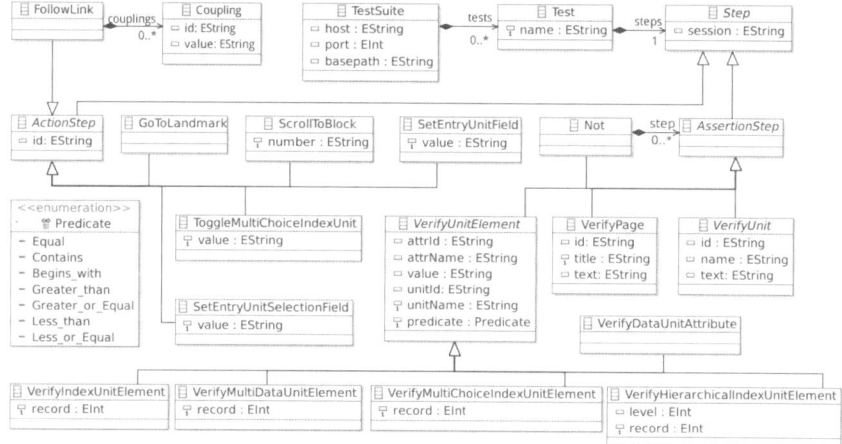

Fig. 6. WebML-Test Metamodel

of a referenced assertion. The WebML-Test metamodel is more complex, as expectable. *ActionSteps* include the activation of links (providing an optional set of correspondent parameter couplings), of landmark elements[3], of input fields, selections and scrolling. *AssertionSteps* allow one to check information about: 1) the current page (i.e. id, title and contained text), 2) currently visualized units (i.e. id, name and contained text), 3) a single element of a currently visualized unit, provided the id of the unit, of the attribute and the numerical coordinates of the record in the table or tree represented by the unit (e.g., to check that the third element of an index contains a given value).

The test models are associated to a default semantics, according to which the test is successful if: 1) it is possible to execute all the *ActionSteps*, 2) no *AssertionStep* evaluates to false. Going back to our case study, the described BPMN test scenario is a single *Test* with this sequence of *Steps* elements:

```
1a. initiate (session='1', id='lane1')
2a. setParameter (session='1', id='title', value='Car Rental')
```

---

[3] Landmarks are global navigation targets, like the home page or the entry pages of main application areas.

```
2b. setParameter (session='1', id='expense', value='50')
3c. followLink (session='1', id='link2')
3a. initiate (session='2', id='lane3')
3b. checkParameter (session='2', id='title', value='Car Rental',
    predicate='equal')
3c. checkParameter (session='2', id='expense', value='50',
    predicate='equal')
4a. setParameter (session='2', id='Receipts status', value='true')
4b. followLink (session='2', id='link4')
4c. followLink (session='2', id='link5')
```

4   Synchronizing Test Representations

The vertical arrows in Figure 4 represent refinement transformations. Transformations in the right column refine the specification of the test case. A BPMN test case, conforming to the BPMN-Test metamodel is translated in one or more WebML test cases, conforming to the WebML-Test metamodel. A model of a WebML test is translated into a Web testing script.

The horizontal arrows represent the synchronization mechanisms between application transformations and test transformations that is the main contribution of this paper. It is important to remark that this kind of synchronization is not always necessary in generic model-driven testing. If the application transformation is *complete*, i.e. it generates automatically the whole target model, and *fixed*, i.e. it does not change over time, then no synchronization mechanism is required. This is a common case in transformation environments. Several applications are based on a single stable transformation that refines an input model, generating a complete output model. Notable examples are compilers, optimizers, analyzers. In all these cases the transformation logic is fixed, and a corresponding fixed transformation can be easily written also for the test cases. In the cases in which the main transformation is not complete (i.e. it is *partial*) or not fixed (i.e. it is *user-defined* or *evolving*), a synchronization mechanism becomes necessary. In Section 4.1 we propose a solution for partial transformations, using the case study BPMN to WebML. Section 4.2 investigates applications with user-defined and evolving transformations using the case study WebML to code.

4.1   Synchronizing Tests with Partial Transformations

Sometimes the main model transformation is *partial*, meaning that it generates only a skeleton of the target model, leaving to the modeler the task of completing the modeling artifact. In these cases, the abstract test case can be easily translated into a skeleton of the concrete test case by a fixed transformation. However, only by means of a synchronization mechanism it is possible to deal with testing the manual additions to the application model.

As exemplified in Section 2, the transformation BPMN to WebML is a case of partial transformation, since the developer may manually intervene on the generated model to add complementary activities to the main workflow. Hence, the transformation between CIT and PIT can't be directly derived by analyzing the CIM-to-PIM transformation. While creating the WebML test we need to take in account also the current state of the WebML model.

In our case study, each *Step* of the BPMN test sequence can be easily translated in one or more *Steps* for testing the generated WebML model. For instance, steps 3b-4b can be transformed automatically into the following steps over the WebML module in Figure 2:

```
3b. verifyEntryUnitElement (session='2', unitID='enu12',
      attrName='title', value='Car Rental', predicate='equal')
3c. verifyEntryUnitElement (session='2', unitID='enu12',
      attrName='expense', value='50d', predicate='equal')
4a. setEntryUnitField (session='2', id='fld12', value='yes')
4b. followLink (session='2', id='ln21')
```

However, manual modifications of the WebML application model could impact the previously generated test set in two ways:

– the test could loose the completeness property, due to the occurrence of new paths in the modified WebML model that would not be subject to test;
– the test could be no longer applicable to the new model, e.g., there could be no link ln21 exiting from the entry unit enu12.

For example, while the above-mentioned test sequence is still applicable to the model in Figure 3, it would not test for errors in the presentation of the list of past expenses. The solution for making the BPMN-Test to WebML-Test transformation aware of manual modifications to the application model is shown in Figure 7. The CIT to PIT transformation is given a composite structure, made of two steps: **T1a.** A first set of *standard CIT to PIT rules* implements the default refinement from BPMN-Test to WebML-Test, following the same logic used for the forward engineering from BPMN to WebML models. These transformation rules match the elements of the BPMN test sequence, retrieve additional information from the BPMN model, if necessary, and apply a default translation to each test step, mirroring the logic in the forward engineering. **T1b.** A second set of *PIT extension rules* implements an algorithm for checking test executability and for extending test coverage to the newly introduced parts of the application model. The algorithm analyzes each test step generated by the standard CIT to PIT rules and checks it with respect to the modified WebML model. If the step is not executable from the modified WebML model or, in case of *followlink* test steps, if the new model presents alternative paths, the algorithm updates the test with a policy that depends on the desired depth of the testing. Otherwise, the step is simply copied to the result. In our prototype, the test update policy chooses non deterministically an alternative link to follow with respect to the non-executable link, or a subset of the newly introduced navigation paths. The algorithm stops when: a) all the BPMN-Test steps have been analyzed (success) or b) there is no way to proceed with the test extension or the number of steps in the test exceeds a threshold (failure).

In the case study, T1a would generate the steps 3b-4b shown above. T1b would copy 3b-4a to the output script, and would start the coverage algorithm for the step 4b, since new alternative paths have been added to the application model, so to add, in at least one of the updated test cases, the path towards the manually added page that shows the expense list.

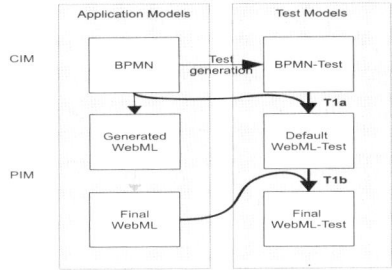

Fig. 7. Transformation framework implementation (BPMN to WebML)

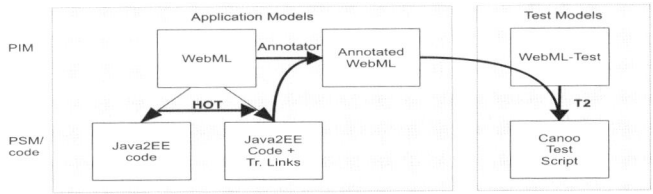

Fig. 8. Adaptation framework to align the PIT to PST transformation

The *PIT extension rules* are able to handle any manual modification to the application model, with the only limitation to elements removal (e.g., the deletion of the User Input unit). Units can instead be repositioned in the model, other units can be interposed between them, and the topology of links can be altered.

4.2   Synchronizing Tests with User-Defined and Evolving Transformations

If the application transformation is *user-defined* or it is *evolving* in time, then adaptation of the PIT to PST transformation is required. This is the typical case of Web applications, in which the PIM can be translated into code in several ways, depending on a number of implementation choices. Notably, on most model-driven Web environments, the implementation transformation depends on the presentation style defined by the graphical designer, which is subject to frequent changes. The code generation process can be seen as a model-to-model transformation that maps an input model at PIM level (e.g., the WebML model of the application) into a an executable model (e.g., the Java2EE code). It is normally a lossy transformation: since its purpose is to produce the code to be actually executed, no extra information is added to the output model and the links between the input and output artifacts are lost.

The transformation from WebML to code is organized into three sub-transformations. The *Layout Transformation* generates a set of JSP pages (one for each page of the WebML model) and miscellaneous elements required by the target platform: Struts configuration (i.e. the controller in the Struts MVC architecture), localization bundles, and form validators. The *Business Logic Transformation* generates a set of XML files (logic descriptors) describing the run-time behavior of the elements of the source model, mainly pages, links, and units. In addition, this transformation produces secondary artifacts, such as the access/authentication logic. The *Persistence Transformation* produces the standard Hibernate artifacts: Java Beans and configuration mapping (one for each entity of the source model) as well as the overall database configuration.

The sub-transformations are based on Groovy. Being the output a set of structured XML and JSP/HTML files, the Groovy generators use a template-based approach: each sub-transformation comprises templates similar to the expected output (e.g., XML or HTML) enriched with scriptlets for looking-up the needed elements of the source model.

The adaptation problem to be solved occurs whenever the code generation produces an implementation with a different way of performing a test step. In this case, also the testing scripts generated from the PIT need to be updated, to automatically align the test session to the updated implementation.

For example, continuing the case study from the previous section, step 4b requires the test to follow a WebML link (ln21) outgoing from an entry unit. The designer may re-generate the application code with a new Groovy template, which alters the presentation of the unit: link 21 could be rendered differently, e.g. as a button instead of an HTML anchor tag. The different rendering could require different activations from the physical test script. For example, in the Canoo WebTest suite, scripts are represented as XML files and links and buttons are activated by specifying different tags, respectively `<clickLink xpath="..."/>` `<clickButton xpath="..."/>`, where the *xpath* attributes is filled by the PIT to PST transformation, in order to point to the correct link or button. In principle, since one cannot make assumptions about the PIM to PSM transformation, which can incorporate any arbitrary code generation rule, the adaptation framework should be able to analyze the code of the transformation itself to detect the new code generation rules. However such an analysis would be remarkably complex. For this reason we propose approach to synchronize the PIT to PST transformation with an evolving PIM to PSM transformation, which relies only on the generated code, and not on the internal structure of the PIM to PSM transformation code. Figure 8 pictorially illustrates the framework.

The key to such a solution is the a posteriori explicitation of the relationship between PIM concepts and the PSM primitive used to render them; this task is performed by an *Annotator* transformation, which enriches the WebML model with the references to the PSM concepts. For the Annotator to remain generic (i.e., not depend on the target implementation platform) another step is required: being able to trace each model concept to the (arbitrary) implementation code produced by the PIM to PSM transformation. This problem is solved by a Higher

Order Transformation (HOT), which automatically weaves traceability links into the PIM to PSM mapping. Therefore, the control flow of the adaptation framework goes as follows: 1) the designer applies the PIM to PSM transformation to generate the code, which may invalidate previous test cases; 2) the framework uses the HOT to mutate the PIM to PSM transformation and produce an augmented PIM to PSM mapping that creates traceability links; 3) the augmented PIM to PSM transformation is executed to produce an augmented implementation code with embedded traceability links; 4) the Annotator transformation uses the PIM (WebML model) and the PSM (J2EE code) augmented with traceability links and produces an annotated PIM model, in which the relationship between PIM concepts and their PSM rendition is made explicit and declarative; 5) the PIT to PST transformation is parametric and exploits the information in the annotated PIM to produce a test case that mirrors the platform dependent primitives used to render the PIM concepts. 6) the test cases automatically generated in this way can be run against the new application implementation.

In Step 2, we exploit our previous work on traceability weaving [15], and we implement an extended version of our Higher Order Transformation (HOT) for traceability. A HOT is a transformation that acts on another transformation, in our case on the transformation used for generating the code. Adding traceability to the generative framework requires preserving the relationship between the elements of the input model and the elements of the output model derived from them. The input of the HOT is the M2M transformation that produces the implementation code. This transformation can be seen as a model, represented by the chosen transformation language (Groovy, in our case study). The output is another transformation, derived by extending the input model with extra elements (additional code generation rules and templates) for producing the traceability links in the implementation code. The HOT must apply to the relevant original transformation rules and produce extended rules such that: 1) they generate the same output elements as the original rules; 2) they add the needed traceability links to the output. The HOT takes only the layout sub-transformation in input, because this is the only one that produces the View elements exercised by the testing script. The traceability links are stored within presentation-neutral, transparent elements (e.g., HTML DIV elements) added to the View artifacts of the output model (namely, the JSP pages).

Once the traceability links are stored into the output code, the *Annotator* parses each element of the WebML model, looks for the associated element in the generated code and adds an annotation to the WebML element (e.g. it would add "type=button" or "type=link" to ln21). Finally, the PIT to PST transformation translates WebML tests into Canoo tests. T2 is parametrized by the element types stored in the annotated WebML model.

The HOT has been implemented using the ATL language and the AmmA [7] framework. To integrate the Groovy language in the transformation framework, a Groovy metamodel has been developed extending the JavaAbstractSyntax metamodel provided by the MoDisco project [1]. The Annotator has been implemented in Java and the PIT to PST transformation is written in ATL.

To summarize, the design of the proposed transformation scheme has the following benefits:

– thanks to the HOT approach the user can freely develop a Groovy template for the website generation;
– the template analysis logic is contained in the HOT and Annotator, and it is kept separated from the test generation logic of T2;
– the template analysis is remarkably semplified by the fact that instead of interpreting the Groovy code, the Annotator has to interpret only the result of this code, i.e. the HTML/JSP.

The main limitation of our current approach is the supposed one-to-one relationship between the PIM and the PSM model (i.e. one PIM element translates into one PSM element, with an arbitrary internal complexity). While this assumption is verified in most WebML applications, an extended version of the algorithm could be advisable for more complex cases.

5   Related Work

The three-layers parallel transformation flow in Figure 4 is first introduced in [13] and the model transformations that compose it are studied in several works, as surveyed in [24]. One of the most popular tasks in this area is test script generation from application requirements, for which an extensive list of references can be found in [14]. In these approaches requirements are modeled by activity diagrams [16], sequence diagrams [9] or natural text [8]. [24] introduces a Functional Requirement Metamodel similar to our BPMN-Test. Our work differentiates from these in being the only one to investigate the automatic synchronization between refinement transformations of application and test cases.

Similar problems to our framework are addressed in the field of model and transformation co-evolution, for instance in [18], [26] and [11]. While some of the design issues are shared with these works, our proposal addresses the peculiar relationship between the model of an artifact and the model of a test case.

Our framework makes use of traceability links to connect a generated model element with its source and avoid the direct analysis of generation code. Transformation frameworks can address traceability during the design of transformations [12], either by providing dedicated support for traceability (e.g., Tefkat [19], QVT [2]), or by encoding traceability as any other link between the input and output models (e.g., VIATRA [25], GreAT [4]). Traceability links may be encoded manually in the transformation rules (e.g., [19]), or inserted automatically (e.g., [2]). A HOT-based traceability system for ATL is already implemented in [17], where the HOT adds to each original transformation rule the production of a traceability link in an external ad-hoc traceability model (conforming to a small traceability metamodel). The approach that we propose is inspired from [17] and can be used to add traceability support to a language like Groovy, that does not provide any built-in support to automatic or manual traceability links.

6 Conclusions

In this paper we have addressed the problem of managing complex model-driven development and testing environments by automatically aligning model transformations. As an application, we have considered the problem of testing Web applications specified at the CIM level with BPMN and at the PIM level with WebML. The proposed framework consists of four "vertical" transformations (CIM-to-PIM and PIM-to-PSM) applied to the forward engineering and to the production of test scripts, which are kept aligned by two "horizontal" transformations that are capable to reinforce integrity after a change of the WebML model produced from the BPMN process diagram and after the update of the WebML-to-Java transformation that yields the executable application. A prototype of the framework has been implemented in Java and ATL. The ongoing and future work will concentrate on the performance validation of the current prototype on very large projects, on its integration with the WebRatio development tool suite, and on the provision of effective mechanisms for evaluating the coverage of a test set with respect to the CIM, PIM and PSM of the application. As a particularly important direction of work, the illustrated framework could be exploited to promote a Test Driven Development approach for MDE.

References

1. MoDisco home page, http://www.eclipse.org/gmt/modisco/
2. QVT 1.0, http://www.omg.org/spec/QVT/1.0/
3. Acerbis, R., Bongio, A., Brambilla, M., Butti, S.: Webratio 5: An eclipse-based case tool for engineering web applications. In: Baresi, L., Fraternali, P., Houben, G.-J. (eds.) ICWE 2007. LNCS, vol. 4607, pp. 501–505. Springer, Heidelberg (2007)
4. Agrawal, A., Karsai, G., Shi, F.: Graph transformations on domain-specific models. Technical report, ISIS (November 2003)
5. Baerisch, S.: Model-driven test-case construction. In: ESEC-FSE Companion '07: 6th Joint Meeting on European SE Conf. and the ACM SIGSOFT Symp. on the Foundations of SE, pp. 587–590. ACM, New York (2007)
6. Baresi, L., Fraternali, P., Tisi, M., Morasca, S.: Towards model-driven testing of a web application generator. In: Lowe, D.G., Gaedke, M. (eds.) ICWE 2005. LNCS, vol. 3579, pp. 75–86. Springer, Heidelberg (2005)
7. Bézivin, J., Jouault, F., Touzet, D.: An introduction to the ATLAS model management architecture. Research Report LINA(05-01) (2005)
8. Boddu, R., Mukhopadhyay, S., Cukic, B.: RETNA: from requirements to testing in a natural way. In: Proceedings of 12th IEEE International Requirements Engineering Conference, vol. 4, pp. 244–253 (2004)
9. Briand, L., Labiche, Y.: A UML-based approach to system testing. Software and Systems Modeling 1(1), 1042 (2002)
10. Ceri, S., Fraternali, P., Bongio, A., Brambilla, M., Comai, S., Matera, M.: Designing Data-Intensive Web Applications. Morgan Kaufmann, USA (2002)
11. Cicchetti, A., Ruscio, D.D., Eramo, R., Pierantonio, A.: Automating Co-evolution in Model-Driven Engineering. In: 12th International IEEE Enterprise Distributed Object Computing Conference, pp. 222–231 (2008)

12. Czarnecki, K., Helsen, S.: Classification of model transformation approaches. In: OOPSLA '03 Workshop on Generative Techniques in the Context of MDA (2003)
13. Dai, Z.R.: Model-driven testing with UML 2.0. Computer Science at Kent (2004)
14. Denger, C.M.M., Mora, M.M.: Test Case Derived from Requirement Specifications. Fraunhofer IESE Report, Germany (033) (2003)
15. Fraternali, P., Tisi, M.: A Higher Order Generative Framework for Weaving Traceability Links into a Code Generator for Web Application Testing. In: Gaedke, M., Grissnikalus, M., Diaz, O. (eds.) ICWE 2009. LNCS, vol. 5648, pp. 273–292. Springer, Hiedelberg (2009)
16. Hartmann, J., Vieira, M., Foster, H., Ruder, A.: A UML-based approach to system testing. Innovations in Systems and Software Engineering (1), 12–24 (2005)
17. Jouault, F.: Loosely coupled traceability for atl. In: European Conference on Model Driven Architecture (ECMDA), workshop on traceability (2005)
18. Lammel, R.: Coupled software transformations. In: First International Workshop on Software Evolution Transformations, Citeseer, p. 3135 (2004)
19. Lawley, M., Steel, J.: Practical declarative model transformation with tefkat. In: Bruel, J.-M. (ed.) MoDELS 2005. LNCS, vol. 3844, pp. 139–150. Springer, Heidelberg (2006)
20. Li, N., Ma, Q.-q., Wu, J., Jin, M.-z., Liu, C.: A framework of model-driven web application testing. In: COMPSAC '06, Washington, DC, USA, pp. 157–162. IEEE Computer Society Press, Los Alamitos (2006)
21. Miller, J., Mukerji, J., et al.: MDA Guide Version 1.0. 1. Object Management Group, 234 (2003)
22. Pretschner, A.: Model-based testing in practice. In: Fitzgerald, J.S., Hayes, I.J., Tarlecki, A. (eds.) FM 2005. LNCS, vol. 3582, pp. 537–541. Springer, Heidelberg (2005)
23. Stahl, T., Voelter, M., Czarnecki, K.: Model-Driven Software Development: Technology, Engineering, Management. John Wiley & Sons, Chichester (2006)
24. Torres, A.H., Escalona, M.J., Mejias, M., Gutiérrez, J.: A MDA-Based Testing: A comparative study. In: 4th International Conference on Software and Data Technologies, ICSOFT, Bulgary (2009)
25. Varró, D., Varró, G., Pataricza, A.: Designing the automatic transformation of visual languages. Sci. Comput. Program. 44(2), 205–227 (2002)
26. Wachsmuth, G.: Metamodel adaptation and model co-adaptation. In: Ernst, E. (ed.) ECOOP 2007. LNCS, vol. 4609, p. 600. Springer, Heidelberg (2007)
27. White, S.A.: Business process modeling notation. Specification, BPMI. org. (2004)

Capture and Evolution of Web Requirements Using WebSpec

Esteban Robles Luna[1,2], Irene Garrigós[3]
Julián Grigera[1], and Marco Winckler[4]

[1] LIFIA, Facultad de Informática, UNLP, La Plata, Argentina
{esteban.robles,julian.grigera}@lifia.info.unlp.edu.ar
[2] Also at Conicet
[3] Lucentia Research Group, DLSI, University of Alicante, Spain
igarrigos@dlsi.ua.es
[4] IRIT, University Paul Sabatier, France
winckler@irit.fr

Abstract. Developing Web applications is a complex and time consuming process that involves different kind of people, ranging from customers to developers. Requirement artefacts play an important role as they are used by these people to perform their daily activities. However, state of the art in requirement management for Web applications disregards valuable features that tend to improve the development process, such as quick validation during elicitation, automatic requirement validation on the final application and useful change management support. To tackle these problems we introduce WebSpec, a requirement artefact for specifying interaction and navigation features in Web applications. We show its use through the development of an example application in the social networking area, and its implementation as an Eclipse plugin.

1 Introduction

It is usual to have multidisciplinary teams (including customers, analysts, developers, QA staff, etc) involved in the development of real world Web applications, making it a complex and time consuming process. Moreover, requirements are susceptible of changing along the development cycle, so it is important to keep them updated and record their changes to reduce risks and time efforts. Many times, the success of a Web project relies on how Web requirements are captured and specified [16].

Several studies [16, 19] in industrial cases have shown the importance of requirements in Web application development. Requirements are generally described in informal documents (e.g. use cases [13]) that are shared by the different stakeholders of the project. However, Web applications tend to evolve in short periods of time [16] and sometimes not having a comprehensive way of handling requirement changes in coherent documents. Therefore, testing against the requirement specification is not feasible [19]. Furthermore, it is sometimes necessary to get deeper in the development or design phases so that customers start to understand their own needs [19].

In this context, capturing requirements should be efficient enough to accomplish the time constraint, without disregarding the interactive nature of Web applications.

B. Benatallah et al. (Eds.): ICWE 2010, LNCS 6189, pp. 173–188, 2010.

Therefore, requirement artefacts have to be easily understood and validated by stake-holders prior to the development, in order to avoid future wastes of time. Moreover, during the development process, the application has to be checked to validate that new requirements have been correctly implemented without "breaking" previous ones. Furthermore, requirement artefacts should help to maintain good quality standards during the development process, which are hard to keep with short time constraints.

In the context of model driven Web engineering approaches [22, 20, 14, 2, 11] the aforementioned concerns are not generally taken into account [7]. As a consequence, Web applications developed with these methodologies share some commonalities with the industrial cases, such as outdated requirements, unfeasibility to test against the requirements and unsuitably to handle fast evolution. Web requirements artefacts (e.g. user interaction diagrams [22], extended use cases [6], etc) capture important aspects of Web applications like navigation; however they are either used to document [13] or to derive the first version of the domain or navigation models [8, 10] and do not consider either evolution or validation (except WebRe [8] which provide test derivation from WebRe models) or even quick validation during the capture phase.

To tackle these problems we present WebSpec, a multi purpose requirement artefact used to capture navigation, interaction and UI (User Interface) features in Web applications. To improve the capturing phase, WebSpec can be used in conjunction with mockups to provide realistic UI simulations, hence improving requirement elicitation. Also, to allow quick requirements' validation in the final application, WebSpec automatically derives a set of interaction tests. Finally, WebSpec enforces change management support which could be used to improve the development cycle by automating structural changes in the application. Summarizing, we show how to:

- Simulate the application using WebSpec and mockups to improve communication between the different stakeholders and reduce elicitation times.
- Derive tests from WebSpec diagrams to reduce requirement validation times.
- Capture requirement changes and use them to semi/automatically upgrade the application and maintain quality standards.

The rest of the paper is structured as follows: in Section 2 we present WebSpec, its concepts and syntax. In Section 3 we show how it is used in different activities in the development cycle by improving requirement's elicitation, helping to automatically validate the requirements and managing their changes. Section 4 briefly shows Web-Spec Eclipse plugin and describes its use in a real application. Section 5 presents related work and finally in Section 6 we conclude and present further work.

2   WebSpec: A DSL to Capture Interactive Web Requirements

WebSpec is a DSL (Domain Specific Language) that allows specifying navigation, interaction and UI aspects in a more formal way than, for example, use cases. A WebSpec diagram has two key elements: *interactions* and *navigations* (Fig. 1).

An *interaction* (the counterpart of a Web page in the requirements stage) represents a point where the user can interact with the application by using its interface objects (widgets). Interactions have a name (unique per diagram) and may have widgets such as: labels, list boxes, buttons, radio buttons, check boxes and panels. Labels define the content (information) shown by an interaction. Interactions are graphically represented with a rounded rectangle which contains the interaction's name and widgets. A WebSpec diagram must have a starting interaction represented with dashed lines.

Fig. 1. WebSpec's basic concepts

A mockup is a sketch of the "possible" application which generally represents UI elements. We can associate interactions with mockups and WebSpec widgets with their concrete UI elements in the mockup to improve the stakeholder's communication during the elicitation phase. There are several tools that could be used to create mockups, such as Balsamiq [1] or plain HTML. WebSpec allows using any of them as long as they provide a unique way to locate the interface elements.

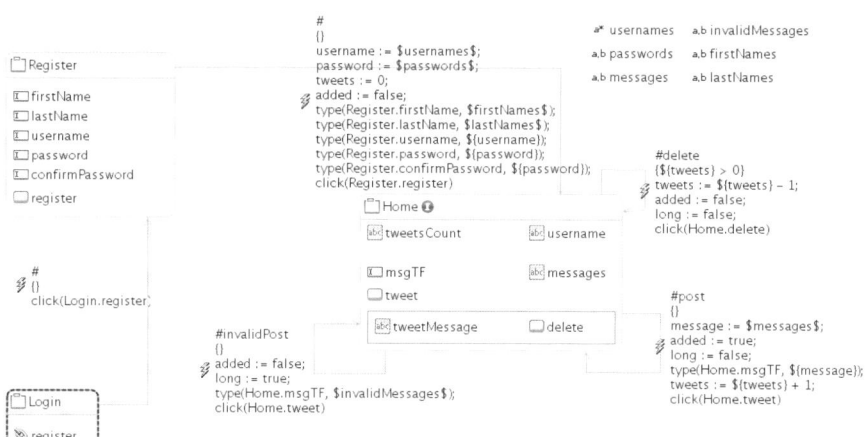

Fig. 2. Tweet Webspec diagram

Invariants are Boolean predicates that must always hold. Every interaction has an invariant that specifies which properties must be satisfied (in case that we do not define one, it is assumed that the invariant is *true*). Fig. 2 shows a simplified diagram of a Twitter-like application that specifies the *post a message* (tweet) requirement and has 3 interactions named: Login, Register and Home. The Home interaction defines an invariant (marked with the I icon near the interaction's name): *Home.username = ${username} && Home.tweetsCount = ${tweets} && ${long} -> Home.messages = "Invalid message"* that states that the contents of the username label must be equal to the username variable (denoted as *${variableName}*) and the contents of the tweetsCount label must be equal to the tweets variable and if the long variable is true then the contents of the messages label must be equal to "Invalid message".

A *navigation* from one interaction to another can be activated if its precondition holds by executing a sequence of actions such as: clicking a button, adding some text in a text field, etc. As well as invariants, preconditions can reference variables previously declared in the diagram. For example, the *delete* navigation (Fig. 2) has the precondition: *${tweets} > 0*. Navigations are graphically represented in the WebSpec diagrams with gray arrows while its name, precondition and actions are displayed as labels over them. Actions are written in an intuitive DSL conforming to the syntax: *var := expr | actionName(arg1,... argn)*. Traditional hyperlink navigation is represented with no precondition (indeed, an always true precondition) and with only one action click (follow) a link widget (see Login to Register navigation in Fig 2). An example of a more complex sequence of actions is the *invalidPost* navigation (Fig. 2):

```
(1) added := false;
(2) long := true;
(3) type(Home.msgTF, $invalidMessages$);
(4) click(Home.tweet);
```

The first 2 sentences (1-2) assign constant values to variables. Then some text generated by the invalidMessages generator (denoted between $) is typed in the msgTF text field (3) and finally the tweet button is clicked (4).

WebSpec allows specifying general properties like "an error must be shown if the user tries to post a message with more that 150 characters" using generators. Following the idea of QuickCheck [3], we extract the data used for specifying interaction requirements into generators. If a property in a WebSpec diagram holds, then it must hold for any element that could be generated by a generator. A generator is a function that can be called from navigation actions (e.g. $invalidMessages$) and generates data. For example, Fig. 2 has 6 generators: usernames, passwords, messages and invalidMessages, firstNames, lastNames. The invalidMessages generator generates strings with size > 150, so when that *invalidPost* navigation is activated, some invalid text will be typed and because the long variable will be true an error message must be display (recall the invariant of the Home interaction) in the messages label.

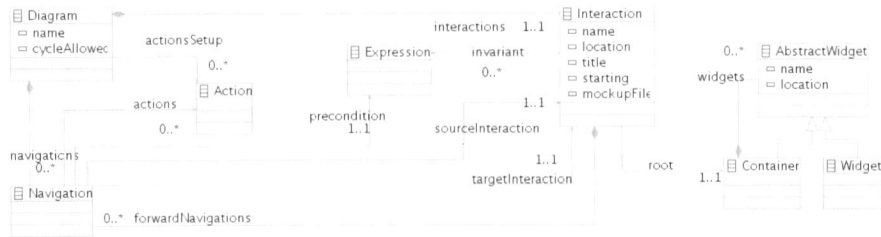

Fig. 3. WebSpec simplified metamodel

For those Web requirements that have strong hidden behaviour (not perceived from an interaction point of view, e.g. send an email), Webspec could be combined with simple notes over the diagram or by linking navigations with use cases or user stories. For example, if an email has to be sent when a user posts a message, we can easily add a note over the *post* navigation.

Finally, WebSpec is formally defined in a metamodel (Fig. 3) that is used to improve the development process as shown in the following section. A diagram has a root object of the class Diagram which contains many Interaction and Navigation instances. An Interaction instance knows its name, forward navigations and associated mockup. A Navigation knows its source and target Interaction and the sequence of Action instances that triggers them. Finally, the interaction knows its root widget Container which can contain many AbstractWidget (Widget or Container) instances.

3   Using WebSpec along the Development Cycle

WebSpec allows specifying interaction requirements for Web applications at a conceptual level without imposing any particular development process. Notwithstanding, WebSpec diagrams can be used at different steps of the development cycle of Web applications. To illustrate this fact, we show in Fig. 4 how WebSpec can be used in the different activities of a test-driven approach like WebTDD [21] and in a methodology using a RUP [15] like process. Simulation (S in Fig. 4) can be used to share design options between stakeholders and validate their requirements in the requirements phase of both kind of processes. Tests generated from the diagrams (TG in Fig. 4) can be used to validate requirements against the final implementation when using a RUP style or to drive the development process in WebTDD. Changes during the development cycles are recorded (CR in Fig. 4) in the requirements phase of both. Finally, semi/automatic upgrades (CA in Fig. 4) using the previously recorded changes can be applied to the application in the development phase of WebTDD and RUP. In the following subsections we show how these features are supported in WebSpec.

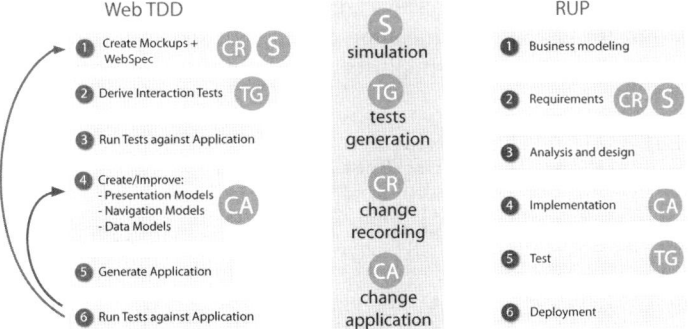

Fig. 4. Using WebSpec in activities of different approaches

3.1 Simulating the Application during Requirements Elicitation

With the aim of improving the requirement elicitation phase, WebSpec diagrams allow the simulation of the resulting application. Simulation is important to bridge the gap between the understanding of customers and designers about requirements thus getting real feedback from them.

Most requirement artefacts [13, 8, 1, 22] require some level of knowledge from customers to be fully understood, causing communication or understanding problems during elicitation. WebSpec is not the exception; understanding a diagram may take some time and require some knowledge of WebSpec's concepts, e.g. variables and interactions. To ameliorate this scenario WebSpec provides some interesting features such as mockup association and formal specification which allows to formally simulating the application to improve the communication between stakeholders during elicitation. We say formally, because different from the simulation provided by tools such as Balsamiq [1], we not only show transitions between the pages but also execute real actions and provide descriptions of what would be the real output of the application directly over mockups. The descriptions provided are generated automatically from the WebSpec diagram and they are easy to understand because they are written in natural language. In this way, from every WebSpec diagram a set of simulations is automatically generated which could be used at any time by customers to understand the meaning of the diagram and suggest changes or improvements to the analyst.

The set of simulations is obtained following the different paths from the starting interaction of each WebSpec diagram. If the diagram has cycles (a path that contains more than one occurrence of an interaction) then we have to prune those paths to obtain finite paths. For example, in the Tweet Diagram (Fig. 2) we can obtain the following paths pruning them (as it is a cycled diagram) to a length of 5 interactions:

Login -> *Register* -> *Home* -> *(post nav)* *Home* -> *(post nav)* *Home*
Login -> *Register* -> *Home* -> *(invalidPost nav)* *Home* -> *(post nav)* *Home*
Login -> *Register* -> *Home* -> *(post nav)* *Home* -> *(invalidPost nav)* *Home*
Login -> *Register* -> *Home* -> *(invalidPost nav)* *Home* -> *(invalidPost nav)* *Home*
Login -> *Register* -> *Home* -> *(post nav)* *Home* -> *(delete nav)* *Home*

Each simulation is created following the sequence of interactions and navigations of the path and data is generated when a generator is referenced inside expressions. The path is transformed into a simulation model (not shown for space reasons) that specifies the simulation steps. A simplified version of the transformation algorithm is shown next:

```
(01) simulation := new Simulation();
(02) for (PathItem item : path.getItems()) {
(03)    if (item.isInteraction()) {
(04)      Interaction interaction = (Interaction) item;
(05)      simulation.openMockup(interaction.getMockup());
(06)      simulation.showPredicate(interaction.getInvariant());
(07)    } else {
(08)      Navigation navigation = (Navigation) item;
(09)      simulation.showPredicate(navigation.getPrecondition());
(10)      for (Action action : navigation.getActions()) {
(11)        simulation.simulateAction(action);
(12)      }
(13)    }
(14) }
```

Line 1 creates the simulation model. For every item (interaction or navigation) in the path (2): if it is an interaction (3) we show the mockup associated with it (5) and show the predicate of its invariant to describe which properties must hold (e.g. "The label should have the value 'John') (6); if the item is a navigation, we show the pre-condition (9) and for every action we simulate it (10-12).

As an example of a simulation we next show a sequence of the simulation steps of the path: **Login** -> **Register** -> **Home** -> *(post nav)* **Home** -> *(post nav)* **Home** generated by the algorithm. For space reasons, we can not show all the steps so we will describe the first 11 steps and show steps 8 through 11 (except step 10 which is equal to step 11 without the label) in Fig. 5.

```
(01) open("loginMockup.html");
(02) click("register", "the user clicks the register button");
(03) open("registerMockup.html");
(04) type("firstName", "John", "the user types 'John'");
(05) type("lastName", "Doe", "the user types 'Doe'");
(06) type("username", "john.doe", "the user types 'john.doe'");
(07) type("password", "aaa", "the user types 'aaa'");
(08) type("confirmPassword", "aaa", "the user types 'aaa'");
(09) click("register", "the user clicks the register button");
(10) open("homeMockup.html");
(11) showDescriptionNearTo("it should contain the text 'John'",
     "username");
```

Line 1 opens the first mockup. Line 2 clicks the register button and line 3 we simulate navigation by opening the mockup associated with the Register interaction. Lines 4-9 execute the actions to move from Register to Home interaction. Specifically, line 8 (Step 8 of Fig. 5) types 'aaa' to the confirm password field and line 9 (Step 9 of Fig. 5) clicks the register button. Line 10 simulates the navigation by opening the mockup associated with the Home interaction and finally line 11 (Step 11 of Fig. 5) shows the label with the condition that must be satisfied according to the filled information. Notice that the algorithm has to use generators in lines 4, 5, 6, 7, 8 to generate data according to the specification of Fig 2 (Register to Home navigation).

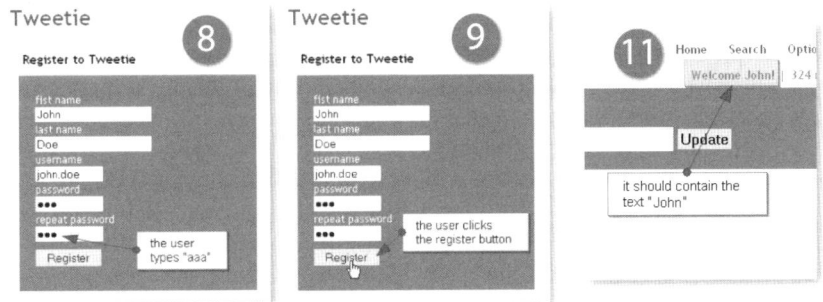

Fig. 5. Simulation steps of the Tweet diagram

Once the requirements elicitation phase is completed we can automatically generate a set of tests that the application must pass as shown in the following subsection.

3.2 Automatic Validation of Requirements

New requirements must be validated to guarantee their correct implementation while previous ones still work as intended. However, it is hard to perform this task in short periods of time thus making it more important to keep requirements updated for the quality assurance team.

A well known way of validating requirements consists in running automated tests (that express the requirements) over the application. If one of these tests fails, then a requirement is not satisfied by the application. In particular, interaction tests play an important role in industrial settings as they execute a set of actions in the same way a user would do on a real Web browser, thus their use is continuously growing [17]. However, in the Web engineering research area their use is recently appearing in approaches like WebTDD [21].

In a similar way we have created the simulations, we build a test suite (a set of test cases) from a WebSpec diagram by following the different paths from the starting interaction. To capture the basic concepts of tests, we have created a metamodel (Fig. 6) which is independent of the technology used. The metamodel contains the Test and TestSuite classes that conceptualize a test and a set of tests. A Test has a sequence of actions: assertions on interface objects or actions performed by the user over the application. Both cases are covered by the TestItem hierarchy.

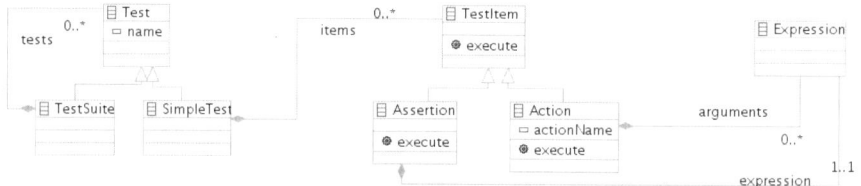

Fig. 6. Test metamodel

To build the test suite, we transform each path into a SimpleTest (see Fig. 6) by executing the following simplified version of algorithm over each path. Similar to simulations, we will use generators to generate data according to the specification when an expression references it. The TestSuite is obtained by simple composition (see the composition relationship in the metamodel of Fig. 6) of the previous SimpleTest instances. More complex scenarios could be manually created by composing different Test suites into a bigger one. Once the TestSuite model is generated, we can translate it to a specific implementation framework such as Selenium [24].

```
(01) test := new SimpleTest();
(02) test.addItem(new OpenURL(applicationURL));
(03) for (PathItem item : path.getItems()) {
(04)    if (item.isInteraction()) {
(05)       Interaction interaction = (Interaction) item;
(06)       test.addItem(new Assert(interaction.getInvariant()));
(07)    } else {
(08)       Navigation navigation = (Navigation) item;
(09)       for (Action action : navigation.getActions()) {
(10)          test.addItem(new Execute(action));
(11)       }
(12)    }
(13) }
```

Line 1 creates the test model and line 2 generates the action to open the application. For each element in the path: if it is an interaction (4), we assert its invariant (6); if it is a navigation (8) we execute the actions that allow us to navigate from one interaction to another one (9-11).

To better illustrate these ideas, let us consider a specific path of the Tweet diagram: *Login* -> *Register* -> *Home* -> *(post nav) Home* -> *(delete nav) Home*. Applying the previous algorithm to the path and deriving a Selenium version of the test gives the next result:

```
(01) selenium.open("http://localhost:8080/index.html");
(02) selenium.click("id=register");
(03) selenium.waitForPageToLoad("30000");
(04) selenium.type("id=firstName", "John");
(05) selenium.type("id=lastName", "Doe");
(06) selenium.type("id=username", "john.doe");
(07) selenium.type("id=password", "wqe4yt24");
(08) selenium.type("id=confirmPassword", "wqe4yt24");
(09) selenium.click("id=register");
(10) selenium.waitForPageToLoad("30000");
(11) assertTrue((selenium.getText("id=username").equals("John"))
(12)     && (selenium.getText("id=tweetsCount").equals("0")));
(13) selenium.type("id=tweetMessage" "@Office");
(14) selenium.click("id=tweet");
(15) selenium.waitForPageToLoad("30000");
(16) assertTrue((selenium.getText("id=username").equals("John"))
(17)     && (selenium.getText("id=tweetsCount").equals("1"))
(18) selenium.click("id=tweetDelete0");
(19) selenium.waitForPageToLoad("30000");
(20) assertTrue((selenium.getText("id=username").equals("John"))
(21)     && (selenium.getText("id=tweetsCount").equals("0")));
```

Line 1 opens the application in the Web browser. Lines 2-3 click on the register link. Lines 4-10 fill the register information (first name, last name, username, password and confirm password) and clicks the register button. Lines 11-12 assert that the labels of the Home page have the values previously filled. Lines 13-15 post a new message to the wall. Lines 16-17 assert the new value that the labels must have after the post are valid. Lines 18-19 click on the delete button of the first message to delete the post. Finally, lines 20-21 assert the values of the labels after the delete operation.

As aforementioned, Web applications tend to change very fast, thus recording requirements changes is important to improve the development process. In the next subsection we show how requirement changes are captured in WebSpec.

3.3 Capturing Requirement Changes

Capturing requirements changes is an important feature to predict their impact in the application. Though some mature requirement artefacts [13] provide extensions to support change management, in the Web engineering field there are not many studies about how requirement changes can be captured and used to improve some part of the development process (see Sect. 5 for details).

In WebSpec, changes are recorded into change objects that group a set of changes. WebSpec can suffer different coarse grained changes, such as the addition or deletion of an interaction or navigation element. These elements can be modified too, by the

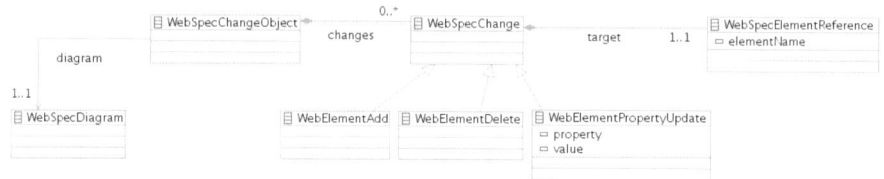

Fig. 7. Change metamodel

addition or deletion of widgets to an interaction, changes in invariants, etc. As for navigations, we can add or delete preconditions, change their source, target, or the actions that triggers them. All these types of possible changes have been represented in the metamodel of Fig. 7. When the user modifies the diagram, a change object is created and the sequence of changes is recorded as instances of these classes.

As an example, let us suppose we want to add a link between the Login interaction (Fig. 2) and a new TermsOfService interaction. The change in the diagram generates a new change object (Fig. 8) which has the following elements: a new interaction (TermsOfService), a new navigation (Login -> TermsOfService), a new link (tosLink) and a new label (the description of the terms of service). To take advantage of capturing changes, we show in the following subsection how to use WebSpec change objects to semi/automatically upgrade the application.

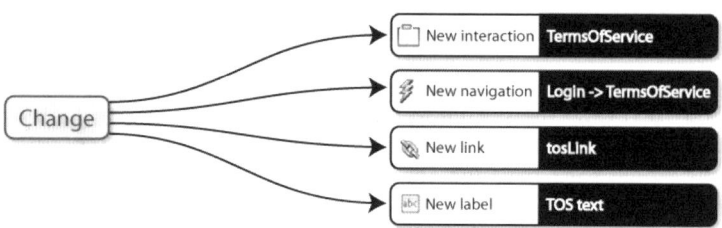

Fig. 8. Change object representing the new Terms of Service functionality

3.4 Using Requirement Changes to Evolve the Application

Though handling requirement changes serves for multiple useful purposes, we will focus on how to semi automatically upgrade the application using them. Since change objects represent changes at the WebSpec level, we decouple the process of upgrading the application by providing different effect handlers. An effect handler is a component responsible of mapping the changes in the diagrams to a concrete technology and storing the trace links between the WebSpec elements and the technology ones. For example, a WebSpec diagram generates a change that can be applied with different effect handlers depending on the underlying technology: Seaside [23], GWT [12], WebRatio [25], etc. Seaside and GWT effect handlers will create/update methods and classes but WebRatio effect handler will produce model transformations in order to update the models.

As an example of the use of effect handlers, we next show how to use the change object of the previous subsection to upgrade the application. We assume that the application is developed with Seaside, so we use the Seaside effect handler.

The effect handler "reads" the change object and suggests actions to the developer. The first change (add the TermsOfService interaction) suggests to create a new class (WATermsOfService) that extends the base class of the Seaside framework (WALayoutPane) (see row 1 of Fig. 9). The developer accepts the proposal and continues with the next change that represents the navigation from Login to TermsOfService interaction. This change refers to behavioral aspects that the effect handler does not handle yet, so it does not propose an action. The two remaining changes involve adding widgets to the interactions. The first one adds a link in the Login interaction; because the effect handler stores the trace link between the interaction and the implementation class, it suggests adding a new method that creates the link to the WALogin class (Row 2). Finally, the effect handler suggests adding a new method to the WATermsOfService to create the new label (Row 3).

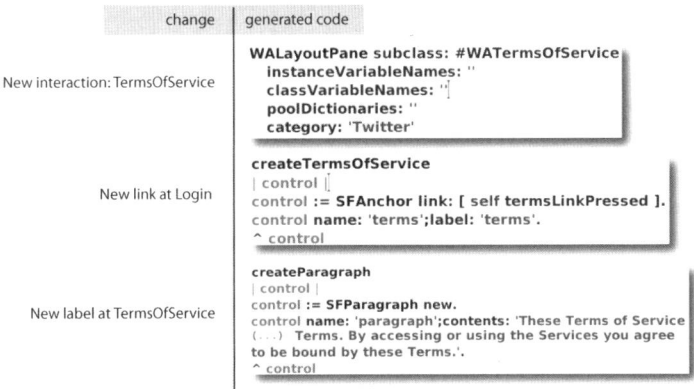

Fig. 9. Semi/automatic upgrades using the Seaside effect handler

4   Implementation

WebSpec has been implemented as an Eclipse plugin using EMF and GMF technologies. The plugin allows the creation of diagrams and the association of interactions with HTML mockups inside the environment. Simulations are implemented using a small extension to the Selenium framework, and JUnit selenium tests are automatically generated from diagrams. Finally, changes are recorded and stored into XML files that could be read by different effect handlers. We have implemented effect handlers for Seaside and GWT. Fig. 10 shows a screenshot of the WebSpec Eclipse plugin.

Using the plugin and following the WebTDD approach, we have successfully implemented a complete application for the Post-graduate area of the College of Medicine in the University of La Plata. We have used GWT, Spring and Hibernate as base technologies for the development process and actively used the generated tests to

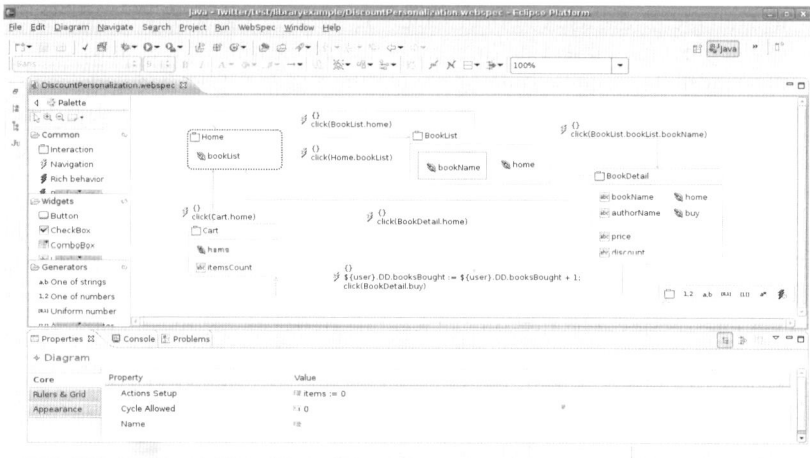

Fig. 10. Webspec Eclipse plugin

check that the application satisfies the requirements in an incremental way. Simulation was used for improving the elicitation of requirements and change objects allowed automating the creation of the structural UI classes of the application.

5   Related work

In the context of Web Engineering, the specification of interaction requirements is a complex task due to some unique characteristics of Web applications such as the need to represent the navigation in information spaces, the need of describing technical constraints related to the information flow (e.g. session management), the rapid evolution of requirements, sensitive communication among developers and the participation of customers in the development process (e.g. marketing experts, editorial board, etc) [26]. In the last years, a large variety of model-based artefacts have been employed to capture Web requirements like UML use cases and sequence diagrams [4], User Interaction Diagrams [22], task models [27], and navigation models [11]. It is also worthy noting a widespread use of paper-based mockups to capture requirements related to the user interface of Web applications [9] which has lead to the development of advanced tools for sketching and storyboarding the user interface of Web applications such as Denim [18] and Balsamiq [1].

In Table 1 we compare the expressiveness power of some artefacts with respect to the concepts for representing Web requirements. As shown in the table, each artefact includes only part of the concepts required to express requirements of Web applications. For example, whilst use cases can be used to represent functional requirements, mockups (either paper-based or supported by tools) are more likely to capture and represent requirements related to the composition of the user interface. Task models allow expressing fine-grained functional requirements including navigation, user transactions and business processes. As can be seen, Web engineering methods have

Table 1. Expressiveness power of requirement artefacts for Web applications

Concept		Artefacts used for representing requirements				
		Use cases (UC)	Task Models	WebRE	WebSpec	Mockups
Behaviour	Navigation	Dependencies between UC	Dependencies between tasks	Navigation	Navigation arrows	Arrows
	Process	Use cases	Tasks,	WebProcess	WebSpec diagram	-
	User interaction	Functional requirements	Interactive tasks	User transaction	Action	-
	Constraints	OCL	Lotus operators	OCL	Precondition	Annotated text
	Information flow	-	Data transfer between tasks	Data transfer in user transaction	Data transfer between interactions	-
Structure	Node / page	-	-	Node	Interactions / navigations	Prototype
	Content	-	-	Content	Widgets	Widgets
	UI composition	-	-	-	Containers	Prototype
	User roles	Actor	Actor	WebUser	-	-

often included more than one artefact for capturing requirements; for example use cases are present in OOHDM [22] in combination with UIDS. Besides, use cases and activity diagrams, WebML [2] uses semi-structured textual descriptions to capture additional information that can hardly be expressed using the former models. Similarly, UWE [14] proposes extended use cases, scenarios and glossaries for specifying requirements and WSDM [6] employs task models using concurrent task trees.

Currently, there is no consensus on which notation(s) should be used to capture and specify Web requirements. In order to provide a more uniform view on the coverage of requirements by each artefact, Escalona and Koch [8] have proposed a metamodel based on WebRE profiles [8]. Its main advantage is the automatic generation of conceptual models (content and navigation models) which automatically satisfy the requirements. Notwithstanding, some requirements such as detailed composition of the user interface and behaviour constraints cannot be fully described with this notation.

In another study, Escalona and Koch [7] have investigated how different Web engineering methods support the capture of requirements. They demonstrated that Web engineering methods do not pay equal attention to requirements. Some methods employ classical notations to deal with Web requirements or ignore this phase of the development process. It is interesting to notice that requirement artefacts might play several roles during the development process: they can act as communication tools (for elicitation requirements with clients), as elements for early specifications (that should be taken into account during implementation phases) and as checklists for assessing if the final implementation complies the initial requirements. Requirement checklists can indeed be employed in regression testing [28] for assessing in a longer term, the evolution of requirements expressed for a single application.

In [5] the authors have investigated the communication role of artefacts and they proposed MoLIC which acts as a kind of blueprint of the application and thus allowing professionals from multidisciplinary backgrounds to share the same understanding of the essence of the application. Other authors however, have investigated how to automate the generation of the system specification from the requirements specification; for example OOWS [20] which extends activity diagrams with the concept of interaction point to describe the interaction of the user with the system. It provides automatic generation of (only) navigation models from the tasks description by means of graph transformation rules. A-OOH [10] considers the i* framework in order to specify the requirements model which is goal-oriented. From this specification, the conceptual models (e.g. domain and navigation models) are generated by means of QVT transformations. Both OOWS and A-OOH approaches are examples of methods that specify requirements and provide code derivation; however the level of detail they provide make them unsuitable as communication tools with clients.

WebSpec supports features that tend to improve the development process when changes appear often and should be implemented fast, in comparison with the afore-mentioned requirement artefacts. It provides a means to describe several of the unique aspects of Web applications (such as navigation and information flow); when used in combination with mockups, it provides animated storyboards to improve the communication between stakeholders. Moreover, they contain enough information to support test generation independently of the development method. Finally, change support and effect handlers allow managing the fast evolution of the application.

6   Concluding Remarks and Further Work

In this paper we have presented WebSpec: a requirement artefact used to capture navigation, interaction and UI features in Web applications independently of the de-velopment process. WebSpec presents several advantages that help to improve the development cycle in short periods of time. We have shown its use in conjunction with mockups to provide a formal simulation of the final Web application, getting real feedback during the requirement elicitation phase. Furthermore, requirements ex-pressed in WebSpec diagrams are easily validated due to the automatic derivation of interaction tests. Finally, it has been shown how keeping diagrams updated contrib-utes to semi/automatically upgrade the application thus improving development times.

This work focuses on interactive requirements of Web applications. In the future we aim at exploring how WebSpec can be used in conjunction with other techniques for expressing non-interactive requirements such as accessibility and usability of Web applications. We are currently working on adding RIA expressiveness to WebSpec, so that RIA features (e.g. autocomplete, hover detail, etc) can be easily specified in the diagrams. Also, we aim to associate WebSpec diagrams to tasks, so we can monitor the progress of a development process. Finally, we are analyzing different alternatives to support the specification of requirements at the domain level which can be seam-less integrated in WebSpec.

References

1. Balsamiq, http://www.balsamiq.com/products/mockups
2. Ceri, S., Fraternali, P., Bongio, A.: Web Modeling Language (WebML): A Modeling Lan-guage for Designing Web Sites. Computer Networks and ISDN Systems 33(1-6), 137–157 (2000)
3. Claessen, K., Hughes, J.: QuickCheck: a lightweight tool for random testing of Haskell programs. In: Proceedings of the fifth ACM SIGPLAN international conference on Func-tional programming, vol. 35, pp. 268–279 (September 2000)

4. Conallen, J.: Building Web Applications with UML, 300 p . Addison-Wesley, Reading (2000)

5. de Paula, M.G., da Silva, B.S., Barbosa, S.D.: Using an interaction model as a resource for communication in design. In: CHI '05 Extended Abstracts on Human Factors in Computing Systems, Portland, USA, April 02-07, pp. 1713–1716 (2005)

6. De Troyer, O., Casteleyn, S.: Modeling Complex Processes for Web Applications using WSDM. In: 3rd Int. Workshop on Web-Oriented Software Technologies, Oviedo, Spain (2003), http://www.dsic.upv.es/~west/iwwost03/articles.htm

7. Escalona, M.J., Koch, N.: Requirements engineering for web applications – a comparative study. J. Web Eng. 2(3), 193–212 (2004)

8. Escalona, M.J., Koch, N.: Metamodeling Requirements of Web Systems. In: Proc. International Conference on Web Information System and Technologies (WEBIST 2006), INSTICC, Setúbal, Portugal, pp. 310–317 (2006)

9. Flannagan, S.: The Paper Version of the Web. In: Deeplinking, http://deeplinking.net/paper-web/

10. Garrigós, I., Mazón, J.N., Trujillo, J.: A Requirement Analysis Approach for Using i* in Web Engineering. In: Gaedke, M., Grissnikalus, M., Diaz, O. (eds.) ICWE 2004. LNCS, vol. 5648, pp. 151–165. Springer, Hidleberg (2009)

11. Gómez, J., Cachero, C.: OO-H Method: extending UML to model web interfaces. In: van Bommel, P. (ed.) Information Modeling For internet Applications, pp. 144–173. IGI Publishing, Hershey (2003)

12. GWT, http://code.google.com/webtoolkit/

13. Jacobson, I.: Object-Oriented Software Engineering: A Use Case Driven Approach. ACM Press/Addison-Wesley (1992)

14. Koch, N., Knapp, A., Zhang, G., Baumeister, H.: UML-Based Web Engineering, An Approach Based On Standards. In: Web Engineering, Modelling and Implementing Web Applications, pp. 157–191. Springer, Heidelberg (2008)

15. Kruchten, P.: The Rational Unified Process: an Introduction, 3rd edn. Addison-Wesley Longman Publishing Co., Inc., Amsterdam (2003)

16. McDonald, A., Welland, R.: Web Engineering in Practice. In: Proceedings of the Fourth WWW10 Workshop on Web Engineering, pp. 21–30 (May 1, 2001)

17. Maximilien, E.M., Williams, L.: Assessing test-driven development at IBM. In: Proceedings of the 25th international Conference on Software Engineering, Portland, Oregon, May 03-10, pp. 564–569. IEEE Computer Society, Washington (2003)

18. Lin, J., Newman, M.W., Hong, J.I., Landay, J.A.: DENIM: finding a tighter fit between tools and practice for Web site design. In: Proceedings of the SIGCHI Conference on Human Factors in Computing Systems, CHI 2000, The Hague, The Netherlands, April 01 - 06, pp. 510–517. ACM, New York (2000)

19. Lowe, D.: Web system requirements: an overview. Journal of Requirements Engineering, 102–113 (2003)

20. Pastor, O., Abrahão, S., Fons, J.: An Object-Oriented Approach to Automate Web Applications Development. In: Bauknecht, K., Madria, S.K., Pernul, G. (eds.) EC-Web 2001. LNCS, vol. 2115, pp. 16–28. Springer, Heidelberg (2001)

21. Robles Luna, E., Grigera, J., Rossi, G.: Bridging Test and Model-Driven Approaches in Web Engineering. In: Gaedke, M., Grissnikalus, M., Diaz, O. (eds.) ICWE 2009. LNCS, vol. 5648, pp. 136–150. Springer, Heidelberg (2009)

22. Rossi, G., Schwabe, D.: Modeling and Implementing Web Applications using OOHDM. In: Web Engineering, Modelling and Implementing Web Applications, pp. 109–155. Springer, Heidelberg (2008)

23. Seaside, http://www.seaside.st/
24. Selenium web application testing system, http://seleniumhq.org/
25. The WebRatio Tool Suite, http://www.webratio.com
26. Uden, L., Valderas, P., Pastor, O.: An Activity-theory-based to analyse Web applications requirements. Information Research 13(2) (June 2008)
27. Winckler, M., Vanderdonct, J.: Towards a User-Centered Design of Web Applications based on a Task Model. In: Proceedings of IWWOST 2005, Porto, Portugal, June 12-13 (2005)
28. Zheng, J.: In regression testing selection when source code is not available. In: Proceedings of the 20th IEEE/ACM international Conference on Automated Software Engineering, ASE '05, Long Beach, CA, USA, November 07-11, pp. 752–755. ACM, New York (2005), doi:http://doi.acm.org/10.1145/1101908.1101997

Re-engineering Legacy Web Applications into Rich Internet Applications*

Roberto Rodríguez-Echeverría, José María Conejero, Marino Linaje,
Juan Carlos Preciado, and Fernando Sánchez-Figueroa

Quercus Software Engineering Group
Universidad de Extremadura, 10003, Cáceres, Spain
{rre,chemacm,mlinaje,jcpreciado,fernando}@unex.es

Abstract. There is a current trend in the industry to migrate its traditional Web applications to Rich Internet Applications (RIAs). To face this migration, traditional Web methodologies are being extended with new RIA modeling primitives. However, this re-engineering process is being figured out in an ad-hoc manner by introducing directly these new features in the models, crosscutting the old functionality and compromising the readability, reusability and maintainability of the whole system. With the aim of performing this re-engineering process more systematic and less error prone we propose in this paper an approach based on separation of concerns applied to the specific case of WebML.

Keywords: Web Models Transformations, Patterns, Rich Internet Applications.

1 Introduction

More and more, traditional Web applications are being migrated to RIAs and, consequently, more and more, traditional Web methodologies are incorporating RIA modeling features [1][10][13][18]. Among these proposals, WebML deserves our attention due it being one of the most promising approaches because of its significant extensions to accomplish RIA features at different levels: client/side processing and storing [1], event handling [17] or presentation [11].

Despite these efforts, the real fact is that the industry is performing this migration in and ad-hoc manner, leading to two different problems: on the one hand, the original model becomes tangled with concerns for different purposes such as distribution or persistence which compromises the readability, understandability and, consequently, maintainability of the system. On the other hand, the adaptations performed related to RIA features are not reusable since they have to be applied again from the scratch in any new migration process. This is a significant drawback since there are quite a few adaptations that recurrently appear in many RIAs, e.g., synchronization patterns needed to work in a disconnected mode.

Precisely, this paper presents a proposal for the systematic re-engineering of Web applications modeled with WebML into RIAs following an Aspect-Oriented approach [4]

---

* This work has been supported by MEC under contract: TIN2008-02985.

B. Benatallah et al. (Eds.): ICWE 2010, LNCS 6189, pp. 189–203, 2010.

in the sense that the RIA features are modeled separately. The main contribution of the proposal is twofold: on the one hand, extending the WebML metamodel to include RIA features and, on the other hand, defining a new systematic model driven re-engineering process, based on model compositions driven by weaving models and pattern instantiations, to perform the weaving between the legacy model (WebML metamodel compliant) and the model describing the RIA related features (extended WebML metamodel compliant). As an additional contribution, several RIA synchronization patterns have been also identified and modeled separately, applying one of them to a running example. The RIA features contemplated are those related with data and business logic distribution and their associated issues, e.g. synchronization, event notifications, communications, etc. The separation approach followed is symmetric in the sense that it takes advantage of the entities and units already defined in WebML. A main objective of the proposal is to define a reusable framework automatically applicable to any re-engineering process within the domain.

The rest of the paper is as follows. Section 2 briefly introduces WebML extensions for RIA. In section 3 we present a motivating example to highlight the problems we want to solve with the proposed approach that is presented in Section 4. Section 5 applies the proposal to the motivating example, while section 6 identifies related works and presents the main conclusions.

2   WebML for Rich Internet Applications Capabilities in Brief

WebML is a language for the high-level description of a Web system consisting of data model, hypertext model, presentation model and personalization model. The application data are modeled using Entity-Relationship (E-R) or UML class diagrams. On top of the data model, WebML allows specifying the *business logic* and the content/containers composition by means of the hypertext model, whose key ingredients include *siteviews, areas, pages, content units, operation units*, and *links*. New entities have been included to treat new challenges like the XML_{OUT} and XML_{IN} ones, which have been used in [12] to marshall and unmarshall data, respectively.

Recently, WebML has been extended to cover RIA features in an approach called WebML for RIA [1]. Extensions proposed for the data model are characterized by two different dimensions: 1) the architectural tier, where client entities and relationships are marked with a "C" label while server ones are marked with a "S"; and 2) the persistence of the data, which is also indicated.

Extensions proposed for the hypertext model are those ones affecting the structural composition of RIAs typing WebML pages into server pages (marked with a "S") and client pages (marked with a "C"). Links are special cases since they could relate entities of client and server pages. In that sense, a link relating two client units is considered as client tier, whereas a link relating a client and a server unit is considered as inter-layer operation chains. An event model to support pulling/pushing RIA capabilities for WebML is also defined in [17]. This event model extends the original WebML data model and the set of units available in the hypertext model (adding the *sent event* and *receive event* ones). The full set of constrains to be accomplished by the model in order to be computable to generate the code are also specified in [1]. Both WebML extensions (i.e., [17] and [1]) are used in our work to specify RIA capabilities plus [12] where XML_{IN} and XML_{OUT} units are introduced to cope with marshalling and unmarshalling data into/from a XML file.

3 Motivating Example

Let us consider a simple e-shop for selling tickets for concerts where a pre-booking mechanism is mandatory in order to avoid situations such as two different users booking the same seat for the same show at the same time.

First, the example is solved using WebML. Then, some RIA features are included and modeled with WebML for RIA. The aim is just illustrating how the resulting model becomes tangled, leading to the two problems identified in section 1.

3.1 The Traditional Ticket e-Shop in WebML

Conceptual Data Model. (Fig. 1): *PreBookings* is specified to avoid booking conflicts (where the *endTime* attribute represents the deadline of a pre-booking). *CartItem* entity stores the different seats selected by a user in a volatile way (note the italic style in the name). Unlike *CartItem*, *PreBooking* is not volatile to ensure that the seats are not booked more than once in the database. The *Order* and *OrderItems* entities manage the information required to make persistent an order.

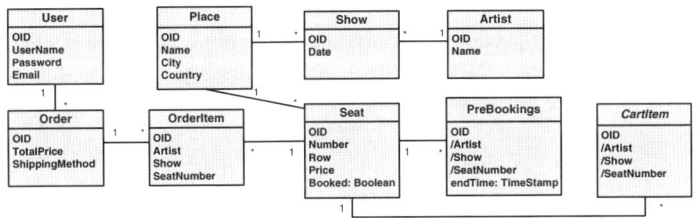

Fig. 1. Data model for the tickets reservation

Conceptual Hypertext Model. (Fig. 2)[1]: When a user accesses to the details of a show from the list of shows (*Events for artist* page), the pre-bookings and availability of seats for this show are displayed (*Booking* page) and they can be added to the cart. By using traditional Web techniques, the free seats are displayed when the user accesses the *booking* page, so this information is not later updated even when other users are performing bookings in that moment.

The operation chains depicted in the excerpt of the hypertext model are responsible for two different actions: i) the addition of a seat to the cart, including the management of the pre-booking (marked *A* with a dashed rectangle) and ii) the generation of an order from the user's cart and the removal of the cart items and the pre-bookings of that user (marked *B*).

---

[1] In the example some units have been omitted for simplicity and space reasons.

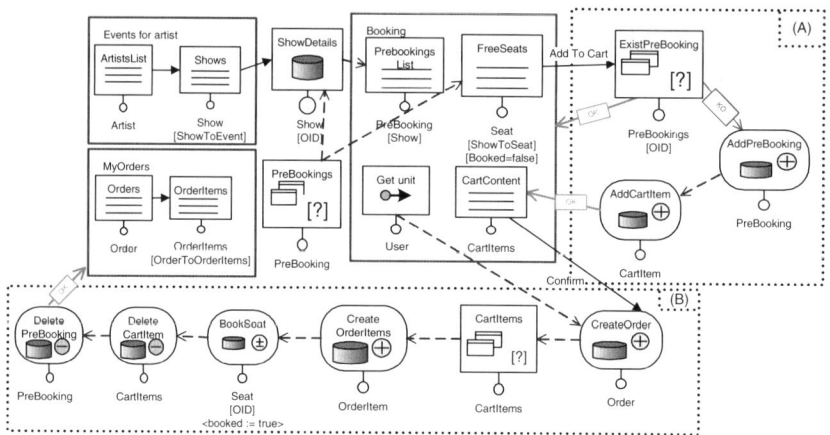

Fig. 2. Hypertext model for the tickets reservation

This solution is good enough to solve our initial problem. However, note some issues arising in this solution: on one hand, the whole process is performed at server side while part of the business logic could be at client-side reducing workloads in the server. On the other hand, the problem of a double pre-booking of a seat could be avoided if the data of the booking page is updated when a different user pre-books a seat. This is only possible with additional synchronous roundtrips in traditional web applications. This synchronous communication imposes processing, rendering and refreshing the whole page (even those parts that have not changed, e.g. cart items).

To solve these issues, RIA features are necessary. Processing and storing capabilities at client side and asynchronous communications come to the scene.

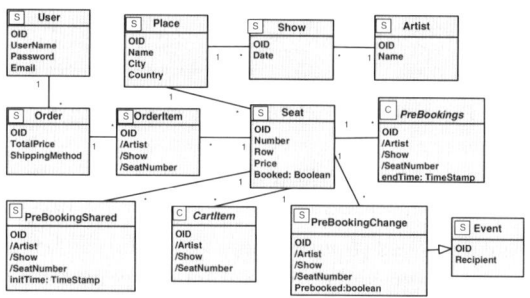

Fig. 3. Data model for the RIA version of the tickets system

3.2   The Ticket e-Shop Revisited Using RIA Capabilities

Conceptual Data Model. (Fig. 3): This data model adds the persistence level and location of the entities. *CartItem* volatile entity is marked as client because each client has a cart and its content is managed at client side. *PreBookings* entity is duplicated following [1] (*PreBooking* volatile at client side and *PreBookingShared* persistent at server side) and two new entities are defined to support event notification according to [17]. For the latter, *Event* stores the information related to each event that is related (through a generalization) with *PreBookingChange* that stores changes in the *PreBooking* or *PreBookingShared* entities.

Conceptual Hypertext Model. (Fig. 4 and Fig. 5): In Fig. 4 the new RIA operations chain (*A* from Fig. 2) is depicted. In this model, the pre-bookings are permanently updated since event is sent whenever a user makes a pre-booking. This event is received by the rest of clients to update their *PreBooking* entity (Fig. 4 right). Using this approach, users may work with their local pre-bookings (e.g., sorting by expiration time) avoiding continuous invocations to the server.

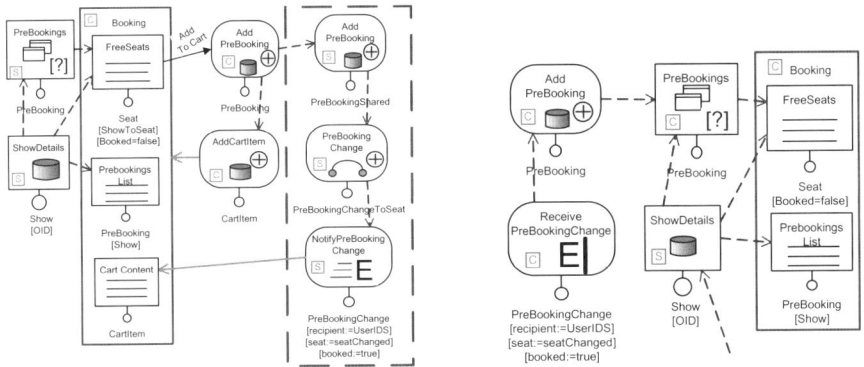

Fig. 4. Adding items to the cart and notifying the rest of users (left-side) and Reception of the PreBookingChange event by clients (right-side)

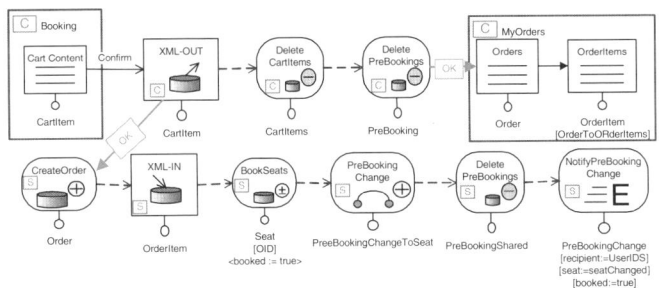

Fig. 5. Client confirms the buying and the order is processed

Regarding to the operation chain *B* (from Fig. 2), now part of the pages (Figure 5) are at client side (i.e., *Booking* and *MyOrders* pages) and also part of the operations related with these content management (i.e., *Delete CartItems* and *Delete PreBookings*). XML_{OUT} is used to collect the items in the cart (pre-booked by the user) at client side and to send this data to the server creating the corresponding order. Server side part of the operation chain also removes the seats from the *PreBookingShared* entity since these seats become booked. This action implies a new event raised by the server to notify the rest of the clients that these seats have been definitively booked. At the client side, the event is received through the *ReceivePreBookingChange* event unit (not shown due to space limitations)

The process used for migrating the system may introduce potential problems that could compromise its applicability:

- Firstly, observe that the original model has been **tangled** with concerns related to distribution, synchronization, event notification or persistence. This situation is harmful for **readability** and **understandability** complicating, thus, the maintainability of the system.
- Secondly, the adaptations performed are **not reusable** since they should be applied again in any new migration process, and these adaptations will be very similar independently of the application context.

4   The Approach

Following the Aspect-Oriented Modeling (AOM) principles [4] three different meta-models are defined: CMM, RMM and WMM. CMM is the meta-model of the model being migrated (core functionality in AOM terminology so it is called Core metamodel). RMM is an extended CMM with RIA features. In our context these RIA features can be considered as aspects in AOM terminology. Finally, WMM is the weaving meta-model. Fig. 6 shows the basic architecture of the approach. The legacy Web application is represented as the model M that conforms to CMM. RIA features are collected in the model M' that conforms to RMM. The composition (weaving) of both models, using the weaving model WM, produces the model M'' that conforms to RMM. M'' represents the resulting model of the migration process from the legacy Web application to the final RIA. In other words, M'' would be a similar solution to that provided in section 3.2, but here obtained following a systematic approach that avoid the two problems identified in previous section.

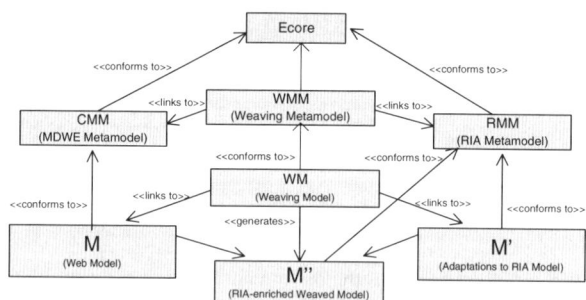

Fig. 6. Weaving of models to obtain the migrated application

4.1   RMM: WebML Metamodel with RIA Features

Among the different WebML metamodel definitions, the work in [16] has been selected due to its completeness and suitability for our approach. This metamodel is enriched with the RIA features introduced in Section 2 as follows (see **Fig. 7**):

- A new attribute *tier* is defined as a member of the metaclass *WebML::ModelElement*. This attribute may hold two different values: *client* and *server*. It indicates the architectural tier of existence for a data entity, or the tier of processing for a hypertext unit.
- New types of units are defined for marshalling data by means of the XML_{IN} and XML_{OUT} metaclasses. They are introduced in the package *ContentManagement* inheriting from *EntityManagementUnit* metaclass.
- And, finally, the units for modeling distributed events, *Send Event* and *Receive Event*, are defined as new kinds of operation units: *Notification Management Unit*.

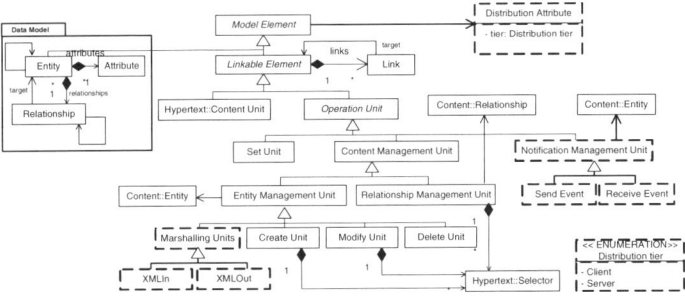

Fig. 7. WebML Extended Metamodel to cope with RIA concepts

4.2 WMM: The Weaving Metamodel

The WMM (Fig. 8) is defined as an extension of the generic model weaver AMW metamodel [5] [6] for the Model Driven Web Engineering [14] (WebML) domain. The composition process consists on the definition of weaving models and their processing by means of generic transformation rules defined in ATL [9]. For the sake of simplicity, we only consider here the extensions needed to process WebML data entities and operation units.

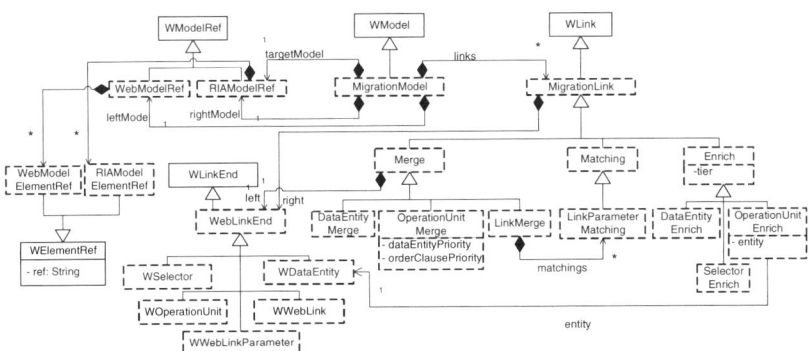

Fig. 8. WMM excerpt (extension of the AMW core metamodel)

As shown in Fig. 8, basically, WMM extends AMW metamodel as follows:

- A new type of weaving model, *MigrationModel*, is defined. It contains references to the woven models: the legacy Web Model (*leftModel*), the RIA model (*rightModel*) and the final model (*targetModel*). It is also composed by a collection (*links*) of *MigrationLink*.
- The metaclass *WLink* has been extended by the definition of three types of migration links: *Merge*, *Enrich* and *Match*. They define how the elements from the source models are woven into the target model.
- The *Merge* metaclass defines a merging connection between a model element from the Web Model and a model element from the RIA model of the same type. Concretely, it may define a link between two data entities, two operation units or two Web links. In case of merging operation units, additional information may be specified to indicate the selection priority of their different components. This information is maintained by the attributes *dataEntityPriority* and *orderClausePriority*. These attributes may hold the values *left* and *right*, indicating which end is selected to compose the target unit. Thus, the *Merge* instances are used to generate a target model element with the selected information of the two source elements.
- The *Enrich* metaclass allows specifying the tier (client or server) of the target model element. In its basic form, this link has only one end that refers to a model element from any of the source models. This metaclass has been extended with three new types, indicating the type of model elements that can be enriched, i.e. data entities, operation units and selectors. As a special case, we highlight the *OperationUnitEnrich* which additionally allows the instantiation of its components, e.g. its related data entity according to the *entity* reference. The Enrich instances of a WM work as annotations of Web elements useful for their transformation to RIA elements.
- The *Matching* metaclass defines a mapping between a model element from the Web Model and a model element from the RIA model. In concrete, the *LinkParameterMatching* defines the parameter matching within a merged link, i.e. it specifies the mapping between the source parameters from the left link and the target parameters from the right link.
- The metaclass *WebLinkEnd* establishes the joinpoint model of the source models. In other words, it defines the set of elements that may be linked by the weaving model (selectors, operation units, data entities, links and link parameters).

4.3 The Composition Process

The current composition process is performed by means of the ATL model transformation language. We have defined a set of ATL rules that generate the final RIA model taking as input the weaving model that links the legacy Web model and the RIA model. These ATL rules are only dependant on the WMM and the CMM. So, once defined for a concrete tuple (wmm, cmm), they can be automatically applied to any migration process conformed to that metamodel tuple.

In order to keep simple the definition and maintainability of this set of ATL rules, we specify the weaving process focusing on the first-class entities of WebML, i.e., entities, units, pages and areas, which drives the composition. In this sense, following the layer decomposition of WebML, we have grouped the ATL rules in two different subsets: one for data entities, and another one for hypertext elements. Here we explain only the rules defined to merge data entities and operation units.

For WebML **data model composition**, two basic rules are defined:

Rule 1. Migrating not merging data entities. They appear in the weaving model as *DataEntityEnrich* instances. They are the data entities from any of the source models that remain barely unchanged on the target model. The application of this rule only produces two kinds of modifications over a data entity: (1) updating the value of its tier attribute; and (2) resolving the target entities of its relationships pointing to a merging data entity (dangling relationships).

Rule 2. Migrating merging data entities. They appear in WM as *DataEntityMerge* instances. The application of this rule produces a new data entity as the result of merging the linked entities of the source models. This new entity takes its name from the source entity of the legacy Web model. It contains the union of the attribute sets contained by the merging entities. It is also composed by all the relationships whose origin was one of the source entities. In this case, it is also necessary to solve the dangling relationships generated, referencing the merged unit in the target model. Finally, the redundancy is removed from the collection of relationships.

For WebML **hypertext model composition** (only operation units), three basic rules are defined:

Rule 1. Migrating not merging operation units. They appear in the weaving model as *OperationUnitEnrich* instances. They are the operation units from any of the source models that remain barely unchanged on the target model. The application of this rule only produces three kinds of modifications over a operation unit: (1) updating the value of its tier attribute; (2) resolving the related data entity according its *entity* reference; and (3) removing its merging outgoing links (processed later by the rule 3).

Rule 2. Migrating merging operation units. They appear in the weaving model as *OperationUnitMerge* instances. In this case, the application of this rule produces a new operation unit as the result of merging the linked units of the source models. This new unit takes its name from the source unit of the legacy Web model. Next, the description of how the different components of a unit are merged is shown:

1. If the units are related to a data entity, the weaving model specifies which one will be used in the final model by means of the *dataEntityPriority* attribute.
2. If they both present selectors, all their conditions are concatenated in the final selector clause, indicating the tier in which they will be processed. In the weaving model, *SelectorEnrich* instances specify the processing tier of each selector.
3. If they both present order clauses, the sorting attributes will be concatenated in the final order clause, according to the sequence specified by the *orderClausePriority* attribute
4. Regarding the link composition, in this case, we only focus on the outgoing links because they are defined as components of the merging units. In this sense, considering the operation unit merging, we only selected the set of not merging outgoing links because their target unit and parameter matching are not modified.

Rule 3. Migrating merging links. They appear in the weaving model as *LinkMerge* instances. In this case, the target unit and the parameter matching of the target link must be solved. The target unit selected is always the one pointed from the right link. And the parameter matching is established according to its collection of *LinkParameterMatch* instances (matching).

Fig. 9 shows a simplified version of one of the ATL rules implemented, in particular, the rule to enrich operation units (rule 1 of hypertext model composition).

```
helper def: entity2rule: Map(CMM!Entity, WMM!MigrationLink) =
  CMM!MigrationLink.allInstances()->iterate(mlk;
    res: Map(CMM!Entity, WMM!MigrationLink) – Map{}|
    res->including(mlk.left,mlk));

rule OperationUnitEnriching {
 from
   oue : WMM!OperationUnitEnrich
 to
   ou : RMM!OperationUnit (
     name <- oue.left.name,
     entity <- thisModule.entity2rule(oue.entity),
     tier <- oue.tier,
     links <- oue.left.links->union(oue.right.links).asSet()
   )
}

rule LinkMigration {
 from
   lnk : CMM!Link
 to
   mlnk: RMM!Link (
     type <- lnk.type,
     target <- thisModule.target2rule(lnk.target)
     )
}
```

Fig. 9. ATL rule to deal with OperationUnitEnrich entities

Although we have only presented adding and merging operations, our approach supports also deletion of elements from the legacy Web application. The simplest form of specifying the deletion of a concrete element is not referencing it from the weaving model.

5 The Motivating Example Revisited

This section illustrates how the approach presented in Section 4 may be applied in a real application (our ticket e-shop example). In particular, this section describes how to model separately a concrete crosscutting concern: synchronization. The behavior of this concern has been defined by means of synchronization patterns so that they are encapsulated into separated models as a highly reusable RIA concern. Those patterns may be stored in model repositories to facilitate their localization and instantiation in future migration processes, reducing costs substantially. Finally, the section shows a particular weaving model (based on the meta-model shown in Fig. 8) used to generate the final RIA system, composing, thus, the original web model with the synchronization patterns.

5.1 Aspect Identification: Synchronization Scenarios

One of the main characteristics of RIA applications is the distribution of data and business between server and client minimizing the communications between them. However, it usually involves synchronization mechanisms to ensure the data consistency at both sides of communication. This implies that many actions related to the synchronization are scattered throughout the system and tangled [20] with the core functionality and business logic of the system (as mentioned in Section 3.3).

In this setting, based on the analysis of many RIA applications (and, in particular, our running example), we have classified the synchronization behavior involved in any RIA application in terms of two different dimensions: source of synchronization (client or server) and matter of synchronization (operation or data). This classification is presented in Fig. 10.

Observe that several kinds of synchronization scenarios may be classified into the same category. As an example, when a client is initialized or reconnected (when offline mode is available), the server must send the data required to run the application to the client. As a different example, when some data have changed at the server tier, the client copy of these data must be synchronized according to these changes. This synchronization is performed in RIA by event notification (e.g., observe the *NotifyPreBookingChange* unit used in the cart confirmation process in our running example, Fig. 5).

The event notification may be also used to synchronize an operation carried out at server tier. In this case, the event stores the data regarding to the operation produced (see also the *NotifyPreBookingChange* unit in Fig. 4 or the task assignation to a user event presented in [17]).

Synchronizations from client to server are also typical in RIA environments (especially in collaborative environments). For instance, when a client reconnects (after working in offline mode) the client must send to the server the data collected in order to be synchronized (e.g., see the cart synchronization using the *XMLOut* unit in Fig. 5). Also in collaborative applications, the replication at server tier of an action performed at client side is very common. This action allows the server (and the rest of clients connected) to keep synchronized (e.g., observe the *addPrebooking* replication when a client adds a seat to the cart in Fig. 4).

5.2 Models for Synchronization Patterns According to RMM

According to the kinds of synchronization identified in previous section, different patterns may be defined to model the behavior of the synchronization concern. In particular, a synchronization pattern has been defined for each type of synchronization. This section presents just one of these patterns: a pattern to model the synchronization carried out when the client must send to the server a set of data (*User Triggered Bulk Data Replication* of Fig. 10)

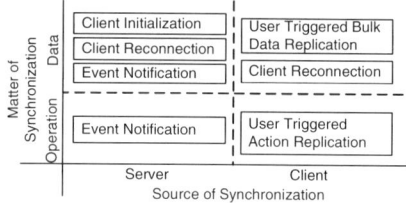

Fig. 10. Scenarios of synchronization based on source and matter dimensions

User Triggered Bulk Data Replication. This pattern happens when a client requires sending a set of data to the server so that these data must be synchronized at client and server tiers. The pattern is defined in terms of the XML_{OUT} and XML_{IN} entities to marshall and unmarshall a set of data (see Fig. 11). As it is suggested by the guidelines defined in [1], this synchronization usually requires the duplication of the data entity which stores the data to be synchronized at client and server tiers (sometimes with different names, e.g. *Order* and *CartItem* entities in our running example).

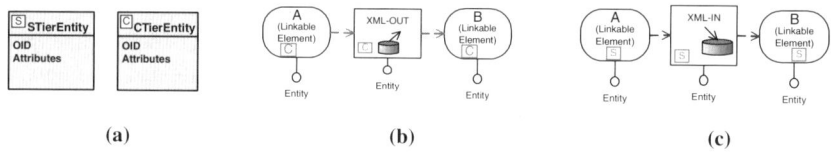

Fig. 11. User Triggered Bulk Data Replication pattern: (a) data entity duplication; (b) marshalling data, (c) unmarshalling data

5.3 Weaving Model

A weaving model, conformed to WMM, is defined to indicate the merging elements that constitute the join points linking the different models. Moreover, this weaving model contains annotations to incorporate additional information for the composition process, such as the distribution tier of different elements of the legacy model. Fig. 12. shows an excerpt of the weaving model for our tickets e-shop.

The weaving model presents the necessary elements to merge the second part (Fig. 11c) of the *User Triggered Bulk Data Replication* pattern with the operation chain *B* from our motivating example. The elements related to the rest of the pattern (Fig. 11a and Fig. 11b) have been omitted due to space limitations. The *OUMerge* instance *V* links the operation unit *Create Order* from the legacy Web model and the linkable element *A* previous to the XML_{IN} unit. In this case, the target unit will be a create unit related to the *Order* data entity (from the left model), as specified by the priority attributes of this instance, and the outgoing link to the XML_{IN} (from the right model), as indicated by a specific *LinkMerge* instance (not shown in the figure). The *OUEnrich* instance *X* indicates that the unit XML_{IN} has to be added to the target model only modifying the data entity related (*entity* reference). The *OUMerge* instance *Z* links the operation unit *Book Seat* and the linkable element *B*. So the target unit will be a modify unit related to the *Seat* data entity and the outgoing link to the *PreBooking Change* unit (both from the left model). Finally, the *LinkMerge* units *W* and *Y* relate the links that must be merged due to the merging of two operation units.

Moreover, two operation units of the chain *B* from the legacy Web Model are deleted, i.e. they do not appear in the final model. These units (selection unit *CartItems* and creation unit *Create OrderItems*) are no longer necessary for the RIA chain *B* redefinition. In this sense, they are not referenced from the weaving model. Observe that the output of the weaving performed (using the ATL rules explained in Section 4.3) is the model M''. The lower part of Fig. 12. shows an excerpt of the application resulting of the composition process, in particular the part obtained by the composition explained by this figure. Note that the models resulting of the whole composition process have been previously presented in the figures of Section 3.2. Thus, the lower part of Fig. 12. shows a snippet of Fig. 5.

The weaving model is the only part that depends on the concrete application. The rules and the ATL transformations generated would be the same for any WebML application. Therefore, the migration of different applications may be performed just defining the corresponding pattern instantiations and weaving models.

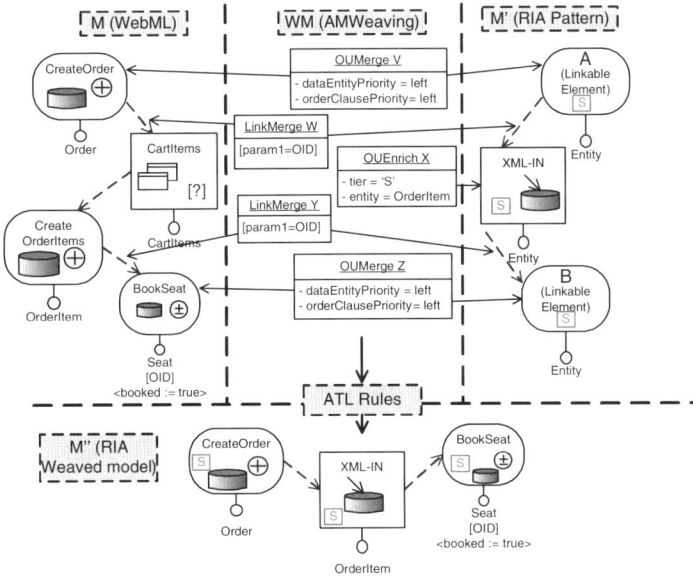

Fig. 12. Snippet of the weaving model

6   Conclusions and Related Works

This paper has presented a systematic model driven re-engineering process, based on a weaving metamodel and composition rules, to perform the migration of traditional Web applications into RIAs. With the proposal here presented this re-engineering process can be barely reduced to the definition of weaving models driving the model composition. Meanwhile the weaving models must be defined specifically for every migration case; the transformation rules are generic and can be applied automatically to any re-engineering process, improving highly the time and effort to carry out such migrations. As an ongoing work, we are studying the application of automatic model matching techniques in order to alleviate the effort of defining the weaving model.

The approach has been presented with the help of a motivating example where synchronization patterns appear as a crosscutting concern that has been separately defined. The patterns ensure the reusability of the migration process to different systems just by defining new weaving models. The applicability of the patterns is not limited to adding RIA features to legacy systems. They may be also applied to new developments where RIA capabilities are added from the beginning. Indeed, similar patterns may be also identified to add different concerns to the system (e.g., to add volatile concerns [7] or context aware information [15].

Crosscutting concerns in Web models have been previously treated in the literature. The approach in [19] allows to clearly decoupling requirements that belong to different concerns. In [15] aspects are introduced, using an asymmetric aspect separation approach which increases the complexity of the obtained models and the learning process. In both approaches RIA issues are not considered.

In [21] a RIA metamodel for OOWS is presented to deal with the new technological challenges arisen with Web 2.0. In [18] an aspect-oriented solution to include RIA issues in OOHDM is presented. However, both works only deal with presentation issues, not dealing with the RIA features here treated.

Although many works have previously introduced the concept of model weaving in WebML, none of them deals with RIA capabilities. In [3] KM3 and Abstract State Machines are introduced defining a Composition Weaving Model to check the consistency among the models of the WebML stack. In [8] a proposal to mix domain specific modeling languages (i.e., WebML and XACML for access control) using a general purpose modeling language (i.e., AMW) is introduced. For the weaving approaches listed above as well as for our work, the WebML metamodel plays a pivotal role. While there is not an official WebML metamodel, many researchers have covered this issue. For example, in [2] a WebML metamodel is presented to transform WebML models to MDA.

Although here applied to WebML, the proposal is generic enough to be applied to other Web modeling languages. The scope of our project is broader than here shown and includes not only other Web modeling languages but also the migration of legacy applications that have not been implemented using a modeling notation. In this case, an important process of reverse engineering and business process reengineering is needed.

References

1. Bozzon, A., Comai, S., Fraternali, P., Carughi, G.T.: Conceptual modeling and code generation for rich internet applications. In: 6th international Conference on Web Engineering ICWE '06. LNCS, vol. 263, Springer, Heidelberg (2006)
2. Brambilla, M., Fraternali, M., Tisi, M.: A metamodel transformation framework for the migration of WebML models to MDA. In: MDWE, CEUR Workshop Proceedings, vol. 389, pp. 91–105. CEUR-WS.org (2008)
3. Cicchetti, A., Di Ruscio, D.: Decoupling web application concerns through weaving operations. Sci. Comput. Program. 70(1), 62–86 (2008)
4. Clarke, S., Baniassad, E.: Aspect-Oriented Analysis and Design: The Theme Approach. Addison-Wesley Professional, Reading (2005)
5. Del Fabro, M., Bézivin, J., Jouault, F., Breton, E., Gueltas, G.: AMW: A generic model weaver. In: Procs. of IDM '05, pp. 105–114 (2005)

6. Del Fabro, M., Valduriez, P.: Towards the efficient development of model transformations using model weaving and matching transformations. Software and Systems Modeling 8(3), 305–324 (2009)
7. Ginzburg, J., Distante, D., Rossi, G., Urbieta, M.: Oblivious Integration of Volatile Functionality in Web Application Interfaces. Journal of Web Engineering 8(1), 25–47 (2009)
8. Hovsepyan, A., Van Baelen, S., Berbers, Y., Joosen, W.: Specifying and Composing Concerns Expressed in Domain-Specific Modeling Languages. In: 47th International Conference, TOOLS EUROPE '09, pp. 116–135 (2009)
9. Jouault, F., Kurtev, I.: Transforming models with ATL. In: Bruel, J.-M. (ed.) MoDELS 2005. LNCS, vol. 3844, pp. 128–138. Springer, Heidelberg (2006)
10. Koch, N., Pigerl, M., Zhang, G., Morozova, T.: Patterns for the Model-Based Development of RIAs. In: Gaedke, M., Grossnilkalus, M., Diaz, O. (eds.) ICWE 2009. 9th international Conference on Web Engineering 2009. LNCS, vol. 5648, pp. 283–291. Springer, Heidelberg (2009)
11. Linaje, M., Preciado, J.C., Sanchez-Figueroa, F.: Engineering Rich Internet Application User Interfaces over Legacy Web Models. IEEE Internet Computing 11(6), 53–59 (2008)
12. Manolescu, I., Brambilla, M., Ceri, S., Comai, S., Fraternali, P.: Model-driven design and deployment of service-enabled web applications. ACM Trans. Internet Technol. 5(3), 439–479 (2005)
13. Meliá, S., Gómez, J., Pérez, S., Díaz, O.: A Model-Driven Development for GWT-Based Rich Internet Applications with OOH4RIA. In: Eighth international Conference on Web Engineering ICWE '08, pp. 13–23. IEEE Computer Society, Los Alamitos (2008)
14. Moreno, N., Romero, J.R., Vallecillo, A.: An Overview of Model-Driven Web Engineering and the MDA. In: Web Engineering: Modeling and Implementing Web Applications, pp. 353–382. Springer, Heidelberg (2007)
15. Schauerhuber, A., Wimmer, M., Schwinger, W., Kapsammer, E., Retschitzegger, W.: Aspect-Oriented Modeling of Ubiquitous Web Applications: The aspectWebML Approach. In: 14th Annual IEEE international Conference and Workshops on the Engineering of Computer-Based Systems, pp. 569–576. IEEE Computer Society, Los Alamitos (2007)
16. Schauerhuber, A., Wimmer, M., Kapsammer, E.: Bridging existing Web modeling languages to model-driven engineering: a metamodel for WebML. In: 2nd international workshop on model driven Web engineering MDWE '06 (2006)
17. Toffetti-Carughi, G., Comai, S., Bozzon, A., Fraternali, P.: Modeling distributed events in data-intensive Rich Internet Applications. In: Benatallah, B., Casati, F., Georgakopoulos, D., Bartolini, C., Sadiq, W., Godart, C. (eds.) WISE 2007. LNCS, vol. 4831, pp. 593–602. Springer, Heidelberg (2007)
18. Urbieta, M., Rossi, G., Ginzburg, J., Schwabe, D.: Designing the Interface of Rich Internet Applications. In: LA-WEB '07. Latin American Web Congress, pp. 144–153 (2007)
19. Valderas, P., Pelechano, V., Rossi, G., Gordillo, S.: From crosscutting concerns to web systems models. In: Benatallah, B., Casati, F., Georgakopoulos, D., Bartolini, C., Sadiq, W., Godart, C. (eds.) WISE 2007. LNCS, vol. 4831, pp. 573–582. Springer, Heidelberg (2007)
20. Van den Berg, K., Conejero, J., Hernandez, J.: Analysis of Crosscutting in Early Software Development Phases based on Traceability. In: Rashid, A., Aksit, M. (eds.) Transactions on AOSD III. LNCS, vol. 4620, pp. 73–104. Springer, Heidelberg (2007)
21. Valverde, F., Pastor, O.: Facing the Technological Challenges of Web 2.0 - a RIAModel-Driven Engineering Approach. In: Vossen, G., Long, D.D.E., Yu, J.X. (eds.) WISE 2009. LNCS, vol. 5802, pp. 131–144. Springer, Heidelberg (2009)

Deriving Vocal Interfaces from Logical Descriptions in Multi-device Authoring Environments

Fabio Paternò and Christian Sisti

CNR-ISTI, HIIS Laboratory, Via Moruzzi 1, 56124 Pisa, Italy
{Fabio.Paterno,Christian.Sisti}@isti.cnr.it

Abstract. Model-based approaches for interactive Web applications have neglected vocal interaction. However, ubiquitous multi-device environments call for better support for such modality. In this paper we present a language for logical descriptions of vocal interfaces along with a transformation for deriving corresponding implementations and show an example application. Such results have been integrated into a multi-device authoring environment.

Keywords: Model-based user interface design, Vocal interfaces, XML-based user interface languages.

1 Introduction

The convergence of telecommunications and the Web is now bringing the benefits of Web technology to the telephone, enabling Web developers to create applications that can be accessed via any telephone, and allowing people to interact with these applications via speech and keypads [16].

In this context the W3C has developed a suite called Speech Interface Framework in which one of the main contribution is VoiceXML 2.0, a language designed for developing vocal interfaces with support for audio dialogues, vocal and DTMF recognition, recording and telephony feature. For this reason we are witnessing the spread of Voice Browsers, which offer Web Based services from any phone.

In a number of contexts of use vocal interaction is important: when the visual channel is busy (e.g. while car driving), for disabled people (e.g. the vision-impaired), or while users are moving. Some possible scenarios are in accessing business (e.g. booking services, airline information, etc.), public (e.g. weather information, news) or personal information (e.g. appointment calendar, telephone list). The vocal features make it suitable to support quick access to information and to interact in a way more similar to that used for communication among humans. Thus, in modern technological ubiquitous settings, the need for supporting vocal interaction is acquiring increasing importance, and the associated technology has considerably improved in terms of efficiency and accuracy, even if some limitations still apply (e.g. performance in noisy environments).

Model-based approaches for interactive systems are characterized by the use of logical languages, which identify and classify user interface elements and ways to compose them according to their effects. They have stimulated interest in particular

B. Benatallah et al. (Eds.): ICWE 2010, LNCS 6189, pp. 204–217, 2010.

with the advent of multi-device user interfaces [7] because they allow designers to better manage the complexity of user interfaces that have to adapt to varying interaction resources. They allow designers to concentrate on logical decisions without having to deal with a plethora of low-level implementation details. However, most model-based approaches have mainly addressed issues related to desktop/mobile adaptation and have neglected other modalities such as vocal interaction. Thus, there is a need for solutions able to facilitate the development of multiple versions of an interactive application, including the vocal one. For this purpose, the ideal solution is to identify a core set of interaction concepts independent of the modality and then refine it for each possible modality in order to account for its specific aspects. For example, the vocal modality in general is linear, not persistent, quicker and more natural in some operations. This implies the need for continuous feedback, and the rendering of short prompts or option lists in order to limit memory efforts.

Stanciulescu [17] indicated some criticisms of the model-based approach in user interface design. Our framework has been developed taking into account such comments and aiming to minimize the negative aspects. One of the highlighted points is the *high threshold*, the designer needs to learn a new language before starting new interface development. To this end, we have created a graphical-based tool to assist designers in developing multiple versions depending on the target platform, which share a common set of abstract concepts, thus reducing the learning effort when they start a new version.

Another problem in model-based approaches may be the *low ceiling*, the user interfaces that can be generated have various limitations due to the excessively abstract design. We address this criticism by proposing a model-based solution using two levels of abstraction, in which the lower level (that refines the higher one) permits good control over the resulting interface without having to know the details of the implementation language. The *unpredictability* of some final results is avoided by furnishing complete documentation of the transformation rules.

On the other hand, model-based approaches offer some advantages in terms of *methodology* (user -centred approach), *reusability* (the various concrete languages are refinements of the same abstract language) and *consistency* (between early design phase and final result).

In the paper after discussing some related work we introduce the proposed approach and present the logical language for vocal interaction, then we describe the transformation from the logical language to VoiceXML, and we show an example application in the museum domain. Lastly, some conclusions are drawn along with indications for future work.

2 Related Work

The problem of designing multi-device interfaces, including vocal ones, has been addresses in some previous work but still needs more general and better engineered solutions. Damask [6] includes the concept of layers to support the development of cross-device (desktop, smartphone, voice) user interfaces. Thus, the designers can specify user interface elements that should belong to all the user interface versions and elements that should be used only with one device type. However, this approach can be useful in developing desktop and mobile versions but does not provide particularly useful support when considering the vocal version, which requires user interface structures profoundly different from the graphical versions. XFormsMM [4] is an attempt to extend XForms in order to derive both graphical and vocal interfaces. In this case the basic idea is to specify the abstract controls with XForms elements and then use aural and visual CSS for vocal and graphical rendering, respectively. The problem in this case is that aural CSS have limited possibilities in terms of vocal interaction and the solution proposed requires a specific ad hoc environment in order to work. For this purpose we propose a more general solution able to derive implementations in the W3C standard Voice XML. Obrenovic et al. [9] have investigated the use of conceptual models expressed in UML in order to then derive graphical, form-based interfaces for desktop or mobile devices or vocal ones. UML is a software engineering standard mainly developed for designing the internal software of application functionalities. Thus, it seems unsuitable to capture the specific characteristics of user interfaces and their software. In [11] there is a proposal to derive multimodal user interfaces using attribute graph grammars, which have a well-defined semantics but limitations in terms of performance. The possibility of deriving vocal interfaces was addressed in [1] but using hardcoded solutions for the transformation and logical descriptions that were unable to describe typical Web2.0 interactions and access to Web services.

A different approach to multimodal user interface development has been proposed in [5], which aims to provide a workbench for prototyping them using off-the-shelf heterogeneous components. In that approach, model-based descriptions are not used and it is necessary to have an available set of previously defined components able to communicate through low-level interfaces, thus making it possible for a graphical editor to easily compose them.

To summarise, we can say that the few research proposals that have also considered vocal interaction have not been able to obtain a general solution in terms of logical descriptions and provide limited support in terms of generation of the corresponding user interface implementations. For example, in [1] the transformations were hard-coded in the Java implementation, while in [11] the transformations were specified using attributed graph grammars, whose semantics is formally defined but have considerable performance limitations.

3   The Proposed Approach to Vocal Interaction

In this paper we present a general logical language for vocal interaction, which is included in an overall environment able to support development of multi-device user interfaces. The associated authoring environment includes a transformation tool able to derive VoiceXML implementations from the logical specifications and satisfies the requirements for multimodal interface generation discussed in previous work [8], such as modality independence, support for specifying hierarchical grouping, etc.

One of our goals is to propose a framework that allows designers to generate vocal interfaces that take into account the challenges and principles for conversational interfaces design identified in [15]. One of these challenges is to *make the interaction feel conversational*. Our approach, in this sense, allows designers to set a number of synthesizing properties (e.g. prosody, volume, tone, speed, etc). Moreover, we support the barge-in technique that allows users to interrupt the speech synthesizer by using their voice. Another point is the problem of recognizing the start/stop of subdialogues corresponding to the grouping. We propose different solutions to communicate this information to the user, such as inserting simple delimiting sounds or prompting short meaningful phrases (e.g. "Main menu").

Another challenge is error recognition: "*One can never be completely sure that the recognizer has understood completely*" [15]. The errors are divided into three categories: rejection errors, occur when the recognizer does not match the user input with any expected utterance; substitution errors, when the platform erroneously recognizes a wrong but legal input; insertion error, if the recognizer accepts a noise as a valid input. Our framework allows different mechanisms to avoid these errors. In the case of rejection errors, a simple solution is to answer with a "*not understand*" feedback. This approach could stimulate user frustration and so, as suggested in [15], we provide designers with the possibility of implementing the *tapered prompting* technique. In this way the error messages that the user receives become more explicit as the number of errors increase.

To reduce the occurrence of the substitution errors, we allow the designers to verify the utterance when necessary. Lastly, the problem of insertion errors is attenuated by including in the generated vocal interfaces two vocal commands to start and stop the dialogue with the platform. In this way, the user can temporarily interrupt the interaction and restart it at his convenience.

4   A Logical Language for Vocal Interaction

MARIA is a recent model-based language, which allows designers to specify abstract and concrete user interface languages according to the CAMELEON Reference framework [2] (Fig. 1 shows an instance of the framework). This language represents a step forward in this area because it provides abstractions also for describing modern Web 2.0 dynamic user interfaces and Web service accesses. In its first version it provides an abstract language independent of the interaction modalities and concrete languages for graphical desktop and mobile platforms [10]. In general, concrete languages are dependent on the typical interaction resources of the target platform but independent of the implementation languages. In this paper we present a concrete language for vocal interfaces, which has been designed within the MARIA framework.

In MARIA an abstract user interface is composed of one or multiple presentations, a data model, and a set of external functions. Each presentation contains: a number of user interface elements (*interactors*) and interactor compositions (indicating how to group or relate a set of interactors); a dialogue model, describing the dynamic behaviour of such elements and connections, indicating when a change of presentation should occur. The interactors are first classified in abstract terms: edit, selection, output and control. Each interactor can be associated with a number of event handlers, which can change properties of other interactors or activate external functions.

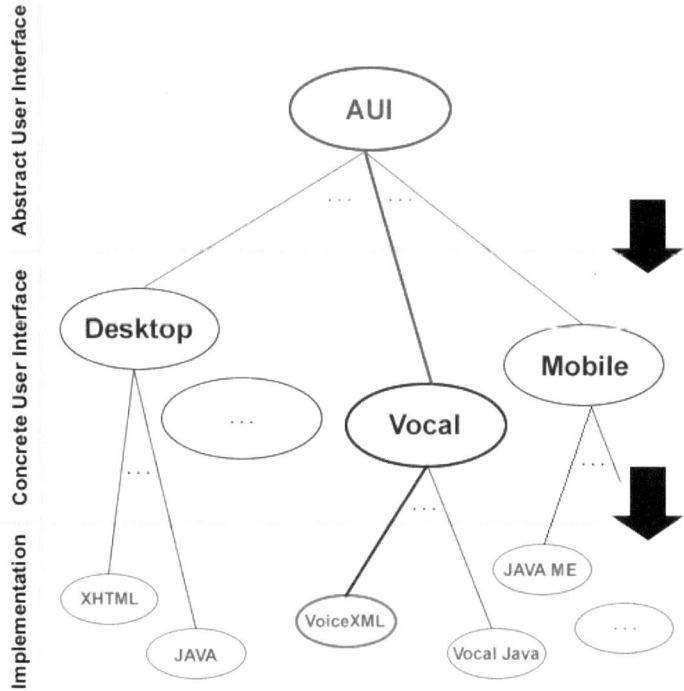

Fig. 1. Possible abstraction levels

While in graphical interfaces the concept of presentation can be easily defined as a set of user interface elements perceivable at a given time (e.g. a page in the Web context), in the case of vocal interfaces we consider a presentation as a set of communications between the vocal device and the user that can be considered as a logical unit, e.g. a dialogue supporting the collection of information regarding a user.

In defining the vocal language we have refined the abstract vocabulary for this platform. This mainly means that we have defined vocal refinements for the elements specified in the abstract language: interactors (user interface elements), the associated events and their compositions.

The refinement involves defining some elements that enable setting some presentation properties. In particular, we can define the default properties of the synthesized voice (e.g. volume, tone), the speech recognizer (e.g. sensitivity, accuracy level) and the DTMF (Dual-Tone Multi-Frequency) recognizer (e.g. terminating DTMF char).

Only-output interactors simply provide output to the user; the abstract interface classifies them into text, description, feedback and alarm output. Refinement of the text element is composed of two new elements: *speech* and *pre-recorded message*. Speech defines text that the vocal platform must synthesize or the path where the platform can find the text resources. It is furthermore possible to set a number of voice properties, such as emphasis, pitch, rate, and volume as well as age and gender of the synthesized voice. Moreover, we have introduced control of behaviour in the event of unexpected user input: by suitably setting the element named *barge in*, we

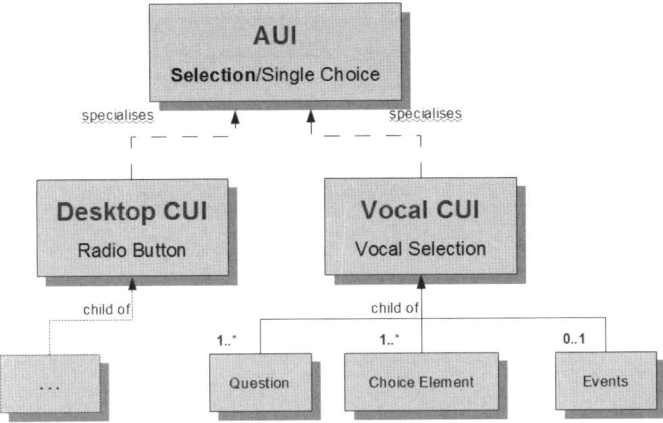

Fig. 2. Specialization example

can decide if the user can stop the synthesis or if the application should ignore the event and continue. As mentioned above, the other element that refines abstract text is pre-recorded message: it defines the path of pre-defined audio resources that must be played. We support the case of missing resources by defining an *alternative content* that can be synthesized when this case occurs. Besides speech and pre-recorded message, the other only-output element is *sound*. This element permits defining the path of a non-vocal audio resource that must be played by the platform. It is also possible to insert a textual description of the sound that could be used as additional information regarding the sound content.

Selection interactors permit performing a selection between a set of elements; the abstract interface distinguishes between *single* and *multiple choice*. In order to support such interactions in vocal context we have introduced the interactor *vocal selection* (see Fig.2). This element defines the question(s) to direct to the user and the set of possible user input that the platform can accept. In particular, it is possible to define textual input (word or sentences) or DTMF input. Depending on the type of selection one or multiple elements can be selected. Semantic differences between this two interactors are made at abstract user interfaces level: for the single choice it is possible to set only one selected element instead, in the case of the multiple choice, more than one user input can be set as selected elements.

The control interactor at the abstract level can be distinguished in activator, to activate a functionality, and navigator, to manage navigation between the presentations. The activator is refined in the vocal language into: *command*, in order to execute a script, *submits*, to send a set of data to a server, and *goto* to perform a call to a script that triggers an immediate redirection. While the navigator is refined into: *goto*, for automatic change of presentation; *link*, for user-triggered change of presentation, and *menu* for supporting the possibility of multiple target presentations.

Edit interactors gather more complex user input. We refined three abstract elements: text edit, numerical edit and object edit. Text edit, from the graphical web context point of view, can be regarded as an editable textual field. In the vocal context

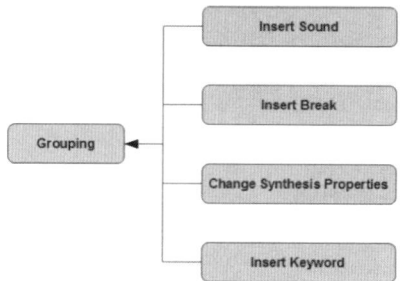

Fig. 3. Grouping refinement

we refined this concept with the *vocal textual input* element, which permits setting a vocal request and specifying the path of an external grammar for the platform recognition of the user input. Numerical edit is an interactor to collect numbers, which are refined into *numerical* input. As in the textual input it is possible to define a request and an external grammar. Moreover, it is also possible to set a predefined grammar, such as date, digits, phone or currency. Finally, we have refined the object edit interactor into a *record* element, which allows specifying a request and storing the user input as an audio resources. It is possible to define a number of attributes relative to the recording, such as *beep* to emit a sound just before recording, *maxtime* to set the maximum duration of the recording, and *finalsilence*, to set the interval of silence that indicates the end of vocal input. Record elements can be used for example when the user input cannot be recognised by a grammar (e.g. a sound).

In the logical language one of the elements that permit the composition of the interactors is *grouping*. From the visual point of view we could refine this concept, for example, into a table element. Group vocal interactor is more complex. We propose four solutions to permit the user to identify the beginning and the end of a grouping (see Fig. 3). Inserting a sound at the beginning and at the end for this purpose can be a good non-invasive solution. Another solution can be inserting a pause, which must be neither too short (useless) nor too long (slow system feedback). Moreover, it is possible to change the synthesis properties such as volume and voice gender. The last possibility is to insert keywords that explicitly define the start and the end of the grouping.

Another substantial difference of vocal interfaces is in the event model. While in the case of graphical interfaces the events are related mainly to mouse and keyboard activities, in vocal interfaces we have to consider different types of events: *noinput* (the user has to enter a vocal input but nothing is provided within a defined amount of time), *nomatch*, the input provided does not match any possible acceptable input, and *help*, when the user asks for support (in any platform specific way) in order to continue the session. All of them have two attributes: message, indicating what message should be rendered when the event occurs, and re-prompt, to indicate whether or not to synthesize the last communication again.

In order to facilitate authoring and editing of logical specifications a graphical editor has been designed and implemented (Fig. 4). It allows editing the various

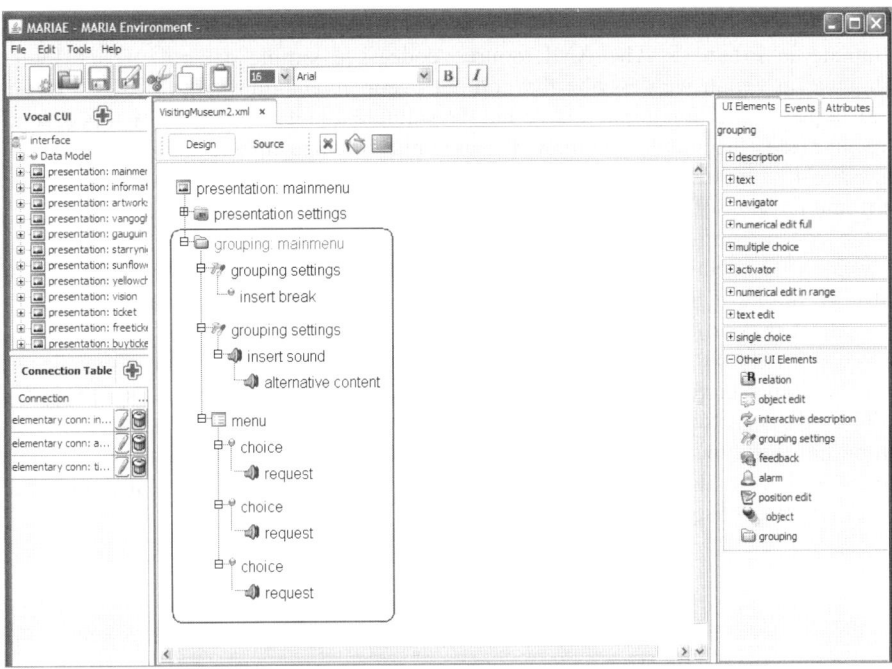

Fig. 4. The graphical editor for the vocal logical language

presentations in the central area. The user interface elements are created by drag-and-drop from the lists in the right frame, which show the elements that can be inserted according to the language specification. In the middle tab in the right frame it is also possible to specify the associated events and attributes. In the left side there is an interactive nested tree view of the presentations and the associated elements. The output of the tool is a logical description of the interface formalized in XML.

5   Transforming Logical Description into VoiceXML Implementations

The current standard for voice browsing implementation is VoiceXML [13]. Thus, this was the first target implementation language from the vocal logical description. Since VoiceXML is also an XML-based language, XSL Transformations (XSLT) [14] seemed the most appropriate technology for implementing such transformation. This language provides a number of constructs for creating mappings among elements of two XML-based languages. However, such mappings are not trivial to create because both languages have a structure that provides constraints about where to locate an element. For example, in VoiceXML a vocal output is implemented differently depending on whether it occurs in a form or in a menu. This has been solved using the "XSLT modes", an XSLT technique to identify which template to use in the transformation when this element occurs. More generally, this mechanism allows the transformation to change template to apply in the mapping depending on the current context.

For the sake of clarity we show a simple example of application of an XSLT template in Figure 5. The XML source code is a simple excerpt of our logical language in which we describe the noinput event. Every time that the *XSLT Engine* finds a match with a noinput source tag the suitable template is called. In this case the template adds the suitable VoiceXML <noinput> tag and then tests if it was set up a message to synthesize. If true a <prompt> VoiceXML tag is added. Note that the attribute *bargein* of the VoiceXML code is forced to be false to prevent interruption of the system communication by any user input. Each presentation is associated with a VoiceXML document. The presentation element has a list of possible default settings. Some of them can be defined once for the entire presentation (e.g. speech recognizer properties); in this case they are opportunely mapped into a <property> VoiceXML. In other cases the properties can be set up multiple times in the presentation (e.g. voice synthesis properties). The policy for defining such properties follows the structure of the specification (bottom-up): first it checks whether local properties have been defined, if not the properties of the surrounding grouping operator, if any, are applied, and lastly, if even these are missing, the properties general to the entire presentation are used.

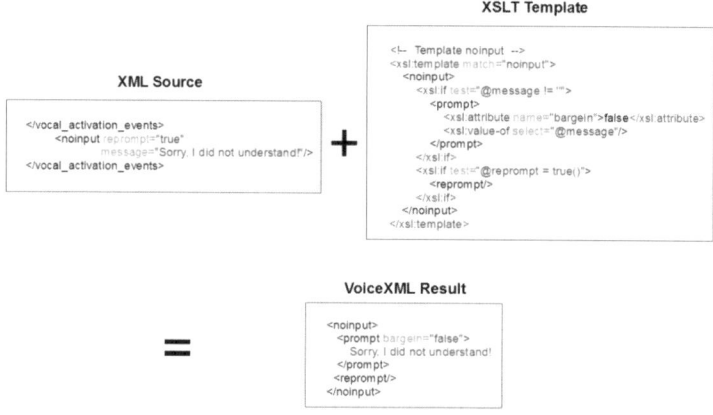

Fig. 5. A small example of XSLT Transformation

SGRS[1] grammars format has been used to specify possible inputs, since that it is what the VoiceXML specification supports. The grammars define the set of possible input that the vocal platform is able to recognize. In the generated implementations we use predefined VoiceXML grammars, when possible, such as date, number and currency in the *numerical edit* element mapping.

In some cases we generate *inline grammars*, which are grammars defined in the transformation process and directly inserted in the VoiceXML code. This is the case of the *link* element that can be activated both by vocal or DTMF command thanks to two expressly created grammars (using the parameters defined in the logical description).

---

[1] Speech Recognition Grammar Specification Version 1.0,
http://www.w3.org/TR/speech-grammar/

Another solution is used in the mapping of the *single vocal selection*: in this case, we know 'a priori' the list of possible inputs, and we can use one VoiceXML *<option>* element for each input. This is equivalent to using a grammar able to recognize the entire list of user input. Figure 5 shows an example of a logical description with a single choice and how its elements are mapped onto VoiceXML elements.

The case of the *multiple vocal selection,* instead, introduces a problem: we do not know how many choices (in the predefined list) the user will make. In our solution the platform asks all the choices one-by-one and the user must answer yes/no to accept/reject each one. This solution may seem verbose but in practise there may be two situations: few choices, in this case the verbosity is negligible; many choices, in this case the verbosity is an advantage because it will reduce the mnemonic effort of the user.

As mentioned in previous sections, *textual edit* is similar to the visual concept of editable text box. It is very hard (if not impossible) to have a grammar able to recognize every kind of user input. We prefer to leave it up to the application developer to implement an *external-grammar* (or to find a pre-built one) that satisfies the possible input (case by case). For example, suppose that the platform asks the name of the user, the developer should build a grammar containing a dictionary of names to provide for recognition.

In the visual context we have some mechanisms to force the input of numbers in a certain range (e.g. a spin box); from the vocal point of view we resolve this problem by carrying out a check, before accepting the user input, using the conditional VoiceXML tags. If the number specified by the user is out of range, we refuse the input and re-prompt the request.

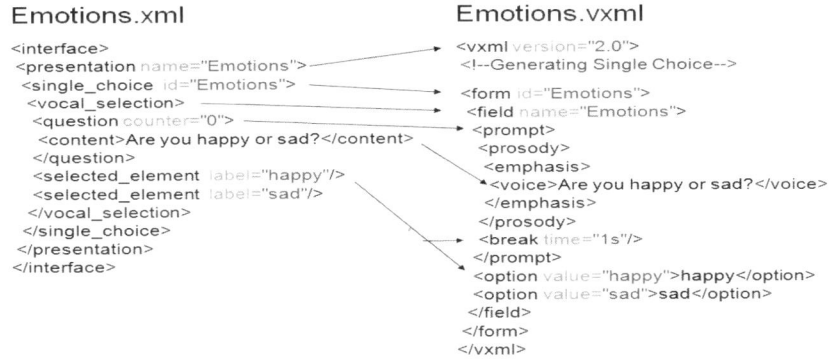

Fig. 6. Example of Logical-to-Implementation transformation

The control interactor command is mapped into a VoiceXML variable that contains the results of the execution of a script that must be defined at presentation level (which corresponds to a VoiceXML document in the implementation). The others control interactors: submit, goto, link, and menu are mapped into corresponding VoiceXML elements. VoiceXML links must be declared externally to the dialogue constructs and thus they are globally activable. Each possible menu choice is transformed into a <choice> VoiceXML element. In this way the grammar for recognizing the user input is automatically provided by the VoiceXML browser.

In the case of textual input the developer has the possibility to specify an appropriate grammar containing the rules for the acceptable inputs. Numerical input can be recognised by predefined grammars. It is also possible to generate in the code the check whether the input satisfies a given range.

6 Application

The VoiceXML code generated by the above transformation has been tested with Tellme.Studio Voice Browser [12] (suggested by W3C) and has passed the validation test integrated in it. The applications have then been used through VoIP access to the vocal server.

Figure 7 shows the structure of some parts of an example application created with our environment. It consists of a virtual museum application. The rectangles represent the presentations. Each of them is a dialogue or a set of dialogues that can be logically grouped together. In general, each presentation supports three basic possibilities: to be executed again, to go to the main menu, and to go to the previous presentation.

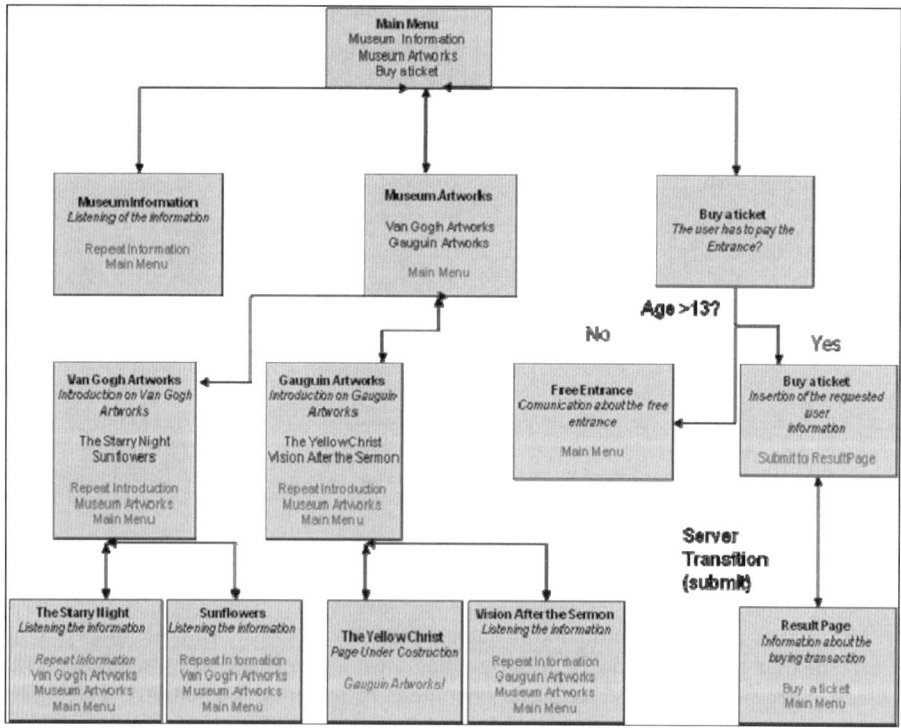

Fig. 7. An Example Application

The main menu allows the user to choose among three options: accessing more detailed information regarding the museum, performing a virtual visit, and buying a museum ticket. In the first case the user can just listen to the detailed information and then return to the main menu. In the second case the user can select an artist and then receive a related general description and then the possibility of accessing the associated artworks. When buying a ticket there are two possibilities depending on the age. In the case of under 13 the application activates a presentation communicating that the entrance is free otherwise the user has to provide personal information and perform the payment. More in details the system ask: *user name*, described logically with a text edit element with associated an external grammar that contain a dictionary of names; *user surname*, described in the same way of the previous one; *credit card type*, described by a single choice with three possible options (Visa, Mastercard, PostePay); *credit card number*, described by a numerical edit with associated a predefined grammar (number); *favoured museum's rooms*, described by a multiple choice with a number of possible options and *further notice*, described by a record element to permit at the user to leave a personal comment.

In the table below we show an example of resulting dialogue in order to better understand the interaction user-platform that we have obtained. We focus in this example only on the module related to the inserting of the user information.

	Input / Output	**Logical Element**
System:	To buy a ticket you have to tell me some information.	Only Output
System:	What is your name?	Text Edit
User:	. . . **(do not respond)**	
System:	Sorry, I did't hear you. What is your name?	No input event
User:	Henry **(with noise)**	
System:	Sorry, I do not understand	No match event
User:	Henry	
System:	What is your surname?	Text Edit
User:	Smith	
System:	Kind of credit card between: Visa, Mastercard and PostePay?	Single Choice
User:	PostePay	
System:	Credit card number?	Numerical Edit
User:	123456789#	
System:	We are interest to know which room you prefer to visit. Are you interested in Van Gogh's room?	Multiple Choice
User:	Yes	
System:	Are you interested also in Leonardo's room?	

User:	Yes	
System:	After the *beep* leave a comment about this services	Record
System:	Beep	
User:	. . . (some second silent length)	
System:	Thanks. Do you want confirm you reservation?	Only Output
User:	Yes	
System:		Submit
System:	Your request has been registered.	Only Output

7　Conclusions and Future Work

This work introduces a novel logical language for vocal interfaces and the associated environment, which allows designers to easily compose vocal interfaces and derive VoiceXML implementations. This has been integrated in an environment for multi-device interface design and development, thus facilitating the implementation of multiple versions adapted to the various target modalities because of the use of a common abstract vocabulary, which is then refined according to the target platforms. This avoids requiring developers to learn a plethora of details of the many possible implementation languages

This result has been validated through the development of a number of vocal applications (one of them is briefly described in the paper), which have been rendered through publicly available voice browsers.

Future work will be dedicated to empirical tests in order to better assess how the development process is facilitated with this approach, especially when multi-device interfaces should be developed (e.g. desktop, mobile and vocal versions of the same application). We also plan to develop an automatic system able to support graphical-to-vocal adaptation and a new logical language for multimodal interfaces able to exploit the language presented in this paper for the vocal part.

Acknowledgments

We gratefully acknowledge support from the EU ServFace Project (http://www.servface.eu).

References

1. Berti, S., Paternò, F.: Model-Based Design of Speech Interfaces. In: Jorge, J.A., Jardim Nunes, N., Falcão e Cunha, J. (eds.) DSV-IS 2003. LNCS, vol. 2844, pp. 231–244. Springer, Heidelberg (2003)
2. Calvary, G., Coutaz, J., Bouillon, L., Florins, M., Limbourg, O., Marucci, L., Paternò, F., Santoro, C., Souchon, N., Thevenin, D., Vanderdonckt, J.: The CAMELEON reference framework. CAMELEON project, Deliverable 1.1 (2002)

3. Edwards, A., Pitt, I.: Design of Speech-Based devices. Springer, Heidelberg (2007)
4. Honkala, M., Pohja, M.: Multimodal interaction with XForms. In: Proceedings ICWE, pp. 201–208 (2006)
5. Lawson, J., Al-Akkad, A., Vanderdonckt, J., Macq, B.: An open source workbench for prototyping multimodal interactions based on off-the-shelf heterogeneous components. In: Proceedings ACM EICS, pp. 245–254 (2009)
6. Lin, J., Landay, J.A.: Employing Patterns and Layers for Early-Stage Design and Prototyping of Cross-Device User Interfaces. In: Proc. CHI, pp. 1313–1322 (2008)
7. Myers, B.A., Hudson, S.E., Pausch, R.: Past, Present and Future of User Interface Software tools. ACM Trans. Comput. Hum. Interact. 7, 3–28 (2000)
8. Nichols, J., Myers, B.A., Higgins, M., Hughes, J., Harris, T.K., Rosenfeld, R., Pignol, M.: Generating remote control interfaces for complex appliances. In: Proceedings ACM UIST'02, pp. 161–170 (2002)
9. Obrenovic, Z., Starcevic, D., Selic, B.: A Model-Driven Approach to Content Repurposing. IEEE Mutimedia, 62–71 (Januray-March 2004)
10. Paternò, F., Santoro, C., Spano, L.D.: MARIA: A Universal Language for Service-Oriented Applications in Ubiquitous Environment. ACM Transactions on Computer-Human Interaction 16(4), 1–30 (2009)
11. Stanciulescu, A., Limbourg, Q., Vanderdonckt, J., Michotte, B., Montero, F.: A Transformational Approach for Multimodal Web User Interfaces based on UsiXML. In: Proc. ICMI, pp. 259–266 (2005)
12. Tellme.Studio Voice Browser, `https://studio.tellme.com/`
13. Voice extensible markup language (VoiceXML) version 2.0, `http://www.w3.org/TR/2009/REC-voicexml20-20090303/7`
14. XSL Transformations (XSLT) Version 2.0, `http://www.w3.org/TR/xslt20/`
15. Yankelovich, N., Levow, G., Marx, M.: Designing SpeechActs: Issues in Speech User Interfaces. In: CHI 1995, pp. 369–376 (1995)
16. Voice Browser Activity (W3C), `http://www.w3.org/Voice/`
17. Stanciulescu, A.: A Methodology for Developing Multimodal User Interfaces of Information Systems. Ph.D Thesis, University of Louvain (2008)

Quality, Quality in Use, Actual Usability and User Experience as Key Drivers for Web Application Evaluation

Philip Lew[1], Luis Olsina[2], and Li Zhang[1]

[1] School of Computer Science and Engineering, Beihang University, China
[2] GIDIS, Web Engineering School at Universidad Nacional de La Pampa, Argentina
philiplew@gmail.com, olsinal@ing.unlpam.edu.ar, lily@buaa.edu.cn

Abstract. Due to the increasing interest in Web quality, usability and user experience, quality models and frameworks have become a prominent research area as a first step in evaluating them. The ISO 25010/25012 standards were recently issued which specify and evaluate software and data quality requirements. In this work we propose extending the ISO 25010 standard to incorporate new characteristics and concepts into a flexible modeling framework. Particularly, we focus on including information quality, and learnability in use characteristics, and actual usability and user experience concepts into the modeling framework. The resulting models and framework contribute towards a flexible, integrated approach to evaluate Web applications. The operability and particularly the learnability of a real Web application are evaluated using the framework.

Keywords: Learnability, quality, usability in use, learnability in use, usability, information quality, user experience.

1 Introduction

Web applications (WebApps), a combination of information content, functionality and services are fast becoming the most predominant form of software implementation and delivery today. Due to these evolutions, *usability, information quality, quality in use* and, in the end, *actual user experience* have taken on increased importance. With the latest ISO 25010 quality standard [8], and other recent work by researchers in the field of *quality in use, usability* and *user experience* such as Bevan [2] and Hassenzahl [7], it is still not totally clear where characteristics such as *information quality*, *learnability in use*, *actual usability* and *user experience* fit in regarding quality modeling.

The recent ISO 25010 standard on quality models, updates and brings previous standards together while delineating three views of quality viz *internal*, *external quality* and *quality in use*. Some researchers including the ISO 9241-11 have suggested that *quality in use* is defined similarly to *usability*. Of particular interest in ISO 25010 is the standard's new breakdown of *quality in use* and *usability* which provides us insight into our framework for this paper. Bevan examined ISO 25010

B. Benatallah et al. (Eds.): ICWE 2010, LNCS 6189, pp. 218–232, 2010.

from the viewpoint of usability and user experience (UX) and drew some interesting relationships regarding usability as performance in use, and satisfaction as it relates to user experience. Hassenzahl´s work in classifying user experience in two categories, hedonic and pragmatic is also useful when examining usability from the *do*, pragmatic viewpoint and the *be*, or hedonic satisfaction viewpoint.

Meanwhile, ISO 25012 [9], a recent standard on data quality, considers data as an entity by itself, and as a separate model. However, extending the number and span of standards runs the risk that nobody will use them [19]. There is a need to integrate *information quality* as part of the overall quality of an application, particularly for WebApps, rather than as a separate entity. Olsina et al [14] made an initiative in this direction as discussed in Section 3.

Rather than rely on separate standards, we propose to augment the ISO 25010 standard to include *information quality* as a characteristic of internal/external quality because this is a critical characteristic of WebApps. Furthermore, the 25010 standard has categorized *learnability* as an internal/external quality subcharacteristic under the *operability* characteristic. We further propose to include *learnability in use* as a characteristic of *usability in use* to account for the learning process and the importance of context of use during learning. Lastly, we propose an integrated means to evaluate the characteristics of *usability in use* as they relate to *user experience*. All these issues will be combined into an integrated framework which embraces the possibility of instantiating different models. This framework and its models, i.e., *internal/external Quality*, *Quality in use*, *actual Usability* and *User experience* (2Q2U, for short) are as compliant as possible with the recent ISO standards, considering other well-known contributions as well.

Ultimately, the specific contributions of this research are: (a) An extension to ISO 25010 internal/external quality model to include *information quality* as a characteristic; (b) An extension to ISO 25010 *quality in use* model to include *learnability in use* as a subcharacteristic of *usability in use;* (c) An integrated and flexible framework for modeling quality requirements, and particularly *quality in use* to relate the concepts of *actual usability* and *user experience;* and (d) An instantiation, for illustration purposes, of the ISO 25010 *external quality* model with subcharacteristics and measurable attributes to evaluate *operability* for WebApps, which can influence *learnability in use*.

Following this introduction, Section 2 reviews recent related work and delineates opportunities for improvements. In Section 3, our proposal of extending the ISO 25010 standard in order to incorporate new characteristics and concepts into a flexible framework called 2Q2U is discussed. In Section 4, we discuss the usefulness of the proposed framework, in the context of evaluating operability and some of its subcharacteristics, e.g. learnability for a WebApp. Section 5 draws our main conclusions and outlines future work.

2 Related Work and Motivation

User experience, a relatively new term, combined with *usability*, *information quality*, and *quality in use*, have all recently come to the research forefront especially for WebApps due to the shift in emphasis to satisfying the end user. Yet based on the current standards and literature reviewed, it can be difficult to understand their relationships and we often observe a lack of consensus in meaning. This section examines the related work with an eye for improvement opportunities.

In the recent ISO 25010 standard, the concept of quality has been broadened from 6 characteristics (ISO 9126-1 [11]) to 8 as shown in Fig. 1.a.

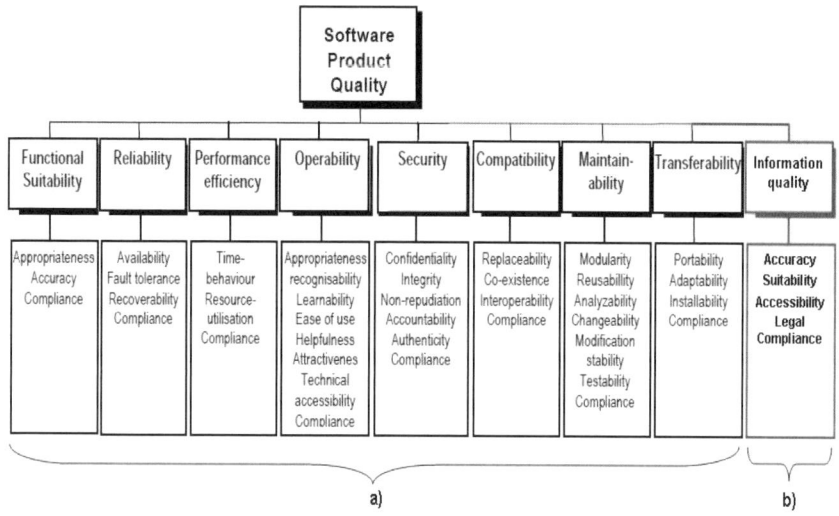

Fig. 1. a) ISO 25010 Internal/External Quality Model; b) Adding Information Quality

The standard is split into 2 quality models. The first is a software internal/external quality model. In comparison with ISO 9126-1, the previous characteristic of *usability* has been renamed as *operability* with broader meaning. For instance some sub-characteristics such as *learnability* among others have remained while new ones such as *technical accessibility compliance* and *helpfulness* were added. *Security* has been added as a separate characteristic, rather than as a subcharacteristic of *functionality* in the former, while other names have changed slightly to enhance descriptiveness. The second model in ISO 25010 depicted in Fig. 2.a refers to *quality in use* and includes previous ISO 9126-1 *quality in use* characteristics while adding others. Note that the *effectiveness* and *satisfaction* characteristics from ISO 9126-1 were imported into this newer standard as subcharacteristics of *usability in use*, while the *productivity* has been renamed as *efficiency in use*. In addition, *flexibility in use* has been added to accommodate different usage contexts including *accessibility in use*. It is worth mentioning that the suffix 'in use' was added to two characteristics and many subcharacteristics.

As can be seen from these figures, *learnability*, an internal/external sub-characteristic of *usability/operability* in ISO 9126-1/25010 has not been moved to *usability in use*. In addition, ISO 25010 does not include information or content quality as a characteristic of either model. Recent ISO's intentions are for data to be addressed by a complementary standard, namely ISO 25012.

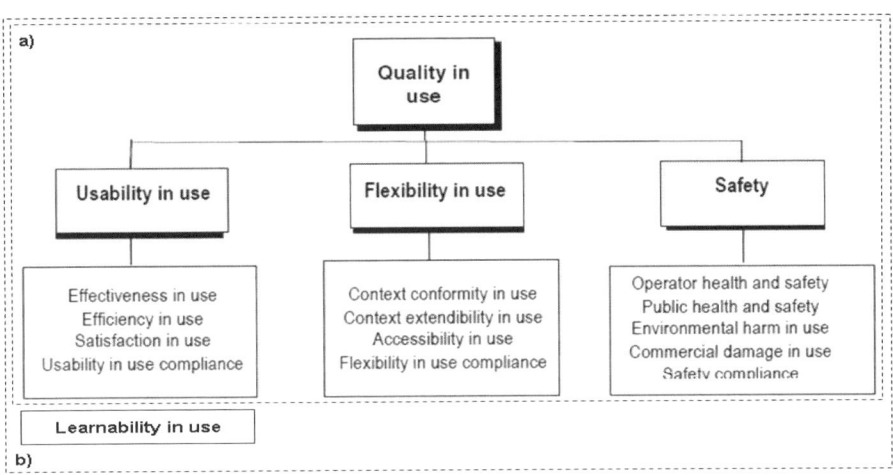

Fig. 2. a) ISO 25010 Quality in Use Model; b) Adding Learnability in Use into Usability in Use

ISO 25012 is a general data quality model intended to be used to establish data quality requirements, and plan and perform data quality evaluations. This standard is intended to be used in conjunction with ISO 25010, but by going to such length to define quality of data, it loses emphasis in using data as information and as a component of a WebApp rather than just data as an entity itself. Information quality as a characteristic was researched by [14] whereby they proposed extending ISO 9126-1 with content quality containing four sub-characteristics including *content accuracy*, *content suitability*, *content accessibility*, and *legal compliance*. We adapt their contribution to the ISO 25010 internal/external quality model as discussed in Section 3.

Learnability has been removed from the usability characteristic of ISO 9126-1 and moved to the *operability* characteristic of ISO 25010 where it is defined as "degree to which the software product enables users to learn its application" borrowing the dialogue principles from ISO 9241-110 [10] regarding suitability for learnability.

For WebApps, users are expected to learn intuitively with no user manual. However, *learnability* in ISO 25010 is solely a product quality, which does not incorporate evaluation of the process of learning and does not model *learnability* in different real contexts of use such as the domain of the system, and its target users. Research made by [17] exemplifies the need to examine learning from different user viewpoints, as learning observed for new users is not necessarily related to continued learning. Furthermore many researchers have determined *learnability* to be linked directly with *usability* as summarized by Abran [1]. Bevan [3] also noted *learnability in use* as part of *usability in use*.

On the other hand, the term UX is becoming more important as evidenced by the definitions by various researchers. To understand the term requires breaking down the word 'user experience' and examining first what experience means.

Experience is a general concept which refers both immediately-perceived events and the wisdom gained in interpretation of events. In the context of UX, it is a sequence of events over time for a user's interaction with the software product. [7]

notes that the time dimension could be either momentary or accumulated and changing over time. The 'moment' perspective does not exclude the accumulation or summary perspective, rather, it adds to it like a continuum.

Examining the 'user' part of the *user experience* concept, [7] characterizes a user's goals into pragmatic, do goals and hedonic, be goals and assumes the interactive product quality is perceived in two dimensions, pragmatic and hedonic. Pragmatic quality refers to the product's perceived ability to support the achievement of tasks such as paying a bill and focuses on the product's utility and usability in completing tasks that are the 'do-goals' of the user. Hedonic quality refers to the product's perceived ability to support the user's achievement of 'be-goals', such as being happy, or satisfied with a focus on self. He also argues that the fulfillment of be-goals is the driver of experience and that lack of usability or inability to complete do-goals may prevent achieving be-goals, but do-goals are not the end goal of the user. Rather the real goal of the user is "to fulfill be-goals such as being autonomous, competent, related to others, stimulated, and popular through technology use." He also states that pragmatic quality enables achieving hedonic quality be-goals and has no value by itself, but only through enabling accomplishment of be-goals. In summary, user experience comes from fulfilling be-goals in the time dimension, at the moment, and in summation over time.

Given that, it's easy to see why UX has become such a buzz word regarding WebApps. Websites and the interactive game industry have combined to change our expectations of software/WebApps. We not only expect them to work and help us get our task done, but also expect them to be pleasant to use and provide satisfaction. Yet, a common standard definition for user experience is still not available. According to Stewart [18], the next revision of ISO 9241-210 defines UX as 'all aspects of the user's experience when interacting with the product, service, environment or facility' and that "it is a consequence of the presentation, functionality, system performance, interactive behavior, and assistive capabilities of the interactive system. It includes all aspects of usability and desirability of a product, system or service from the user's perspective". Stewart states that *usability* depicts a narrower concept than *user experience* and simply focuses on systems being easy to use. In his words, "Easy to use is not enough" as exemplified through the IPod's market dominance through more than just ease of use.

Satisfaction in use, as noted by Bevan, correlates to Hassenzahl's hedonic goals whereas *usability in use* and its do-goals are related to a user's pragmatic goals. In summary, *usability in use*, *satisfaction in use*, and *user experience*, need clearer relationships in order to model and evaluate them. After a reviewing the related work, we found possible opportunities for improvement, as we discuss in the next Section.

3   2Q2U Models and Framework: Proposal and Discussion

The aim of the proposed models and framework is twofold: first, adding characteristics to extend the ISO 25010 standard and; second, add two new concepts, *actual usability* and *actual UX*, to which characteristics and subcharacteristics can be related and new models created in a flexible way. Regarding the extension of the ISO 25010 standard, as shown in Figures 1.b and 2.b, we added the following:

- *Information quality,* defined as the degree to which the software/WebApp provides accurate, suitable, accessible and legally compliant information. This new characteristic becomes part of the internal/external quality model.
- *Learnability in use* defined as the degree to which specified users can learn efficiently and effectively while achieving specified goals in a specified context of use. This new subcharacteristic becomes part of the *usability in use* characteristic in the *quality in use* model.

The ISO 9126-1 standard stated "evaluating product quality in practice requires characteristics beyond the set at hand". Hence, the revised ISO 25010 increased the total number of characteristics from 6 to 8. As software applications continue to change, we suggest including the above mentioned characteristics accordingly.

Second, regarding the addition of two new concepts, *actual usability* and *UX*, we hereby provide the following definitions:

- *Actual Usability*: the degree to which specified users can achieve specified goals with effectiveness in use, efficiency in use, learnability in use, and accessibility in use in a specified context of use. Note: *Actual usability* is measured and evaluated in a real operational environment where real users perform actual specified tasks.
- *Actual User Experience*: the degree to which specified users can achieve actual usability, safety, and satisfaction in use in a specified context of use. Note: *Actual UX* is evaluated not only by measures and indicators of user performance –as in *actual usability*-, but also by means of satisfaction instruments.

Subsections 3.1 through 3.3 further develop the aspects above and then relate them together in our flexible framework in section 3.4, which models characteristics of usability in use to bring together the concepts of *usability* and *user experience*.

3.1   Adding Information Quality

In our proposal to extend ISO 25010, we assume that the software quality models, definitions, and concepts in the standard were intended for application to software products as a whole and therefore are also applicable to WebApps, a type of software application. Like any software product, building WebApps involves different development stages, from inception and development to operation and maintenance. Thus we should be able to use the same ISO *internal* and *external quality* and *quality in use* models for WebApps with the same eight prescribed quality characteristics (and their subcharacteristics) for internal and external quality requirements, and the three characteristics for *quality in use* requirements, but some other considerations might be taken into account.

As highlighted elsewhere [5] the very nature of WebApps is a mixture of contents and functions. Therefore we argue that the eight internal/external quality characteristics (see Figure 1.a) are not well suited, nor were they intended to specify requirements for *information quality*.

We intentionally use the term 'information' to differentiate from 'data'. Data comes from attribute measurements, facts, formula calculations, etc. and are often organized and represented in databases. On the other hand, information is the meaningful interpretation of data for a given purpose and context. Given that a WebApp is very often content oriented and intended mainly to deliver information, the central issue is the ability to specify the *information quality* for WebApps from the internal and external quality viewpoints. This viewpoint is also supported by ISO 9241-110 which relates characteristics of presented information to its dialogue principles. Therefore, we propose augmenting the internal/external quality model with the *information quality* characteristic, with *content accuracy, content suitability, content accessibility,* and *content legal compliance* as subcharacteristics, shown in Fig. 1b. Definition for each characteristic is done by Olsina *et al* in [14].

It may be argued that *information quality* should be added as a *quality in use* characteristic. However, it can be evaluated as an internal/external quality characteristic by measuring its attributes. In addition, when designing tasks for *quality in use,* for example for evaluating *efficiency* and *effectiveness in use,* content and functions are embedded in the task design itself rather than as attributes of the software/WebApp. *Satisfaction in use* questionnaires can also address *information quality* with specific questions related to its sub-characteristics [4].

3.2   Adding Learnability in Use

As mentioned earlier, *learnability* solely as a product quality does not incorporate evaluation of the learning process or different usage contexts. More specifically, we examine learning context to strengthen our reasoning to include *learnability in use* as a *usability in use* subcharacteristic, from the user group type and time viewpoints.

Regarding the former, the learning objectives and therefore behavior of different user groups have an impact on the learning process as novice users behave differently than expert users [17]. Ease of learning depends on the user group type and the task being attempted. As an extreme, a quality requirement characteristic to minimize the necessary learning time, or to make the learning time equal to zero depends entirely on who the user is, and what tasks they are trying to do. As another example of how user group types behave differently based on their background and task at hand, requirements for users who are trying to re-learn a task can be difficult to model as a product characteristic. Grossman *et al* [6] also noted several other user group types including: i) Level of experience with computers; ii) Level of experience with interface; iii) Level of related domain knowledge; iv) Experience with similar software. Therefore, the dimension of user group types and its influence in *learnability* is of paramount importance.

Regarding the time aspect, learning from different user viewpoints such as initial learning and continued learning as researched by [17] are not necessarily correlated. So, measuring the learning of users must be done in the time dimension as the time delta between initial learning and continued learning has an influence on the *learnability* of the software in a real context. Many *learnability* measures focus on initial *learnability*. As Nielsen states: "One simply picks some users who have not used the system before and measures the time it takes them to reach a specified level of proficiency in using it" [13]. However, continued *learnability* requires assessing performance over time using a constant user group.

Some may argue that *effectiveness in use* and *efficiency in use* either combined or solely, can constitute *learnability in use*. However, software that is easy to learn is not always efficient to use and vice versa. A WebApp´s design evaluated highly as part of the *learnability* external quality characteristic, may lead to less efficient procedures.

Learnability therefore depends on the domain of the software, its target users and tasks at hand. Hence it cannot solely be determined by inspection, as an internal or external quality characteristic. Bevan also related *learnability in use* as a sub-characteristic of *usability in use*, as discussed in Section 2. Given the aforementioned reasons, and recalling our definition as "the degree to which specified users can learn efficiently and effectively while achieving specified goals in a specified context of use", we argue that *learnability in use* should be added to the *usability in use* characteristic as a subcharacteristic. Moreover, if we agree on including *learnability in use* into the model, we can combine, for instance, the following measurable attributes [12]:

- *Assisted learning time:* Time for user from a specific user group type to learn to complete a specific task in a specified time, plus the amount of instructional guidance time if needed.
- *Relative user learning efficiency:* The efficiency of an ordinary user as compared to an expert measuring the time required to reach a predefined threshold ratio.

Other attributes such as *task time deviation, error rate over time, learning efficiency* and so forth can be supplemented and used in conjunction with external attribute learnability measurements in order to draw relationships between *quality in use* and *learnability*.

3.3 Modeling Actual Usability and User Experience

As mentioned in Section 2, these two concepts are mainly derivatives through the [2, 7] works. In Bevan´s work relating and explaining factors contributing to system *usability* and *UX* he defines 4 characteristics of *usability in use*: i) *Effectiveness and productivity in use* ii) *Learnability in use* iii) *Accessibility in use* and iv) *Safety in use*. He also matches *usability* to performance in use measures equivalent to those characteristics related to the pragmatic 'to-do' goals of the end user.

Measures of *UX* are noted by Bevan as being composed of *satisfaction in use* as equivalent to achieving pragmatic and hedonic goals, with its subcharacteristics as specified by ISO 25010 including *pleasure, likability, comfort,* and *trust.* Hassenzahl further elaborates on hedonic goals as: "fulfilling the human needs for autonomy, competency, stimulation (self-oriented), relatedness, and popularity (others-oriented) through interacting with the product or service". He further states that pragmatic quality facilitates satisfaction of be-goals. That is, be-goals are not dependent on, but facilitated by do-goals; i.e. a user could be satisfied even if do-goals are not satisfied. For example, if a user cannot buy a product online efficiently (slowly with mistakes), but ends up buying what they like, then they may achieve their be-goals and be very satisfied.

Thus, achieving *UX* is influenced through satisfaction of both *usability in use* (pragmatic goals) and *satisfaction in use* (hedonic goals). Ultimately, the *actual usability* and *UX* definitions given in the introduction of Section 3 are based on this rationale. The above concepts and relationships are shown in Figure 3 and further explained in the following section in the context of our proposed framework.

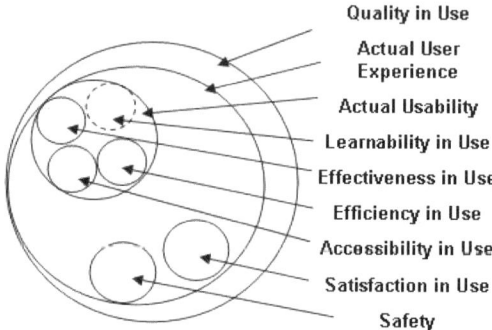

Fig. 3. Relationships of Quality in Use, Actual Usability, and Actual User Experience

3.4 Specifying and Using the Proposed Framework

Now, given the aforementioned definitions and relationships, our 2Q2U framework for modeling nonfunctional requirements for *internal/external Quality*, *Quality in use*, *actual Usability* and *User experience* can be specified and generalized to flexibly meet the evaluator needs to represent these calculable concepts. Figure 4 shows a basic Venn diagram to represent this viewpoint.

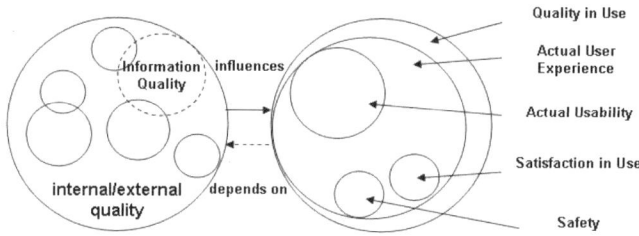

Fig. 4. Proposed 2Q2U modeling framework

The particular 2Q2U models can in turn be instantiated relying on the INCAMI (*Information Need, Concept model, Attribute, Metric and Indicator*) nonfunctional requirement specification component. As detailed in by Olsina et al [16], the *InformationNeed* allows evaluators to establish the evaluation *purpose*, user *viewpoint*, *Entity*, *focus*, and *CalculableConcept* such as external quality, actual usability, user experience, among others. As such, our proposed framework can be used flexibly to generate *ConceptModels* by choosing *CalculableConcepts* (characteristics in the ISO terminology) depending on the specific *InformationNeed*. We can then combine *Attributes* to characteristics or subcharacteristics to fully instantiate the selected model, resulting in a requirements tree, for further measurement and evaluation purposes.

2Q2U is in line with the intention of the ISO 25010 standard where tailoring the model is encouraged for relative importance of characteristics depending on the situation at hand: *"It is not practically possible to measure all internal and external*

.... Similarly it is not usually practical to measure quality in use for all possible user-task scenarios. The relative importance of quality characteristics will depend on the product and application domain. So the model should be tailored before use..."

Regarding the above, we just examine –for space reasons- the actual UX model (subset of quality in use) including our proposed learnability in use subcharacteristic in the light of our *actual usability* concept. To this aim, we modeled four characteristics, namely, *efficiency in use*, *effectiveness in use*, *learnability in use* and *accessibility in use* as shown in Fig. 5.

As per our discussion that *actual usability* is related to satisfying the do goals of the end user while completing real specified tasks, we include *satisfaction in use* as part of *actual UX* rather than in *actual usability*. Fig. 5 depicts our model composition as part of the framework with *actual usability* and *actual user experience* as defined and discussed above.

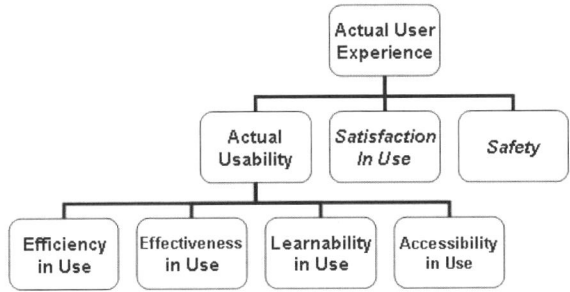

Fig. 5. Model composition representing Actual Usability and Actual User Experience

Note from Fig. 5 that *actual usability* is not a prerequisite for *actual UX*, but rather has influence as one out of three characteristics. Both concepts involve the temporal component, not just UX [7] but also *actual usability*. Note that *safety* is a *quality in use* characteristic defined by ISO 25010 as "Acceptable levels of risk of harm to people, business, data, software, property or the environment in the intended contexts of use". *Satisfaction in use* and *safety* are in italic to denote their hedonic nature. *Safety* is depicted as a hedonic characteristic of *actual UX* because often it contributes to the user´s emotional needs for security and trust rather than a characteristic which satisfies a do-goal. Lastly, *accessibility in use*, defined in ISO 25010 as the "degree of usability in use for users with specified disabilities" is modeled as part of *actual usability*.

4   Example of Evaluating WebApp Operability

As the contributions mentioned in the Introduction Section, we developed a flexible modeling framework that uses to a great extent the latest ISO quality models, but also we have enlarged their models to include new characteristics and concepts, taking into account other well-known contributions in the discipline. As discussed above, our outcome is a set of new related models, assembled using the 2Q2U framework, which ties together well-known terms: quality, usability and user experience. Moreover the 2Q2U framework can be flexibly inserted and used with the so-called INCAMI nonfunctional component. In turn, we can use INCAMI to design the measurement, evaluation, and analysis.

INCAMI and its methodology have been used in different case studies. Recently, we performed a practical case by evaluating *external quality* requirements for a shopping cart, followed by *Web Model Refactoring* [15] for improvements. As stated by ISO 25010, for the end user, *quality in use* results mainly from *functional suitability, reliability, operability* and *performance efficiency*. Thus, for illustration purposes, we have applied the same methodology, taking into account the ISO 25010 *operability* characteristic because subcharacteristics and attributes related to *operability* as well as *information quality* could affect *quality in use* and *usability in use* including not only *efficiency in use* and *effectiveness in use* but also *learnability in use*. This will be addressed in our future work.

In Table 1, we now examine some of the subcharacteristics of *operability*, namely, *learnability, ease of use*, and *helpfulness* to measure and evaluate external quality requirements for software/WebApps. We strive to maintain consistency and alignment with the ISO 25010 quality models, while recognizing that there are many possibilities for other dimensions and attributes. In addition, these subcharacteristics have been modeled for general evaluative purposes for use with all WebApps. Some attributes associated to subcharacteristics could be more applicable to some domains than others and therefore receive higher relative weighting.

Table 1. Definition of used operability characteristics, subcharacteristics and attributes

Operability Characteristics and *Attributes*	Definition
1 Learnability	Degree to which the software/WebApp enables users to learn its application.
1.1 Predictability	Degree to which the software enables users to predict its interactions, functionality or content. <u>Note:</u> By being able to predict the consequences of an action, users can operate the software with minimal unintended consequences and fewer errors.
1.1.1 Action determinability	Degree to which the software /WebApp enables the user to predict what his action will do. <u>Note:</u> For instance, the user can evaluate potential inputs showing the result before changes are applied.
1.1.2Predictive textual anchor information	Degree to which the textual link provides users meaningful anchors or contextual information in order to help predict the target destination.
1.2 Feedback Suitability	Degree to which mechanisms and information regarding the success, failure or awareness of actions is provided to users to help them interact with the application. <u>Note:</u> Users need to know what might happen given the options available. Feedback about system states relieves users from having to remember these states, thereby making learning easier.
1.2.1Task Progress feedback suitability	Degree to which users are made aware of what they are doing for a specific task, function, or process. <u>Note:</u> For example, display progress in current process with number of steps completed and how many remaining to complete the task, or please wait, system is processing.

Table 1. (*continued*)

1.2.2 Navigability feedback completeness	Degree to which users are made aware of past, current, and possible locations while performing a navigation-oriented task.
1.2.3 Entry form feedback awareness	Degree to which users are made aware of the correctness or incorrectness of data entries.
1.2.4 Error message appropriateness	Degree to which meaningful error messages are provided upon invalid operation so that users know what they did wrong, what information was missing, or what other options are available. Note: This also relieves users from learning error recovery methods.
2 Ease of Use	Degree to which the software/WebApp makes it easy for users to operate and control it.
2.1 Controllability	Degree to which users can initiate and control the direction and pace of the task until task completion.
2.1.1 Permanence of main controls	Degree to which main controls are consistently available for users in all appropriate screens.
2.1.2 Stability of main controls	Degree to which main controls are in the same location in all appropriate screens.
2.2 Error Tolerance	Degree to which if, despite input errors, the intended result may be achieved with either no, or minimal, corrective action by the user.
2.2.1 Invalid Action Forgiveness	Degree to which users are allowed to attempt invalid actions without negative consequences.
2.2.2 Error Recovery Support	Degree to which the software/WebApp provides support for error recovery. Note: For instance, cursor is automatically positioned at the location where correction is required.
3 Helpfulness	Degree to which the software product provides help that is easy to find, comprehensive and effective when users need assistance.
3.1 Help Suitability	Degree to which the software/WebApp provides appropriate help given the users, their experience and context of help when required.
3.1.1 Context-sensitive help availability	Degree to which the software/WebApp provides context sensitive help depending on the user profile and goal, and current interaction.
3.1.2 Help Appropriateness	Degree to which the software/WebApp provides traditional online help with a structure that is easily readable and searchable. Note: For example, a top down hierarchy with hyperlinks for more detail for easier reading, semantic search, and advanced search.

Table 1 shows the selected characteristics, subcharacteristics and attributes *(in italic)*. The purpose of the evaluation was to understand the *external quality* level of the *operability* characteristic for filling new prescriptions (the evaluated *entity*) of a pharmacy WebApp. For confidentiality reasons, we do not disclose the company name and site, but are closely cooperating with the company to recommend improvements in case low satisfaction of these requirements were achieved. Table 2 shows an excerpt of the whole current evaluation.

Note in table 2 that for each attribute of the requirement tree has a metric to quantify it. For example, for *Error recovery support* attribute (coded 2.2.2), users should not have to search to find their mistake to correct it. So we designed a direct metric whose scale specifies four categories considering an ordinal scale type, namely: (0) *None,* does not support at all; (1) *Partially*, sometimes there is support but not always; (2) Complete support but only partial controllability, and (3) *Complete*, always support with complete controllability.

Table 2. External quality requirements for new prescription filling for Operability; EI = Elementary Indicator; P/GI = Partial/Global Indicator

Operability Characteristics and *Attributes* (External Quality)	Measure	EI value	P/GI value
Global Quality Indicator			62.8%
1 Learnability			66.7%
1.1 Predictability			75.0%
1.1.1 Action determinability	1	50.0%	
1.1.2 Predictive textual anchor information	2	100.0%	
1.2 Feedback Suitability			58.3%
1.2.1 Task progress feedback suitability	2	66.7%	
1.2.2 Navigability feedback completeness	1	33.3%	
1.2.3 Entry form feedback awareness	1	33.3%	
1.2.4 Error message appropriateness	2	100.0%	
2 Ease of Use			75.0%
2.1 Controllability			50.0%
2.1.1 Permanence of main controls	1	50.0%	
2.1.2 Stability of main controls	1	50.0%	
2.2 Error Tolerance			75.0%
2.2.1 Invalid action forgiveness	1	50.0%	
2.2.2 Error recovery support	3	100.0%	
3 Helpfulness			46.7%
3.1 Help Suitability			46.7%
3.1.1 Context-sensitive help availability	1	33.3%	
3.1.2 Help appropriateness	3	60.0%	

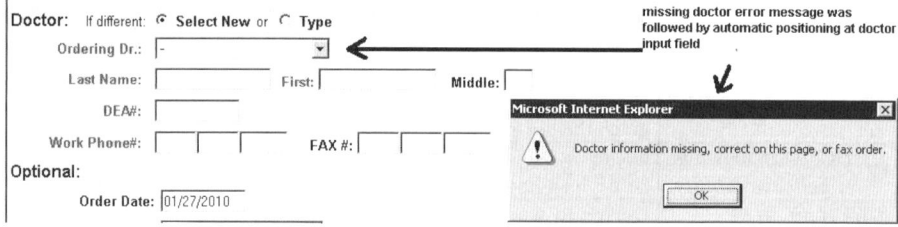

Fig. 6. Screenshot exhibiting error recovery support in a pharmacy prescription WebApp

As can be seen from table 2, its measurement resulted in 3 because it was clear to the user what the error was and the cursor was automatically placed at the error location throughout the task. However, the measure by itself does not have meaning so we must design an elementary indicator to interpret the level of satisfaction met. Therefore, a new scale transformation and decision criteria for acceptability ranges needs to be defined. In our study, we used three acceptability ranges in a percentage scale: a value within 40-70 (marginal –yellow) indicates a need for improvement actions; a value within 0-40 (unsatisfactory –red) means changes must take place with high priority; a score within 70-100 indicates a satisfactory level –green- for the analyzed attribute. Table 2 shows an elementary indicator value of 100% for the 2.2.2 attribute, but for instance resulted in 50% for the 1.1.1 attribute taking into account that a measure value of 1 mapped to 50%. This attribute would be totally suitable if the measure is 2. In the current state for the new prescription filling WebApp, users should not have to guess the results of their actions especially if they are infrequent users and improvement is needed.

Expanding the analysis as shown in table 2 can enable evaluators to understand the application's current external quality state and make recommendations for improvements. Note that the above model is just a possible instantiation of the external quality model represented in the 2Q2U framework (recall Figures 1 and 4). Going forward, we can also use the framework to model characteristics of *quality in use* and perform evaluation by extending the above study. By doing this with real users executing real tasks, we could derive relationships between *learnability* and *quality in use* characteristics such as *learnability in use* in order to determine how making improvements at the external quality level can affect *quality in use*. Regarding this, characteristics defined for *learnability in use* in section 3.2 can be measured with relationships drawn to the external quality *learnability* characteristics above.

5 Conclusions and Future Work

In this paper, we have developed a framework for modeling Quality, Quality in use, Usability and User experience (2Q2U) for Web applications, which in turn can be instantiated by using the INCAMI nonfunctional requirement specification component. In doing so, we have provided reasoning for and recommendations for adding 2 characteristics to extend the ISO 25010 standard, namely *information quality* and *learnability in use*. We have also characterized and described two new concepts, *actual usability* and *actual user experience* to bring to light their relationships while demonstrating using the framework.

To illustrate the applicability of the proposed approach, an example of external quality evaluation has been presented, in which the importance of taking into account *operability* was highlighted. 2Q2U offers not only an integrated framework for modeling requirements for quality, quality in use, actual usability and user experience, but also a consistent, and flexible way for representing calculable concepts (characteristics) and subconcepts which can then be used in conjunction with INCAMI for measurement and evaluation.

Another manuscript will thoroughly discuss the subcharacteristic, *learnability in use* covered in Section 3.2, including the further definition of attributes, metrics, and indicators for user group types performing specific tasks in real environments. Ongoing research is focused on further utilizing the 2Q2U framework when modeling and understanding the relationships among *internal/external quality, quality in use, actual usability* and *UX*. This concern has often been neglected in the literature, but may help evaluators to make sound design recommendations and ultimately better decision-making for improving the *user experience* as a whole.

Acknowledgments. Thanks to the support by Nat'l Basic Res. Prog. of China (973 project-2007CB310803) and Nat'l Natural Science Foundation of China (No. 60773155), and PAE-PICT 2188 project at UNLPam, Science and Technology Agency, Argentina.

References

1. Abran, A., Surya, W., Khelifi, A., Rilling, J., Seffah, A., Robert, F.: Consolidating the ISO Usability Models. In: 11th annual Int'l Software Quality Management Conference (2003)
2. Bevan, N.: Extending quality in use to provide a framework for usability measurement. In: Proc. of HCI Int'l 2009, San Diego, California, USA (2009)

3. Bevan, N.: Classifying and selecting UX and usability measures. In: Proc. of the 5th COST294-MAUSE Workshop on Meaningful Measures: Valid Useful User Experience Measurement (2008)

4. Covella, G., Olsina, L.: Assessing Quality in Use in a Consistent Way. In: ACM proceedings, Int'l Congress on Web Engineering (ICWE '06), SF, USA, pp. 1–8 (2006)

5. Ginige, A., Murugesan, S.: Web Engineering: An Introduction. IEEE Multimedia 8(1), 14–18 (2001)

6. Grossman, T., Fitzmaurice, G., Attar, R.: A survey of software learnability: metrics, methodologies and guidelines. In: Proceedings of the 27th Int'l conference on Human factors in computing systems, pp. 649–658 (2009)

7. Hassenzahl, M.: User experience (UX): towards an experiential perspective on product quality, IHM. In: Proc. of the 20th Int'l Conference of the Assoc. Francophone d'Interaction Homme-Machine, vol. 339, pp. 11–15 (2008)

8. ISO/IEC CD 25010 Software engineering – Software product Quality Requirements and Evaluation (SQuaRE) – Quality model and guide (2009)

9. ISO/IEC 25012 Software engineering – Software product Quality Requirements and Evaluation (SQuaRE) – Data quality model (2008)

10. ISO 9241-110, Ergonomics of human-system interaction, Part 110: Dialogue principles (2006)

11. ISO/IEC 9126-1. Software Engineering— Software Product Quality—Part 1: Quality Model (2001)

12. Lew, P., Zhang, L., Wang, S.: Model and Measurement for Web Application Usability from an End User Perspective. In: Proc. of QAW 2009: 1st Quality Assessment in Web Workshop held at Int'l Congress on Web Engineering (ICWE '09), CEUR Workshop Proceedings, San Sebastian, Spain, vol. 561 (2009) ISSN 1613-0073, http://ceur-ws.org

13. Nielsen, J., Levy, J.: Measuring usability: preference vs. performance. Comm. of the ACM 37(4), 66–75 (1994)

14. Olsina, L., Sassano, R., Mich, L.: Towards the Quality of Web 2.0 Applications. In: Proc. of 8th Int'l Workshop on Web-oriented Software Technology (IWWOST 2009) held at Int'l Congress on Web Engineering (ICWE09), San Sebastian, Spain, vol. 493, pp. 3–15, CEUR (ceur-ws.org) (2009) ISSN 1613-0073

15. Olsina, L., Rossi, G., Garrido, A., Distante, D., Canfora, G.: Web Applications Refactoring and Evaluation: A Quality-Oriented Improvement Approach. Journal of Web Engineering 4(7), 258–280 (2008)

16. Olsina, L., Papa, F., Molina, H.: How to Measure and Evaluate Web Applications in a Consistent Way. In: Rossi, Pastor, Schwabe, Olsina (eds.) Springer book: Web Engineering: Modeling and Implementing Web Applications, ch. 13, pp. 385–420 (2007)

17. Santos, P.J., Badre, A.N.: Discount learnability evaluation. Graphics, Visualization & Usability Center, Georgia Institute of Technology (1995), ftp://ftp.gvu.gatech.edu/pub/gvu/tr/1995/95-30.pdf (accessed by 20/12/2009)

18. Stewart, T.: Usability or user experience - what's the difference? System concepts (2008), http://www.usabilitynews.com/news/article4636.asp (accessed by 20/12/2009)

19. Vaníček, J.: Software and Data Quality. In: Proc. Conference of Agricultural Perspectives XIV, Czech University of Agriculture in Prague, vol. 52 (3), pp. 138–146 (2005)

Interfaces for Scripting: Making Greasemonkey Scripts Resilient to Website Upgrades

Oscar Díaz, Cristóbal Arellano, and Jon Iturrioz

ONEKIN Research Group, University of the Basque Country,
San Sebastián, Spain
{oscar.diaz,cristobal.arellano,jon.iturrioz}@ehu.es
http://www.onekin.org/

Abstract. Thousands of users are streamlining their Web interactions through user scripts using special *weavers* such as *Greasemonkey*. Thousands of programmers are releasing their scripts in public repositories. Millions of downloads prove the success of this approach. So far, most scripts are just a few lines long. Although the amateurism of this community can partially explain this fact, it can also stem from the doubt about whether larger efforts will pay off. The fact that scripts directly access page structure makes scripts fragile to page upgrades. This brings the nightmare of maintenance, even more daunting considering the leisure-driven characteristic of this community. On these grounds, this work introduces *interfaces for scripting*. Akin to the *JavaScript* programming model, *Scripting Interfaces* are event-based, but rather than being defined in terms of low-level, user-interface events, *Scripting Interfaces* abstract these DOM events into *conceptual events*. Scripts can now subscribe to or notify of *conceptual events* in a similar way to what they did before. So-developed scripts improve their change resilience, portability, readability and easiness to collaborative development of scripts. This is achieved with no paradigm shift: programmers keep using native *JavaScript* mechanisms to handle *conceptual events*.

Keywords: Greasemonkey, JavaScript, Maintenance, Web2.0.

1 Introduction

Traditional adaptive techniques permit to adjust websites to the user profile with none (a.k.a. adaptive techniques) or minimum (a.k.a. adaptable techniques) user intervention [13]. No design can provide information for every situation, and no designer can include personalized information for every user. Hence, traditional customization techniques do not preclude the need for do-it-yourself (DIY) approaches where users themselves can locally modify websites for their *own* purposes.

A popular client-based DIY technology is *JavaScript*, using special plugins such as *Greasemonkey (GM)* [1]. A *GM* script resides locally, and it has a scope, i.e. the websites to be subject to scripting (identified through URL patterns). *GM* silently watches whether the current URL matches the URL patterns, and if so, *locally* executes the script that leads to on-the-fly changes on the current page. Scripts can be uploaded to

B. Benatallah et al. (Eds.): ICWE 2010, LNCS 6189, pp. 233–247, 2010.

script repositories such as *userscripts.org*. With thousands of members and scripts, *userscripts.org* registers hundreds of downloads everyday! These remarkable figures stem from both the usefulness of scripts created by anyone, and the easiness of installation (a two-click process). This success evidences how scripting is moving from being a solitude practice to become a community phenomenon where laymen can enjoy scripts even if ignorant about *JavaScript.*

Unfortunately, current scripting practices do not scale up. Scripts directly access the structure of the page being rendered (i.e. the DOM tree). If the page changes, all the scripting can fall apart. And the page can change due to either upgrades on the website or changes made by previously enacted scripts. The problem is that websites are reckoned to evolve frequently, and the number of simultaneously enacted scripts tends to increase. This brings the nightmare of maintenance, even more daunting considering the altruistic, leisure-driven characteristic of many script programmers. *Our base premise is then that the maintenance burden is hindering GM scripters from becoming a mature community, not so in size but on the complexity of the scripts available.*

On these grounds, this work aims at shielding scripts from changes in the scripted pages. This implies separating the stable part of the script from the one more exposed to page changes. The former stands for the "mod logic", i.e. the code that supports the additional functionality provided by the script. The unstable part corresponds to code that weaves this mod logic to the current page, i.e. the code that consults/updates the scripted page.

We propose to encapsulate the fragile part of current scripts through interfaces. *Interfaces for scripting* encapsulate the current realization of HTML pages in terms of the *concepts* these pages convey. Now, scripts can subscribe to *conceptual events* (rather than subscribing to low-level, UI events) and notify of *conceptual events* (rather than directly modifying the page). This approach does not imply so much a change in the programming model but in the development methodology. Scripts keep using handlers but now upon *conceptual events* rather than UI events. The difference rests on the two-stage process. A first script implements the interface (so-called *Class Script*). A second script supports the mod logic on top of this interface (so-called *Mod Script*). In this way, the mod logic is decoupled from the concrete realization of the concepts this logic acts upon. Page changes only impact the *Class Script.*

From the perspective of the *Mod Script*, this approach accounts for change resilience (i.e. mod scripts are sheltered from changes in the website), readability (i.e. mod scripts are described in terms of *conceptual events* rather than HTML scraping), portability (i.e. mod scripts work for any website providing the same interface), and "collaborativeness" (i.e. the *Class Script* and the *Mod Script* can be developed by different users).

We regard as main contributions of this work, first, the rationales for bringing interfaces into the scripting realm, backed by examples taken from the *GM* community. And second, a non-disruptive approach that keeps current scripting practices without changes in neither the *JavaScript Engine* nor *GM*. Although the study is conducted for *Greasemonkey*, the approach can be generalized to any other *weaver*. Readers are encouraged to download the solution from *http://userscripts.org/users/61033.*

The paper begins by stating the problem, i.e. (1) script tight coupling to page rendering and, (2) companion script collision. Section 3 introduces *interfaces for scripting*. Its realization entails the introduction of three types of scripts: *Interface Scripts*, *Class Scripts* and *Mod Scripts* which are the subject of sections 4, 5 and 6, respectively. Section 7 revises the approach. Related work and conclusions end the paper.

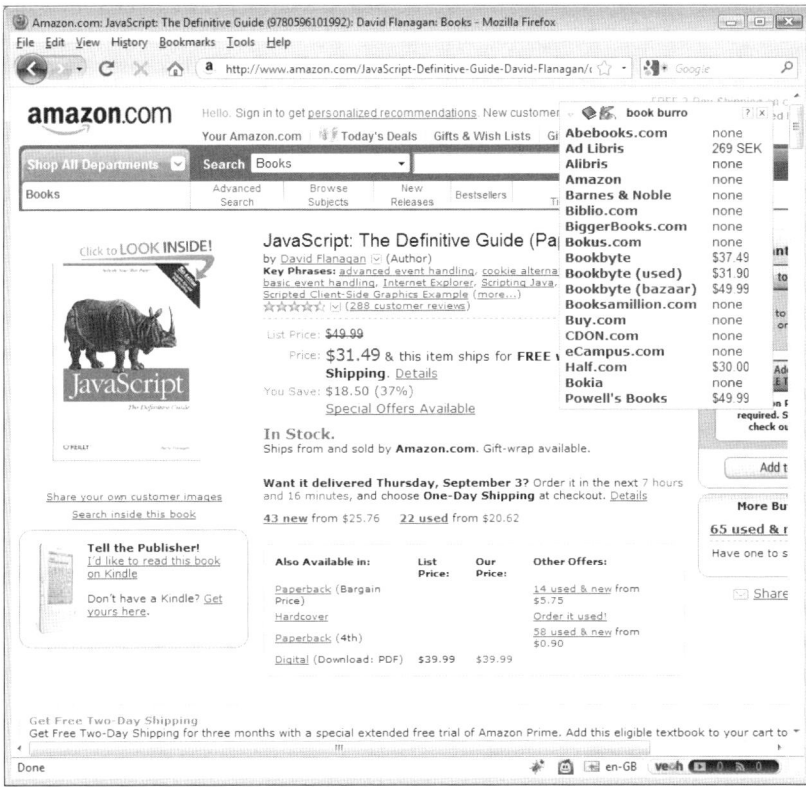

Fig. 1. *Amazon* enhanced with the *BookBurro* script: prices for the current book at other online bookshops are shown at the top-right side

2   Problem Statement

Greasemonkey (GM) is a *Firefox* extension that permits end users to install scripts at the client that make on-the-fly changes to the underlying HTML page structure (a.k.a. DOM tree) [2]. This is known as a *weaver*. This work focuses on *Greasemonkey* as a *weaver* for *Firefox*. Besides *Firefox*, *weavers* are available for Internet Explorer (e.g. *IE7Pro* or *Turnabout*), Opera (e.g. *User javascript*), Safari (e.g. *SIMBL + GreaseKit*) and natively supported in Google Chrome.

Table 1. Rationales for upgrading *BookBurro* along a year period. (source: changelog in *Book-Burro* script)

Reason	Times
Perfective maintenace	22
Adding new bookshops	18
Upgrades on bookshop websites	3
Remove sites	1
Corrective maintenance	13
Centralize scraping data	1
Bugfix	3
Improve code understandability	2
Improve UI	5
Add extra functionality (AJAX, caching)	2
Adaptive maintenance	4
Weaver-based no backward compatibility	3
Port to other browser weavers	1

Weavers permit scripts to act upon Web pages at runtime. Pages are perceived as DOM trees[1]. The script is triggered by User Interface events (UI events) on this DOM tree (e.g. *load*, *click*). Event payloads provide the data that feed script handlers which in turn, update the DOM tree. Scripts are written in *JavaScript*. For instance, a popular script, *BookBurro*[2], embeds price comparison in *amazon.com* web pages. On loading, the script retrieves the book's ISBN being rendered, and next, embeds a panel with the price of this book at other online bookshops, e.g. *Buy, BN, Powell, Half*, etc (see Figure 1). At install-time, scripts are associated with URL patterns that denote the pages to which the script applies. You can keep the script for yourself or upload it into a script repository (e.g. *userscripts.org*) for others to download. *BookBurro*, with more than 18,684 downloads, illustrates this ripple benefit.

Millions of downloads and thousands of uploads at *userscripts.org* provide anecdotal evidence of the profound impact that user scripting is having on a large number of users. Unfortunately, this practice is being hindered by maintenance burdens.

Next subsections provide two real scenarios where *BookBurro* is confronted with changes in either the scripted pages or the companion scripts. These scenarios are far from being just an academic exercise but they reflect similar settings as those faced up by *BookBurro* during its lifetime. Indeed, we examined the 18 different versions of *BookBurro* to assess the main reasons for the upgrades. Table 1 depicts the findings.

2.1 Upgrades on Scripted Pages

BookBurro embeds price comparison in *amazon.com* web pages from distinct online bookshops. The script is outlined in Figure 2 (left side). The process goes as follows:

- interacting with a page triggers UI events (e.g. *load*),
- the script can *react* to this event by triggering a handle (e.g. *"init"*, line 20). The association between event and handler (a.k.a. event listener) are achieved through the *addEventListener()* function (line 35),

---

[1] The Document Object Model (DOM) is a platform- and language-independent standard object model for representing HTML or XML documents as well as an Application Programming Interface (API) for querying, traversing and manipulating such documents.

[2] *BookBurro* is available at *http://userscripts.org/scripts/source/1859.user.js*

```
 1  // --UserScript==
 2  // @name          Simplified version of Book Burro
 3  // @namespace      http://overstimulate.com/userscripts/
 4  // @description    Book prices from various book stores
 5  // @include        http://www.amazon.com/*
 6  // ...
 7  // ==/UserScript==
 8
 9
10  //MOD-LOGIC
11 ⊟function createBookBurroPanel(isbn) {
12    var priceListHTMLPanel=document.createElement("div");
13    //A panel with a price list of selected book from
14    //distinct bookstores is created and assigned to the
15    //priceListHTMLPanel variable
16    ...
17    return priceListHTMLPanel;
18 ⌊}
19
20 ⊟function init() {
21    //HTML SCRAPING
22    var isbn=document.evaluate(
23      "//body[@class='dp']"+
24      "//td[@class='bucket']/div[@class='content']/ul/li[4]",
25      document,null,
26      XPathResult.UNORDERED_NODE_SNAPSHOT_TYPE,null
27      ).snapshotItem(0).innerHTML.match(
28      /[0-9]{10}/i )[0];
29    //MOD-LOGIC CALL
30    var bookBurroPanel=createBookBurroPanel(isbn);
31    //HTML INJECTION
32    document.body.appendChild(bookBurroPanel);
33 ⌊}
34
35  window.addEventListener("load",init,true);
```

```
 1  // ==UserScript==
 2  // @name          Interface-aware version of Book Burro
 3  // @namespace      http://overstimulate.com/userscripts/
 4  // @description    Book prices from various book stores
 5  // @include        http://www.amazon.com/*
 6  // ...
 7  // ==/UserScript==
 8 ⊟var doc=window.document;
 9
10  //MOD-LOGIC
11 ⊟function createBookBurroPanel(isbn) {
12    var priceListHTMLPanel=doc.createElement("div");
13    //A panel with a price list of selected book from
14    //distinct bookstores is created and assigned to the
15    //priceListHTMLPanel variable
16    ...
17    return priceListHTMLPanel;
18 ⌊}
19
20 ⊟function init(loadBookOcc) {
21    //EVENT PARAMETER RETRIEVE
22    var book=loadBookOcc.currentTarget;
23    var isbn=book.getElementsByTagName("isbn").item(0).nodeValue;
24    //MOD-LOGIC CALL
25    var bookBurroPanel=createBookBurroPanel(isbn);
26    //EVENT DISPATCH FOR HTML INJECTION
27    var appendChildBookOcc=doc.createEvent(
28      "ProcessingEvents");
29    appendChildBookOcc.initProcessingEvent(
30      "appendChildBook",book,bookBurroPanel);
31    doc.dispatchEvent(appendChildBookOcc);
32 ⌊}
33
34
35  doc.addEventListener("loadBook",init,true);
```

Fig. 2. Two versions of the *BookBurro* script: traditional (left side) vs. interface-aware (right side)

– a handler can access any node of the page (using DOM functions such as *"document.evaluate()"* in lines 22-26), and create HTML fragments (e.g. *bookBurroPanel* in lines 11-18),
– a handler can also *change* the DOM structure at wish by injecting HTML fragments. In the example, a *bookBurroPanel* is injected at a point identified by an *XPath* expression on the underlying DOM structure (i.e. the injection point). A DOM functions are used for this purpose (e.g. *"document.body.appendChild (bookBurroPanel)"* in line 32).

Scripts can do any change on the underlying page. But this freedom has a downside: makes the script bound to the actual page structure. If *Amazon* website is upgraded, all the scripting can fall apart. Back to our sample case, *BookBurro* first needs to retrieve the book's *ISBN* from the current page, and next, injects the *bookBurroPanel* at a certain location. This is normally achieved through *XPath* expressions (line 23-24). If the underlying page changes, *XPath* expressions could no longer recover/identify the right DOM portion which could make *BookBurro* stop working properly.

Recovering from these failures can be classified as perfective maintenance. For *BookBurro*, this accounts for 22 changes (see Table 1). This includes not only upgrades on consulted pages but also the need for *BookBurro* to run in bookshops other than *Amazon* (see user discussion thread at *http://userscripts.org/topics/15357*). These changes are not always straightforward which leads to delays on meeting these petitions by *BookBurro* programmers, more to the point if we consider that most programmers tend to do it altruistically. The problem is that websites are reckoned to evolve frequently, and programmers might not have the time to keep the script updated.

2.2   Changes in Companion Scripts

Scripts and scripted pages can exhibit an M:N relationship: a script can be applied to different pages, and a page can be the target of distinct scripts. *BookBurro* illustrates the first case where the very same script provides its mod (i.e. the *bookBurroPanel*) to different websites (*e.g. Amazon, Buy, BN...*). But, these websites can also be the substrate for other scripts. *Amazon* is a case in point. At the time of this writing, 261 scripts are reported to be available for *Amazon* at *userscripts.org*. If you are a regular *Amazon* visitor, it would be more than likely you have different scripts installed. These scripts (i.e. the companion set) will be enacted simultaneously when you visit *Amazon*. It is important to notice that script enactment is not in parallel but in sequence, i.e. scripts are launched in the order in which they were installed. This implies that the first script acts on the raw DOM tree, the second script consults the DOM once updated by the first script, and so on.

The problem is that programmers develop scripts from the raw DOM, being unaware of changes conducted by other companion scripts. This can end up in a real nightmare where code developed by different authors with different aims, is mixed up together with unforeseen results. Even worse, the final DOM tree can even be dependent on the order in which scripts are intermingled! This is not unusual for popular websites that enjoy a large set of scripts. The larger the set of (companion) scripts installed, the higher the likelihood of clashed. And the number of scripts available is steadily increasing which will likely lead to an increase in the number of scripts in each *user* installation. We then perceive "this DOM-based interaction model" as a main stumbling block for scripting to scale up. As learnt from previous experiences in Software Engineering, the approach is to abstract the way scripts are developed by moving away from "the implementation platform" (basically, the DOM document, and the UI events) to a more abstract platform. This is the aim of the *interfaces for scripting*.

3   The Big Picture

So far, scripts act directly upon DOM trees. We strive to abstract away from the DOM tree and the UI events through *Scripting Interfaces*. Interface reckons to hide "design decisions in a computer program that are most likely to change, thus protecting other parts of the program from change if the design decision is changed" [12]. *Scripting Interfaces* aim at shielding scripts from layout/presentation decisions *"that are likely to change"* in current HTML pages.

Scripting Interfaces characterise pages in terms of the concepts these pages convey, hiding the circumstantial representation of these concepts in HTML. For instance, *"Bookmark"* could be a concept for *del.icio.us*, *"Book"* for *amazon.com*, *"Article"* for *acm.org*, etc. Now, scripts do no longer access directly the DOM tree but through the interface: you can subscribe to *loadBook* (rather than the DOM event, *load*) and obtain book data through the event payload rather than scraping the DOM tree; you can publish the event *appendChildBook* to add an HTML fragment as a child of a *Book* (rather than using an *XPath* expression). The right side of Figure 2 shows the *BookBurro* script but now supported through a *Scripting Interface*. The mod logic is the same (lines 11-18). Differences rest on (1) page scraping being substituted by event payload recovering (lines 22-23) and (2), injection points described by the point where a conceptual event occurred (lines 27-31) rather than an *XPath* expression.

Figure 3 outlines the main notions of the problem space (in bold). **"User-Scripts"** act upon **"BaseWebsites"** but rather than accessing websites directly, scripts are now specified in terms of a **"ScriptingInterface"**. Programming languages clearly distinguish between the interface and the realization of this interface (a.k.a. *class*). Likewise, *Scripting Interfaces* are implemented through **"ScriptingClasses"** which implement interfaces based on websites.

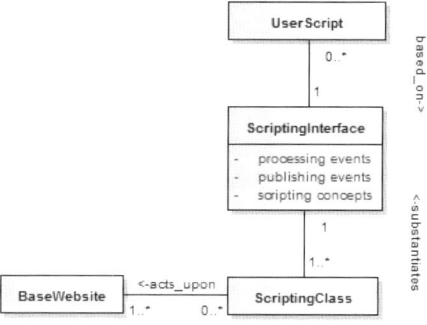

Fig. 3. The Problem Space

How are *Scripting Interfaces* described? Interfaces are commonly specified in terms of operations defined upon data types. However, *JavaScript* favours event-based programming, i.e. listeners are associated to UI events. Unlike operations, listeners are not explicitly called but *triggered* when the associated event occurs. Akin to the *JavaScript* approach, *Scripting Interfaces* are to be described in terms of events rather than operations, but they will act upon *concepts* rather than DOM nodes.

A (scripting) concept *is a data compound whose rendering is liable to be enhanced as a unit.* This approach resembles that of *microformats* [10] (e.g. *hCalendar*). However, there exist two main differences. First, *microformats* are designed to be widely applicable, i.e., they are targeted at general-purpose agents. By contrast, scripting is website-specific, i.e. each website can have its own concepts. Second, *microformats* are server-based, i.e. they are annotated by the site owner. Conversely, scripting is client-based, i.e. it is up to the scripter to decide which the concepts of interest are. While *microformats* aim at re-using existing HTML tags to convey metadata, "scripting concepts" can be website-specific.

Additionally, components distinguish between the provided and the required interfaces to differentiate between services facilitated or necessitated by the component, respectively. Likewise, a *Scripting Interface* encapsulates a DOM tree, and offers a set of services to "read" and to "write" this DOM tree. From the interface perspective, the "read" part realizes the required interface as the set of events the interface realization just signals but leaves to scripts their processing (a.k.a. *Publishing Events*). This basically implies abstracting away from UI events (i.e. those rised when manipulating HTML elements) to *conceptual events* (i.e. those signalled when acting upon *concepts*). For instance, *amazon.com* can publish events on book loading (e.g. *loadBook*). Scripts can now subscribe to *loadBook* rather than listening to the UI event, *load*. From this perspective, *Publishing Events* are to concepts what UI events are to DOM nodes: means to operate on the underlying structure. But while DOM nodes are implementation-dependant, concepts are design notions, more stable under website upgrades.

```
// ==UserScript==
// @name          Scripting Interface for Books
// @include       *
// ==/UserScript==
var bookInterface={
 "ScriptingConcepts":[{
  "conceptId":"Book",
  "attributes":[{"attributeId":"title","type":"string"},
               {"attributeId":"author","type":"array"},
               {"attributeId":"isbn","type":"string",
                "pattern":"/[0-9]{10}/"},
               {"attributeId":"price","type":"number"}]
 }],
 "PublishingEvents":[{"id":"loadBook",
                      "payloadType":"Book",
                      "uiEventType":"load",
                      "cancelable":false}],
 "ProcessingEvents":[{"id":"appendChildBook",
                      "payloadType":"HTMLDivElement",
                      "operationType":"appendChild",
                      "targetConcept":"Book"}]
}
```

Fig. 4. The *bookInterface* script

As for the "write" part, scripts directly modify the DOM structure, coupling the script to the current page implementation. This coupling is now eliminated by identifying the place to be modified in terms of concept occurrences rather than through DOM nodes: the *Processing Events*. *Processing Events* are to concepts what DOM operations are to DOM nodes: means to operate on the underlying structure. For instance, Figure 2 (right side) shows the new version of *BookBurro*. Now, the location to place the *BookBurro* panel is specified in terms of event dispatching: *append-ChildBook* (line 31). It is up to the *Scripting Interface* to map this event to the specific physical location (i.e. DOM node). From this perspective, *Processing Events* are the means to operate on the encapsulated realization of concepts. *Processing events* then realize the provided interface[3].

So far, the main notions of the problem space have been introduced. Next, we move to the solution space by addressing how previous concerns are engineered. *Our main requirement is non-disruptiveness from current practices.* The implications are twofold:

– from a programmer perspective, this implies *Mod Scripts* to be developed in a similar way to traditional scripts. This entails native *JavaScript* mechanisms to be used to notify/subscribe *conceptual events*,
– from the user perspective (i.e. users who install scripts), non-disruptiveness implies minimum diverge with current practices for script installation.

As a result, our proposal is *uniquely* based on traditional *GM* scripts where no plugin for neither *Firefox* nor *Greasemonkey* is required. Specifically, three types of *GM* scripts are introduced: **Interface Scripts,** which *specify* the interfaces; **Class Scripts,** which *implement* interfaces for a given base websites, and finally, **Mod Scripts** which built the mod logic on top of an interface. Next sections delve into the details.

4   Interface Scripts

Interfaces for scripting are specified through *Interface Scripts* using *JSON* (*JavaScript Object Notation*). *JSON* is a text format which is less verbose than XML, and whose

---

[3] The terminology of "processing events" and "publishing events" is widely used for event-based components such as portlets [9].

syntax is familiar to *JavaScript* programmers [3]. *JSON* document structure can be constrained through a *JSON Schema* specification [4] (much like what *XML Schema* provides for *XML*). An interface is a collection of *"ScriptingConcepts"*, *"PublishingEvents"* and *"ProcessingEvents"* instances (see Figure 4 for an example).

Concepts have a *conceptId* and a set of attributes. Each attribute has an *attributeI*d, a *type* and other features that constraint the set of possible values (e.g. *min, max, pattern,* etc). The sample case includes *Book* as a concept with four attributes: *title, author, isbn* and *price,*

Publishing Events are described through (1) the event payload, (2) when the event arises and (3), whether it can be cancelled or not. The event payload i.e. the type of parameter the event conveys (*"payloadType"* property), corresponds to one of the interface's *Concepts*. This concept is communicated at loading time, mouseover time, etc as specify by the *"uiEventType"* property whose values are taken from the W3C's DOM Level 2 Events specification [5]. Finally, the *"cancelable"* property mimics the namesake property available for *JavaScript* events whereby an event is liable to be called off by a handler so that the occurrence is no longer propagated to other handlers. The *bookInterface* exhibits *loadBook* as a publishing event to be raised when a page containing a *Book* is loaded,

Processing Events are specified through (1) the event payload and (2), how the signalled event is going to be processed. *Processing Events* carry a piece of *HTML markup*, i.e. its payload type is described along the W3C's DOM Level 2 HTML and Style specification [5] (e.g. *HTMLParagraphElement*). In this way, the *Scripting Interface* can restrict the type of the markup to be injected to those that can be safely injected (e.g. if the concept is realized through an *HTMLTableElement <table>*, only *HTMLTableRowElement <tr>* could be permitted). As for the processing mode, it specifies *what* is affected (i.e. the *"targetConcept"* property), and *how* is affected (i.e. the *"operationType"* property). The latter through a reference to the W3C's DOM Level 2 Core operations [5] (e.g. *appendChild*). The *bookInterface* provides an example: *appendChildBook* is raised when *HTMLDivElement* fragments are to be added as children of a *Book* concept.

Interface Scripts just contain the description of an interface. Additionally, they can be uploaded into script repositories so that a URL is generated. For instance, *bookInterface* can be found at *userscripts.org* where the following URL was generated: *http://userscripts.org/scripts/source/60315.user.js* . This is important for others to univocally refer to this script, e.g. *Class Scripts.*

5   Class Scripts

A *Class Script* implements an interface based on a specific website. A *Class Script* contains mappings that indicate how interface concepts are obtained from the circumstantial representation of those concepts in a concrete website. Figure 5 shows the *bookAmazonClass* script that implements the *bookInterface* for the *Amazon* website along the following characteristics:

```
 1  // ==UserScript==
 2  // @name        Scripting Class for Amazon Books
 3  // @include     http://www.amazon.com/*
 4  // @require     http://userscripts.org/scripts/source/60315.user.js
 5  // @require     http://userscripts.org/scripts/source/60318.user.js
 6  // ==/UserScript==
 7  var bookAmazonClass={
 8    "baseWebsite":"http://www.amazon.com/*",
 9    "implements":"http://userscripts.org/scripts/source/60315.user.js",
10    "scrapers":[
11    {"scrapedConcept":"Book","XPath":"//body[@class='dp']",
12      "attributeScrapers":[
13      {"scrapedAttribute":"title",
14       "XPath":"//span[@id='btAsinTitle']"},
15      {"scrapedAttribute":"author",
16       "XPath":"//div[@class='buying']/span/a"},
17      {"scrapedAttribute":"isbn",
18       "function":function(book){
19          var isbnNode=document.evaluate(
20            ".//td[@class='bucket']/div[@class='content']/ul/li[4]",book,
21  null,XPathResult.UNORDERED_NODE_SNAPSHOT_TYPE,null).snapshotItem(0);
22          return isbnNode.innerHTML.match(/[0-9]{10}/)[0];}},
23      {"scrapedAttribute":"price",
24       "function":function(book){
25          var price=document.evaluate(".//b[@class='priceLarge']",book,
26  null,XPathResult.UNORDERED_NODE_SNAPSHOT_TYPE,null).snapshotItem(0);
27          return price.innerHTML.match(/[0-9]+(\.[0-9]+)?/)[0];}}]}]}
28  }
29  window.registerScriptingClass(bookAmazonClass);
```

Fig. 5. The *bookAmazonClass* script

- *"baseWebsite"*, which holds a URL expression for the base website (e.g. *www.amazon.com/**) [4],
- *"implements"*, which keeps the URL of the *Interface Script* whose interface is being realized. This URL is obtained from *userscripts.org* (see later),
- *"scrapers"*, which contains a tripplet *<scrapedConcept, expression, attributeScrapers>* that indicates that *scrapedConcept* is to be located by applying the *expression* to the *baseWebsite* page, and its properties obtained by applying *attributeScrapers*. The latter is a set of pairs *(attribute,expression)* that denotes that *attribute* can be obtained by applying *expression* on the *baseWebsite* page. Expressions can be either functions or *XPath* expressions. In the latter case, *XPath* expressions for attributes are relative to the location of the concept. In the example, *XPath* is used to identify *"title"* and *"author"* while *JavaScript* functions are needed to extract *"isbn"* and *"price"*. It is worth noticing, how the function to extract the ISBN coincides with the one embedded in the traditional script of Figure 2 (left side, lines 22-28).

Greasemonkey allows for scripts to have *require* dependencies (specified through the *@require* comment tag). At install-time, *Greasemonkey* will download and keep a locally cached copy of the required files. This facility is used for *Interface Scripts* to be simultaneously downloaded with the installation of the *Class Scripts* so that consumers of *Class Scripts* do not have to install the interface separately. Figure 5 shows this

---

[4] This URL expression should coincide with the one specified at the time the *Class Script* is installed.

dependency through the *@require* tag in line 4[5]. Also, this facility is used to import the library that manages *conceptual events* and initializes the environment. This library is shared among all *Class Scripts* and its description is omitted due to paper length restrictions. Finally, the *registerScriptingClass* instruction (line 29), which is defined inside this library, makes the environment aware of this class.

Class Scripts can be uploaded to *userscripts.org* and installed as traditional scripts. Once a *Class Script* is successfully installed, the environment will generate *conceptual events* in the very same way that traditional UI events (see later). Now, it is the turn of *Mod Scripts* to capitalize on these *conceptual events*.

6   Mod Scripts

This section addresses the definition of scripts based on *conceptual events*. Native *JavaScript* mechanism is used to notify/subscribe *conceptual events* with no variations w.r.t. traditional script development. This is the most important characteristic to ensure non-disruptiveness with current *JavaScript* practices.

Notification of Processing Events. *JavaScript* follows an event-based approach where listeners can be associated with DOM-based events. An event is a happening of interest. Event types include: *MouseEventTypes* (e.g. *click, mouseover, mousemove...*), *UIEventTypes* (e.g. *DOMFocusIn, DOMFocusOut* and *DOMActivate*), *MutationEventTypes* (e.g. *DOMSubtreeModified, DOMNodeInserted*) and *HTMLEventTypes* (e.g. *load, change*). Operations are available for creation of event occurrences (e.g. *createEvent("MouseEvents")*), assigning the payload to an occurrence (e.g. *initMouseEvent("eventInstance", "eventParameters")*), or raising the event manually (e.g. *dispatchEvent(eventOccurrence)*). The following code simulates a click on a checkbox:

```
var ev=document.createEvent("MouseEvents");
var cb=document.getElementById("checkbox");
evt.initMouseEvent("click", true, true, window, 0, 0, 0, 0,
    0, false, false, false, false, 0, null);
cb.dispatchEvent(ev);
```

The snippet illustrates the pattern for dispatching an event occurrence: [*createEvent*, obtain DOM node, *initMouseEvent*, *dispatchEvent* on this node]. This is standard *JavaScript* code.

Conceptual events mimic this pattern. Back to our running example, a *bookBurroPanel* (i.e. an HTML fragment) is to be injected as a child of a *Book*. For this case, the pattern goes as follows: [*createEvent*, obtain concept, *initProcessingEvents*, *dispatchEvent* on this concept]. The code follows (the complete script can be found at Figure 2):

```
var ev=document.createEvent("ProcessingEvents");
var book = loadBookOcc.currentTarget;
```

---

[5] This *@require* tag is only read at install-time, and the remote file is not monitored for changes. *Class Scripts* are not aware of changes made to interfaces once installed.

```
evt.initProcessingEvent("appendChildBook", book,
    bookBurroPanel);
doc.dispatchEvent(ev);
```

The only difference with traditional scripting is that now injection points are not DOM nodes but the current concept. This current concept is to be obtained through *Publishing Events*.

Subscription to Publishing Events. *JavaScript* achieves event subscription by registering a listener through the *addEventListener* method. An example follows:

```
function init(...){...}
var cb=document.getElementById("checkbox");
cb.addEventListener("click",init,true);
```

This code associates the script function *init()* with the occurrence of clicks on a *checkbox* node (a.k.a. the event target). From then on, a click on a checkbox will cause *init()* to be enacted. Since most *JavaScript* events are UI events, event occurrences are generated while the user interacts with the interface, raised by the *JavaScript Engine*, and captured and processed through script functions.

Subscription to *conceptual events* is accomplished in the very same way: associating a listener. For instance, instruction (line 35 in Figure 2 (right side)) *"doc.add EventListener("loadBook",init,true)"* adds a listener to the *loadBook* event, i.e. occurrences of *loadBook* will trigger the *init()* function. The difference rests on listeners being associated to the whole document (i.e. variable *doc*) rather than to DOM nodes (e.g. a *checkbox*). This highlights the fact of events happening on a document of books rather than on DOM nodes that are the circumstantial representation of these books.

7 Discussion

Change resilience. We advocate for traditional scripts to decouple the stable part (i.e. the mod logic which is realized through *Mod Scripts*) from the unstable part (i.e. the logic that reads/writes the DOM which is supported through *Class Scripts*). In so doing, website upgrades will only impact *Class Scripts*. *Mod Scripts* are sheltered from the circumstantial realization of concepts in the website. For example, if *Amazon* decides to add more product details (e.g. other online bookshops include the book format: "printed" or "electronic"), *bookAmazonClass* needs to be rewritten but the *BookBurro* script itself is preserved. More to the point, the extreme change is moving to a different site altogether. For instance, *BookBurro* is initially thought to work upon *Amazon*. However, it can be made available for your favourite on-line bookshop as long as appropriate *Class Script* realizing *bookInterface* are provided for the target bookshop. This can be regarded as improving the portability of the script.

Script interferences. Companion scripts refer to those scripts that are simultaneously executed, then acting upon the very same DOM tree. The problem is that programmers develop scripts from the raw DOM, being unaware of changes conducted by other companion scripts. So far, scripts are enacted sequentially based on the time they were installed. This implies changes made by the first script *to be visible* to ulterior scripts. Two types of dependencies arise: read dependency (a script can accidentally read data written by a previous script), and write dependency (the injection point can be displaced by the writing of a node made by a previous script). As a result, the very same set of scripts can deliver different outcomes depending on the time they were installed.

Scripting Interfaces alleviate these problems. Read dependencies are obviated by making script changes transparent to other scripts. Scripts can only access the raw DOM, i.e. the DOM sent by the web server, previous to being updated by the scripts. Implementation wise, this is achieved by imposing *Class Scripts* (i.e. those accessing the DOM) to be executed before any *Mod Script* (i.e. those updating the DOM). Since *Class Scripts* are the first to be executed, the raw data is first captured as event payloads. Then, *Mod Scripts* take their data from these payloads. Therefore, there is no risk of a *Mod Script* reading data introduced by other *Mod Script*.

As for write dependencies, they are avoided by describing the writing point (i.e. the injection point) in terms of concepts (e.g. the location where a book is rendered) rather than directly addressing the DOM nodes. The injection point becomes logical rather than physical. If two *appendChild* scripts acts upon the very same concept, the "interference" is only noticeable in the order in which the children are rendered.

Readability. Both subscription and notification of *conceptual events* is easier to understand that their code counterparts (i.e. HTML scraping with XPath expressions). Additionally, both *Interface Scripts* and *Class Scripts* are *JSON* structures, hence, more declarative than *JavaScript* code. Besides improving legibility, this makes scripts liable to be automatically generated and processed. Indeed, *Class Scripts* can be automatically obtained using *MashMaker*'s data extractors [7] where XPath expressions are easily obtained by directly clicking on the page rendering (see section on related work).

Collaborative development. Compared with traditional practices, this proposal makes explicit a three-stage process during script development: interface definition, interface realization and mod-logic implementation. At first glance, this could look cumbersome compared with current practices where a unique script is needed. This additional effort can payoff in the following scenarios: the website suffers frequent updates, the very same mod logic is reused for distinct websites, or the page scraping part is complex enough to advice for separation of concerns. Additionally, this separation of concerns promotes collaborative development of scripts. For instance, a community interested in a certain topic (e.g. book price comparison) can provide interfaces that isolate this topic from its realization in distinct websites and then, releases *Class Scripts* for other programmers to capitalize upon (e.g. through *userscripts.org*). This resembles the genesis of *microformats* where *microformat* tags are first introduced, and later, become *ad-hoc* standards as the rest of the community adopts them.

8   Related Work: Scraping, Scripting, Mashuping

Web scraping is the process of automatically converting Web resources into a specific structured format. For instance, *Piggy Bank* is a plugin for Web browsers that "lets Web users extract individual information items from within Web pages and save them in Semantic Web format" [8]. Both the extraction technique and the target format serve to characterize scripting tools [11].

Mashuping overcomes scraping by addressing not only data extraction but also how this data is combined in novel ways [14]. For the purpose of this work, mashup approaches can be classified as compositional and "customizational" based on the role played by the source applications. Compositional approaches are akin to *integration* efforts to built *new applications* out of existing resources. This is, most mashup examples aggregate data coming from different sources to conform a new application in its own right, detached from the source websites. *Yahoo Pipes* is a case in point [6]. By contrast, "customizational" approaches focus on a given application which is then leverage using mashups. The mashup can only be understood that by referring to the customized website. *MashMaker* is one of the few examples. According to its authors, *MashMaker* augments "the familiar web browsing interface that the user already uses to browse data, and enhances this with mashed up information" [7].

Scripting aligns with customizational mashup efforts. *Greasemonkey* scripts also act within the realm of Web pages. The difference stems for the target audiences. *MashMaker* is oriented towards end users. Do-it-yourself is a main tenant of the mashup movement. The downside is expressiveness. *MashMaker* amendments can only occur at the time the page is load. By contrast, scripts can be attached to any DOM event. *MashMaker* amendments tend to be gadgets as reusable components which end users can easily plug into the target application. By contrast, scripts can attach any HTML fragment where deletions are also possible (e.g. removing banners). Basically, any *MashMaker* amendment can be achieved through scripting but not the other way around. Notice however, that the very same amendment that takes some few minutes to complete in *MashMaker*, could become a lengthy scripting effort[6].

This also supports the rationale for this work, i.e. *Scripting Interface* as the means to preserve the costly development of scripts. Indeed, our work strives to detach scripts from the underlying Web pages. Events are the means to realize both data extraction (i.e. publishing events) and data injection (i.e. processing events). By contrast, loose coupling is not a priority in *MashMaker*. *MashMaker* couples data extractors (i.e. the counterpart of *Class Scripts*) and gadgets (i.e. the counterpart of the *Mod Script*). The data is extracted as a raw structure which should coincide with the parameters of the gadget to be plugged. In practice, this implies that each gadget has its own data extractor. More to the point, the location where the gadget is to be injected is limited to the place the data is extracted. This is in contrast to *Scripting Interfaces* where the very same *Mod Script* can inject its output at different locations raising distinct *Processing events*.

Is it possible to obtain the best of both worlds, i.e. the easiness of *MashMaker* and the flexibility of scripting? As an attempt, we were able to obtain *Class Scripts* out of *MashMaker* data extractors. This permits *Greasemonkey* programmers to resort to *MashMaker* to obtain their *Class Scripts*, and move down to *JavaScript* to code their own amendments without being limited to reusing existing *MashMaker* widgets.

---

[6] Of course, if you have to program your own *MashMaker* gadgets then this is a complete different matter.

9 Conclusions

This work introduces *interfaces for scripting* to shelter user scripts from changes in the underlying Web pages. These interfaces are realised through standard scripts that generate *conceptual events* from UI events. The approach is aligned to *JavaScript* practices (event-based) and supported through standard *Greasemonkey* scripts. No plugin is required. Preliminary experiments suggest that this approach does not convey main disruptions for programmers while improving both change resilience and readability of scripts. The presumption is that improving change resilience will lead to greater user implication and more sophisticated scripts.

Next follow-ons include how to engineer sites to be more user-script friendly, with the possibility of providing the class/interface part directly from the site. At this regard, the possibility of automatically generating the class/interface parts using widgeting tools such as *MashMaker*, has so far being very encouraging.

Acknowledgements. This work is co-supported by the Spanish Ministry of Education, and the European Social Fund under contract TIN2008-06507-C02-01/TIN (MODE-LINE), and the Avanza I+D initiative of the Ministry of Industry, Tourism and Commerce under contract TSI-020100-2008-415.

References

1. Greasemonkey Homepage, http://www.greasespot.net/
2. Greasemonkey in Wikipedia, http://en.wikipedia.org/wiki/Greasemonkey
3. JSON (JavaScript Object Notation),http://json.org/
4. JSON Schema, http://json-schema.org/
5. W3C DOM Level 2, http://www.w3.org/DOM/DOMTR#dom2
6. Yahoo Pipes, http://pipes.yahoo.com/pipes/
7. Ennals, R., Brewer, E.A., Garofalakis, M.N., Shadle, M., Gandhi, P.: Intel Mash Maker: join the web. SIGMOD Record (2007)
8. Huynh, D., Mazzocchi, S., Karger, D.R.: Piggy Bank: Experience the Semantic Web Inside Your Web Browser. In: Gil, Y., Motta, E., Benjamins, V.R., Musen, M.A. (eds.) ISWC 2005. LNCS, vol. 3729, pp. 413–430. Springer, Heidelberg (2005)
9. Java Community Process (JCP). JSR 168: Portlet Specification Version 1.0 (2003), http://www.jcp.org/en/jsr/detail?id=168
10. Khare, R., Çelik, T.: Microformats: a pragmatic path to the semantic web. In: The 15th International Conference on World Wide Web (2007)
11. Kushmerick, N.: Languages for Web Data Extraction. In: Encyclopedia of Database Systems, p. 1595. Springer, Heidelberg (2009)
12. Parnas, D.L.: On the Criteria To Be Used in Decomposing Systems into Modules. Communications of the ACM 15, 1053–1058 (1972)
13. Magoulas, G.D., Chen, S.Y.(eds.): Adaptable and Adaptive Hypermedia Systems. IRM Press (2005)
14. Yu, J., Benatallah, B., Casati, F., Daniel, F.: Understanding Mashup Development. IEEE Internet Computing 12, 44–52 (2008)

Context-Aware Interaction Approach to Handle Users Local Contexts in Web 2.0

Mohanad Al-Jabari[1,*], Michael Mrissa[2], and Philippe Thiran[1]

[1] PReCISE Research Center, University of Namur, Belgium
[2] Université de Lyon, CNRS
Université Lyon 1, LIRIS, UMR5205, F-69622, France

Abstract. Users sharing and authoring of Web contents via different Web sites is the main idea of the Web 2.0. However, Web users belong to different communities and follow their own semantics (referred to as local contexts) to represent and interpret Web contents. Therefore, they encounter discrepancies when they have to interpret Web contents authored by different persons. This paper proposes a context-aware interaction approach that helps Web authors annotate Web contents with their local context information, so that it becomes possible for Web browsers to personalize these contents according to different users' local contexts.

1 Introduction

Web users usually belong to different communities and follow their local contexts for representing and interpreting Web contents. A *local context* (or context, for short) refers to the shared knowledge of a community such as a common language and common local notations and conventions (keyboard configurations, character sets, and notational standards for measure units, time, dates, durations, physical quantities, prices [3,4,14]. As a consequence, the same real world concepts might be represented and interpreted in different ways by different *Web authors and readers*. Such concepts are referred to in the following as Context-Sensitive Contents, or *CSCs*). For example, the concept of *price* could be represented using different currencies (e.g., Euro, US Dollar) and according to different price formats. Also, *date and time* concepts could be represented using different time zones and according to different formats. This situation leads to several *discrepancies* Web readers encounter on the Web, as they (need to) follow their contexts to interpret these *CSCs*.

Recently, the Web 2.0 has revolutionized the way information is designed and accessed over the internet. In our previous work [2], we advocated that the heart idea of the Web 2.0 lies into sharing and authoring of Web contents via different Web users and sites. Also, we illustrated a set of use cases that Web 2.0 sites/services provide. Indeed, Web 2.0 sites enable users not only browsing the Web but also creating and updating Web contents (i.e., they can act

---

* Supported in part by the Programme for Palestinian European Academic Cooperation in Education (PEACE).

B. Benatallah et al. (Eds.): ICWE 2010, LNCS 6189, pp. 248–262, 2010.

as *Web authors*). Several authors can author and update the same Web contents (e.g., *wiki*). In addition, Web 2.0 sites/services can aggregate Web contents from several sites and display them as a single Web page. Contents aggregation may happen both on client-side (e.g., using RSS) or on server-side (e.g., blogs' aggregations on sites like *Technorati*). Therefore, we concluded that Web contents in a *single Web page* are represented according to *several* authors' contexts, and the discrepancies Web readers encounter increase accordingly.

One possible solution is to annotate *CSCs* with semantic metadata (i.e., authors' contexts), so that it becomes feasible for Web browsers to adapt the former to different users' contexts [6,10]. This paper proposes a context-aware approach to resolve the discrepancies Web users encounter when they interact with Web 2.0 sites. This approach is an extension to our previous work [2,3] and consists of: (1) an evaluation of several design alternatives to adapt *CSCs* to different users' contexts; (2) a model that describes how *CSCs* are annotated with context information; (3) a context-aware architecture that shows how our approach works seamlessly with Web technologies; and (4) an annotation process that details how Web authors are assisted to specify their contexts and to annotate *CSCs* with a suitable context information.

The rest of the paper is structured as follows. Section 2 introduces a motivating example and evaluates the design alternatives. Section 3 summarizes a semantic model proposed in [3] to represent *CSCs* together with context information. Section 4 illustrates the annotation of *CSCs* based on the semantic model. Section 5 presents our proposed architecture and Section 6 details the annotation process. Section 7 introduces a prototype as a proof-of-concept. Section 8 discusses related work, and Section 9 concludes the paper.

2 Motivation and Design Alternatives

2.1 Motivating Example

This section presents an example to illustrate the discrepancies that Web users could encounter when they interact with Web 2.0 sites, as shown in Figure 1. This example considers several authors and readers from different communities. Also, it considers several tasks (i.e., T1-T7) performed on different Web 2.0 pages in sequential manner as follows:

– A British author creates and publishes a length and a date contents on page A (T1). After that, an American author browses the contents of page A (T2), and then updates the date content created by the British author to *07/09/2009* and publishes it again (T3).
– A Canadian author (from French speaking community) browses the contents of page B and deletes the date content *2009-09-11* (T4). We consider the page B's contents were created by this author. Next, the length and date contents from pages A and B are aggregated to page C (T5).
– An Italian reader browses the date contents that are automatically aggregated, via RSS engine, from pages A and B (T6). Finally, a French reader browses the date and length contents that are aggregated to the page C (T7).

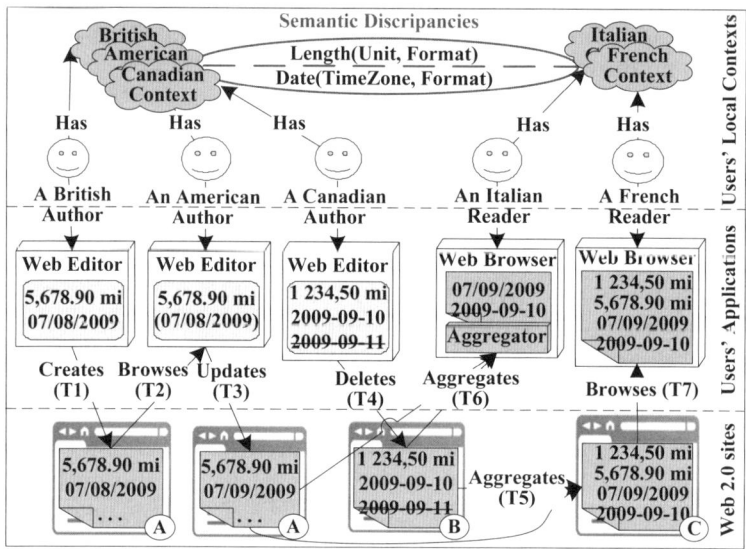

Fig. 1. Web 2.0 contents' sharing and the implicit use of users' local contexts

It is obvious that the date and length contents are represented in different ways by different authors. For example, the British author implicitly uses the British context[1] in T1. In contrast, Web readers usually (need to) interpret these contents according to their contexts. For example, as the French reader uses the Meter unit and the French length format (e.g. 1 234,50), he is responsible to adapt the length from Mile to Meter and to French length format. The problem is similar with respect to the date content 07/09/2009 which is updated at task T3. It is not obvious whether the American author updates the date content according to his context or according to the British context of the original author. Even if the ambiguity is resolved (i.e., he uses his context), the French reader might misinterpret this date as the 7th of September (following the French format) instead of the 9th of July (following the American format). Finally, several time zones are implicitly used by different users for representing and interpreting the date contents.

To conclude, the local context is clearly part of the $CSCs'$ semantics. Also, the discrepancies that rise do not relate to the $CSCs$ themselves, but rather to the contexts of Web users who represent and interpret them.

2.2   Design Alternatives

To resolve the aforementioned discrepancies, there is a need to adapt $CSCs$ from their multiple authors' contexts to their readers' contexts. To do so, there are several design alternatives, each of which has its own strengths and limitations. In the following, we evaluate three alternatives with respect to Web 2.0 use cases.

---

[1] Mile unit, British notation (e.g. 1,234.50) and date format (dd/mm/yyyy).

Adaptation to a Standard Local Context

The first alternative imposes a standard, unified context for all Web sites. Then, each CSC needs to be annotated with a standardized machine interpretable version (MV). The latter is generated by adapting the value of CSC from the author's local context to the standard context at creation and update time. Additionally, there is a need to adapt the MV into different human-readable versions according to different readers' contexts. In practice, we can use Microformat specifications and Microformat's *abbr design pattern*[2] for generating MV and annotating CSC. For instance, the date content above can be annotated with an MV date based on the ISO 8601 date/time standard.

This alternative allows $CSCs$ from several sites to be aggregated seamlessly as they are annotated with unified MVs. However, it violates the "do not repeat yourself" (DRY) design principle [1]. Indeed, each CSC needs to be represented twice (in the text and in the MV), and therefore needs to be maintained twice. For instance, when an author updates an annotated CSC, then both versions need to be updated. In addition, it lacks flexibility and may not satisfy the requirements of all Web sites. For example, most Europe countries use a tax system called VAT, while different states in the USA use different tax systems. Finally, $CSCs$ need to be adapted twice: one from the author's version to MV version and from the latter to the reader's version.

Adaptation to a Single Page Local Context

The second alternative is to identify a local context for each page. In this setting, each CSC is adapted from an author's context to a page's context at creation and update time, and adapted to different readers' contexts at browsing time.

This alternative does not violate the DRY principle and does not impose a standard context. Moreover, Web contents in a single Web page are homogeneously represented. However, $CSCs$ need to be adapted many times. These adaptations are often not necessary. For instance, assume the British author above needs to update the date content he created before. To this end, this date needs to be adapted to the author's context, since it was adapted from the author's to the page's A contexts at creation time. Also, it needs to be adapted from the author's to the page's A contexts after update. Furthermore, when the date and length contents are aggregated from pages A and B to the page C, then other unnecessary adaptations are needed to adapt the aggregated contents according to the context of page C.

Annotation of a CSC with Author's Local Context

The final alternative is to annotate $CSCs$ with authors' contexts at creation and update time and to adapt the annotated $CSCs$ to the readers' contexts at browsing time. For instance, annotating the above date and length contents with several authors' contexts, and adapting them to the French contexts at task T7.

---

[2] See `http://microformats.org/wiki/`

This alternative does not violate the DRY principle and does not impose a standard context. Furthermore, it preserves the initial Web contents as they were submitted to the Web page, which may be useful for their in-depth understanding or analysis. Also, it optimizes the number of required adaptations of $CSCs$. Finally, context information like *date format* can be made visible to readers in case $CSCs$ cannot be adapted to readers' contexts. However, when an author needs to update a CSC created by another author, the Web editor must take care to update the annotation too (hidden from the user).

2.3   Discussion

Our proposal is to adopt the third design alternative, since it is the best tradeoff with respect to the context representation flexibility, the DRY principle, and the number of $CSCs$ adaptations, as summarized in Table 1 below:

Table 1. Evaluation summary of the design alternatives

Design Alternatives	Standard Context	Single Page's Context	Multiple authors' Contexts
Context Rep. Flexibility	No	Yes	Yes
DRY Principle Compliance	No	Yes	Yes
Number of Adaptations	Twice	Many	Once

However, there are several techniques to annotate $CSCs$, and annotation of $CSCs$ is still a complex process. Indeed, we should consider that authors often do not know the relations between $CSCs$ and local context information. They also do not have theoretical and technical knowledge about the annotation process. As a consequence, we identify several objectives to be addressed in the rest of the paper as follows:

1. Identify the relations between $CSCs$ (e.g., date) and context information (e.g., date format and time zone) at the conceptual (meta) level.
2. Evaluate and optimize different annotation techniques, and illustrate how to annotate $CSCs$ with context information.
3. Assist authors (e.g., British author) to specify their contexts and annotating each type of $CSCs$ (e.g., date) with suitable context information (e.g., date format = "dd/mm/yyyy", date style = "short", and time zone = "UTC").

3   Semantic Model of Web Contents

To address the first objective presented in Section 2.3, we proposed a semantic representation model in our previous work [3]. This model builds on a local context ontology and uses the notion of semantic object to represent each type of CSC together with suitable context information, as summarized in the following.

3.1   Local Context Ontology

As already mentioned, *CSCs* such as date, time, price, physical quantity, phone number, and address are represented and interpreted in different ways by different users. To annotate *CSCs* with authors' context, we specify a set of local context attributes in an ontology called *local context ontology*. These attributes are mainly related to *country and community*. Indeed, each country has a set of local conventions such as currency, tax, measure system, etc. Also, each country has many cities, sometimes located in different time zones. In addition, one country may have one or more communities (e.g., French and Dutch speaking communities in Belgium). Each community usually relies on a common natural language and a set of common conventions such as writing formats of the above *CSCs*. More details about local context ontology and *CSCs* are given in [3].

3.2   Semantic Object

A *semantic object* provides a way to represent each *CSC* together with one or more context attributes. Basically, a semantic object *SemObj* is a triple $\langle S, V, C \rangle$. S represents the real world concept that the *CSC* adheres to, V is the physical representation (the value) of the *CSC*. C specifies the local context of *SemObj*. This context is represented as a finite set of context attributes. Also, context attributes themselves are represented as semantic objects, which may also have context attributes. This provides a recursive means for context description. Also, we categorized context attributes into two subsets: static and dynamic. Static attributes are the minimum context attributes that are used to describe the context of a semantic object and hence, their values must be specified explicitly. Dynamic attributes can be inferred from other context attributes that belong to that semantic object (See Table 2 below). However, Web authors can explicitly specify the values of one or more dynamic attributes if required, thus overriding the inference results. The motivation behind dynamic attributes is twofold. First, some dynamic attributes such as currency exchange rate could have dynamic value, and therefore cannot be statically stored. Second, it simplifies the specification of context information. For example, it is easier for users to specify the city instead of the time zone of this city.

To illustrate the notion of semantic object, Figure 2 represents the date from our scenario updated by the American author during Task T3 as a semantic object (See Figure 1). *Date* refers to the *date* concept S. '*07/09/2009*' represents the value V. Finally, *Context* represents the set of context attributes C. Here, the date *SemObj* has *DateStyle* as *static* attribute, and the *DateFormat* and *TimeZone* as *dynamic* attributes. The other context parts further describe the context of other semantic objects (i.e., DateFormat and TimeZone). The *DateFormat* value is inferred from the *country*, *language*, and *dateStyle* and the *TimeZone* value is inferred from the *country* and *city* static attributes.

Table 2 summarizes the relations between *CSCs* and static/dynamic context attributes. These relations are mainly derived from the *W3C Internationalization* initiatives that helped us to build *CSCs* [9]. As we will see in Section 6, these relations are utilized to extract the context attributes of semantic objects.

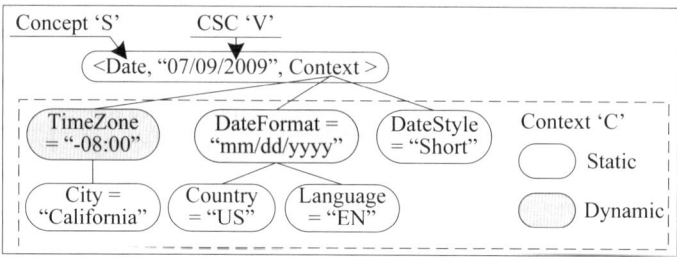

Fig. 2. Sample of date semantic object

Table 2. Relations between $CSCs$ and context attributes (the value of a dynamic attribute can be inferred if the value(s) of attribute(s) between brackets are known)

Context-Sensitive Contents ($CSCs$)	Context Attributes	
	Static	Dynamic
Date/Time	Date style	Date format(Country, Language, Date style) Time zone(Country, City)
Price	VAT included	Currency(Country), VAT rate(Country) Currency Exchange Rate (Date issued) Price format(Country, Language)
Physical Quantity	Measure unit, scale Error percentage	Measure System(Country), unit prefix(scale) Quantity format(Country, Language)
Telephone Number		Country calling code(Country) International prefix(Country) Phone format(Country, Language)
Address		Address format(Country, Language)

4 Semantic Annotation of Web Contents

Representing a CSC as semantic object requires annotating it with metadata (i.e., a concept S and a set of context attributes C). This section initially gives an insight on the different annotation techniques, namely external and internal annotation. Then, it details the advantages of internal annotation and RDFa technology for the purpose of this paper.

4.1 Annotation Techniques: External vs. Internal

Document annotation can be external and internal. With external annotation, metadata is represented and stored in an external annotation document. Then, these metadata refers to a part of a document (typically an XHTML tag) that is annotated using a pointing language such as Xlink and Xpointer. For instance, the value V of the aforementioned date semantic object can be annotated using this technique as follows. The concept S and the context information C are represented inside an RDF statement as a set of RDF attribute-value pairs. The subject of this statement is an Xpointer (e.g., /pageA#Xpointer(html/body/div[2])) which refers to the second *div* element. The latter represents the date value V inside the annotated document (page A)[3]. The main motivations behind external annotation are twofold. First, it provides a way to annotate already-published HTML documents without changing them and to annotate new ones without introducing new elements into their document-type definition (DTD). Second, one annotation document can be reused for annotating specific parts of multiple Web documents. This is useful for annotating (already-published) parts that have common semantics and structures such as calender events and products data [6,7,10].

However, since Xpointers refer to annotated elements based on their paths in the annotated document, this technique requires additional work to synchronize the annotation document with the annotated document. Furthermore, external annotation leads to problems when aggregating contents. First, references to the document structure are modified; and second, aggregation may imply context changes, and as the external annotation files are attached to the original contents, they should not be updated when the context changes due to aggregation in different contexts.

On the contrary, internal annotation stores Web contents and metadata together in the same document. Metadata is embedded as an XHTML attribute of the document element (e.g., XHTML tag) that delimits the Web contents to be annotated. As we will see in the next section, internal annotation remains simple, and the aggregation of annotated contents does not require additional synchronization, since metadata are directly embedded in the document. Then, as long as the document is accessible for editing, both Web content and annotation can be edited, deleted, and/or aggregated without any problems.

However, internal annotation requires additional work for annotating already-published Web documents. Indeed, each XHTML tag that represents a Web content to be annotated should be edited separately to embed the annotation, leading to a redundancy problem. For instance, if two or more date contents are created by one author (may be in the same page), then the same context information needs to be provided in all the corresponding XHTML elements.

To conclude, external annotation faces significant limitations with respect to creation/delete, update, and aggregation of Web contents and metadata. Therefore, we adopt internal annotation, which despite its redundancy drawback, remains the best tradeoff with respect to the Web 2.0 use cases as it eases the creation/delete, update, and aggregation tasks. We rely on the RDFa[4] syntax in order to annotate our documents. We give a short introduction to RDFa in the following before detailing our architecture.

4.2   RDFa-Based Internal Annotation

RDFa provides an annotation syntax to express RDF statements in XHTML documents. It relies on a collection of XHTML attributes such as *about*, *property*,

---

[3] More details available on W3C annotation note : http://www.w3.org/TR/annot/

[4] http://www.w3.org/TR/xhtml-rdfa-primer/

and *content* to embed RDF statements in XHTML. Also, it provides processing rules to extract these statements from XHTML [1,3].

This section shows how we utilize RDFa for annotating *CSCs* as semantic objects and illustrates how it works seamlessly with the Web 2.0 use cases. To keep the paper self-contained, we annotate the date contents presented in Section 2.1 as semantic objects and track the tasks to perform, as shown in Figure 3. We represent the concepts S and the context attributes C as RDF statements and localize them using RDFa syntax.

In page A, the *namespaces* inside HTML tag represent the URLs of RDF constructs and the *XHTML+RDFa* version of XHTML DTD. The RDFa attribute $about = `\#D1'$ identifies a date *ScmObj*. The RDFa attribute *property='dc:date'* represents the date concept S and the date contents 07/09/2009 represents the value V of the date after being updated by the American author[5]. In the inner $\langle span \rangle$ tags, the set of RDFa *property* attributes represent the date context attributes, and the RDFa *content* attributes represent the values of the context information related to the American author, as illustrated in Figure 2. Also, the date content *2009-09-10* in page B is annotated with the context information related to the Canadian author in similar way. Next, the annotated dates from pages A and B are aggregated like *"copy and past"* to the page C. It is worth noting that when the American author updates the date contents on page A (i.e., T3), the annotations are updated according to his context too.

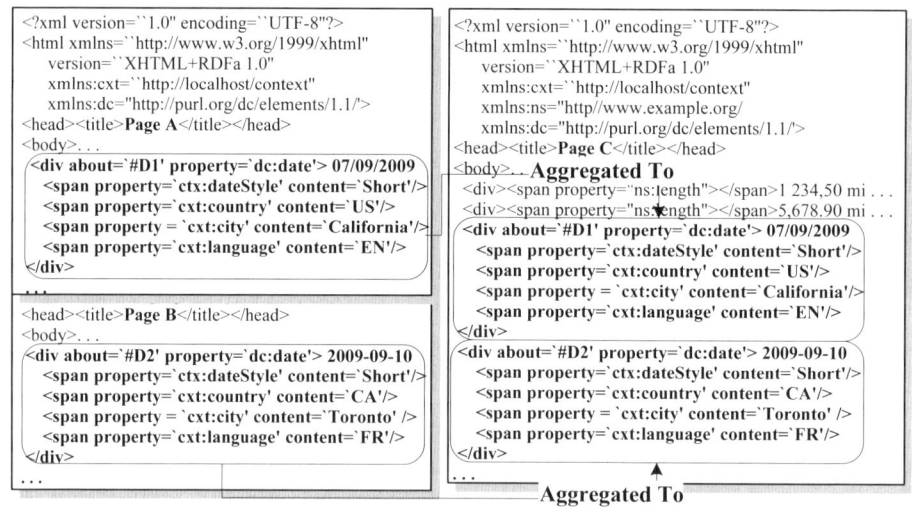

Fig. 3. Internal annotation using RDFa

5   Architecture

In this section, we present our proposed architecture to illustrate how our approach can be deployed and work seamlessly with the existing Web technology

---

[5] We do not present the original date value created by the British author for brevity.

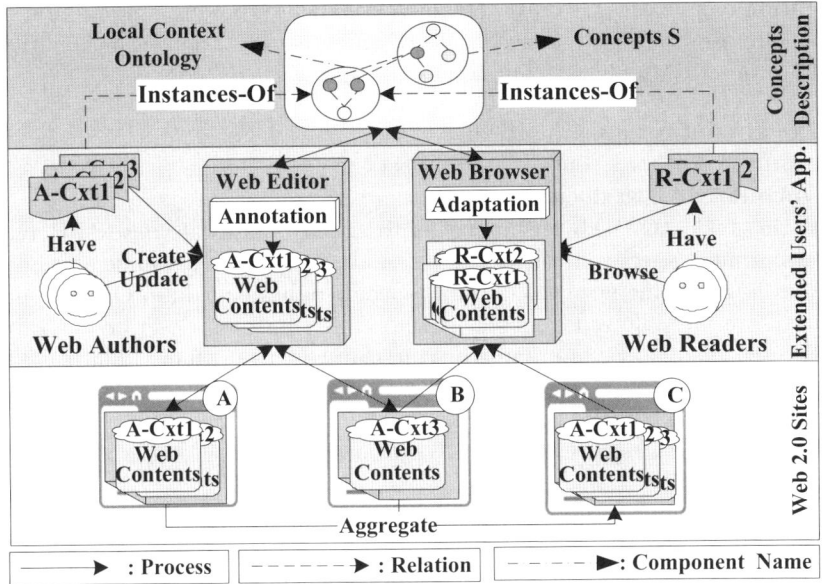

Fig. 4. A general architecture overview

stack. This architecture adds a layer called *concept description* and extends both traditional Web editors and Web browsers with an annotation engine and an adaptation engine respectively, as shown in Figure 4.

The *concept description* layer illustrates the relation between $CSCs$ and context attributes at a conceptual level as described in Section 3. The role of this layer is to provide the necessary vocabularies to specify users' local contexts (i.e., A-Cxts and R-Cxts) and concepts S. We consider that Web authors and readers agreed on these common vocabularies.

The role of the annotation engine is to assist authors for annotating $CSCs$ with their context. Web authors need to specify their context (i.e., A-Cxts) and the type of the CSC to be annotated. Then, the annotation engine interactively annotates the specified CSC with context attributes, which now forms a semantic object $\langle S, V, C \rangle$. Section 6 discusses the annotation process in more details with the extended Web editor. The adaptation engine also allows Web readers to specify their contexts (i.e., R-Cxts). Then, it adapts each annotated CSC from authors' to a reader's context. The output is semantically equivalent to the annotated CSC, but it is represented according to the reader's context. The adaptation process of the extended Web browser are out of this paper scope (See [3] for more details).

6   Annotation Process

This section details our vision on how to interactively accomplish the annotation process. Basically, this process illustrates the role of the aforementioned annotation engine during a Web author/extended Web editor interaction (Figure 5). Our annotation process consists of one pre-annotation task (i.e., Task 1) and four annotation tasks as follows:

1. **Local context specification**
   Input 1: Author's context attributes C
   Output 1: A-Cxt document
   In this task, the Web author needs to specify his local context. Here, the author must specify static attributes, and one or more dynamic attributes if there is a specific need for unusual/specific dynamic values. Other dynamic attributes are inferred from static context attributes as described in Table 2. The specified values are then stored in the A-Cxt document. This task (pre-annotation task) is thus performed prior to content annotation.

2. **Context attributes extraction**
   Input 2: A concept S, a content V, Context ontology, A-Cxt document
   Output 2: S, V, C
   During content creation or update, the author needs to select (i.e., highlight) the target value V of CSC to be annotated, and then selects a concept S. Upon selection of S and V, this process extracts the corresponding context attributes from the context ontology based on the concept S and its relations with these attributes (See Table 2). After that, the value of static and specified dynamic context attributes are extracted from the A-Cxt document. S, V, and C are utilized as inputs to Task 3.

3. **Annotation creation**
   Input 3: S, V, C
   Output 3: $SemObj = \langle S, V, C \rangle$
   Using the inputs received from the Task 2, this process annotates V as a $SemObj$ as follows. First, it builds a semantic object instance $SemObj$ from S, V, and C. Second, it generates the XHTML+RDFa representation of this $SemObj$. Finally, in the extended Web editor interface, it replaces the value V with the generated $SemObj$. Note that, an author can repeat this task and the previous one in order to annotate other $CSCs$.

4. **Annotation testing**
   Input 4: $SemObjs$ in the editing interface, A-Cxt document
   Output 4: Tested $SemObjs$
   During authoring and annotation process, this task scans Web contents in the background and checks the generated SemObjs. To this end, a semantic object instance called $testerSemObj$ is built for each generated $SemObj$. The $testerSemObj$ takes the concept S and the context C of the generated $SemObj$ as parameters and generates a test value TV. Then, the generated $SemObj$ is compared with the tester $SemObj$ as follows. First, the value V is compared with the value TV and provides a warning message if the former does not comply with TV (like a smart tag in Microsoft Word). For example, if an author, by mistake, annotates a length CSC with a date concept, then the generated $SemObj$ is highlighted and provides a warning message (e.g., this content is not a date). Second, it compares V with the context attributes

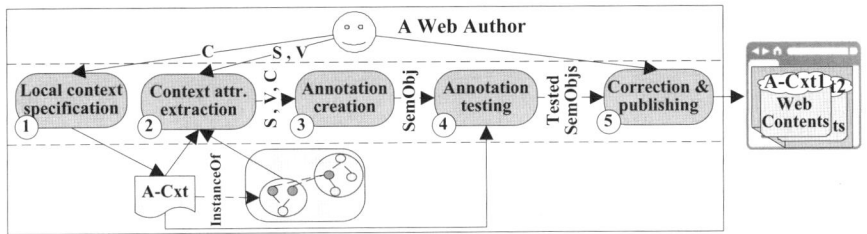

Fig. 5. Details of the annotation process

stored in the A-Cxt. For example, comparing a date content with a date style attribute and provides the corresponding warning message if V violates this attribute.

5. **Correction and publishing**
   Input 5: Tested $SemObj$
   Output 5: Annotated $CSCs$ in a Web page
   In this task, an author needs to correct the highlighted $SemObjs$. For example, he needs to correct the annotation of the length CSC, in the above example, with the length concept (instead of the date concept). Finally, the author publishes the annotated Web contents.

The above tasks provide the means to annotate $CSCs$ in an interactive and easy manner. First, context attributes that are specified in the A-Cxt document can be reused to annotate Web contents at different authoring times. Second, Task 2 enables authors annotating $CSCs$ as easy as formatting text in word processors [12]. Furthermore, authors do not need to know the relations between the $CSCs$ and the local context attributes, since the context attributes and their values are extracted automatically based on the concept S. At the same time our approach is flexible as advanced authors still have the possibility to override dynamic inferred attributes in the annotation. Third, creation of semantic annotation, in Task 3, hides the technical complexities of the RDFa syntax. In addition, it reduces the annotation efforts and the number of annotation errors that could be performed by Web authors. Finally, testing semantic objects, in Task 4, assists authors into correctly authoring $CSCs$ and therefore reduces the potential errors that they could perform.

The annotation process can be extended in one of the following directions. First, it can be extended with information extraction, together with semantic-aware auto-complete interface [8]. Information extraction is used to match typed $CSCs$, at authoring time, with one of the concepts S based on predefined concept patterns for example. If the matching task succeeds, then the semantic auto-complete interface will recommend this concept to the author. This increases the willingness of authors for annotating $CSCs$. Also, it helps authors knowing which $CSCs$ need to be annotated. Second, providers (i.e., Web sites designers or administrators) can relate concepts S and context attributes C to *annotation templates*, such as event and product templates. Based on an author's context, these templates are generated upon an author's request, and the typed (filled) $CSCs$ are annotated accordingly. This is useful for annotating structured contents [11]. Third, the annotation testing task can reuse the information extraction technique to check if there is a CSC that matches one of the concept S. If so, then it provides a warning message to the author in order to confirm or deny this matching. This enhances the annotation results, since authors could forget annotating some $CSCs$.

7   Prototype

This section presents a prototype as a proof-of-concept of our approach. Basically, the proposed prototype demonstrates the annotation process presented above. To this end, we use an HTML form to allow authors specify and store their contexts information into A-Cxts. The latter and the context ontology are implemented using RDF/XML syntax, and existing concepts from published ontology (e.g., Dublin Core) are reused as concepts S. We also extends the TinyMCE^{TM}WYSIWYG editor[6] with a concept menu. The latter allows authors to select concepts S for annotating $CSCs$ typed in the editor. As a back end, we use Java^{TM}APIs to extract context attributes and their values for each concept S. Finally, Javascript is utilized to annotate the value V with the selected concept S and the extracted context attributes C as semantic object. Figure 6 presents a screenshot of our prototype that illustrates how the date content from our scenario during Task T3 can be annotated as semantic object.

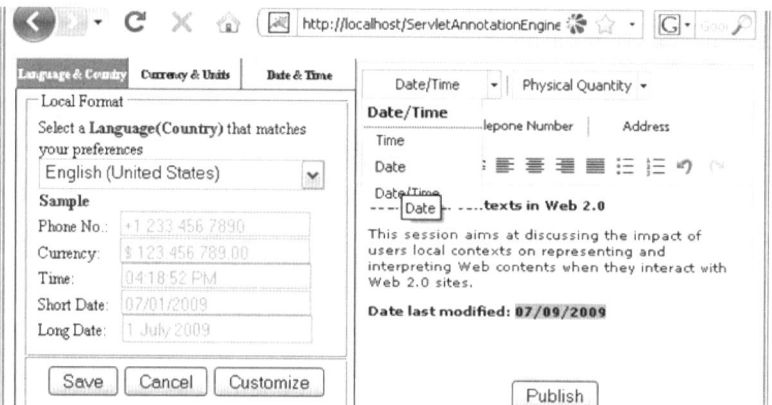

Fig. 6. A screenshot of the extended Web editor

8   Related Work

This section discusses existing works related to Web annotation and users' local contexts. The term *Web annotation* has been used to refer to a process of

---

[6] http://tinymce.moxiecode.com/

adding metadata to a Web document (e.g., XHTML document) or to metadata itself [6,10]. In practice, annotation is performed manually by authors or auto-matically by software application. Manual annotation incurs several problems such as usability, error proneness, scalability, and time consumption. Automatic annotation usually relies on information extraction techniques and/or machine learning to identify and annotate Web contents based on a natural language knowledge base (e.g., WordNet). However, automatic annotation faces concept disambiguation problems. Indeed, many Web contents have overlapping seman-tics and an annotation application cannot correctly recognize them all [5,15].

Interactive annotations have been commonly used recently [5]. Amaya is a W3C annotation framework that enables authors to add and share notes about Web contents. These notes are created and updated by authors, represented us-ing RDF schema, associated with contents using XPointer, and can be stored on authors' machines or on a remote annotation server [10]. Saha [15] is another interactive annotation system. It aims at enabling authors to enrich Web docu-ments with ontological metadata and uses the latter as a semantic index. Saha uses a set of predefined ontological schemas and associates them with annotated documents using Xpointer.

In the domain of the Web 2.0, several systems have relied on interactive an-notation and on RDFa. For instance, Luczak-Roesch and Heese [12] propose a system that enables authors to annotate Web contents and publish them as RDF-based linked data using RDFa. Also, semantic wikis such as SweetWiki[7] and OntoWiki[8] allow authors to annotate wiki contents with RDF-based prede-fined metadata (e.g., FOAF) using RDFa.

Local context has been acknowledged as an important issue [4,13,14]. However, a few approaches have used annotations for handling users' contexts. Further-more, most existing approaches rely on two assumptions. First, Web contents in a single Web page are represented according to one context only. Second, con-text information such as users' preferences or users' device capabilities are ac-quired into a context model, and then different contents are provided to different users. For example, transcoding systems annotate Web contents with transcod-ing metadata to define contents' roles. Then, annotated contents are adapted (e.g., restructured, summarized, or deleted) based on users' device capabilities or based on special requirements of users (i.e., visually impaired) [7].

To conclude, to our best knowledge, none of the existing approaches enable authors to annotate Web contents with their local contexts, so that Web browsers can handle local contexts' discrepancies. We advocate that our approach is one step further towards enriching the semantics of Web contents. Several works can be extended with this approach.

9   Conclusion

Today, the Web has evolved to a new era characterized by authoring and shar-ing of Web contents via different Web users and sites, known as Web 2.0. This

---

evolution leads to misunderstandings of Web contents as users from different communities use their own local contexts to represent and interpret these contents. This paper proposes a context-aware interaction approach that enables authors enriching Web contents with context information as easily as formatting text. Accordingly, it becomes feasible for Web browsers to personalize annotated contents according to different users' contexts. We present an architecture and a prototype to show how our approach works seamlessly with the Web technology stack. As a future work, we plan to set up an experiment to evaluate the practical feasibility of our approach from both authors' and readers' perspectives.

References

1. Adida, B.: hGRDDL: Bridging microformats and RDFa. J. Web Sem. 6(1) (2008)
2. Al-Jabari, M., Mrissa, M., Thiran, P.: Handling users local contexts in web 2.0: Use cases and challenges. In: AP WEB 2.0 International Workshop. CEUR Workshop Proceedings, vol. 485, pp. 11–20 (2009)
3. Al-Jabari, M., Mrissa, M., Thiran, P.: Towards web usability: Providing web contents according to the readers contexts. In: Houben, G.-J., McCalla, G., Pianesi, F., Zancanaro, M. (eds.) UMAP 2009. LNCS, vol. 5535, pp. 467–473. Springer, Heidelberg (2009)
4. Barber, W., Badre, A.: Culturability: The merging of culture and usability. In: The 4th Conference on Human Factors and the Web (1998)
5. Corcho, Ó.: Ontology based document annotation: trends and open research problems. IJMSO 1(1), 47–57 (2006)
6. Handschuh, S., Staab, S.: Authoring and annotation of web pages in CREAM. In: WWW, pp. 462–473 (2002)
7. Hori, M., Kondoh, G., Ono, K., Hirose, S., Singhal, S.K.: Annotation-based web content transcoding. Computer Networks 33(1-6), 197–211 (2000)
8. Hyvönen, E., Mäkelä, E.: Semantic autocompletion. In: Mizoguchi, R., Shi, Z.-Z., Giunchiglia, F. (eds.) ASWC 2006. LNCS, vol. 4185, pp. 739–751. Springer, Heidelberg (2006)
9. Ishida, R.: Making the World Wide Web world wide. W3C internationalization activity, http://www.w3.org/International/articlelist/ (last accessed: February 19, 2010)
10. Kahan, J., Koivunen, M.-R., Hommeaux, E.P., Swick, R.R.: Annotea: an open RDF infrastructure for shared web annotations. Computer Networks 39(5) (2002)
11. Kettler, B.P., Starz, J., Miller, W., Haglich, P.: A template-based markup tool for semantic web content. In: Gil, Y., Motta, E., Benjamins, V.R., Musen, M.A. (eds.) ISWC 2005. LNCS, vol. 3729, pp. 446–460. Springer, Heidelberg (2005)
12. Luczak-Roesch, R.H.M.: Linked data authoring for non-experts. In: Proceedings of the WWW '09, Workshop Linked Data on the Web, LDOW 2009 (2009)
13. Reinecke, K., Bernstein, A.: Culturally adaptive software: Moving beyond internationalization. In: Aykin, N. (ed.) HCII 2007. LNCS, vol. 4560, pp. 201–210. Springer, Heidelberg (2007)
14. Troyer, O.D., Casteleyn, S.: Designing localized web sites. In: Zhou, X., Su, S., Papazoglou, M.P., Orlowska, M.E., Jeffery, K. (eds.) WISE 2004. LNCS, vol. 3306, pp. 547–558. Springer, Heidelberg (2004)
15. Valkeapää, O., Alm, O., Hyvönen, E.: An adaptable framework for ontology-based content creation on the semantic web. J. UCS 13(12), 1835 (2007)

Rethinking Microblogging:
Open, Distributed, Semantic

Alexandre Passant[1], John G. Breslin[1,2], and Stefan Decker[1]

[1] Digital Enterprise Research Institute, National University of Ireland, Galway
`firstname.lastname@deri.org`
[2] School of Engineering and Informatics, National University of Ireland, Galway
`john.breslin@nuigalway.ie`

Abstract. In order to break down the walls that lock-in social data and social networks, new paradigms and architectures must be envisioned. There needs to be a focus on the one hand on distributed architectures — so that users remain owners of their data — and on the other hand on means to semantically-enhance their content — so that it becomes more meaningful and interoperable. In this paper, we detail the anatomy of SMOB, a distributed semantic microblogging framework. In particular, we describe how it achieves the previous objectives using Semantic Web standards (including RDF(S)/OWL, RDFa, SPARQL) and Linked Data principles, as a consequence rethinking the microblogging experience and, more generally, providing Linked Social Data as part of the growing Linking Open Data cloud.

Keywords: Social Web, Semantic Web, Linked Data, Microblogging, Distributed Systems.

1   Introduction

Founded in 2006, Twitter[1] defined the foundations of a now well-known phenomena: microblogging. While blogs let people openly share their thoughts on the Web, microblogging goes further by enabling real-time status notifications and micro-conversations in online communities. While it is mainly recognised as a way to provide streams of information on the Web, it can be used in various settings such as Enterprise 2.0 environments. The simplicity of publishing microblogging updates, generally shorter than 140 characters, combined with the ubiquitous nature of microblog clients, makes microblogging an unforeseen communication method that can be seen as a hybrid of blogging, instant messaging and status notification. Moreover, by considering microblog content as being information streams, new real-time applications can be imagined in the realm of citizen sensing [18].

So far, most of the current research around microblogging focuses on studying and understanding its communication patterns [10] [11]. However, its technical

---

[1] `http://twitter.com`

B. Benatallah et al. (Eds.): ICWE 2010, LNCS 6189, pp. 263–277, 2010.
© Springer-Verlag Berlin Heidelberg 2010

and engineering aspects still have to be studied, especially regarding the need for more openness in Social Web applications, argued by various manifestos such as "A Bill of Rights for Users of the Social Web"[2]. This would enable means to make microblogging integrated more closely with other Social Web applications, as well as with other data from the Web, for instance to identify information about a particular topic or event. In this paper, we discuss the motivations, the architecture and the use of SMOB — `http://smob.me` —, an open-source framework for open, distributed, and semantic microblogging[3]. In particular, our contributions include:

- an ontology stack to represent microblogs and microblog posts, using and extending popular vocabularies, hence making microblog data integrated with other efforts from the Semantic Web realm;
- a distributed architecture, based on hubs interacting together using SPARQL and SPARQL/Update, and which could also be used in a broader context of real-time RDF data replication; and
- interlinking components, enhancing existing practices such as `#tags` and providing the means to make microblogging interlinked with any resource from the growing Linking Open Data (LOD) cloud[4].

The rest of the paper is organised as follows. In Section 2, we discuss different engineering issues of microblogging systems, from which we derive three main requirements to enhance the microblogging experience. Based on these requirements, Section 3 describes the anatomy of SMOB, focusing on our main contributions and their applicability to other components of the Social Semantic Web. We then describe the system in use (Section 4), demonstrating how the previous principles enhance the microblogging experience in terms of content publishing and data reuse and discovery. We discuss related work in Section 5, and finally conclude the paper.

2   Issues with Current Microblogging Services and Requirements for a Richer User Experience

2.1   Lack of Machine-Readable Metadata

A first issue regarding microblogging services concerns the lack of metadata exposed from these applications, in a way that could be easily reused. Twitter has adopted microformats[5] for describing lists of followers (and followees), but this does not provide the means to match one username with other profiles that she or he may have in other platforms. Relatedly, users must create a new profile

---

[2] `http://opensocialweb.org/2007/09/05/bill-of-rights/`
[3] While the first release of SMOB has been designed mid-2008, this paper describes its recent 2.x series, redesigned with a completely new architecture and introducing new paradigms for semantic microblogging.
[4] `http://richard.cyganiak.de/2007/10/lod/`
[5] `http://microformats.org/`

— and fill in related details — on each platform they join. While OpenID could be used as a first step to solve this, there are still some issues with regards to maintaining attributes (profile pictures, etc.) across different websites. Furthermore, posts themselves do not provide enough fine-grained metadata (creation date, topics being addressed, recipient(s), etc.), beyond the use of RSS elements. Such a lack of metadata makes it difficult for microblogging to be interoperable with other systems, and it is also difficult for the data to be efficiently queried (*e.g.* to retrieve all microblog posts written by one's friend in the last six days). Consequently, we identified a first requirement to enhance microblogging systems:

– *R1: machine-readable metadata* — in order to make microblogging more interoperable, not only between microblogging applications but also considering the Social Web at large, there is a need for more machine-readable metadata about (1) microblog posts (time, author, etc.), (2) their content (topics, etc.) and (3) their authors (name, depiction, etc.).

2.2   Microblogs as Closed-World Data Silos

A second issue is that microblogging services act as closed worlds similar to most Web 2.0 applications. To that extent, the current centralised architecture of microblogging systems raises an important issue in terms of data portability, since data is locked into a particular platform, and cannot be automatically reused in other applications. While subscribing to RSS feeds in order to get a local copy of one's data can be an alternative, it does not provide the means to gather a complete dump of, *e.g.* last year's Twitter activity in a particular domain, unless one crawled the feed since the beginning. Therefore, our second requirement is the following:

– *R2: decentralised architecture and open data* — in order to solve the walled-garden issue of current microblogging systems, new decentralised architectures must be provided, so that everyone can setup her or his own platform and claim ownership on the published data, while at the same time making it openly available for other applications that may require it.

2.3   Lack of Semantics in Microblog Posts

Finally, in addition to the previous metadata concerns, microblog posts themselves do not carry any semantics, making their querying and reuse difficult. Twitter users have adopted hashtags (`#tag` patterns included in microblog posts to emphasise particular words, now officially supported by the service, as is the `@user` pattern), but their semantics are not readily machine-processable, thus raising the same ambiguity and heterogeneity problems that tagging practices cause [14]. Someone interested in the Semantic Web would have to follow various tags to get an accurate description of what is happening in that realm. Especially, she or he will have to consider the different tag variations caused by the sparse expertise levels and backgrounds of users, as raised by [7] in Delicious, *i.e.* subscribing to #RDF, #SPARQL, etc. This consequently leads to our third requirement:

- *R3: data-reuse and interlinking* — in order to enhance semantic descriptions of microblog posts and their content, they must be interlinked with existing resources using meaningful and typed relationships, going further than the traditional usage of ambiguous and heterogeneous hashtags and hyperlinks.

3   Anatomy of SMOB

Based on the aforementioned requirements, we engineered SMOB — Semantic MicrOBlogging —, an open-source microblogging framework providing an open, distributed and semantic microblogging experience. In particular, the system completely relied on state-of-the-art Semantic Web and Linked Data [3] technologies to achieve this goal. Therefore, SMOB offers a combination of these technologies with existing Social Web paradigms and tools (microblogging, information streams), leading to what is generally known as the Social Semantic Web [5]. To enable such integration and solve the aforementioned issues, we have engineered a system relying on several components. In particular:

- *R1: machine-readable metadata* is achieved thanks to lightweight *ontologies* and RDFa markup, enabling common semantics and standard representations to model microblog posts (and microblog services) and their metadata and consequently providing interoperable descriptions of microblog posts. Thus, posts can be exchanged not only between SMOB applications but with any service capable of consuming RDF(S)/OWL data;
- *R2: decentralised architecture and open data* is achieved thanks to *distributed hubs*, spread across the Web and exchanging information (posts and subscriptions) based on the previous ontologies and a sync protocol (based on SPARQL/Update over HTTP). These hubs also act as end-users publishing and browsing interfaces for microblog posts;
- *R3: data-reuse and interlinking* is achieved thanks to *interlinking components*, so that microblog posts can be interlinked with resources from the Web, and in particular those from the aforementioned Linking Open Data cloud, by turning #tags into URIs identifying Semantic Web resources. Thus, it allows (1) microblog posts to become more discoverable, by being linked to existing resources using the Linked Data principles [2]; and (2) microblogging and the Social Web to join the Linking Open Data Cloud, and not exist as an isolated subset.

In addition, thanks to these different components, the boundaries between microblogging and other Social Semantic Web applications become (voluntarily) weaker. Microblog content can indeed be immediately mashed-up and integrated between various applications supporting the same standards. This emphasises the object-centric [13] nature of many online conversations and social networks, enhancing them with a real-time component. For example, it provide means to integrate microblog updates with blog posts or RSS feeds mentioning the same object (research topic, project, etc.), this object being identified by its own URI.

Technically, as our goal was to make the system as easy as possible to deploy, SMOB only requires a LAMP — Linux, Apache, PHP and MySQL — environment. This also emphasises how Semantic Web and Linked Data technologies can be brought to end users and developers thanks to simple frameworks and engineering practices linking these technologies to object-oriented paradigms. In our case, we relied on the ARC2 PHP framework[6], which provides a simple toolkit to build RDF-based applications in PHP. The different objects (posts, users, etc.) and views (lists of posts, map views, etc.) have been mapped to SPARQL and SPARQL/Update queries using the aforementioned ontologies in order to be generated and saved in the local database, so that the underlying RDF(S)/OWL structure is directly mapped to PHP objects.

3.1   The SMOB Ontologies Stack

In order to semantically-enhance microblogging services and microblog posts by providing more fine-grained metadata (R1), there is a need for models representing these services and their related content. This entails the need for ontologies representing:

- users, their properties (name, homepage, etc.), their social networking acquaintances and contextual information about themselves (geographical context, presence status, etc.);
- microblogging services and microblog posts, including common features such as #tags and replies (@user).

In order to model user profiles, we naturally relied on FOAF — Friend of a Friend [6] — as it provides a simple way to define people, their main attributes and their social acquaintances. Moreover, FOAF is already widely used on the Web, providing SMOB users with a way to reuse their existing profiles in their microblogs, thus also enabling distributed profile management and authentication. As we mentioned earlier, one major issue with current Web 2.0 services is the need to create a new profile each time one wants to join a new website. Given the popularity and the number of new services appearing regularly on the Web, this quickly becomes cumbersome and leads to what some have called "social network fatigue". We have addressed this issue by letting users create a SMOB account simply by specifying their FOAF URI. User profiles are then retrieved from these URIs (assuming they follow the Linked Data principles), so that name of an author, depiction, etc. can be stored and updated in third-pary websites, but are however integrated in microblog updates. In addition, the original profile can be any any RDF serialisation, for example in the form of an RDFa-annotated profile (Fig. 1). Furthermore, each post is linked to this author (using foaf:maker), every author being uniquely identified on the Web

---

[6] http://arc.semsol.org

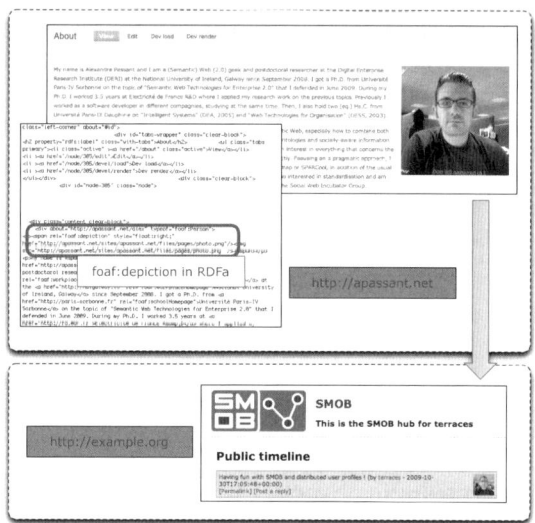

Fig. 1. Distributed and user-owned profiles

thanks to her of his URI. Then, for each microblog post, additional information about its author can be immediately obtained by dereferencing its author's URI. Moreover, we also used OPO — Online Presence Ontology [20] — to describe users' presence information, such as their geolocation aspects. Finally, we can benefit from these FOAF-based user profiles for authentication purpose, since SMOB hubs rely on FOAF-SSL [21] to enabling distributed authentication.

Regarding the modelling of microblogs and microblog posts, we relied on and extended SIOC — Semantically-Interlinked Online Communities [4]. Notably, we introduced two new classes to the SIOC Types module[7] to model these services and their data: (1) `sioct:Microblog` and (2) `sioct:MicroblogPost`. We also introduced two additional properties: (1) `sioc:follows`, to express following / follower notifications (the same property being used for both, benefiting of the RDF graph model), and (2) `sioc:addressed_to`, to represent whom a given post is intended for.

Finally, a further aspect concerns the enhancement of tagging practices, notably to bridge the gap between `#tag`s as simple keywords and URIs identifying Semantic Web resources. We rely on MOAT — Meaning Of A Tag [16] — to do so, since it provide a model to represent the meaning of tags using Semantic Web resources, such as identifying that in a particular context, the tag `apple` is used to identify "apple, the fruit" (identified by `dbpedia:Apple`) but "'Apple, the computer brand" (`dbpedia:Apple_Inc.`) in another one.

Combined together, these ontologies form a complete stack to represent the various elements involved in microblogging applications (Fig. 2). We did not want to provide a new unique and huge ontology, but rather defined this combination

---

[7] `http://rdfs.org/sioc/types#`, prefix `sioct`

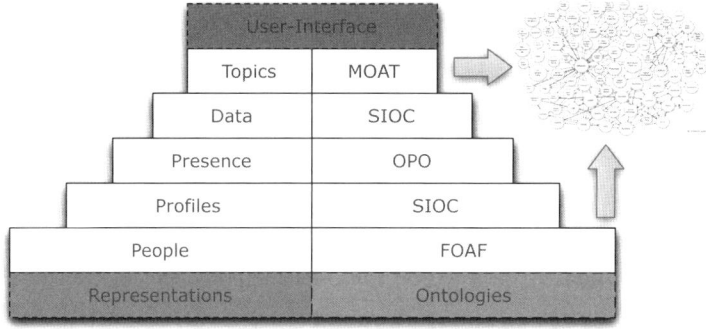

Fig. 2. The SMOB ontologies stack

of lightweight ontologies [8] to fit with existing applications and provide a coherent framework to represent the different artifacts of paradigms appearing in the Social Web. Therefore, this stack is not specific to SMOB but can be referred to as a more global ontologies stack for the Social Semantic Web and can be reused in any applications relying on similar modeling requirements. Each post created with SMOB is then modelled using this stack and made available on the Web as RDFa-annotated content, in the author's hub. The following snippet of code (Fig. 3, prefixes omitted) shows an example of a microblog post represented via SMOB with this ontologies stack, here in Turtle.

```
<http://example.org/smob/post/20091101-1755> a sioct:MicroblogPost ;
    sioc:content "Drinking #coffee with @milstan in #Torino" ;
    sioc:has_creator <http://apassant.net/smob/owner> ;
    foaf:maker <http://apassant.net/alex> ;
    sioc:has_container <http://apassant.net/smob> ;
    sioc:addressed_to <http://ggg.milanstankovic.org/foaf.rdf#milstan> ;
    moat:taggedWith <http://dbpedia.org/resource/Coffee> .
<http://example.org/smob/presence/20091101-1755> a opo:OnlinePresence ;
    opo:customMessage <http://example.org/smob/post/20091101-1755> ;
    opo:currentLocation <http://sws.geonames.org/2964180/> .
```

Fig. 3. Representing microblog posts using the SMOB ontologies stack

3.2 Distributed Hubs and Synchronisation Protocols

In order to fulfill our second requirement (R2: decentralised architecture and open data), we designed an architecture based on distributed microblogging hubs that communicate with each other to exchange microblog posts and notifications. That way, there is no centralised server but rather a set of hubs that contains microblog data and that can be easily replicated and extended. Hubs communicate with each other via HTTP thanks to SPARQL/Update (the Update part of SPARQL, currently being standardised in the W3C[8]). We rely in

---

[8] http://www.w3.org/TR/sparql11-update/

particular on a subset of SPARQL/Update, namely the `LOAD` clause, in order to publish items from one hub to another (Fig. 4). When creating a new microblog post on a SMOB hub, the workflow is the following:

- the post is immediately stored in the local RDF store, and published in an RDFa-enabled page at its own URI, e.g. `http://example.org/post/data`, using the aforementioned vocabularies;
- a SPARQL query identifies all the poster's followers from the local store, as well as the URL of their hub;
- a SPARQL/Update query is sent (*via* HTTP `POST`) to each hub so that they aggregate the newly-created RDF data.

To avoid hijacking, only posts whose URI corresponds to the URI of a followee can be loaded in remote stores, so that only trusted data (*i.e.* generated by a followee) is aggregated. A SPARQL/Update pre-processor is used to provide this additional level of security on top of SPARQL endpoints. Future improvements may include FOAF-SSL or OAuth[9] to further address this issue.

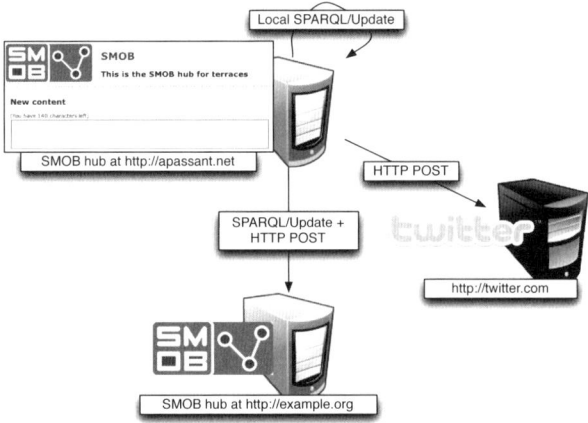

Fig. 4. Communication between SMOB hubs using SPARQL/Update

Using this workflow, posts are immediately broadcasted to every follower as soon as the content is created. This push approach (contrary to a pull one where hubs would regularly fetch followees' information) is similar to what is provided by the OStatus protocol[10]. However, while OStatus sends all information about the posts (using Atom feeds), our approach directly sends a SPARQL/Update `LOAD` query to each follower's hub. Such a query only contains the URI of the post to be loaded, this URI being dereferenced when it is loaded in the follower's RDF store. That way, we rely on "self-descriptive" posts as each URI

---

[9] `http://oauth.net/`
[10] `http://ostatus.org/`

identifying a microblog post can be dereferenced and delivers a full set of RDFa information about itself. This nicely illustrate the use of Linked Data principles to transmit rich status updated (since they can contain geolocation information, semantically-enhanced tagging, etc.) in the context of distributed microblogging.

In addition, each hub features a triggering approach which loads followees' FOAF profiles each time one of their post is loaded into the store. Then, if a user edits his or her depiction, this will be forwarded to each hub following as soon as a new post is created, without any additional intervention, and without having to send this information in the post itself, thanks to the FOAF-based approach that we earlier described.

A similar approach is used with regards to the followers and followees subscriptions. A bookmarklet is provided to let anyone become a follower of another user when browsing that user's hub. The subscription is registered in the follower's hub by adding a `"<user-uri> sioc:follows <remote-user-uri>"` triple, while the same triple is included in the remote store. Thus, that both parties are instantaneously aware of this new relationship, which can be then used in the aforementioned publishing protocol when new updates are published.

3.3   Integrating Microblogging in the Linking Open Data Cloud

As we mentioned in our third requirement (R3: data-reuse and interlinking), our goal was to make microblog posts more discoverable by linking them to existing resources on the Web. In particular, our vision is to make microblogging updates linked to resources from the Linking Open Data cloud. While such interlinking is already provided by reusing existing FOAF profiles, there is a need to go further and link to relevant data to make sense of the content of microblog updates.

To achieve such goal, we extended the common `#tag` practice in order to turn these tags into machine-readable identifiers, *i.e.* URIs identifying Semantic Web resources. In particular, we aim at relying on URIs from the Linking Open Data cloud, since it provides identifier for various things, from generic topics in DBpedia (the RDF export of Wikipedia) to drug information information in LODD [12]. Our approach therefore focuses on extending the genuine `#tag` practice by turning tags into identifier to resources from the LOD cloud, enabling interlinking between tagged content and these resources. Then, `#tags` are not simple tags anymore, but provide links to uniquely identified and structured resources. In addition to the interlinking, this practice also solves the ambiguity and heterogeneity issues of tagging. Indeed, by linking a microblog post initially tagged with `#apple` to `dbpedia:Apple_Inc.`, one can identify that this post is about the computer brand (and not the fruit nor the record label), and could also benefit from the links existing from (and to) the related URI to identify that this is a message about computers (since `dbpedia:Apple_Inc.` and `dbpedia:Computer_hardware` are linked in DBpedia).

Fig. 5. The SMOB publishing interface and its interlinking components

4   SMOB in Use: Publishing and Discovering Semantically-Enhanced Microblog Posts

4.1   Publishing and Interlinking Microblog Posts

SMOB is available under the terms of the GNU/GPL license at `http://smob.me` and can be simply setup on any LAMP environment. Its user interface is somehow similar to existing microblogging applications, but when writing posts, a parser interprets the content entered by the user (using JQuery and a set of regular expressions) in order to identify if a know pattern has been used, such as `#tag`, `@user` or `L:loc` (generally used for geolocation purposes). In addition, each pattern is mapped to a set of wrappers that query existing services in real-time to suggest relevant URI(s) for the resource they may refer to.

Default wrappers for `#tag` include Sindice (the Semantic Web index[11]) or DBpedia, whilst also letting people write their own wrappers. The mappings are then modelled using MOAT and exposed in the microblog posts as RDFa, so that they can be used for querying as we shall see next. Regarding `@replies`, we have mainly relied on Twitter and on other SMOB hubs that the user interacts with so that we can link to a user's FOAF profile when responding to a message. Finally, for the `L:location` patterns, we use the GeoNames service[12] that provide URIs for more than six million geographic entities. Interestingly, not only cities or countries are referenced, but also various places such as hotels, which offers an interesting level of granularity for representing this information.

---

[11] `http://sindice.com`
[12] `http://geonames.org/export`

From the suggested URIs, users can then decide which one is the most relevant (Fig. 5), that choice being saved for further reuse. As soon as the post is saved, informations are stored in the local RDF store and can be browsed in the user interface using RDFa. Furthermore, the mappings are saved for next posts and can be shared between hubs so that one's mapping can be provided to followers to enhance the interlinking process. While this process of manual interlinking from #tags may sound complex, we recently demonstrated the usefulness of MOAT in a corporate context, showing how users could benefit from the system to improve information retrieval [16] and showed that users are willing to do the additional effort of assigning resources to their tags.

As new wrappers can be created, it may be useful to build these in enterprise contexts if people want to refer to their own knowledge bases. For example, a tag such as #p:1453 could automatically be linked to the URI of the corresponding project, identified by its ID (1453). It therefore provides a use-case for enhancing microblogging in Enterprise 2.0 environments [15], following some of our previous work on Semantic Enterprise 2.0 [17].

SMOB can also be used as a Twitter client, relying on the Twitter API to do so. In that case, not only SMOB content is cross-posted to Twitter, but Twitter data is integrated in SMOB hubs and translated into RDF using the previous ontologies. Thus, Twitter data can be queried in a similar way to native SMOB data (using SPARQL) providing another way to make existing microblogging data enters the Linking Open Data cloud.

4.2   Geolocation Mash-Ups

In order to let users define geolocation information, SMOB enables a deep integration with GeoNames[13]. In addition to the webservice used to map the L:loc patterns, SMOB provides an autocompletion field so to users can define their current location. The auto-completion is based on the GeoNames webservice, and its JSON answers are interpreted on runtime to fill the location textbox. The main interest of GeoNames in our context is that each location has its own URI and has a description available in RDF. Thus, each time a new post featuring location information is created (or posted to a hub), the GeoNames URI corresponding to the current location (linked to the post with OPO, see previous example) is dereferenced and the data is integrated in the SMOB hub. We therefore benefit from any related information, such as the location of the feature, its inhabitants, its parent zone, etc. Consequently, posts can be geolocated in real-time, as seen in Fig. 6, and new features can be provided from the querying side, for instance identifying all content posted in a given country, while the only information available in the annotation refers to a city.

4.3   Data Discovery and Querying with SPARQL

In addition, SMOB hubs can be queried using SPARQL, either directly (via their local endpoint) or distributively, as each hub pings Sindice when new content

---

[13] http://geonames.org

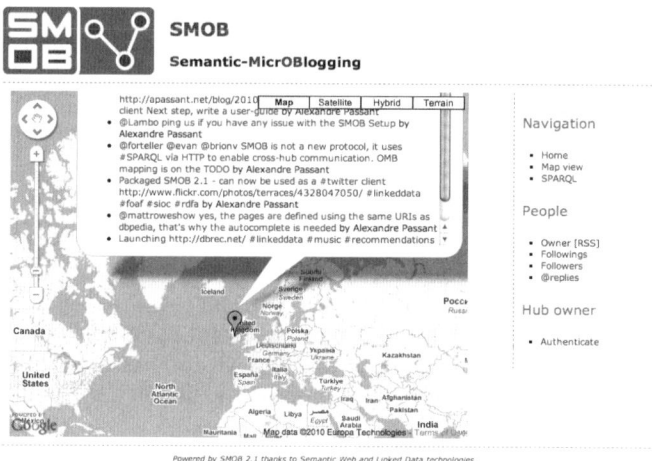

Fig. 6. Real-time geolocation of microblog posts with SMOB and GeoNames resources

is created. Then, using a library such as the Semantic Web Client Library [9], which uses an approach to enable SPARQL query over the Web by traversing RDF graphs, one can discover hubs and microblog content that are distributed on the Web.

```
SELECT DISTINCT ?post ?author
WHERE {
    ?post a sioct:MicroblogPost ;
        sioc:has_creator ?author ;
        moat:taggedWith [ ?related dbpedia:Italy ] .
}
```

Fig. 7. Example of advanced SPARQL query retrieving SMOB data

As an example, the SPARQL query in Fig. 7 (prefixes omitted) identifies posts about Italy and their author, even if the tag `#italy` was not initially used in the post, but by beneficing of (1) the interlinking from the original post to the URI identifying a given location (*e.g.* `dbpedia:Milan`) and (2) the existing links between `dbpedia:Italy` and this location in DBpedia.

Additionally, it can provides means to subscribe to topic-based or geolocation-based feeds, without having to rely on various subscriptions feeds. Moreover, if the links in DBpedia are enriched, the feeds will be automatically updated with new resources.

5   Related Work

Since the first release of SMOB mid-2008, the first semantic microblogging platform developed, various related projects have emerged. Microblogging platforms generating semantic data (*i.e.* represented using RDF(S)/OWL) include smesher[14] and StatusNet[15] (formerly Laconica). smesher is a semantic microblogging client with local storage, that integrates with Twitter and Identi.ca (another popular microblogging website, powered by StatusNet). StatusNet publishes both FOAF (describing people) and SIOC data (as SIOC-augmented RSS feeds for users and groups), and allows users to create friend connections across installations. It also uses the OpenMicroBlogging (OMB) protocol[16] for client-server communication (currently redesigned as OStatus). However, these applications do not provide interlinking with the Linked Data cloud, focusing only on representing the containers or structures using semantics, but not on linking its content to existing resources, hence leaving microblogging isolated from the ongoing LOD initiative. In addition, the semantics exposed in StatusNet are relatively light and do not take into account particular features of microblogging (such as modelling @user message recipients).

Moreover, in order to extend current microblogging systems, various syntax extensions have been proposed, including MicroTurtle[17], microsyntax[18], nanoformats[19], Twitterformats[20] and TwitLogic [19]. These syntaxes however generally require specific ways of formatting microblog posts, that may be not be widely used and are restricted to particular niches of users.

In addition, one can also implement semantic capabilities on top of existing systems. SemanticTweet[21] provides exports of user profiles from Twitter using FOAF, and the Chisimba Twitterizer[22] provides microblog data using open formats, but the original content is still stored in a closed system. We also recently designed a system translating Twitter streams to RDF data (including some of the vocabularies used in the SMOB ontologies stack) in real-time[23] so that it can be used in Semantic Web applications using streamed SPARQL extensions such as C-SPARQL [1] However, many of these applications and syntaxes rely on the Twitter infrastructure, and do not offer the distributed and open architecture that SMOB provides.

6   Conclusion

In this paper, we detailed the architecture of SMOB, a framework for an open, distributed and semantic microblogging experience. In particular, we focused on the issues of existing microblogging systems and detailed how we designed and used (i) a set of ontologies combined together to represent metadata related to

---

[14] http://smesher.org/

[15] http://status.net/

[16] http://openmicroblogging.org/

[17] http://buzzword.org.uk/2009/microturtle/

[18] http://www.microsyntax.org/

[19] http://microformats.org/wiki/microblogging-nanoformats

[20] http://twitterformats.org/

[21] http://semantictweet.com

[22] http://trac.uwc.ac.za/trac/chisimba/browser/modules/trunk/twitterizer

[23] http://realtimesemanticweb.org

microblogs and microblog posts, associated with RDFa markup to model microblog data, (ii) a distributed architecture to make microblogging more open and let users claim their data, combined with a synchronisation protocol that can be reused in other distributed Social Semantic Web applications, and (iii) interlinking components to make microblogging — and more generally the Social Semantic Web — part of the ongoing Linking Open Data cloud and offer new querying and mash-up capabilities.

In future work, we may consider how to not only link to existing content, but to extract new information from microblog updates, following practices generally used in Semantic Wikis, and enabling in this case streamed statements creation and notification that could provide new possibilities in terms of real-time information monitoring.

Acknowledgements

The work presented in this paper has been funded in part by Science Foundation Ireland under Grant No. SFI/08/CE/I1380 (Líon 2). We would like to thanks Uldis Bojārs and Tuukka Hastrup for their involvement in SIOC and in the first version of SMOB as well as Milan Stankovic and Philippe Laublet for their work on OPO.

References

1. Barbieri, D.F., Braga, D., Ceri, S., Valle, E.D., Grossniklaus, M.: Continuous Queries and Real-time Analysis of Social Semantic Data with C-SPARQL. In: Proceedings of the Second Social Data on the Web Workshop (SDoW 2009). CEUR Workshop Proceedings, vol. 520, CEUR-WS.org (2009)
2. Berners-Lee, T.: Linked Data. Design Issues for the World Wide Web, World Wide Web Consortium (2006), http://www.w3.org/DesignIssues/LinkedData.html
3. Bizer, C., Heath, T., Berners-Lee, T.: Linked Data — The Story So Far. International Journal on Semantic Web and Information Systems (IJSWIS) 5(3), 1–22 (2009)
4. Breslin, J.G., Harth, A., Bojārs, U., Decker, S.: Towards Semantically-Interlinked Online Communities. In: Gómez-Pérez, A., Euzenat, J. (eds.) ESWC 2005. LNCS, vol. 3532, pp. 500–514. Springer, Heidelberg (2005)
5. Breslin, J.G., Passant, A., Decker, S.: The Social Semantic Web. Springer, Heidelberg (2009)
6. Brickley, D., Miller, L.: FOAF Vocabulary Specification. Namespace Document, FOAF Project (September 2, 2004), http://xmlns.com/foaf/0.1/
7. Golder, S., Huberman, B.A.: Usage patterns of collaborative tagging systems. Journal of Information Science 32(2), 198–208 (2006)
8. Gómez-Prez, A., Corcho, O.: Ontology languages for the Semantic Web. IEEE Intelligent Systems 17(1), 54–60 (2002)
9. Hartig, O., Bizer, C., Freytag, J.-C.: Executing SPARQL queries over the web of linked data. In: Bernstein, A., Karger, D.R., Heath, T., Feigenbaum, L., Maynard, D., Motta, E., Thirunarayan, K. (eds.) ISWC 2009. LNCS, vol. 5823, pp. 293–309. Springer, Heidelberg (2009)

10. Huberman, B.A., Romero, D.M., Wu, F.: Social networks that matter: Twitter under the microscope. Computing Research Repository (CoRR), abs/0812.1045 (2008)
11. Java, A., Song, X., Finin, T., Tseng, B.: Why We Twitter: Understanding Microblogging Usage and Communities. In: Proceedings of the Joint 9th WEBKDD and 1st SNA-KDD Workshop 2007 (2007)
12. Jentzsch, A., Zhao, J., Hassanzadeh, O., Cheung, K.-H., Samwald, M., Andersson, B.: Linking Open Drug Data. In: Linking Open Data Triplification Challenge 2009 (2009)
13. Knorr-Cetina, K.D.: Knorr-Cetina. Sociality with objects: Social relations in post-social knowledge societies. Theory, Culture and Society 14(4), 1–30 (1997)
14. Mathes, A.: Folksonomies: Cooperative Classification and Communication Through Shared Metadata (2004)
15. Mcafee, A.P.: Enterprise 2.0: The Dawn of Emergent Collaboration. MIT Sloan Management Review 47(3), 21–28 (2006)
16. Passant, A., Laublet, P., Breslin, J.G., Decker, S.: A URI is Worth a Thousand Tags: From Tagging to Linked Data with MOAT. International Journal on Semantic Web and Information Systems (IJSWIS) 5(3), 71–94 (2009)
17. Passant, A., Laublet, P., Breslin, J.G., Decker, S.: SemSLATES: Improving Enterprise 2.0 Information Systems using Semantic Web Technologies. In: The 5th International Conference on Collaborative Computing: Networking, Applications and Worksharing. IEEE Computer Society Press, Los Alamitos (2009)
18. Sheth, A.: Citizen Sensing, Social Signals, and Enriching Human Experience. IEEE Internet Computing 13(14), 80–85 (2009)
19. Shinavier, J.: Real-time SemanticWeb in = 140 chars. In: Proceedings of the WWW 2010 Workshop Linked Data on the Web (LDOW 2010), CEUR Workshop Proceedings. CEUR-WS.org (2010)
20. Stankovic, M.: Modeling Online Presence. In: Proceedings of the First Social Data on the Web Workshop. CEUR Workshop Proceedings, vol. 405. CEUR-WS.org (2008)
21. Story, H., Harbulot, B., Jacobi, I., Jones, M.: FOAF+TLS: RESTful Authentication for the Social Web. In: First International Workshop on Trust and Privacy on the Social and the Semantic Web (SPOT 2009) (June 2009)

A Web-Based Collaborative Metamodeling Environment with Secure Remote Model Access*

Matthias Farwick[1], Berthold Agreiter[1], Jules White[2], Simon Forster[1],
Norbert Lanzanasto[1], and Ruth Breu[1]

[1] Institute of Computer Science University of Innsbruck, Austria
{matthias.farwick,berthold.agreiter,simon.forster,
norbert.lanzanasto,ruth.breu}@uibk.ac.at
[2] Electrical Engineering and Computer Science Vanderbilt University,
Nashville, TN, USA
jules@dre.vanderbilt.edu

Abstract. This contribution presents GEMSjax – a web-based meta-modeling tool for the collaborative development of domain specific languages. By making use of modern Web 2.0 technologies like Ajax and REST services, the tool allows for simultaneous web browser-based creation/editing of metamodels and model instances, as well as secure remote model access via REST, which enables remote model modification over a simple HTTP-based interface. This paper describes the complex technical challenges we faced and solutions we produced to provide browser-based synchronous model editing. It further explains on the XACML-based access control mechanisms to provide secure remote access to models and model elements. Additionally, we highlight the usefulness of our approach by describing its application in a realistic usage scenario.

1 Introduction

Nowadays, there exist a multitude of different modeling tools for a variety of purposes. The wide adoption of model-driven techniques further stimulates the creation of domain specific modeling languages. Furthermore, in today's global-ized economy engineering teams are often geographically dispersed to cut down costs, bring together expertise, or to explore new markets. However, to allow such teams to collaborate in an efficient manner, specific tools are needed which support the collaborative way of working. Modeling is a well-established task in software engineering and enterprise architecture today. In this context, modeling tools are used, for example, to communicate software architecture or to model the IT-landscape of large organizations. Contemporary research studies have shown evidence that complex projects conducted by such virtual teams are less

---

* This work was partially supported by the Austrian Federal Ministry of Economy as part of the Laura-Bassi – Living Models for Open Systems – project FFG 822740/QE LaB.

B. Benatallah et al. (Eds.): ICWE 2010, LNCS 6189, pp. 278–291, 2010.

successful than geographically concentrated teams [5]. The difficulty of sharing knowledge in dispersed teamwork has been identified as one of the main reasons for such failures [4,3]. Therefore, collaborative tools have to be created that aid the experts in sharing knowledge, and working together on solutions. Another fact which is mostly not respected by current modeling tools are concepts that allow feeding back information from the real world into models, thus keeping the model (partly) in–sync with the real world. To cope with the two requirements of collaborative modeling and remote model access, we present GEMSjax[1], a web-based metamodeling tool leveraging the full technological capabilities of Web 2.0. GEMSjax has been built from the ground up having collaborative model-ing and bidirectional information flow in mind, e.g. by supporting concurrent browser-based modeling and collaborative chat.

The tool allows for developing metamodels from which web-based graphical editors for the corresponding domain specific language (DSL) are generated. These generated editors run in the same browser-based graphical environment, hence there is no need for users to install software for participating in the mod-eling process. Many of the features, such as look-and-feel customization via stylesheets, dynamic styles reflecting attribute changes, as well as the remote model access, are inspired by the Generic Eclipse Modeling System (GEMS [13]). Apart from the metamodeling capabilities, the tool employs a novel approach to remote model access by providing a REST-based (REpresentational State Trans-fer [6]) remote interface for model elements. The interface can be used to easily integrate GEMSjax models into other applications and to provide live updates of model elements by remote applications. These updates can be used to keep the model partly in–sync with what it represents in the real world. For example, in a model of a server landscape, the remote interface can be used to signal when a new server is added to the system or when a server is down. A malfunctioning server could be, for example, drawn in red, or, in case of the addition of a new server, a new model element can be added to the model remotely.

We faced the following implementational and conceptual challenges because of the web-based nature of our solution: i) browser-based manipulation of mod-els, ii) synchronization of simultaneous model changes by different clients, iii) access control management of the remote interfaces. To provide the browser-based manipulation of models we created a client-side representation for them. In order to tackle the synchronization problem, we made use of a bidirectional push protocol. In order to cater for adequate security protection of the interfaces to models, we developed a sophisticated access control architecture based on the eXtensible Access Control Markup Language (XACML [9]). This contribution focuses on our solutions to the aforementioned challenges.

The remainder of this paper is structured as follows. The next section mo-tivates the need for GEMSjax by giving a short introduction on metamodeling and summarizing the requirements. Section 3 outlines a typical usage scenario that requires the features of GEMSjax. In Section 4 we provide a high-level

---

[1] GEMSjax is a short form of GEMS(A)jax, where GEMS is the Eclipse-based tool whose features GEMSjax brings to the web via Ajax (see Section 4).

overview of the challenges we faced and solutions we created to provide browser-based graphical model editing. After that, Section 5 focuses on the secure REST model interface and the XACML-based architecture to enforce access control on models and model elements. The final section concludes the paper by referring to related tools and pointing out directions of future work.

2   Motivation

To justify the need for the tool presented here, we first cover background in metamodeling as this is the main foundation this work builds on. After that, we identify the requirements for such at tool, to prepare the reader for the expected solution realized by GEMSjax.

2.1   Metamodeling

Today, software engineers have several possibilities in order to create models for a given domain. The predominant choice is still the use of General Purpose Modeling Languages (GPML) like the Unified Modeling Language (UML)[2]. The meta-model of such languages predefines their syntax and partially their semantics. In order to customize, e.g. UML, one needs to utilize the stereotyping mechanism, so that customized models remain compatible to their base language. With this mechanism one can apply new semantics to meta-elements. However, the underlying syntax remains static and needs to be manually adjusted via the use of the Object Constraint Language (OCL). Restricting the metamodel in such a way is very cumbersome and error-prone, leading to inconsistent models. Another alternative is creation of a Domain Specific Modeling Language (DSML). Opposed to a GPML, a DSML is specifically created to model selected domain aspects. Here, a metamodel is created from a generic meta-metamodel that specifies the syntax and semantics of the desired language, thereby only describing what is allowed in the language, and implicitly prohibiting everything that is not specified. With UML, on the other hand, everything that is not explicitly prohibited is allowed to be modeled. The comparison of the two above-mentioned modeling processes is shown in Figure 1.

Therefore, UML is more suitable for application areas that are closely related to the core competence of UML, like Software Engineering, where the syntax and semantics only need to be slightly adapted. DSMLs however, allow for a precise language definition *without* the need to restrict a large metamodel. Furthermore, by using well-designed DSMLs, the modeling process can be considerably speeded up compared to GPMLs like UML [8,7].

2.2   Requirements Analysis

Before we start with the description of the tool and the problems we faced, we first identify its requirements. As many design decisions are motivated by the

---

[2] http://www.uml.org/

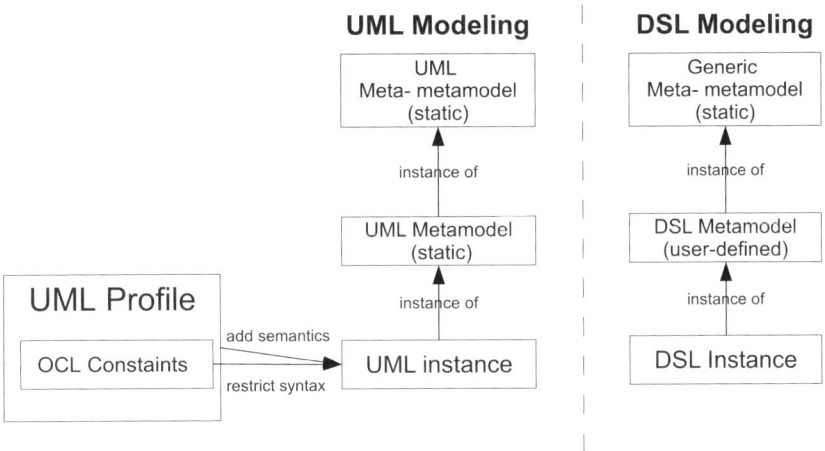

Fig. 1. UML Profiling vs. Domain Specific Modeling

requirements, this section clarifies why specific approaches have been selected. For a quick overview, Table 1 shows all requirements in a condensed representation.

Today's economy becomes more and more globalized, e.g. companies have subsidiaries distributed all over the world. This leads to the need for synchronization and effective collaboration over distance. We want to provide a tool supporting this collaboration. Collaboration should be as easy as possible for the partners, i.e. instant availability without the need for installing software packets (*R1*). Furthermore, it is likely that due to the geographical dispersion, heterogeneous environments will be encountered. Because of this, our tool should be platform-independent wherever applicable (*R2*). With GEMSjax we want to provide a tool that helps domain experts to express their knowledge as directly as possible. Domain-specificity is aimed at targeting exactly the questions of the domain, without blurring the important information by side information. For this reason, a further requirement to GEMSjax is the ability to create domain specific languages via metamodeling (*R3*). As already mentioned, an additional requirement to our tool is the possibility to support a collaborative way of modeling. This collaboration should allow several domain experts to view and edit models simultaneously (*R4*).

A known problem of models is that they are often created in the design phase only, and not maintained anymore at later stages. However, if the information captured in the models changes later on, the model and the real world are out-of-sync. This is certainly not desired, and should be avoided to derive greater benefit from models, cf. [2]. Such updates are not necessarily executed by humans, e.g. an information system which has just been booted up can update its status in the model autonomously. For this reason, GEMSjax aims to provide a way for updating models remotely, even without using a graphical modeling tool (*R5*). Naturally, this interface needs to be protected against unauthorized usage.

Table 1. GEMSjax requirements

R1	Instant availability for users, without installing large packages.
R2	Platform independence.
R3	Ability to create domain specific languages.
R4	Simultaneous model access.
R5	Secure remote model API for querying/manipulating models.

3   Usage Scenario

The IT-landscape of the large global enterprise Example Corp. is distributed over several continents, technologically heterogeneous, and poorly documented, since it evolved over several decades and was changed by many different individuals. Additionally, it is often not clear to the management of Example Corp. how each (technical) component (e.g., servers, information systems, etc.) in the IT-landscape contributes to the core business of the enterprise. Therefore, the CIO decides to start an Enterprise Architecture Management (EAM) effort, in order to document and model the global IT-landscape, standardise used technology (e.g. only Apache Tomcat version 6 should be in use), plan infrastructure change, and to analyze how each component contributes to the business goals of the company. It is also decided that in the long run, the IT-landscape model should be coupled with the actual run-time infrastructure in order to always have an up-to-date view of the infrastructure.

The first step in the EAM initiative of Example Corp. is the *metamodel defini-tion* (R3) of the enterprise's IT-landscape and the business functionalities (e.g. selling hotel bookings, selling cars). *Due to the distributed nature of Example Corp. not all stakeholders of the company's IT can gather for a physical meeting to discuss the metamodel. Therefore, a web-based meta-modeling solution like GEMSjax is chosen (R4).* The IT-responsibles at each data center meet in the virtual modeling environment, where they can collaboratively create the enterprise metamodel, and communicate via a chat to discuss ideas. Furthermore, the dynamic graphical appearance of each model element is defined. For example, a server, whose workload reaches a certain threshold dynamically appears in red.

After several weeks of discussion, an agreement on the metamodel is found that is capable of expressing all necessary IT and business assets of the company. An instance of this metamodel is created in GEMSjax and the stakeholders at each site insert their IT-infrastructure in the global model. Where interfaces between two data centers exist, these model elements are also collaboratively created in GEMSjax.

Finally, a satisfying representation of the enterprise IT-landscape is created. However, this representation is hard to keep up to date without unjustifiable manual labor. *In order to reduce this workload, some of the infrastructure devices are equipped with agents that communicate their existence and state to the GEMSjax model via its REST interface (R5).* The high security standards of Example Corp. are met by the XACML-based access control engine of GEMSjax.

4   The GEMSjax Metamodeling Tool

To enable distributed, collaborative (meta-)modeling, needed for the usage scenario described before, we developed GEMSjax, a browser-based graphical model editor. The web-based nature of the tool allows for instant availability and platform independence, as everything needed on the client-side is a web browser with Javascript capabilities (R1 & R2). GEMSjax provides a graphical modeling view for Eclipse Modeling Framework (EMF)[3] models. EMF is an Eclipse project that provides the means to work on top of a structured data model, include code generation and manipulation interfaces. GEMSjax is written in Java and the client-side browser editor is compiled from Java into Javascript and HTML using the Google Web Toolkit (GWT)[4]. The server-side of GEMSjax is built on Java servlets. Figure 2 shows its web interface with the classical modeling tool setup. The left side shows the metamodels and model instances specific to the logged-in user. The modeling canvas is located in the center and has nested tabs for metamodels and different views for them. Model elements can be dragged from the palette on the right onto the canvas. Each model instance has its own chat where users, which have the right to view or edit the model, can post messages. Attributes of a selected model element can be viewed and edited in the center-bottom panel. Table 2 summarizes the key features of GEMSjax to give the reader a quick overview.

The remainder of this section provides a high-level overview of the challenges we faced and solutions we created to provide browser-based graphical model editing.

4.1   Client-Side Manipulation of EMF Models

One of the key challenges of developing GEMSjax was establishing a method for building a client-side in-memory representation of a server-side EMF model. EMF models, themselves, cannot be transported to the client via HTTP and loaded into memory as Javascript (a future alternative might be the recent Eclipse project proposal JS4EMF[5]). To address this challenge, we created a generic Javascript object graph for representing EMF models in memory on the browser. The client-side memory representation manages structural constraints in the model, such as allowed parent/child relationships or valid associations between model elements. The reason for checking such constraints on the client-side instead of on the server is to avoid one roundtrip. This enhances the user-experience because immediate feedback is provided without the need to wait for a server response on each action. If, in the future, complex constraints are introduced, e.g via OCL, these should be evaluated on the server-side to ensure model consistency. Our solution allows for leveraging server-side computation power and reusing existing constraint libraries which are currently non-existent for client-side Javascript.

---

[3] http://www.eclipse.org/modeling/emf/
[4] http://code.google.com/webtoolkit/
[5] http://www.eclipse.org/proposals/js4emf/

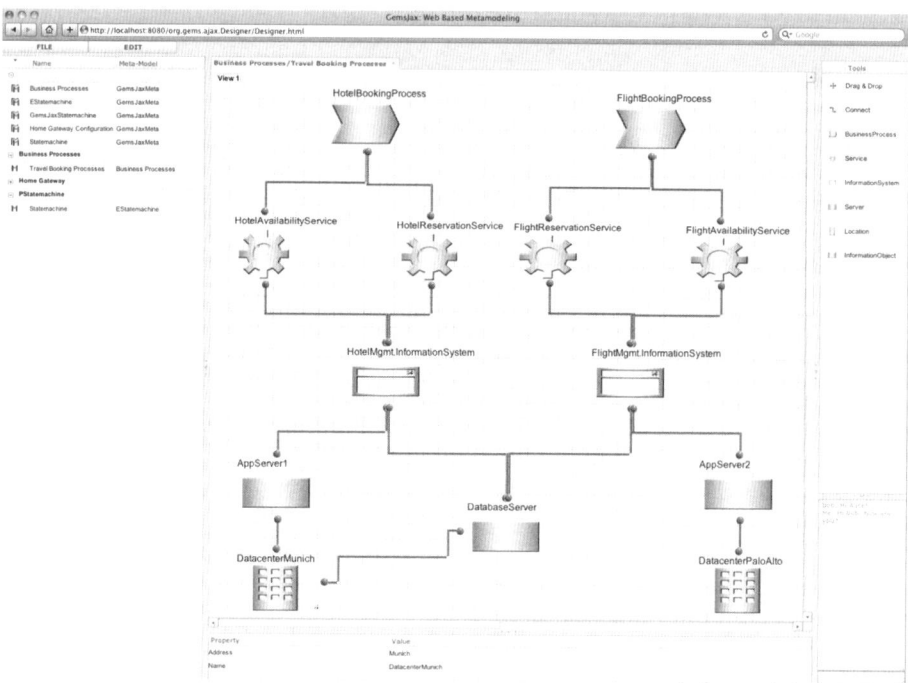

Fig. 2. Screenshot of the GEMSjax web-interface

When a model is first opened, a Java servlet on the server reads in the EMF model from disk and translates the model into an equivalent JavaScript Object Notation (JSON) representation that is sent to the client to be loaded into memory using our generic Javascript object graph framework. The server-side also associates a unique ID with each EMF and client-side Javascript object in order to provide a mapping from EMF objects to Javascript objects and vice-versa. The structural rules from the Ecore metamodel of the EMF model are also extracted into a JSON representation and transported to the client-side to enforce basic structural constraints, such as allowed containment relationships.

4.2   Bidirectional Client/Server Model Synchronization

For implementing the simultaneous model access (R4), a bidirectional client/server synchronization scheme is used. HTTP is designed for a request/response style communication between a client and remote server. A key challenge we faced was determining how to *push* changes triggered in the server-side EMF model to the client-side Javascript representation. In situations where a model edit originated from the client-side, any updates, such as triggered model transformations, performed on the server can be returned in the HTTP response

Table 2. GEMSjax key features

Key Feature	Description
Metamodeling	Typical metamodeling possibilities: (abstract-) classes, attributes, connections, inheritance
Model Instances	Instantiate metamodel in same web interface
Customization	Instance customization via stylesheets
Dynamic styles	Dynamic styles of model elements depending on attribute values
Remote Interface	Model modification via REST interface
Access Control	Role Based Access Control of graphical and REST interface
Collaborative Modeling	Simultaneous modeling
Chat	One chat instance per model instance
Export	Exports EMF models, e.g. for model transformation

to the client. However, changes in the model that originate in the server *outside* of an HTTP response to a client request, such as time-based triggers, incoming REST API calls, or edits from modeling collaborators, cannot be pushed to the browser using standard HTTP approaches.

To address this limitation of standard HTTP request/response architectures, GEMSjax uses the bidirectional HTTP push protocol Cometd[6] to allow updates from the server to be delivered to the client browser. GEMSjax uses an event bus built on top of the Cometd protocol to communicate changes between clients and the server. Synchronization between clients editing the same model is maintained by broadcasting model edits on the event bus to each connected client. Each client compares the state change in incoming events to the model to determine if the event represents a duplicate change. A priority scheme is used to reconcile and rollback conflicting changes. Also, information about which elements are currently selected by other modelers can be pushed to the other clients to avoid conflicting changes a priori. However, the effectiveness of our conflict avoiding/resolution approach needs further investigation.

The sequence diagram in Figure 3 shows an example of the bidirectional communication with two clients connected to the server. When a client connects to the server, it first gets the most current version of the model by calling the `getModelPackage(modelID: String)`-method. Note that all calls originating from clients are asynchronous calls, s.t. the behavior is non-blocking. For every client to receive updates that originate from a different source than itself, it makes a call to `receiveModelChange()`. This call opens a connection to the server and leaves it open until either an update has to be delivered to the client, or a timeout occurs. In our example, Client1 updates the model (step 7) and afterwards Client2 receives the updated version (step 9) as response of the call in step 6. After client2 receives the update it re-opens the connection to the server in the final step 10. Note that this technique avoids frequent polling. Furthermore, updates are delivered to all clients immediately after the server changes the status of a model.

---

[6] `http://cometdproject.dojotoolkit.org/`

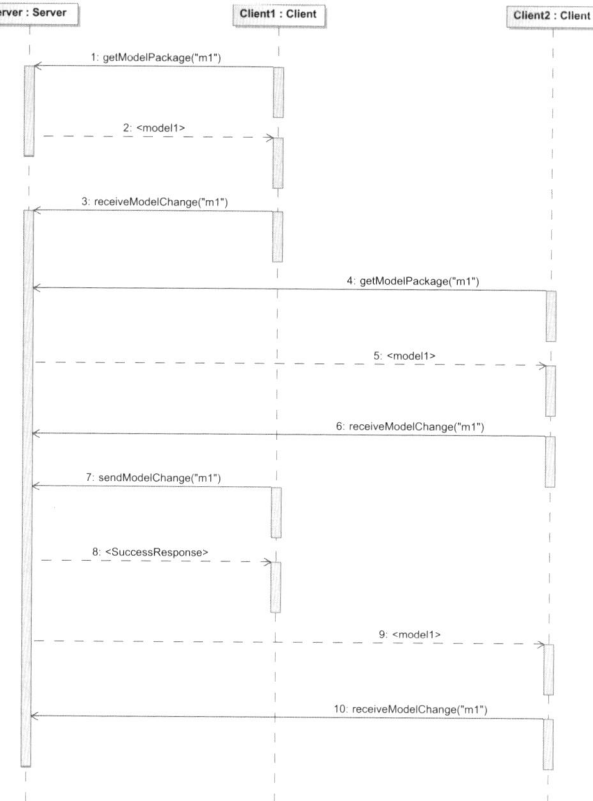

Fig. 3. Client-push protocol example with two clients connected to the server

4.3   (Meta-)modeling Lifecycle

The metamodel representations in GEMSjax are managed on the server-side. Figure 4 shows the lifecycle of the involved metamodels and their corresponding instantiations. On top of the figure is the GEMSjax metamodel which is predefined as an Ecore metamodel (1).

In the metamodeling step an EMF instance (2) of the GEMSjax Ecore meta-metamodel is created according to the requirements of the new DSML. This realizes our requirement R3. Each time the metamodel is saved on the client side (web–browser) the model is serialized to its EMF representation on the server. Once the metamodel is finished, it is transformed again to act as an Ecore metamodel (3) for new DSML model instances. This is achieved via JET templates[7] that transform the EMF representation of the metamodel into an Ecore metamodel. Finally, instances of that DSL-specific Ecore metamodel can be created with the client (4).

---

[7] `http://wiki.eclipse.org/M2T-JET`: JET is a template-based code generation framework, which is part of the Eclipe Model to Text (M2T) project.

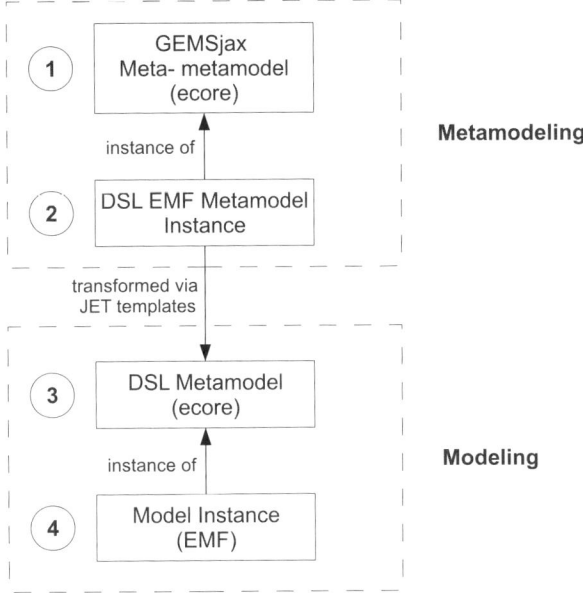

Fig. 4. Image showing the modeling lifecycle from the GEMSjax metamodel to a domain specific model instance

5   Secure Remote Model Access

As mentioned earlier, one of the key-features of GEMSjax is the easy-to-use REST interface of the models. Using this interface allows developers to integrate information of models in their own tools or update the models, because the HTTP protocol is available in practically any programming environment. This way, models can become actual configuration artifacts and represent the real state of a system.

5.1   The REST Interface

The REST API treats models as hierarchical resources that can be accessed and manipulated via the basic HTTP operations GET, PUT, POST and DELETE. It uses the hierarchical property of URIs to describe containment of model elements. The *GET* operation is used to retrieve information about model elements, *PUT* to create new elements, *POST* and *DELETE* to change and delete elements respectively. Table 3 exemplifies the usage of the simple interface calls.

By specifying return mime-types the user can choose the return type of the operation, e.g. XML or JSON. This allows for a flexible client implementation. Input values for the PUT and POST commands are transmitted via simple key-value pairs in the body of the messages. Responses conform to the HTTP specification, e.g. by returning *201 (Created)* for a PUT request containing the URI of a newly created model resource.

Table 3. Examples of REST HTTP Commands

Operation	HTTP Command and URL
Get list of available models	GET http://.../gemsjax/models
Get attributes of an element	GET http://.../gemsjax/modelID/.../elementID/attributes
Get all meta element names of model	GET http://.../gemsjax/modelID/meta
Create new element	PUT http://.../gemsjax/modelID/.../metaType
Update Element	POST http://.../gemsjax/modelID/.../elementID
Delete Element	DELETE http://.../gemsjax/modelID/.../elementID

5.2   Access Control Architecture

To constrain access to model elements we incorporate an *Access Control Layer* in our architecture (cf. Fig. 5). With each request a user needs to provide credentials in the form of a username/password combination over HTTPS. As mentioned before, there are two possibilities to access GEMSjax models – the *REST interface* and the graphical interface of the modeling tool (*GWT RPC interface*). Calls via the REST API are first translated to the actual action to be taken on the model by the *URI Mapper*. After that, requests are forwarded to the *Request Handler* which provides a model interface independent of the access method. These actions could immediately be executed on the model, however the request first has to pass the Access Control Layer.

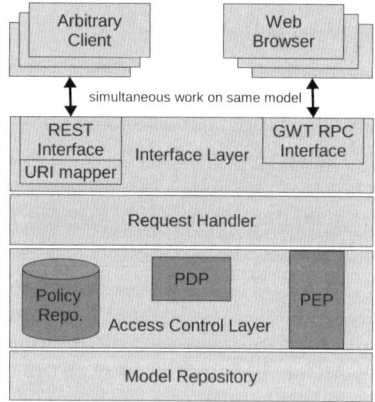

Fig. 5. Access Control Architecture

The access control layer is based on the XACML Target Architecture (cf. [9]), since it is a proven access control architecture with several open-source implementations. In an XACML architecture, requests are generally first intercepted by the *Policy Enforcement Point* (PEP). It forwards the request context (e.g. user credentials, requested action) to the *Policy Decision Point* (PDP). The PDP decides on the access rights according to policies stored in a policy repository. This decision is then enforced by the PEP. A typical example for an XACML

```
<Policy PolicyId="PolicyId123" RuleCombiningAlgId="permit-overrides">
 <Target><Subjects><AnySubject/></Subjects>
   <Resources><Resource>
     <ResourceMatch MatchId="isSubURI-or-equal">
      <AttributeValue DataType="anyURI">
       http://.../gemsjax/exampleModelID</AttributeValue>
     <ResourceAttributeDesignator DataType="anyURI" AttributeId="resource-uri"/>
    </ResourceMatch></Resource></Resources>
   <Actions>
     <ActionMatch MatchId="string-equal">
      <AttributeValue DataType="string">update</AttributeValue>
      <ActionAttributeDesignator DataType="string" AttributeId="action" />
     </ActionMatch>
   </Actions></Target>
 <Rule Effect="Permit" RuleId="updateRule">
  <Target />
  <Condition FunctionId="string-equal">
   <Apply FunctionId="string-one-and-only">
    <SubjectAttributeDesignator DataType="string" AttributeId="role" />
   </Apply>
   <AttributeValue DataType="string">admin</AttributeValue>
  </Condition>
 </Rule>
</Policy>
```

Fig. 6. An example XACML policy

policy in GEMSjax is depicted in Figure 6. First, a rule combining algorithm is set in the initial $< Policy >$ element. This algorithm governs the final result of the policy evaluation depending on the evaluation of $< Rule >$ elements. In this case a $PERMIT$ evaluation of a rule overrides the $DENY$ result of a previous rule. The following $< Target >$ element specifies to which subject, resource, and requested action the policy applies. Here, it applies to actions of the type update, on the model resource with ID exampleModelID. Finally, the $Rule$ element states that for the user with the role admin access should be permitted.

Depending on the PDP's decision, the Policy Enforcement Point will forward the action to the *Model Repository* or block access. The creator of a metamodel is in charge of granting permissions to registered users of the system. This entails that GEMSjax enforces a combination of Role-based Access Control (RBAC) and Discretionary Access Control (DAC)[8].

Note that the XACML policies can be very fine grained down to the attribute level, and are very flexible, e.g. to allow another group of users to update the model it suffices to add another rule element to the policy. Also policies can be used for complex requirements like delegation of rights, if, e.g., a user wants to invite an expert that only has read access to the model for consulting purposes. The REST interface to GEMSjax and the access control architecture described here, form the realization of requirement R5.

6   Related Work and Conclusion

In this work we presented GEMSjax a web-based metamodeling tool for the collaborative development of domain specific languages. We highlighted its

---

[8] In Discretionary Access Control the owner of an object has the right to delegate permissions on an object to other users.

usefulness in a usage scenario and described the technological challenges and our solutions for implementing a metamodeling tool in a web enabled manner. We also presented a novel approach to secure (remote) model access, based on REST and XACML.

Related literature describes several collaborative modeling approaches, none of which combines all three key features of GEMSjax: the secure REST model interface, web-based DSL definition, and simultaneous collaborative modeling.

In his work on COMA [10], Rittgen describes a collaborative modeling tool for UML models, that provides a voting mechanism to achieve consensus on a model. Opposed to our work this tool does not provide the means to create domain specific languages and has no means to enforce access control on models. The commercial modeling tool MagicDraw Teamwork Server [1] allows for collaborative modeling by providing a locking mechanism for models but does not provide a REST interface and strictly operates on UML models. Many of the features of GEMSjax are inspired by GEMS [13], however GEMS is not browser-based, does not support collaborative modeling, and does not provide security for its RPC interface. The Eclipse-based Graphical Modeling Framework (GMF)[9] also allows for graphical metamodeling, but is relatively more complicated to configure and does not provide web-based and collaborative modeling. The research prototype SLIM, presented in [12], uses similar technology to GEMSjax and also aims for a lightweight collaborative modeling environment. However, it is only able to represent UML class diagrams, which contradicts to our requirement to create domain specific languages. Moreover, SLIM does not offer a remote model API which eliminates the usage of the tool where updates should not be made by human users only.

Our future work in this area will include further development of the tool as well as experiments. Specifically, we will work on an Eclipse integration, that will enable to use all Eclipse features while editing GEMSjax models in the Eclipse browser tab. Collaborative modeling environments expose a number of further issues to solve, like locking or versioning. Future work will also comprise the integration of existing solutions for such problems (e.g. [11]). We will investigate on the integration of model voting, concurrency issues and code generation techniques. We also plan a collaborative modeling experiment with dispersed teams of students in the US and Austria to get further insight on the effectiveness of our collaborative modeling approach.

References

1. Blu Age: MagicDraw TeamWork Server (2009),
   http://www.bluage.com/?cID=magicdraw_teamwork_server
2. Breu, R.: Ten principles for living models - a manifesto of change-driven software engineering. In: 4th International Conference on Complex, Intelligent and Software Intensive Systems, CISIS-2010 (2010)
3. Conchúir, E.O., Ågerfalk, P.J., Olsson, H.H., Fitzgerald, B.: Global software development: where are the benefits? Commun. ACM 52(8), 127–131 (2009)

---

[9] http://www.eclipse.org/modeling/gmf

4. Cramton, C.D.: The mutual knowledge problem and its consequences for dispersed collaboration. Organization Science 12(3), 346–371 (2001)
5. Cramton, C.D., Webber, S.S.: Relationships among geographic dispersion, team processes, and effectiveness in software development work teams. Journal of Business Research 58(6), 758–765 (2005), http://www.sciencedirect.com/science/article/B6V7S-4BM92C5-4/2/60cbbf7fa88eb389c6e745f355acca58
6. Fielding, R.T.: Architectural Styles and the Design of Network-based Software Architectures. Ph.D. thesis, University of California, Irvine, Irvine, California (2000)
7. Frank, U., Heise, D., Kattenstroth, H., Fergusona, D., Hadarb, E., Waschkec, M.: ITML: A Domain-Specific Modeling Language for Supporting Business Driven IT Management. In: DSM '09 (2009)
8. Luoma, J., Kelly, S., Tolvanen, J.: Defining Domain-Specific Modeling Languages: Collected Experiences. In: Proceedings of the 4th OOPSLA Workshop on Domain-Specific Modeling, DSM '04 (2004)
9. OASIS: eXtensible Access Control Markup Language (XACML) Version 2.03. OASIS Standard (February 2005)
10. Rittgen, P.: Coma: A tool for collaborative modeling. In: CAiSE Forum, pp. 61–64 (2008)
11. Schneider, C., Zündorf, A., Niere, J.: CoObRA-a small step for development tools to collaborative environments. In: Proc. of the Workshop on Directions in Software Engineering Environments (WoDiSEE), Edinburgh, Scotland, UK (2004)
12. Thum, C., Schwind, M., Schader, M.: SLIM – A Lightweight Environment for Synchronous Collaborative Modeling. In: Schürr, A., Selic, B. (eds.) MODELS 2009. LNCS, vol. 5795, pp. 137–150. Springer, Heidelberg (2009)
13. White, J., Schmidt, D.C., Mulligan, S.: The Generic Eclipse Modeling System. In: Model-Driven Development Tool Implementer's Forum at the 45th International Conference on Objects, Models, Components and Patterns (June 2007)

Carbon: Domain-Independent Automatic Web Form Filling

Samur Araujo, Qi Gao, Erwin Leonardi, and Geert-Jan Houben

Delft University of Technology, P.O. Box 5031, 2600 GA Delft, The Netherlands
{s.f.cardosodearaujo,q.gao,e.leonardi,g.j.p.m.houben}@tudelft.nl

Abstract. Web forms are the main input mechanism for users to supply data to web applications. Users fill out forms in order to, for example, sign up to social network applications or do advanced searches in search-based web applications. This process is highly repetitive and can be optimized by reusing the user's data across web forms. In this paper, we present a novel framework for domain-independent automatic form filling. The main task is to automatically fill out a correct value for each field in a new form, based on web forms the user has previously filled. The key innovation of our approach is that we are able to extract relevant metadata from the previously filled forms, semantically enrich it, and use it for aligning fields between web forms.

Keywords: Auto-filling, auto-completion, concept mapping, web forms, semantic web.

1 Introduction

Current applications on the Web show a high degree of user interaction. A large amount of the data that users input into web applications, is supplied through web forms. This process of filling out forms is highly repetitive and can be optimized by intelligently reusing the user's data across web forms, basically following the observation that web forms from applications in a similar domain demand the same data from a user. As a simple example, most sign-up forms require the user's email and first name.

Recently, auto-filling and auto-completion emerged as techniques to assist the users in reusing their data for filling out web forms. **Auto-filling** is a mechanism for automatically filling out web forms. It exists as tools, in web browsers, and when the user visits a web page containing a form, auto-filling can be triggered by a simple mouse click. Google Toolbar Auto-fill [6] is one of the tools available, however it is the simplest form of auto-filling, as it only works in sign-up forms that demand the user's personal data. Another tool is the Firefox Auto-fill Forms plug-in [1] that is also limited to the user's personal data. However, it allows the user to extend the pre-configured fields. Both approaches demand the user to fill out an application-proprietary form containing some basic fields (e.g., name, address, zip code, etc.) before they can be used to automatically fill out web forms. The Safari browser has an auto-fill feature that reuses data from previously filled out forms for automatically

B. Benatallah et al. (Eds.): ICWE 2010, LNCS 6189, pp. 292–306, 2010.

filling out different forms with this data. However, its matching mechanism is only based on the *string matching* of field names. **Auto-completion** is a feature provided by many applications for suggesting a word or phrase that the user wants to enter without the user actually entering it completely. Most of the browsers have native support for auto-completion. It stores values that have been filled before and based on this history it recommends values for a field that has been already visited. However, this mechanism demands user interaction for each field that needs to be filled out.

In this paper, we propose and validate a new concept-based approach for automatically filling out web forms re-using data from previously filled out forms.

Manipulating (the code behind) web forms is not a trivial task, due the high heterogeneity among them. Web forms have different shapes, different numbers of fields, different labels, different representation for values, and different purposes. To be able to derive proposed values for a new form from the knowledge obtained from the already filled out forms, a concept-based structure is used. This helps to represent the knowledge from the filled out forms at a conceptual level and exploit that conceptual level to reason about the proposal for values for fields from the new form. Figure 1 shows an example of the translation of a simple string representing a field name to a meaningful concept in some known vocabulary, for example, WordNet for forms in English. With these concepts we can thus construct a conceptual model with which it is then possible to connect (the concepts from) the target form and its fields to (the concepts from) the data gathered from the previously filled out forms. An essential step in that process is the mapping of 'target' concepts to 'source' concepts. Several options are possible here and we will choose in this paper a semantic-based approach to show the feasibility. Figure 2 shows a mapping between two form fields using a path of concept relations.

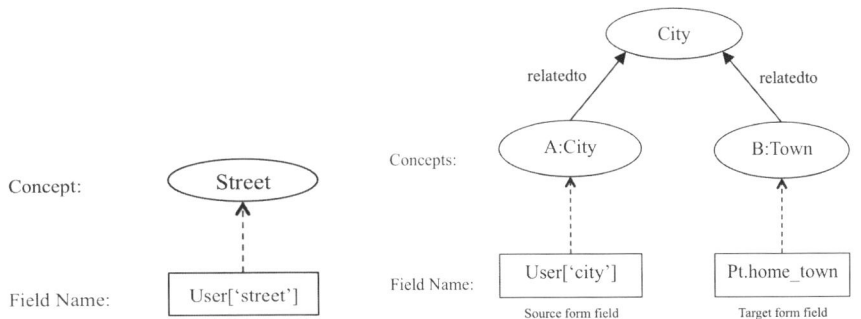

Fig. 1. Example Translation   **Fig. 2.** Example Mapping

The concept-based structure and mappings can thus be used to suggest values for a target form to be filled out automatically. In this respect, we see two main notions that exist in current techniques, auto-completion and auto-filling, which we mentioned before. The main difference between them is that *auto-completion* guesses a value from a fixed and defined set of values (e.g. rows in a database column) based on what the user types and only applies for text fields (not for select, checkbox or radio button fields), while *auto-filling* guesses a value based on the concept that the field represents and applies to all types of fields. Both techniques are instantiations of a *general* (meta) model for automatically filling out forms that discerns the following aspects:

- **Mapping technique:** Mapping can use string or semantic-based techniques.
- **Data source:** Proposals for values to be filled out can be made on knowledge gathered through forms that have been filled out previously, from a predefined list of values extracted from a database (e.g. list of cities), or it can be gathered by explicitly forcing the user to fill out an application-specific form, like in the case of Google Toolbar Autofill.
- **Field identifier:** For matching fields the name of the INPUT tag can be used or the label of the field (the latter is not always available), for example a field can have the label *"First Name"* and the name *"F_name"*.
- **User-intervention:** User interaction can be required as in the case of completion or it can be omitted in the fully automatic case.
- **Domain-dependence:** Approaches can be using domain-specific intelligence, like Firefox Autofill, or be domain-independent.

Existing *auto-filling* approaches are typically string-matching-based tools. They do not consider the concepts associated with the field label and/or field name. Consequently, they are not able to exploit available background knowledge. A simple example is the use of synonym relationships between concepts (e.g., city is the synonym of town).

In this paper, we present Carbon, a domain-independent automatic web form filling approach that exploits the semantic of concepts behind the web form fields. Carbon uses semantic techniques to accumulate and connect the data from previously filled out forms to the target form to be filled out. Carbon goes beyond the state of the art in *auto-filling* since it can be applied for auto-filling any kind of form in the English language.

This paper is organized as follows. After this motivation, we discuss the related work in Section 2. In Section 3, we present a deep look at web forms, with an evaluation of the distribution of field labels in web forms. Section 4 presents our approach for auto-filling forms. Section 5 elaborates on our Carbon implementation in details by describing the architecture and data model for the implementation. We then present an experimental validation of the approach, with a study of the performance of the approach in Section 6. Finally, Section 7 concludes this paper.

2   Related Work

Form Extraction. Extracting labels from web forms is a challenging and non-trivial problem. In [17], Raghavan et al. presented HiWE (Hidden Web Exposer) that used a *layout-based* information extraction technique to find candidate labels and a set of heuristic rules to compute and rank the distance between form elements and candidate labels. In [8], He et al. presented a technique for extracting labels from web forms by capturing the *textual layout* of elements and labels, and based on their position the relationships between elements and labels are determined. Recently, Nguyen et al.

presented LABELEX for extracting element labels from web form interfaces by using *learning classifiers* [14] and they introduced the idea of mapping reconciliation.

Auto-completion. Auto-completion is a feature found in web browsers, e-mail clients, text editors, query tools, etc. Based on the user's input, it *predicts* a word or phrase that a user wants to type before the user types the *whole* word or phrase. One of the early auto-completion facilities was command line completion in the Unix Shell. In [2], the authors introduced an auto-completion feature for full-text search using a new compact *indexing* data structure called HYB to improve the response time of this feature. In most *web browsers*, an auto-completion feature is available for completing URLs, suggesting search terms, and for auto-completing form fields. In the context of web form filling, the web browser typically reuses previously inputted form data for the prediction of values for fields to be filled out in a new form. Recently, Hyvönen et al. [9] generalized the idea of syntactic text auto-completion onto the *semantic* level by matching the input string to ontological concepts.

Auto-filling. In spite of the usefulness of auto-completion in helping users to fill out forms, auto-filling is a step further and often more suitable for this task because it does not require *explicit* user intervention. With an auto-completion tool, a user has to type at least one letter for each field (and thus helps the tool with the proposal of values), where with auto-filling the only action needed is to signal the system to start its process of finding suggestions; note that tools can allow the user to confirm or adjust the given proposal. A number of auto-filling tools [1,6,18] require the users *beforehand* to fill out a predefined "form" (outside the use of any form-based application) that will act as the source of data for later when the tools attempts to automatically fill out a target form. Note that the predefined form usually has a very small number of fields that are typically related to standard *personal* information [18]. Even though some of the tools have the capability of adding extra fields and defining rules [1,18], the users have to explicitly define them beforehand. This task is not trivial and is often perceived as cumbersome by many users. Another tool, called iMacros [10], *records* the interactions between a user and a web page when she fills out a form and then generates macros for these interactions. These macros are then later used to automatically fill out the *same* form. While this tool is useful if the user wants to fill out one kind of form with different values (e.g., in the process of searching for products using various keywords), it *cannot* be used to automatically fill out *different* forms. In [21], a framework called iForm was presented for automatically filling out web forms using data values that are implicitly available in a text document given by the user as input.

Syntactical Matching and Semantic Matching. A host of works [11,13,19,22] has addressed the problem of measuring *syntactical* similarity between strings, and in many approaches for predicting form values such work is used. They measured how similar two strings are by computing the minimum number of *operations* needed to transform one string into the other [13,19], by computing the numbers of matches and transpositions, or by indexing the strings by their pronunciations [11]. Semantic matching is a technique used to identify information that is *semantically* related. Typically, this is achieved through concept-based structures that represent semantic relations between concepts, and the matching process then tries to find connections between concepts based on such relations. The semantic relations are obtained from ontologies such as WordNet, DBpedia, etc. In [3,4,5,15] the problem of mapping and aligning ontologies was addressed. Tools exist [12,16] that are specifically built to measure semantic similarity between concepts. WordNet::Similarity [16] is a freely available software package that implements a variety of semantic similarity and relatedness measures based on information from the lexical database WordNet[1]. The DBpedia Relationship Finder [12] is a tool for exploring connections/relationships between objects in a Semantic Web knowledge base, specifically DBpedia[2].

3 Web Forms

Web forms are the main input mechanism for users to enter data for web sites and web applications. They occur in different shapes, different numbers of fields, different labels, different values representations (e.g., 'Brazil', 'BR' or 'BRA'), and different purposes (e.g., for signing up, searching or commenting). Forms in HTML are by far the most used ones on the web, but there also exist forms in others standards such as XFORMS[3] and Adobe Flash[4]. Our focus in this paper is on forms expressed in HTML (also XHTML).

In this section we consider the format of forms and report on a study into the nature of forms, reporting how the form elements are distributed and shared between forms.

3.1 Format of Data in HTML Forms

A web form is a set of HTML INPUT tags enclosed by a tag HTML FORM that when rendered in the browser, allows users to enter data in a HTML page. After a user has filled out a web form, it can be submitted to a server application for further processing. Each INPUT field in a form is associated with a name (e.g., "user['First_Name']"), a value (e.g., "John"), and a type (e.g., checkbox, textarea, text, select, radio button or button). Also, a form field can be associated to a human-readable label (e.g. "First name"). In spite of the fact that the HTML specification defines the tag LABEL to represent a human-readable label, it is not widely used by HTML programmers and most forms represent labels in an alternative way. However, even that is not always the case, as form fields can come without label, can have labels in a position hard to detect by machines, or can have meaningless names. In the last case, it is hard to process it for auto-filling, due to the fact that we cannot map an opaque field (with a meaningless name such as "$er32") using any matching strategy.

3.2 Nature of Web Forms

For the challenge of filling out web forms automatically, the first step is to "understand" how similar web forms can be. One way to measure this is by looking for the field labels, and how these labels (or concepts) are distributed and shared between forms. For this purpose, we conducted a study using a typical dataset on web forms.

---

[1] http:// wordnet.princeton.edu
[2] http://dbpedia.org/About
[3] http://www.w3.org/TR/2009/REC-xforms-20091020/
[4] http://www.adobe.com/products/flashplayer/

3.2.1 Experimental Setup

When we consider web forms in a specific domain, there are two characteristics that are worth further investigation. One phenomenon is that the same labels occur distributed in different forms. For example, in the book domain a label named "author" occurs in many different forms. Another characteristic is that different labels can be connected to a single concept. For example, both of the labels "postal code" and "zip code" can be associated with one concept representing the "postal code". Identifying and exploiting these two phenomena, we have the capability of a significant reduction from labels to concepts.

In an experiment to evaluate these characteristics mentioned above, we adopted the TEL-8 dataset [20], a manually collected dataset, which contains a set of web query interfaces of 447 web sources across 8 domains in the Web, including Airfares, Automobiles, Books, Car Rentals, Hotels, Jobs, Movies, and Music Records. We note that they are typical domains on the Web and that this dataset captures the structures of forms on the Web nowadays. For each domain, we gathered all labels from TEL-8 files and we grouped them into sets of distinct labels. Then we categorized these distinct labels by mapping them to concepts, manually. Afterwards, we investigated the distribution of labels and concepts in the forms for each domain and analyzed the reduction from labels to concepts.

3.2.2 Experimental Results

The results show the distribution of forms, (distinct) labels and concepts and also the average number of labels and concepts in the forms, for each domain (see Table 1). In Table 1 we see the number of forms per domain and the number of labels that are contained in all those forms. We see also how many of those labels are distinct labels. Further, we see how many concepts are associated with these distinct labels.

Table 1. Nature of Web Forms

	Airfares	Auto mobiles	Books	Car Rentals	Hotels	Jobs	Movies	Musics Records
forms	65	47	84	25	39	49	73	65
labels	332	505	402	189	269	284	455	421
distinct labels	168	241	175	126	197	221	287	183
concepts	132	164	116	110	149	162	213	134
reduction rate	60%	68%	71%	42%	45%	43%	53%	68%

We define a measure named "Reduction Rate" to represent the reduction from the labels to the concepts in the web forms due to the conceptual mapping. Given one specific domain, let N_c be the number of concepts and N_l be the number of labels. The reduction rate is defined as follows:

$$ReductionRate = \frac{N_l - N_c}{N_l}$$

The reduction rate ranges from 42% to 71% and is 55% in average. The higher this number is, the fewer concepts we need to cover all the fields in one domain. In Table 1, we can see that the reduction rate is composed of two parts. The first part of the reduction is from labels to distinct labels, due to the fact that labels are repeated among different forms. The second part is caused by the application of concept mapping that groups distinct labels into clusters of concepts that they represent. The intelligent exploitation of these two reduction steps is the main motivation for our approach.

4 Conceptual Framework for Form Auto-Filling

The main idea behind our approach for automatically filling out web forms is to extract data from previously filled out forms and propose them for the new form. Thus we exploit the observations we made in the study presented above. The use of data from previously filled out forms, increases, progressively, the variety of forms that can be automatically filled, since, once a user manually has filled out a form in a specific domain (e.g., social networks or hotels), this data is then available as a solid basis for automatically filling out other forms in the same domain.

This approach allows Carbon to go beyond of the state of the art of available auto-filling tools by exploiting the semantic overlap of knowledge contained in the previously filled out forms.

4.1 Extracting Concepts from Web Forms

We represent a form field by a conceptual structure containing the field name, the label, the type, the values, the URL of the page containing the form, the domain of the URL, the universal unique identifier of the form, and the update date. The process of auto-filling forms starts with the instantiation of this conceptual structure by extracting metadata from web forms that the user fills out. Afterwards this metadata is stored and enriched for further use. The next step is to instantiate a similar structure for a target form, and subsequently to map this instance to the previously obtained knowledge structure, to obtain suggestions for values for empty fields. The mapping between form representations can exploit any attribute of the knowledge structure, but in the implementation for the experiment described in this paper, we just exploited the attribute representing the form field's name.

As we typically do not have access to (the database behind) the web application, we choose to collect the form field metadata in real-time - i.e. after a user fills out the form and before submitting it. We extract this metadata from the DOM[5] (Document Object Model) tree of the HTML page that contains the form. The DOM is a programming API for documents that has a logical structure that represents a document as a tree of objects. By navigating through the HTML DOM tree we can access any web page element and their attributes, including form input fields. Carbon processes the DOM tree, extracting name, value, and type of the HTML INPUT tags. The Carbon version used in this experiment does not implement a label extraction strategy, however this can be easily plugged into the architecture.

The extraction intelligence applied for previously filled out forms and for an (empty) target form is logically similar, but obviously occurs at distinct moments.

---

[5] http://www.w3.org/DOM/

4.2 Mapping Web Form Concepts

The main idea behind Carbon is to connect fields at the conceptual level. For example, Carbon can connect the (different) concepts "city" and "town" using WordNet, a large lexical database for English, for bringing two words together that represent synonyms.

Carbon starts by mapping a field name string to atomic syntactic elements or lexical words in WordNet. Carbon splits the string by eliminating all non-letter characters, e.g., the string "reg.name[last]" is split into "reg", "last" and "name". Afterwards, Carbon builds a tree of prefixes and suffixes for each of the thus split strings and looks for English terms in the WordNet database that match these substrings. So, in the example it only retrieves (the concepts for) the terms "last" and "name". By exploiting WordNet collations (sequences of words that go together, such as "zip code"), Carbon also retrieves the term "last name". Carbon uses WordNet synsets (a set of word or collation synonyms) to retrieve synonyms of the found terms. So, Carbon maps the field name string to all WordNet terms thus found, e.g., "reg.name[last]" is mapped to the WordNet terms "name", "last", "last name", "surname" and "family name".

Like this, Carbon can use this knowledge to propose values for a target form field.

5 Auto-Filling Web Forms with Carbon

Carbon has two main parts: Carbon Client and Carbon Server. Figure 3 shows an overview of the interactions between the Carbon Client and Carbon Server.

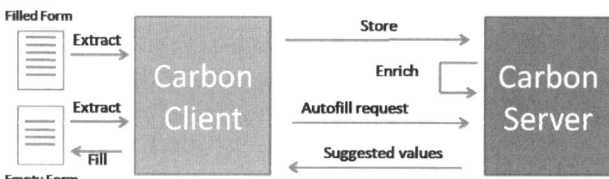

Fig. 3. Client-Server Interactions.

Carbon Server is a semantic application that stores and enriches metadata about web forms in order to recommend values in the process of automatically filling forms. Carbon Client is a browser extension that processes web pages extracting relevant metadata from previously filled out web forms, extracts relevant metadata from an empty target form, and automatically suggests values for the empty form's fields.

5.1 Carbon Client

Carbon Client was implemented as a Greasemonkey[6] script. Greasemonkey scripts can be easily added to the Firefox browser and can be enabled or disabled with a simple mouse click. Carbon uses it to have access to the DOM tree of the HTML page

---

[6] http://www.greasespot.net/

that the user visits. The communication between Carbon Client and Carbon Server is done via the *XMLHttpRequest*[7] object.

Since Carbon uses previously filled out forms for the purpose of auto-filling an empty form, Carbon Client extracts metadata about all forms that the user fills out. For doing this, for every page that the user visits Carbon Client accesses the *DOM tree*, searches for all input elements, and adds to them a *Javascript*[8] *Onblur DOM event*. An *Onblur* event triggers a function when the user moves the focus from an HTML INPUT element to another. At this moment, Carbon Client extracts the metadata (name, value, and type) of the input tag that triggered the event and sends, together with the page URI, such information to Carbon Server via an *AJAX*[9] *request*. Once the request has been received, Carbon Server processes it and stores it.

When the user visits a page with an empty form, Carbon Client can be triggered to automatically fill out the form. This triggering can occur upon the explicit request of the user or automatically whenever the user visits a page with an empty form. At that moment, for each field, Carbon Client extracts metadata, and sends, with the page URI, an *autofill* request to Carbon Server. Carbon Server retrieves suggestions for values that fit those fields. Carbon Client uses these values to fill out the empty fields.

5.2   Carbon Server

Carbon Server plays a main role in Carbon as it is responsible for storing and enriching metadata about web forms. Also, Carbon Server implements the logic behind the mapping of form fields.

Carbon Server is a web application that was implemented using the Ruby on Rails[10] framework. As storage technology we use Sesame[11], an RDF (Resource Description Framework)[12] open source framework. We also use ActiveRDF[13], a Ruby library for manipulating RDF data following the object-oriented paradigm. Figure 4 gives an overview of the Carbon Server implementation architecture.

Carbon Server uses 3 different data models: a Form Data Model, a Configuration Data Model, and an Enriched Data Model. All these models are instances of the Carbon Ontology that we will not detail in this paper due to space limitations. Using RDF as the representation model, Carbon can be easily extended to include any relevant metadata about web forms. Also, RDF is the foundation of the Semantic Web, and a lot of open data is being published in this environment, such as WordNet that Carbon uses in the process of mapping form fields. The main benefit of the use of such a model is that Carbon can consume data from any external source in the Semantic Web, and thus provides a flexible and extensible environment that allows for the definition of new rules and their application in the conceptual mapping of fields. For example, Carbon could use dictionaries in other languages, in addition to the English version of WordNet that is already used in the process of mapping fields.

---

[7] http://www.w3.org/TR/XMLHttpRequest/
[8] https://developer.mozilla.org/en/About_JavaScript
[9] http://en.wikipedia.org/wiki/Ajax_%28programming%29
[10] http://rubyonrails.org/
[11] http://www.openrdf.org/
[12] http://www.w3.org/RDF/
[13] http://www.activerdf.org/

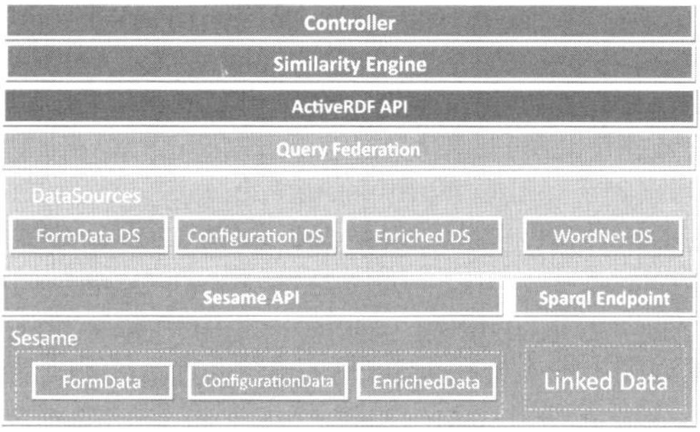

Fig. 4. Carbon Server Implementation Architecture

The Form Data Model stores the conceptual structures of forms and fields, for all previously filled out forms. For example, the concept for the field with the label "Last Name", the name "reg.lastName", the type "text", and the value "Donald" is represented as follows in the Carbon Ontology using the RDF notation N3[14].

```
@prefix carbon: <http://www.carbon-autofill.org/> .
@prefix rdf:    <http://www.w3.org/1999/02/22-rdf-syntax-ns#>  .
@prefix rdfs:   <http://www.w3.org/2000/01/rdf-schema#>   .
<carbon:7382> <rdfs:type> <carbon:Field>   .
<carbon:7382> <carbon:fieldLabel> 'Last Name'   .
<carbon:7382> <carbon:fieldId> <carbon:7382>   .
<carbon:7382> <carbon:originalFieldName> 'reg.name[last]'   .
<carbon:7382> <carbon:name> 'regnamelast' .
<carbon:7382> <carbon:formUri>
<https://secure.gettyimages.com/register/>   .
<carbon:7382> <carbon:urlDomain> <https://secure.gettyimages.com>   .
<carbon:7382> <carbon:updated_at> '1254390507.71' .
<carbon:7382> <carbon:fieldValue> 'Donald ' .
```

Following this data model, all *form data* is stored in a Sesame repository called FormData.

The Enriched Data Model stores extra knowledge about instances of the Form Data Model. It can store any meta-knowledge about the conceptual structures; in the implementation that we used in the current experiment it stores concepts that are related to fields' names and labels. The example below shows the enrichment related to a form field labeled "Last Name":

```
@prefix carbon: <http://www.carbon-autofill.org/> .
@prefix wordnet: <http://www.w3.org/2006/03/wn/wn20/instances/>   .
<carbon:7382> <carbon:relatedto> <wordnet:synset-last_name-noun-1> .
<carbon:7382> <carbon:relatedto> <wordnet:synset-surname-noun-1>   .
<carbon:7382> <carbon:relatedto><wordnet:synset-family_name-noun-1> .
```

Here, the WordNet synset consisting of the words or collations "last name", "surname" and "family name" is associated to the concept 7382 that represents the form field labeled "Last Name". The property *relatedto* defined in the Carbon

---

[14] http://www.w3.org/DesignIssues/N3Resources

Ontology was used to denote this association. Following this data model, all *enriched data* is stored in a Sesame repository called EnrichedData.

The Configuration Data Model allows to extend the knowledge from the Form Data Model and the Enriched Data Model. For example, the concepts "street" and "address" are not synonyms in the WordNet vocabulary, so Carbon will not map these two. Even another semantic matching strategy such as WordNet::Similarity[16], will give them a low degree of similarity. However, the designer can add extra triples in this repository (using Carbon's named graph), see the example below, which extends the WordNet synsets enabling Carbon to perform better for this example.

```
@prefix carbon: <http://www.carbon-autofill.org/> .
@prefix wordnet: <http://www.w3.org/2006/03/wn/wn20/instances/>  .
<wordnet:synset-address-noun-1><wordnet:containsWordSense>
<wordnet:wordsense-street-noun-1> .
```

The triple above connects these two concepts using the property "containsWordSense" which is the main relation exploited by Carbon to determine the similarity among concepts.

6   Evaluations

In this section, we present two sets of evaluations. In the first set of experiments we want to see how many concepts in a new form have been found in the previously filled out forms. We define this as concept completeness. In the second set of experiments, we examine the effectiveness of Carbon in terms of precision and recall. Both sets of experiments use the TEL-8 (see Section 3.2) data set with Airfares, Automobiles, Books, Hotels, Jobs, and Movies domains.

6.1   Concept Completeness: Definition and Evaluation

We first formally define the notion of concept completeness. Given a set CS of concepts related to fields in a set S of filled out forms and a set Cf of concepts related to fields in a new form f, the concept completeness of form f given a set of filled forms S is defined as follows:

$$Completeness(f,S) = \frac{|C_S \cap C_f|}{|C_f|}$$

The value of concept completeness ranges from 0 to 1. If it equals 1, then it means that all the concepts in the new form are completely covered by the concepts in the previously filled out forms. If the concept completeness equals 0, then the previously filled forms are not useful in filling out the new form.

For all the forms in a domain *d* in the TEL-8 data set, we did the following:

1) We mapped the fields in all forms to WordNet concepts. In this set of experiments, we used the field labels that were extracted and available in the data set and mapped them to WordNet concepts.

2) For each form f in domain d, we determined all possible subsets of forms in domain d with size r, and used these subsets as sets of source forms. Then, for each subset S_r, we computed the *Completeness(f,S_r)*.

3) Finally, we computed the concept completeness average grouped by d and r.

Table 2 shows the concept completeness for 6 different domains based on the concept mapping performed by human experts (denoted by "Experts") and by our Carbon Server (denoted by "Carbon"). The results of the Experts act as benchmark for Carbon. Note that r denotes the number of forms that a user has previously filled out and in this table r ranges from 1 to 6.

Table 2. Concept Completeness Evaluation Results

r	Airfares		Automobiles		Books		Hotels		Jobs		Movies	
	Carbon	Experts	Carbon	Experts	Carbon	Experts	Carbon	Experts	Carbon	Experts	Carbon	Experts
1	25.21%	49.24%	19.45%	26.27%	34.07%	45.97%	16.53%	21.51%	7.92%	15.01%	11.64%	17.83%
2	40.59%	64.85%	29.57%	39.42%	48.10%	60.05%	27.91%	34.67%	14.66%	26.13%	19.86%	29.29%
3	50.37%	71.50%	35.84%	47.22%	54.90%	66.05%	36.05%	43.33%	20.46%	34.57%	25.91%	37.07%
4	56.93%	75.32%	40.34%	52.60%	58.90%	69.59%	42.11%	49.47%	25.49%	41.15%	30.54%	42.65%
5	61.61%	78.00%	43.89%	56.70%	61.69%	72.15%	46.83%	54.13%	29.91%	46.42%	34.24%	46.88%
6	65.14%	80.10%	46.86%	60.03%	63.88%	74.17%	50.67%	57.88%	33.83%	50.74%	37.30%	50.23%

We see that the concept completeness of a new form becomes higher if a user has filled out more forms previously. For Carbon, the concept completeness of the "Airfares" domain increases from 25.21% to 65.14%. For Experts, it increases from 49.24% to 80.10%. On average the concept completeness of Carbon and Experts for $r = 6$ is 50.73% and 62.31%, respectively. This means that even though a user has only filled out 6 forms, the concepts in the filled forms can cover 50-62% of the concepts in a new form. The uncovered concepts in the new form are usually application-specific, meaning that those concepts occur only in a small number of forms.

We also see that the increment rate becomes slower as the user fills out more forms: the more forms a user fills out, the fewer new concepts can be discovered. For example, the increment rate of the concept completeness for Carbon in the "Hotels" domains drops from 11% when r=2 to 4% when $r = 6$. A similar result is also revealed for Experts. Checking all results in the 6 domains, we found the increment in all domains to become rather small (all below 5%) when r reaches 6. Considering the increase of computation complexity and the decrease in increments for the concept completeness, we therefore set the maximum value of r to 6.

We also observe how the concept completeness of human experts is (expectedly) always higher than Carbon's. On average, the concept completeness of Carbon reaches almost 74% of that of the human experts. It shows that, while there is still room for improvement, the ability of this Carbon version to map labels and concepts is quite close to the one of human experts.

6.2 Effectiveness: Performance Measures and Dataset

In our second evaluation, precision and recall are used as the performance measures [7]. Precision expresses the proportion of retrieved relevant fields among all the retrieved fields, while recall expresses the proportion of retrieved relevant fields from the total relevant fields:

$$Precision = \frac{\{Relevant\} \cap \{Retrieved\}}{\{Retrieved\}} \qquad Recall = \frac{\{Relevant\} \cap \{Retrieved\}}{\{Relevant\}}$$

The basic idea of this experiment is to compare the forms filled by the human experts according to users' profiles to those filled by Carbon automatically. In detail it is divided into four steps described as follows:

1) We randomly chose 10 forms for each domain as the test set.
2) We collected 6 user profiles according to the real personal information, for each domain. These profiles will be used as the facts for filling the forms.
3) In order to evaluate the performance of Carbon, we needed to set up a benchmark for the evaluation by filling out 10 forms for each domain according to these user profiles. This was accomplished by human experts.
4) For each domain we then chose one available form as a target form. Based on the results of the first set of experiments, another 6 forms were selected randomly as the source forms, representing the forms that users have filled out formerly. Then Carbon filled the target form automatically according to the knowledge extracted from these source forms. Repeating this process for 8 different target forms, we calculated the *precision* and *recall* for the domain.

6.3   Effectiveness Evaluation Results

Table 3 summarizes the results of the effectiveness evaluation on Carbon in terms of precision and recall for each domain. The precision ranges from 0.54 to 0.81 and on average is 0.73. The recall ranges from 0.42 to 0.61 and on average is 0.53. We can see that the recall is less than ideal. There are two explanations for this recall result. One is that when we use the field names instead of the labels in the experiments, some of the field names, e.g. "inp_ret_dep_dt_dy", "DEST-1", etc., are not meaningful enough to be parsed to words and mapped to concepts. Sometimes the field name is meaningful, but Carbon fails to map it to related concepts. Taking the example of the Airfare domain that has the lowest recall, among all the 77 relevant fields in this domain, 18 fields are meaningless. If we ignore these fields and re-calculate the recall, it will increase to 0.55 from 0.42. Furthermore, we observe that in those cases where there are labels they are more meaningful than the field names, and despite that these labels could improve Carbon's precision and recall, using the names we have shown the capability of the semantic matching approach for auto-filling forms.

Table 3. Precision and Recall Results

	Airfares	Automobiles	Books	Hotels	Jobs	Movies	Avg
Precision	0.54	0.73	0.75	0.79	0.81	0.73	0.73
Recall	0.42	0.61	0.60	0.53	0.59	0.47	0.54

7   Conclusion

Filling out forms is an essential aspect of many web applications and many users are confronted with a large degree of repetition in this process across applications. Tools exist to help users in this process, but an optimization step is not only welcome but also feasible if we are able to integrate form data across web forms. In this paper, we presented a novel framework for domain-independent automatic form filling. The main challenge behind the approach is to provide good suggestions for the values to be used for each field in a new form to be filled out, and to do so based on the web forms the user has previously filled out. We have approached this challenge with a number of innovative steps in which we are able to extract relevant metadata from the previously filled forms, semantically enrich this metadata, and use it for aligning fields between web forms. We have also given details of experimental validations of the approach. First, to describe the nature of the problem and challenge, we have done a study of the distribution of field labels in web forms. Second, we have conducted a study of the performance of the approach with the Carbon implementation.

As our focus in this paper is to show how to exploit the semantics of concepts behind the web form fields and to use semantic-based techniques to automatically fill out web forms, several usability and privacy issues could not be discussed in this paper. Amongst them are web form domain resolution, encrypted transfer of data and the management of form data for multiple users. The first issue can effectively be resolved by plugging a web page classifier into Carbon's architecture. Thus, Carbon is able to select data from previously filled forms that are classified in the same domain as the target form. The second and third issues can also be addressed, by adding an encryption and authentication system into Carbon's architecture, respectively.

In the continuation of this research, we study how a hybrid approach that uses a combination of auto-filling and auto-completion performs in terms of effectiveness, based on the observation that auto-completion could be exploited over the enriched knowledge structure that was created over the form data stored in the user's history.

References

1. Autofill Forms – Mozilla Firefox Add-on, http://autofillforms.mozdev.org/
2. Bast, H., Weber, I.: Type Less, Find More: Fast Autocompletion Search with a Succinct Index. In: The Proceedings of SIGIR 2006, Seattle, USA (August 2006)
3. Bouquet, P., Serafini, L., Zanobini, S.: Semantic Coordination: A New Approach and an Application. In: Fensel, D., Sycara, K., Mylopoulos, J. (eds.) ISWC 2003. LNCS, vol. 2870, pp. 130–145. Springer, Heidelberg (2003)
4. Doan, A.H., Domingos, P., Halevy, A.Y.: Learning to Match the Schemas of Data Sources: A Multistrategy Approach. Machine Learning 50(3), 279–301 (2009)
5. Doan, A.H., Madhavan, J., Domingos, P., Halevy, A.Y.: Learning to Map between Ontologies on the Semantic Web. VLDB Journal, Special Issue on the Semantic Web 12(4), 303–319 (2003)
6. Google Toolbar Autofill, http://toolbar.google.com/
7. Han, J., Kamber, M.: Data Mining: Concepts and Techniques. Morgan Kaufman, San Francisco (2001)

8. He, H., Meng, W., Yu, C.T., Wu, Z.: Automatic Extraction of Web Search Interfaces for Interface Schema Integration. In: the Proceedings of WWW 2004 - Alternate Track Papers & Posters, New York, USA (May 2004)

9. Hyvönen, E., Mäkelä, E.: Semantic Autocompletion. In: Mizoguchi, R., Shi, Z.-Z., Giunchiglia, F. (eds.) ASWC 2006. LNCS, vol. 4185, pp. 739–751. Springer, Heidelberg (2006)

10. iOpus Internet Macros, http://www.iopus.com/

11. Knuth, D.E.: The Art of Computer Programming. Sorting and Searching, vol. 3, pp. 394–395. Addison-Wesley, Reading (1973)

12. Lehmann, J., Schüppel, J., Auer, S.: Discovering Unknown Connections - the DBpedia Relationship Finder. In: The Proceedings of CSSW 2007, Leipzig, Germany (September 2007)

13. Levenshtein, V.I.: Binary Codes Capable of Correcting Deletions, Insertions, and Reversals. Soviet Physics Doklady 10(8), 707–710 (1966)

14. Nguyen, H., Nguyen, T., Freire, J.: Learning to Extract Form Labels. In: the Proceedings of VLDB 2008, Auckland, New Zealand (August 2008)

15. Noy, N.F., Musen, M.A.: PROMPT: Algorithm and Tool for Automated Ontology Merging and Alignment. In: The Proceedings of AAAI/IAAI 2000, Austin, USA (July-August 2000)

16. Pedersen, T., Patwardhan, S., Michelizzi, J.: WordNet: Similarity - Measuring the Relatedness of Concepts. In: The Proceedings of AAAI/IAAI 2004, San Jose, USA (July 2004)

17. Raghavan, S., Garcia-Molina, H.: Crawling the Hidden Web. In: The Proceedings of VLDB 2001, Rome, Italy (Septmeber 2001)

18. RoboForm, http://www.roboform.com/

19. Smith, T., Waterman, M.: Identification of Common Molecular Subsequences. Journal of Molecular Biology 147(1), 195–197 (1981)

20. TEL-8 Query Interfaces, http://metaquerier.cs.uiuc.edu/repository/datasets/tel-8/

21. Toda, G.A., Cortez, E., de Sá Mesquita, F., da Silva, A.S., de Moura, E.S., Neubert, M.S.: Automatically Filling Form-based Web Interfaces with Free Text Inputs. In: the Proceedings of WWW 2009, Madrid, Spain (April 2009)

22. Winkler, W.E.: The State of Record Linkage and Current Research Problems. Statistics of Income Division, Internal Revenue Service Publication R99/04 (1999)

Scalable and Mashable Location-Oriented Web Services

Yiming Liu and Erik Wilde

School of Information
UC Berkeley

Abstract. Web-based access to services increasingly moves to location-oriented scenarios, with either the client being mobile and requesting relevant information for the current location, or with a mobile or stationary client accessing a service which provides access to location-based information. The Web currently has no specific support for this kind of service pattern, and many scenarios use proprietary solutions which result in vertical designs with little possibility to share and mix information across various services. This paper describes an architecture for providing access to location-oriented services which is based on the principles of *Representational State Transfer (REST)* and uses a tiling scheme to allow clients to uniformly access location-oriented services. Based on these *Tiled Feeds*, lightweight access to location-oriented services can be implemented in a uniform and scalable way, and by using feeds, established patterns of information aggregation, filtering, and republishing can be easily applied.

Keywords: Web Services, Location-Oriented Services, REST, Loose Coupling.

1 Introduction

The *mobile Web*, the mobile access to Web-based resources, has gained a lot of momentum and attention. The increasing availability of sophisticated devices for mobile Web access (smartphones and netbooks) means that an increasing share of Web users access the Web from mobile settings, and in many cases, these users are interested in localized services. Complementing this, the increasing usage of the mobile Web produces an increasing amount of localized data ("location trails" of users and networked objects), which allows novel and personalized access to services based on this localized data. In summary, location-orientation on the Web has seen a sharp incline in interest and sophistication, and it is likely that this development will continue for the foreseeable future.

An increasing share of the services provided over the Web, either in the form of Web pages for browser-based UIs, or in the form of Web service APIs for use by applications, use or support location in some form, but the Web itself still is a location-unaware information system [7], which means that in many cases, the services or information made available are hard to repurpose and reuse. A lot

B. Benatallah et al. (Eds.): ICWE 2010, LNCS 6189, pp. 307–321, 2010.

of location-oriented services nowadays use Web technologies, but in many cases, the spatial component of service design is service-specific, and thus cannot be easily reused or recombined across services. A first change in that landscape is the W3C's current work towards a geolocation API [19], which allows scripting code to access a client's location services, but this API is still in its draft stage, and only covers the case where location information should be made available on the client.

In the general area of *Geographic Information System (GIS)* research, Web-oriented access to GIS systems has seen only little advances beyond the traditional model of client-based access to a centralized GIS server. *Web GIS* [1] are often simply regarded as Web interfaces for GIS systems, which means that they replace other client-side technologies with browser-based access to a single GIS. The *Open Geospatial Consortium (OGC)* has released a number of standards for accessing GIS systems using Web services, but all of these specifications are based on RPC models of interaction, and thus do not apply principles of Web architecture to GIS systems.

In this paper, a more loosely coupled architecture [17] of accessing GIS services is described, which can be used to provide access to any kind of geospatial service. It is based on the principles of *Representational State Transfer (REST)*, and thus is designed around interlinked resource representations served through a uniform interface, instead of using the RPC model of a sophisticated method set provided through a single endpoint.

2   Related Work

The *Open Geospatial Consortium (OGC)* has defined the *Web Map Service (WMS)* [11] and *Web Feature Service (WFS)* [13] as two standards for accessing map imagery and map features using Web services. Both of these standards are based on SOAP, which means that instead of exposing map imagery and map features as Web resources, they expose functions that can be remotely called to request map imagery and features. The RESTful approach of our work, on the other hand, exposes information in the interlinked fashion of Web resources, and thus clients do not need to support SOAP or the specific set of functions defined by WMS/WFS. We claim that the decision between RPC-style Web services and RESTful Web services [18] should be based on the expected use cases, and that map imagery and even more so map features should be provided in a way which allows "serendipitous reuse" [22].

The idea of tiled access to map information in general has been proposed very early and goes back to the approach of using *Quadtree* [20] structures for organizing spatial data. Popular Web mapping services such as *Google Maps* are using tiled access to map imagery for scalability reasons, so that the vast amount of map imagery can be efficiently stored, served, and cached. Beyond these specific implementations, the *Tile Map Service* [16] proposes a general scheme for how services can expose map imagery through a tiling scheme. However, APIs for location-based services as Google Maps, Flickr do not use the tiling concept to organize and deliver spatial information. Proprietary APIs for location-base services are typically designed in ad-hoc fashion, with different types of access capability. The simplest instantiations, such as Flickr, allow queries by point and radius, while some services may allow user-specified bounding boxes.

To the best of our knowledge, no scheme so far has been proposed that combines the tiled model of spatial information representation with a RESTful architecture for accessing the resources organized in this fashion.

3   Feeds as RESTful Web Services

Starting from the principle of loose coupling [17], a popular and well-established method for implementing RESTful access to collections of resources is *Atom* [10], a language for representing feeds. Extending the idea of the feed as a single representation of a collection, the model of interaction with feeds can be extended to cover queries and other interaction mechanisms [24], so that feeds and interactions with feeds turn into a more feature-rich model of interaction with resource collections. This model, however, is still a read-only model, but the *Atom Publishing Protocol (AtomPub)* [5] can be used to extend it into a model that also allows write access to collections.

While feeds satisfy the criteria for RESTful resource access (URI-based access, self-describing representations, interlinked information resources), additional feed-based capabilities for filtering, querying, and sorting collections are not yet standardized. Some feed-based services such as Google's *GData* or Microsoft's *OData* introduce their own extensions to add this functionality to feed-based services, but none of these so far has reached critical mass to be accepted as a new standard. In this paper, we focus on describing an access model to spatial information that builds on feeds to define a lightweight interaction model with spatial data that provides access to spatial information services. This model lays the groundwork for establishing scalable and open interaction patterns with these services. We highlight some of the issues for a more flexible and customizable access to these services in the final parts of the paper, but leave these issues for further work.

4   The Tiled Feed Model

In this section, we present a REST-based model for delivering mashable location-oriented services, based on multiple potential data sources. The design goals for this model are to address four areas of concern in the design of location-oriented Web services:

Remixability. Much of the geospatial information on the Web remains stand-alone and service-specific. Application scenarios involving multiple sources of geospatial information — now becoming common with highly localized and personalized services — require ad-hoc integrations with each of the desired data sources in turn. One of our primary goals in this model is to facilitate the creation of novel location-oriented services that reuse and remix available geospatial data.

Loose coupling. Much as we seek to liberate geospatial data from the constraints of service-specificity and tight coupling, clients of our model should
   not be tightly bound a specific system of our own design.
Scalability. With the rapid growth of location-oriented services and users, a
   model that scales cheaply and efficiently, using existing technology and know-
   how is particularly desirable.
Ease of Deployment. A driving factor of the rapid growth of the Web rests
   with ease of development of Web sites via HTML. In a similar vein, many
   individuals and organizations may have geospatial data that would be valuable if made available as open Web services, and more valuable if easily
   remixed with data from other entities. The development of applications that
   use existing data and services should be made as easy as possible.

Our solution is called *Tiled Feeds*, geospatial data feeds based atop various geographic information systems. Tiled feeds can be published by an individual or
organization with first-hand geospatial information, a commercial data provider,
or even motivated third-parties, much like conventional RSS or Atom feeds. As
with conventional news or blog feeds, client applications may use one or more
tiled feeds to build, mix, and visualize spatial information as they wish.

4.1   Tiled Feed Architecture

Tiled feeds consist of, as the name implies, spatial tiles and data feeds. In this
model, the world is partitioned into standard sets of tiles, at varying levels of
resolution. Each tile is then published as an Atom feed with simple spatial extensions. Entries within the feed represent geospatial features that are located within
that tile. The entries are represented in *Keyhole Markup Language (KML)* [15]
or *Geographic Markup Language (GML)* [12], which are standard markup languages for representing geospatial information features, such as points, lines,
polygons, etc.

Tiles. In the tiled feed system, tiles represent a standardized spatial unit for
access to feeds. Tiling is a well-known technique for spatial division and aggregation of georeferenced data. Size of the tiles indicates level of resolution — larger
tiles offer less detail but more coverage of an area, and vice versa for smaller
tiles. Repeated tiling of the same area provides different levels of resolution, or
"zoom levels", for a tiled map. For example, a map may be divided into four
large tiles at the first level of detail, and 16 smaller tiles at the second level.
   There are many possible tiling schemes, but we adopt the basic tiling approach
used in most Web-based mapping services [21,4]. The base map is a map of the
world, in WebMercator or EPSG:3857 projection [2]. We subdivide this base
map into smaller and smaller tiles, using a conventional quadtree-based [20]
tiling algorithm.
   In our canonical tiling scheme for tiled feeds, the world map is recursively
divided into four smaller tiles. The top-level tile is set to be the entire base map.
This constitutes zoom level 0. We then divide the top-level tile into four equal

square tiles, creating the tiles for zoom level 1. Each of the four new tiles in level 1 can be divided again, in four, to create 16 tiles in level 2. This process continues for each new level of resolution desired. We recommend a maximum of 20 levels, the same provided by most mapping services. Due to the properties of the Mercator projection[1], we restrict the base map to latitudes between -85.05112878 and 85.05112878 — the same restrictions used in both Google Maps and Bing Maps. This also conveniently creates a square map, which yields square tiles simplifies tile-based distance computations.

For identification and linking purposes, an addressing value is assigned to each tile. Both the tiling and the addressing scheme are illustrated in Figure 1, with the world recursively divided and each area numbered 0 to 3. In essence, the top-left tile, covering the top-left sector of its parent tile, is always given a value of 0. The top-right tile is always given a value of 1, the bottom-left given 2, and the bottom-right is given 3. Thus, the *tile key*, a hash string that uniquely identifies a particular tile, is given by the concatenation of the addressing values of its parent tiles. For example: the key "0-1" identifies the top-right tile of the top-left tile of the base map. The length of its *tile key* represents the zoom level of the tile.

Fig. 1. Quadtree-based tiling and addressing scheme

Tile Feeds. Each tile is represented by an Atom feed, called the *tile feed*. The tile feed carries metadata information and descriptions similar to an ordinary Atom feed. The tile is referenced by a unique URI. For example, the tile representing the entire world would have an ID of `http://example.com/feed/top`. The northwestern tile at zoom level 1 would have an ID of `http://example.com/feed/0`. The tile feed links to neighboring, contained, and containing tiles, for RESTful navigation of the map.

However, the tile feed is extended with some simple spatial properties. In particular, a tile feed has the following specific properties:

- The `atom:id` element contains the URI for the tile.
- Four `atom:link` elements point to the neighboring tiles. Links with the relations of *north*, *south*, *east*, and *west* point to the URIs of neighboring tiles in the four cardinal directions. If there is no tile in that direction[2], the corresponding link element is omitted.
- An `atom:link` element points to the containing parent tile. A link with the relation of *up* points to the URI of the tile one zoom-level up, which contains the current tile.

---

[1] Singularities are present on both poles. The line at the top of a Mercator map, is in actuality a point — the north pole. The same applies for the line at the bottom of a Mercator map.

[2] For example, the tile 0 has no neighboring tile to its north.

– Four `atom:link` elements point to the tiles at the next zoom level, contained within the current tile. A link with the relation of *down-northwest* points to the URI of the tile at the northwestern quadrant, the *down-northeast* relation for the northeastern quadrant, *down-southwest* for the tile at southwestern quadrant, and *down-southeast* for the tile at the southeastern sector.

With these extensions, subscribing clients of the tile feed can navigate the world using the provided links, load neighboring tiles, and retrieve information as needed. Each individual `atom:entry` in the tile feed consist of a resource available within this tile, representing some GIS feature and its attributes, written in KML or GML. For read-only feeds, these entries are refreshed when its back-end data source is updated. For writeable feeds, AtomPub may be used as a standard means for posting, editing, and deleting resource items within the feeds, creating more dynamic tile feeds.

Features and attributes are provided by the underlying geographic information system. The system may be, for example, geographic information databases, third-party location-oriented services, or standardized GIS web services. We discuss the methods for integration with these data sources in more detail in Section 5.

A tile feed may support paging, and does so according to the standardized feed paging mechanism [8]. A specific page is requested with a *page* query parameter in its URI, with a value equals to some positive integer representing desired page number. In this case, additional `atom:link` elements will point to the URIs of the previous, next, first, and last pages, as directed in the feed paging specification. More advanced operations, such as filtering, sorting, or querying, may also be supported; we discuss this in more depth in Section 5.

Implementation. We have implemented proof-of-concept tiled feeds using freely available geospatial datasets and geospatial-aware relational databases. The popular PostgreSQL RDBMS can be augmented for GIS *geometry* columns and spatial query functions using the PostGIS[3] add-on. GIS features were loaded into PostgreSQL tables, and a thin service layer is written to serve GIS features in the tiled feed format as described. We created, as tiled feeds, the locations and addresses of 861 Amtrak[4] stations in the United States, the addresses and species of 64,318 street trees maintained or permitted by the city of San Francisco, and 25,928 magnitude 3.0+ earthquakes occurring in the United States in 2008.

An excerpt from the feed is provided in Listing 1. As noted in the previous section, the feed, in addition to resource entries and standard Atom metadata, also contains navigational extensions pointing to neighboring tile feeds, as well as the feeds of its parent tile and child tiles.

The Amtrak tiled feed effectively describes points of access for inter-city passenger rail service in the United States, and is useful as an example of an information collection covering a large area. The San Francisco trees collection, on the

---

[3] Available at `http://postgis.refractions.net/`
[4] Amtrak is the sole intercity passenger rail service in the United States.

```
<?xml version="1.0" encoding="UTF-8"?>
<feed xmlns:fh="http://purl.org/syndication/history/1.0" xml:lang="en-US"
    xmlns="http://www.w3.org/2005/Atom">
  <id>http://tfserver/tiles/02301021</id>
  <link type="text/html" rel="alternate" href="http://tfserver/tiles
      /02301021"/>
  <link type="application/atom+xml" rel="self" href="http://tfserver/tiles
      /02301021"/>
  <title type="text">Tile 02301021</title>
  <updated>2009-11-04T11:10:46-08:00</updated>
  <author>
    <name>TileFeed Generator</name>
  </author>
  <link type="application/atom+xml" rel="http://tfserver/tiledfeeds/
      relation/north" href="http://tfserver/tiles/02301003"  />
  <link type="application/atom+xml" rel="http://tfserver/tiledfeeds/
      relation/south" href="http://tfserver/tiles/02301023"/>
...
  <link type="application/atom+xml" rel="http://tfserver/tiledfeeds/
      relation/down-southwest" href="http://tfserver/tiles/023010212"/>
  <link type="application/atom+xml" rel="http://tfserver/tiledfeeds/
      relation/down-southeast" href="http://tfserver/tiles/023010213"/>
  <fh:complete/>
  <entry>
    <id>http://tfserver/items/287</id>
    <link type="text/html" rel="alternate" href="http://tfserver/items/287"
        />
    <title>ACA</title>
    <updated>2009-11-04T11:10:44-08:00</updated>
    <content type="application/vnd.google-earth.kml+xml">
      <kml xmlns="http://www.opengis.net/kml/2.2">
        <Placemark>
          <name>ACA</name>
          <description>100 I Street, Antioch-Pittsburg, CA</description>
          <Point>
            <coordinates>-121.815132000194,38.0180849997628</coordinates>
          </Point>
        </Placemark>
      </kml>
    </content>
  </entry>
  ...
```

Listing 1. An example, unpaged tile feed containing Amtrak stations for the tile 02301021, a tile covering northern California

other hand, is a demonstration of a highly localized yet relatively dense dataset. Such collections are potentially useful for community-based location-oriented services, such the creation of a municipal dashboard for local government or a neighborhood environmental program.

We use a typical page-caching pattern for optimizing feed services. When a tile feed is requested, its feed is generated, and a static file containing the entire feed is saved to disk. As long as the underlying information is not updated, the file is not regenerated, and the feed service simply serves the static file for every subsequent request for the tile. The feed itself supports standard HTTP ETag based caching, for reader-level caching.

Our service implementation supports 20 zoom levels for the entire world, which yields $\sum_{i=1}^{20} 4^i$ tiles. However, for typical datasets, only a small number of tiles are in service. While Amtrak covers much of the continental United States, its stations are distributed unequally and mostly in coastal or urban areas. A similar effect occurs for the earthquake feed, where tiles of seismically active areas in the western United States tend to have data. When a tile without data is requested, the requesting client is simply presented with a standard HTTP 404 response. The response, too, can be cached unless a new feature is created within that tile.

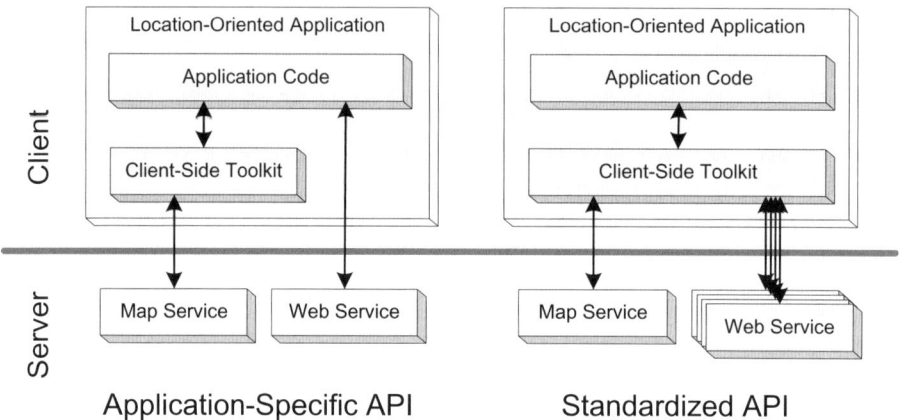

Fig. 2. Architectural comparison between a conventional location-aware service built upon tightly-coupled, application-specific APIs, and a service built upon tiled feeds

Discussion. Tile feeds are accessed by clients in the same manner as ordinary Atom feeds. Compared with current client-server architectures for location-oriented applications, the tiled feed model creates a standardized interface for interacting with location-oriented services. An architectural comparison diagram is provided in Figure 2.

The tiled feed design accomplishes the four design goals we initially laid out for access to geospatial information. For a typical location-oriented application, tiled feeds provide access to multiple resources in an easy, uniform way, providing for easy remixability. For example, currently, a mobile application for travelers may write code to interact with several APIs: one for flight updates and traffic information, another for restaurant reviews, etc. The application must also download map visualizations via a specific mapping service such as Google Earth, Bing Maps, or standard services like the *Web Map Service (WMS)* [11]. In contrast, with a tiled feed, the application can simply include a tiled feed reader and subscribe to the tile feeds for the areas needed. Remixing data — such as traffic conditions, restaurant reviews, and map visualization — is as simple as subscribing to two different tiled feeds for the same tile.

Feed accesses are loosely coupled, much as Atom feeds themselves are loosely coupled from their internal systems and representations. Interactions with the tiled feeds take place via standard HTTP requests to resources and resource collections. This also provides advantages in simple scalability and deployment, in that existing techniques for scaling Web servers and feeds (as opposed to GIS-specific servers and optimizations) can be directly applied to scaling tiled feeds.

As an example, tile feeds can be cached as typical feeds. At client level, entity tag-based caching [3] for RSS and Atom-based feeds, can be used with no modification for a tile feed. Since tile feeds corresponds to unique URIs, conventional Web caches and caching proxies can be used for multi-level caching. At server level, simple page caching techniques can be used to generate and serve static feeds. This enables feed servers to satisfy a large number of requesting clients at very little additional expense. If no data is available for a particular tile — an expected outcome for many data sources covering only specific locations — simple HTTP 404 responses, or redirects to the closest available tile, can be used. It would then be up to the tiled feed client to handle error responses and decide whether to follow redirects to another tile.

The tiled feed model, as described, presents distinct advantages over existing methods for accessing and using geospatial information. As Atom feeds, the data presented is conveniently consumable via standard tools. Clients can choose the tiles that interest them, at any particular level of geographic resolution — thus avoiding the complexity and overhead of geospatial queries. Further, the tile model provides for simple horizontal scalability. By mediating queries for geospatial data into predictable, RESTful patterns, techniques for typical HTTP- and feed-based content delivery can be directly applied for these geospatial feeds as well. Perhaps most importantly, tiled feeds can be easily remixed and combined. Much as Web-based news and blogs are delivered via standard feeds, processed by a variety of reader software, and syndicated across many web sites, geospatial data in tiled feeds can also be consumed in this manner. Remixability also lowers the barrier to reusing data, and encourages the creation of information dashboards and mashups.

5   Publishing Tiled Feeds

A tiled feed publisher may publish data from many types of geospatial information sources. We describe three implementation scenarios of tiled feeds atop existing data sources, including geospatial databases, proprietary location-oriented services, and standard GIS services. A diagram illustrating the three particular scenarios is presented in Figure 3.

5.1   Geospatial Databases

Much geospatial data exist in GIS-specialized relational databases and static files. Both consumer and enterprise-grade relational databases now possess geospatial information storage and querying capabilities, implementing and extending the OpenGIS *Simple Feature Access* [14] to varying degrees. Static geospatial datasets, often provided as *shapefiles* of vector or raster data, can be imported to these databases using widely available tools, as we have done in our implementations.

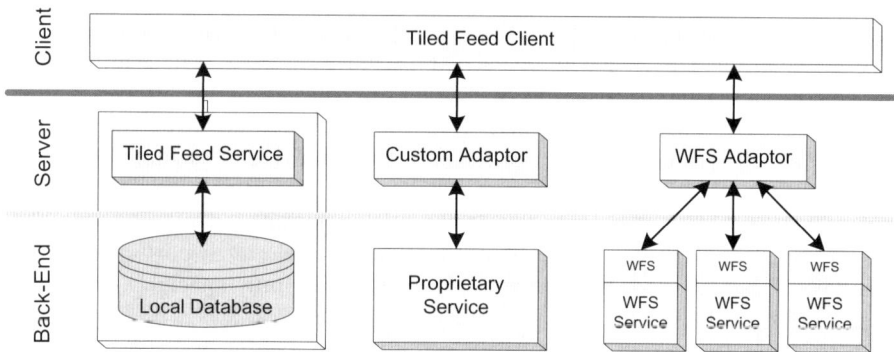

Fig. 3. Implementations of Tiled Feeds atop various geospatial data sources

A tiled feed implementation has the greatest potential capability and flexibility when directly integrated with such a spatial database, at the cost of a tight coupling. The feed generator would be able to create any tile feed on request, and provide grouping and filtering capabilities cheaply. For example, listing all features and associated attributes in a given tile is a trivial spatial SQL query.

Due to the tight coupling, the publisher of the tiled feed must have direct access to the database, and must create a tiled feed service based on the data. As tiled feed services are simply Atom feeds with spatial extensions, this should be relatively painless. GIS server software such as GeoServer[5], which integrate with geospatial databases to provide data access interfaces, may be also extended to publish tiled feeds, so as to reduce custom implementation work.

5.2   Proprietary Location-Oriented Services

Tiled feeds can also be implemented atop existing location-oriented services with APIs. The growth of location-aware devices has prompted many Web applications, such as Flickr or Twitter, to provide geotagged data. A tiled feed can be created from these proprietary APIs via a software adapter that queries the underlying API for information within the requested tile boundaries.

Adapted tiled feeds based on proprietary APIs may have widely varying levels of capability and flexibility, depending on the spatial queries offered by the underlying API. There is significant potential for mismatches in desired capability and offered capability. For example, the Flickr API offers radius-based search for photos; the query takes as input a geographic coordinate and a set of query terms. Obviously this query is not fully compatible with the square geometries of tiles and multiple zoom levels of the tiled feed model. Furthermore, advanced features such as filtering or paging entries, or writing new entries into the tiled feed, depend upon support from the underlying API.

---

[5] Available at http://geoserver.org/

There are potential fixes. Algorithmic approaches may be used to mask some spatial querying mismatches, such as using geometric approximations of the tile using a crafted radius-based query. Adapting existing sources of data also enables a wider variety of spatial data to become available as tiled feeds, and promotes data reuse and remixing. It is also more likely that third parties would be able to offer tiled feeds based on these APIs.

5.3   Standardized Services

Somewhere in the middle of the spectrum of potential capability are tiled feed implementations based on standard GIS web services. The *Web Feature Service (WFS)* [13], an *Open Geospatial Consortium (OGC)* standard, provides a standardized interface to geospatial data. The Web Feature Service defines a set of operations for querying, creation, update, and deletion of geographic features contained within a geospatial dataset. Results can be encoded in different formats, including the Geography Markup Language [12] and ordinary key-value pairs. The aforementioned GeoServer software is a reference implementation of the WFS, and many other GIS support WFS as well.

A tiled feed model can be implemented atop standardized GIS services such as WFS with relative ease, given the rich set of operations defined by these standards. Further, only one tiled feed adapter need be implemented, which then can be used to consume all WFS-compliant services. WFS-based tiled feeds may also support advanced features such as querying or sorting of tile entries, which are supported by the underlying API.

There still exists potential for capability mismatch between the tiled feed service and the underlying WFS or other standard API. Depending on the geospatial dataset, spatial features provided by WFS queries may be difficult to group into higher-level features. For example, a WFS service may return simple line geometries to establish the boundaries of a plot of land, which the tiled feed adapter must assemble into a polygon for its KML or GML resource entry. The WFS may also provide more advanced features than a tiled feed adapter requires, such as complex, multi-dimensional geometries that may be incompatible with a feed or feed reader. Such mismatches may be reconciled via additional metadata, or incompatible features may be excluded altogether from tiled feeds.

6   Experimental Client

To experiment with tiled feeds and assess the consumption of tiled feed-based services in both application development and practical use, we also implemented a prototype mobile tiled feed client using the iPhone SDK (shown in Figure 4).

The tiled feed client prototype has three views. In the first, the user adds tiled feeds as services to be consumed. The process is similar to adding feeds to a typical feed reader. Each feed can be hosted by a different organization, backed by a different geospatial dataset.

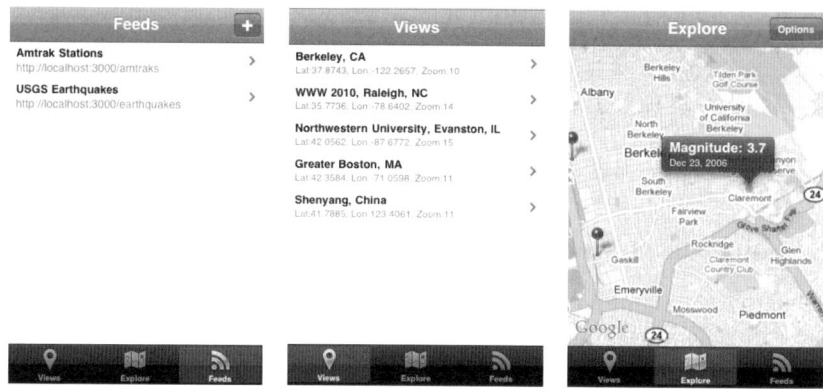

Fig. 4. A basic tiled feed reader. In (a), the user is consuming two location-based services as tiled feeds — locations of nearby Amtrak train stations, and recent M3+ earthquakes from the USGS, respectively. In (b), the user has subscribed to five locations of interest, including Berkeley, CA. In (c), the user selected the Berkeley location, is shown a mashup of Amtrak services and earthquakes at the location of Berkeley, CA.

Second, he identifies specific locations that interest him, such as "two blocks around UC Berkeley campus", or "Vienna, Austria". From this view of all "subscribed locations', he can immediately jump to a location on the map view. The tiled feed reader will retrieve features for that tile from all subscribed services and draw a mashup of the data in the map view. Alternatively, he can explore the world map manually, and the reader will load features for the current visible tile in the viewport from all services. As he navigates around the map, tiles are loaded and unloaded as needed — tiles with no features return 404 from the server, and are simply ignored by the client. At any particular location, the user may choose to add the current map view as a location of interest, adding it to the list of subscribed locations.

Views at specific locations may be personalized further. In the consumption of location-based services, users' information needs are highly context-sensitive. Depending on the usage, data useful at one location is not useful at another [6]. The tiled feed reader adapts to these user contexts; if a user is not interested in data from a particular service for a location, he can turn the service off for that location. For example, if a user is subscribed to a location-based service that provides traffic data, but is interested in the Los Angeles area solely for earthquake and air pollution information, he can turn off the (potentially data-heavy) traffic data for that location, but keep the earthquake and air pollution data on the map view.

This client development experiment demonstrates the significant advantages of the tiled feed model. From the user perspective, there is no central silo or any single point of failure — anyone with access to a geotagged dataset may publish it as a tiled feed. Creating mashups of geospatial services is easy, by simply using two or more services at a given location. From the developer perspective, creating location-aware applications using tiled feed services required minimal code, mostly in the presentation layer to draw the maps, markers, and overlays. We used a standard Atom feed reader component, slightly extended to read the spatial properties from Tiled Feeds; no ad-hoc API adapters were required for adding additional services.

Both the server and the client caches retrieved tiles. Standard caching approaches yields significant benefit. The majority of accesses hits cache rather than triggering new geospatial queries.

In all, the tiled feed model makes information remixing from multiple data sources easy, and encourages the creation of in-context, personalized views and mashups of geospatial data.

7   Future work

With large amounts of geospatial data, tiles feed may contain tens of thousands of features in a given tile at average zoom level. Bandwidth consumption and server load increases significant at this scale, and tiled feed clients on constrained devices are often unable to process or visualize such amounts of data. As GIS datasets and wrapped geospatial APIs often contain large amounts of data, tiled feed servers and clients should be able to properly handle such tiles, if requested.

One solution involve clustering or aggregation on the server side. Aggregate features, identified by a special collection type that marks them as clustered versions of individual geospatial features. The tiled feed client is given the option of dereferencing aggregates into their component features.

Tiled feed filtering and sorting capability are also useful in this context. Feed querying in general remain an open problem [23], as the Atom standard does not adequately cover the query use case. There has been some prior work, including the aptly named *Feed Item Query Language (FIQL)* [9]. However, there are significant mismatches between FIQL and the set of operations desirable in a tiled feed query language. FIQL is largely interested in enabling the retrieval of markup elements in a feed, in a relatively domain-agnostic way. In contrast, tiled feeds may require queries against specific geospatial features contained within a feed, which would require some understanding of GML or KML datatypes.

It is also desirable for clients to autodiscover the query capabilities supported by a particular tile feed (otherwise, human intervention or configuration would be required to read new tiled feeds). For example, certain feed servers may not be able to execute complex queries like `GROUP BY` aggregation, or it may not be able to sort against a particular key. There is no appropriate mechanism to publish this information in FIQL. The design of an appropriately powerful query language for feeds in general, and tiled feeds in particular, is left for future work.

We are also working on more extensive user evaluations of tiled feed client programming (beyond our current simple client) and tiled feed-based applications, as well as quantitative scalability measurements of tiled feed server systems, to explore the advantages and drawbacks of the tiled feed model.

8   Conclusions

We have presented an architecture for providing uniform, lightweight access to location-oriented services, using feeds and geospatial tiling. Based on the principles of *Representational State Transfer (REST)*, tiled feeds is designed around interlinked resource representations served through a uniform interface, allowing loosely coupled and scalable access patterns. Tiled feeds can be easily created from existing sources of data and consumed by standard clients, reducing application complexity and easing application development. Established patterns of information aggregation, filtering, and republishing on the Web can be applied to tiled feeds, allowing the creation of novel information dashboards and mashups using geospatial data.

References

1. Di Martino, S., Ferrucci, F., Paolino, L., Sebillo, M., Tortora, G., Vitiello, G., Avagliano, G.: Towards the Automatic Generation of Web GIS. In: Samet, H., Shahabi, C., Schneider, M. (eds.) 15th ACM International Symposium on Geographic Information Systems, pp. 57–64. ACM Press, Seattle (November 2007)
2. EPSG: 3857, WGS84 (2009),
   `http://www.epsg-registry.org/report.htm?type=selection\&entity=urn:`
   `ogc:def:crs:EPSG::3857\&reportDetail=short\&style=urn:uuid:`
   `report-style:default-with-code\&style_name=OGP%20Default%`
   `20With%20Code\&title=EPSG:3857`
3. Fielding, R.T., Gettys, J., Mogul, J.C., Frystyk Nielsen, H., Masinter, L., Leach, P.J., Berners-Lee, T.: Hypertext Transfer Protocol | HTTP/1.1. Internet RFC 2616 (June 1999)
4. Google: Map Overlays (2009),
   `http://code.google.com/apis/maps/documentation/overlays.html#Google_`
   `Maps_Coordinates`
5. Gregorio, J., de Hóra, B.: The Atom Publishing Protocol. Internet RFC 5023 (October 2007)
6. Kaasinen, E.: User Needs for Location-Aware Mobile Services. Personal and Ubiquitous Computing 7(1), 70–79 (2003)
7. Kofahl, M., Wilde, E.: Location Concepts for the Web. In: King, I., Baeza-Yates, R. (eds.) Weaving Services and People on the World Wide Web, pp. 147–168. Springer, Heidelberg (2009)
8. Nottingham, M.: Feed Paging and Archiving. Internet Draft draft-nottingham-atompub-feed-history-11 (June 2007)
9. Nottingham, M.: FIQL: The Feed Item Query Language. Internet Draft draft-nottingham-atompub-fiql-00 (December 2007)
10. Nottingham, M., Sayre, R.: The Atom Syndication Format. Internet RFC 4287 (December 2005)
11. Open Geospatial Consortium: OGC Web Map Service Interface. OGC 03-109r1, Version 1.3.0 (January 2004)
12. Open Geospatial Consortium: OpenGIS Geography Markup Language (GML) Encoding Specification. OGC 03-105r1, Version 3.1.1 (February 2004)
13. Open Geospatial Consortium: Web Feature Service Implementation Specification OGC 04-094, Version 1.1.0 (May 2005)

14. Open Geospatial Consortium: OGC Simple Feature Access. OGC 06-103r3, Version 1.2.0 (October 2006)
15. Open Geospatial Consortium: OGC KML. OGC 07-147r2, Version 2.2.0 (April 2008)
16. Open Source Geospatial Foundation: Tile Map Service Specification (2009), http://wiki.osgeo.org/wiki/Tile_Map_Service_Specification
17. Pautasso, C., Wilde, E.: Why is theWeb Loosely Coupled? A Multi-Faceted Metric for Service Design. In: Quemada, J., León, G., Maarek, Y.S., Nejdl, W. (eds.) 18th International World Wide Web Conference, pp. 911–920. ACM Press, Madrid (April 2009)
18. Pautasso, C., Zimmermann, O., Leymann, F.: RESTful Web Services vs. "Big" Web Services: Making the Right Architectural Decision. In: Huai, J., Chen, R., Hon, H.W., Liu, Y., Ma, W.Y., Tomkins, A., Zhang, X. (eds.) 17th International World WideWeb Conference, pp. 805–814. ACM Press, New York (April 2008)
19. Popescu, A.: Geolocation API Specification. World Wide Web Consortium, Working Draft WD-geolocation-API-20090707 (July 2009)
20. Samet, H.: The Quadtree and Related Hierarchical Data Structures. ACM Computing Surveys 16(2), 187–260 (1984)
21. Schwartz, J.: Bing Maps Title System (2009), http://msdn.microsoft.com/en-us/library/bb259689.aspx
22. Vinoski, S.: Serendipitous Reuse. IEEE Internet Computing 12(1), 84–87 (2008)
23. Wilde, E.: Feeds as Query Result Serializations. Tech. Rep. 2009-030, School of Information, UC Berkeley, Berkeley, California (April 2009)
24. Wilde, E., Marinos, A.: Feed Querying as a Proxy for Querying theWeb. In: Andreasen, T., Bulskov, H. (eds.) FQAS 2009. LNCS, vol. 5822, pp. 663–674. Springer, Heidelberg (2009)

A Flexible Rule-Based Method for Interlinking, Integrating, and Enriching User Data

Erwin Leonardi[1], Fabian Abel[2], Dominikus Heckmann[3], Eelco Herder[2], Jan Hidders[1], and Geert-Jan Houben[1]

[1] Delft University of Technology, P.O. Box 5031, 2600 GA Delft, The Netherlands
`{e.leonardi,a.j.h.hidders,g.j.p.m.houben}@tudelft.nl`
[2] L3S Research Center, Appelstrasse 9a, 30167 Hannover, Germany
`{abel,herder}@l3s.de`
[3] German Research Center for Artificial Intelligence, Saarbrucken, Germany
`heckmann@dfki.de`

Abstract. Many Web applications provide personalized and adapted services and contents to their users. As these Web applications are becoming increasingly connected, a new interesting challenge in their engineering is to allow the Web applications to exchange, reuse, integrate, interlink, and enrich their data and user models, hence, to allow for user modeling and personalization across application boundaries. In this paper, we present the Grapple User Modeling Framework (GUMF) that facilitates the brokerage of user profile information and user model representations. We show how the existing GUMF is extended with a new method that is based on configurable derivation rules that guide a new knowledge deduction process. Using our method, it is possible not only to integrate data from GUMF dataspaces, but also to incorporate and reuse RDF data published as Linked Data on the Web. Therefore, we introduce the so-called Grapple Derivation Rule (GDR) language as well as the corresponding GDR Engine. Further, we showcase the extended GUMF in the context of a concrete project in the e-learning domain.

Keywords: user modeling, user data integration, personalization, semantic enrichment, knowledge derivation.

1 Introduction

Nowadays, numerous Web applications provide adapted and personalized contents and services to their users. To be able to provide such contents and services, these applications explicitly or implicitly collect data about their users and their behavior. Explicit user data collection approaches rely on asking the user directly, for example, by using a survey form or by asking the user to give ratings to certain products. Implicit approaches imply the observation of the users' behavior: Web applications log and monitor the user behavior in order to construct a user model fitting with the personalization goals of the application. So, a key concern in developing such adaptive Web applications is to model the users and their behavior for achieving the personalization and adaptation goal of the applications. At the same time, these Web

B. Benatallah et al. (Eds.): ICWE 2010, LNCS 6189, pp. 322–336, 2010.

applications are becoming increasingly connected. This creates the interesting challenge of performing user modeling and personalization across application boundaries. It requires approaches allowing various Web applications to exchange, reuse, interlink, and integrate user data. On the one hand, the ability of exchanging, reusing, interlinking, and integrating the user models allows applications to enhance and broaden their user models with additional data. In addition, it is particularly essential for a better integration and cooperation between the applications. On the other hand, it helps users to get the content and services that suit their needs and situations and to syndicate these services. As different applications may represent the same information in different ways, using different syntactic and semantic, the Web applications have to ensure interoperability of the user data in order to be able to exchange, reuse, and integrate user data. Consequently, addressing the interoperability issue is essential when developing interoperable adaptive Web applications.

In essence, there are two ways to ensure interoperability between two applications and their user models: the *shared format* approach [5,20,22] and the *conversion* approach [6]. The shared format approach involves a lingua franca, an agreement between all parties on a common representation and semantic. An alternative approach, which is more flexible, involves conversion between the different applications' user models. Conversion allows for flexible and extensible user models, and for applications to join into a platform. Moreover, in contrast to a shared format approach, conversion is suitable for "open-world user modeling", which is not restricted to one specific set of systems [6].

Furthermore, we observe that there is a growing effort known as Linking Open Data[1] to make data interlinked and openly accessible on the Web by following the principles of Linked Data [7]. This effort opens opportunities to unlock a huge potential of data, including the user data. By *reusing* this interlinked data (such as DBpedia[2] and GeoNames[3]), various relationships between data can now be derived and discovered, and thus make data more meaningful and richer. Note that this data is published as RDF and accessible through a SPARQL endpoint. Nevertheless, the distributed nature of the RDF data sources creates a new interesting problem, that is, the problem of integrating RDF data from multiple distributed data sources. There are two possible solutions for this problem: *data centralization* and *query federation* [8,9,10]. The first approach provides a query service over a collection of data copied from different sources on the Web, while the second approach executes queries only on selected datasets that are part of the collection. This observation leads us to investigate how this distributed interlinked data can be *reused* and be beneficial for the purpose of exchanging, integrating, and enriching user data in the interoperable adaptive Web applications.

In this paper, we present the Grapple User Modeling Framework (GUMF) that facilitates the brokerage of user profile information and user model representations. We show how the existing GUMF [11] is extended with a new flexible rule-based method that enhances the reasoning capability of GUMF by allowing the applications

---

[1] http://esw.w3.org/topic/SweoIG/TaskForces/CommunityProjects/ LinkingOpenData
[2] http://dbpedia.org/
[3] http://geonames.org/

to specify a "*recipe*" that guides the new knowledge deduction process in the distributed setting using a rule language called Grapple Derivation Rule language (GDR). GDR extends GUMF with the flexibility for applications to flexibly define configurations that guide the user data integration and enrichment processes. Also, with GDR the applications are able not only to integrate data from GUMF dataspaces, but also to incorporate and reuse linked data published on the Web. Without GDR performing such processes are more complex and may not be efficient. To validate this, the implementation of the GUMF extended with GDR is applied in the GRAPPLE project[4] for user data in the e-learning domain.

The rest of this paper is organized as follows. Section 2 discusses the related work. We briefly introduce GUMF in Section 3. In Section 4, the Grapple Derivation Rule language (GDR) and the GDR Engine are presented. We also elaborate how GUMF is extended with GDR. Section 5 showcases the extended GUMF in the e-learning domain in the context of a concrete project. Finally, Section 6 concludes our discussion.

2   Related Work

In the user modeling research field, a host of approaches have been delivered to address the user model interoperability problem. There are basically two approaches: the *shared format* approach and the *conversion* approach. In the first approach, a common language for a unified user profile (*a lingua franca*) is needed. Examples of this approach are the General User Model Ontology (GUMO) [20] and Composite Capability/Preference Profiles (CC/PP)[5]. This approach is easily exchangeable and interpretable as there is no syntactic and semantic heterogeneity issue to be addressed [20]. However, this approach is not suitable for open and dynamic environments, such as the Web, as it is impractical and in many cases impossible to enforce Web applications to follow the *lingua franca* [21]. The conversion approach is more flexible and suitable for open and dynamic environments [6]. In this approach, a technique has to be developed for converting a user model of one application to another application. It should deal with the problem of syntactic and semantic heterogeneity. The potential drawbacks of this approach are that it is possible that some information is lost during the conversion process, and that it is possible that models are simply incompatible. It is also possible that the mappings are incomplete because required information in one model is not available in the other model.

Furthermore, the Grapple Derivation Rule language builds upon existing rule languages such as the Rule Markup Language (RuleML) [18] defined by the Rule Markup Initiative. RuleML is a markup language developed to express both forward (bottom-up) and backward (top-down) rules in XML for deduction, rewriting, and further inferential-transformational tasks. RuleML itself covers the entire rule spectrum, from *derivation rules* to *transformation rules* to *reaction rules*, and thus can specify queries and inferences in Web ontologies, mappings between Web ontologies, and dynamic Web behaviors of workflows, services, and agents. The

---

[4] GRAPPLE is the acronym for an EU FP7 STREP Project denoting "Generic Responsive Adaptive Personalized Learning Environment" http://www.grapple-project.org/
[5] http://www.w3.org/Mobile/CCPP/

Semantic Web Rule Language (SWRL) [15] is a proposal for a Semantic Web rules-language that is based on a combination of the OWL DL and OWL Lite sublanguages of the OWL Web Ontology Language [16,17] with the Unary/Binary Datalog RuleML sublanguages of the Rule Markup Language [18]. Rules are of the form of an implication between an antecedent (body) and consequent (head). The intended meaning can be read as "whenever the conditions specified in the antecedent hold, then the conditions specified in the consequent must also hold". The observation that there are currently many "rules languages" in existence in the web community lead to the Rule Interchange Format (RIF) which is a standard in development within the W3C Semantic Web Activity [19]. GDR is different from the existing rule languages at least for the following reasons. Firstly, it provides definitions of premise and consequent at the level of Grapple statements that constitute the lingua franca when interacting with GUMF. Secondly, it allows the integration of knowledge using multiple distributed data sources published as Linked Data on the Web.

To deal with the distributed nature of data sources published on the Web as RDF data, recently, there has been much research on the subject of integrating different RDF graphs into a single RDF graph and the related problem of querying distributed RDF data sources that were integrated into a single virtual RDF data source. Langegger et al. present in [10,12] the *SemWIQ* system that has a mediator-wrapper architecture and allows the integrated data to be queried with a subset of SPARQL and implements and optimizes these queries by translating them to an algebra called ARQ2. The notion of *networked graphs* is introduced by Schenk et al. in [13] where they discuss the problem of integrating different RDF graphs by defining SPARQL-based integration rules between them. The problem of optimizing a query that queries different external RDF data sources is discussed by Zemanek et al. in [14] which concentrates on minimizing communication cost by using semi-joins. The same problem is addressed by Hartig et al. in [9] which focuses on the subproblems of efficiently finding the data sources related to the query during query execution and efficiently executing the queries by using an iterator-based pipeline approach in its query evaluation plans. Finally, the *DARQ* system, described by Quilitz et al. in [8] allows the integration of distributed RDF data sources into a single virtual RDF data source by specifying which data is to be found in which external data source. It uses query-rewriting and cost-based query optimization to obtain efficient distributed query evaluation plans.

3   GUMF

The Grapple User Modeling Framework (GUMF) [11] enables systems to benefit from the multi-faceted user data traces that are distributed across different Web systems. GUMF provides generic user modeling functionality that is adaptable to the requirements of the individual systems that utilize it: it aggregates, contextualizes and models user data so that systems can easily incorporate the data without having to solve interoperability issues such as schema mapping. Further, GUMF together with its plug-ins feature reasoning capabilities for deducing new information about users from their profile and activity data. In the context of the afore mentioned GRAPPLE project, GUMF is applied to provide user modeling functionality across e-learning application boundaries and thus it connects learning management systems such as Moodle, AHA!, and CLIX. In the remainder of this section we present the architecture and components of GUMF in more detail.

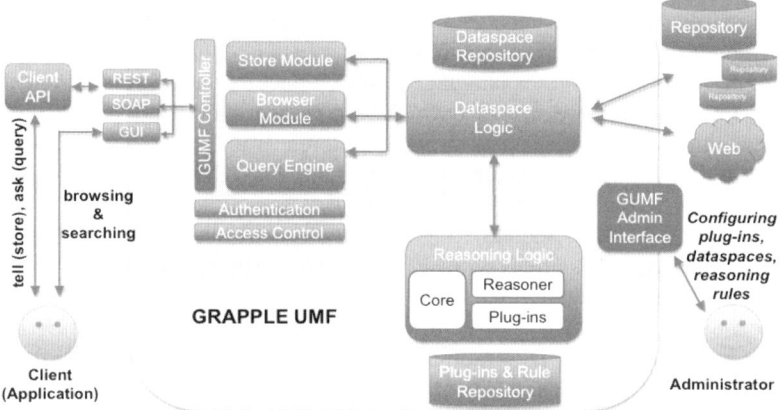

Fig. 1. GUMF Architecture

3.1 Architecture and Building Blocks

GUMF can be considered as an intelligent storage and reasoning engine that provides uniform access to distributed heterogeneous user data. Fig. 1 FigFshows its architecture. The blue elements at the top provide the essential, generic functionality of the framework; the purple components at the bottom provide generic as well as domain-specific plug-in and reasoning functionality.

Client applications can access GUMF either via a RESTful or SOAP-based API. Further, there is a Java Client API that facilitates development of GUMF client applications. Client applications mainly approach GUMF to store user information (handled by the *Store Module*) or to query for information (handled by *Query Engine*). By default, user profile information is modeled by means of Grapple statements (see below) that constitute the lingua franca when interacting with GUMF. Grapple statements are basically reified RDF statements about a user, enriched with DCMI metadata[6] for describing provenance details. The current GUMF implementation supports SPARQL [4] and SeRQL [2] queries as well as a pattern-based query language – *Grapple Query* language – that exploits the Grapple statement structure to specify what kind of statements should be returned.

Queries are executed on so-called dataspaces (*Dataspace Logic*) that logically bundle data that is possibly distributed across different sources on the Web, as well as offer reasoning functionality provided by different reasoners and plug-ins of the *Reasoning Logic*. Dataspaces thus go beyond the notion of namespaces as they explicitly denote a set of things (e.g. data, reasoning rules, data aggregation plug-ins, schema mapping rules), on which an operation – such as a query, store or reasoning

---

[6] http://dublincore.org/documents/dcmi-terms/

operation – should be performed. In more detail, such dataspaces represent the part of GUMF that a certain client application is managing and responsible for, i.e., its own workspace. The Administrator of a GUMF client application can configure dataspaces and plug-ins via the *GUMF Admin Interface* (see Fig. 1F). Activating or deactivating plug-ins and adjusting plug-ins and reasoning rules directly influence the behavior of dataspaces. Inspired by Web 2.0 practices, a key principle of GUMF is that dataspaces can be shared across different client applications. Therefore, clients can subscribe to other dataspaces, as long as the administrator of the dataspace approves them. When subscribed to a dataspace, the client is allowed to query it. However, it might still not be allowed to access all statements that are made available via the dataspace, as fine-grained access control functionality can be embedded in the dataspaces as well.

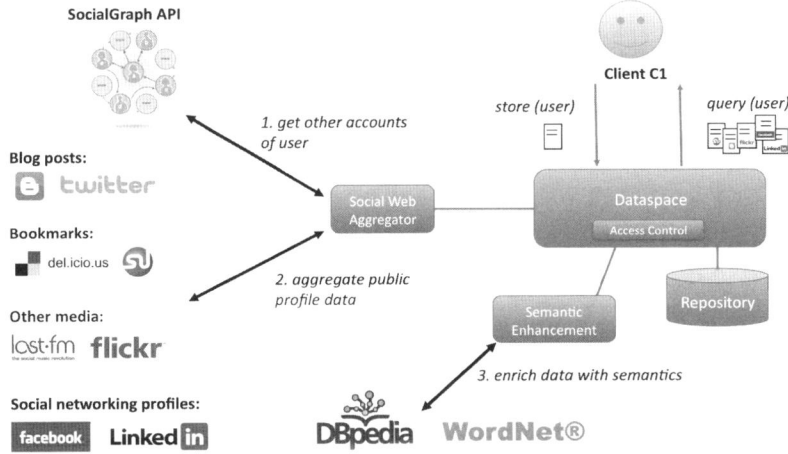

Fig. 2. User Modeling with GUMF Dataspaces

3.2 User Modeling with Intelligent Dataspaces

User modeling functionality of GUMF is embedded into dataspaces. In [1] we implemented the user modeling components that are applied to enrich data stored by client applications as depicted in Fig. 2. Client C1 stores information about a user in a dataspace and more precisely in the repository associated with the dataspace. C1 might for example report that a new user registered to the system. Information about the user is internally modeled by means of Grapple statements, for example, C1 stores that a new user whose name is "Bob Myers" registered to C1. Fig. 3 shows the corresponding statement in RDF/XML syntax.

Grapple statements are subject-predicate-object bindings enriched with metadata. They not only describe the actual statement, i.e. Bob's (*gc:subject* = *http://bob.myopenid.com*) name (*gc:predicate* = *http://xmlns.com/foaf/0.1/name*) is "Bob Myers" (= *gc:object*), but also additional details such as the creator of the statement (*gc:creator*), the time when the statement was created (*gc:created*) or the degree to which the statement holds for the subject (*gc:level*)[7]. Storing a Grapple statement might trigger some plug-ins embodied into the dataspace. In Fig. 2, the *Social Web Aggregator* [1] obtains other accounts the user has via the Social Graph API[8]. Given these mappings, the plug-in gathers – if available – public profile data about the user from the corresponding platforms: tag-based profiles from Delicious, StumbleUpon, Last.fm, and Flickr, social network profiles from LinkedIn and Facebook, and blog posts from Twitter and Blogspot. The aggregated profile data is then enriched with semantic annotations (*Semantic Enhancement* in Fig. 2). In particular, the elements of the tag-based profiles [3] are mapped to DBpedia URIs that specify the semantic meaning of the tags and WordNet[9] categories are applied to cluster the profile [1]. Hence, based on the rather basic Grapple statement, which is listed in Fig. 3, GUMF gathers the distributed profile traces of the user so that the client can exploit a rich profile the next time it is querying the dataspace (cf. Fig. 2).

```
<?xml version="1.0" encoding="UTF-8"?>
<rdf:RDF xmlns:gc="http://www.grapple-project.org/grapple-core/"
         xmlns:rdf="http://www.w3.org/1999/02/22-rdf-syntax-ns#"
         xmlns:xsd="http://www.w3.org/2001/XMLSchema#">
<gc:Statement rdf:about="http://grapple-project.org/2010-01-28-526341">
  <gc:subject redf:resource="http://bob.myopenid.com"/>
  <gc:predicate rdf:resource="http://xmlns.com/foaf/0.1/name"/>
  <gc:object>Bob Myers</gc:object>
  <gc:level rdf:datatype="xsd:double">1.0</gc:level>
  <gc:created rdf:datatype="xsd:dateTime">
    2010-01-28T00:09:20.621+02:00
  </gc:created>
  <gc:creator rdf:resource="http://grapple-project.org/client/1"/>
</gc:Statement>
</rdf:RDF>
```

Fig. 3. Grapple statement: Bob's name is Bob Myers

The components that are plugged into dataspaces come in different flavors: Some plug-ins are black-box components while others are rule-based and are thus highly flexible. In [1], an example of black-box plug-ins is presented and in [11] a rule-based plug-in that is limited to integrate data *only* within a single Grapple dataspace is discussed. In the next section, we introduce the GDR language that extends and enhances the reasoning capability of GUMF and enables developers and administrators to create such flexible, rule-based dataspace plug-ins that are capable of integrating user data from multiple Grapple dataspaces and data published as Linked Data on the Web.

4 GDR

In this section, we elaborate in details on the Grapple Derivation Rule language (GDR) that enables GUMF to provide a flexible way of defining plug-ins by allowing

---

[7] Note that gc:creator and gc:created are sub-properties of dc:creator and dc:created as defined by DCMI.
[8] http://socialgraph.apis.google.com
[9] http://wordnet.princeton.edu

the applications to specify a "*recipe*" for integrating and enriching user data. We also discuss the GDR Engine that processes a GDR rule and derives new Grapple statements. Finally, we present how GUMF is extended with GDR.

4.1 GDR Definition

In the human readable syntax, a GDR rule has the form: $a \Rightarrow c$, where a and c are the antecedent and consequent of the rule, respectively, where a is a conjunction of premises written $p_1 \wedge \ldots \wedge p_n$. The premises of a GDR rule are classified into two types: *dataspace premises* and *external source premises*. A dataspace premise describes conditions over a Grapple dataspace in the form of a pattern-based Grapple Query. An external source premise specifies conditions in the form of triple patterns over an external data source accessible through a SPARQL endpoint. The consequent describes the Grapple statements that will be derived if all the premises are hold. It specifies the subject, predicate, and object properties of the Grapple statements, and optionally the level properties. A GDR rule also has extra information such as name, description, and creator. Variables are indicated using the standard convention of prefixing them with a question mark (e.g., ?x). The GDR rule is formally defined as following.

Definition 1. [Dataspace Premise] A dataspace premise d is a 2-tuple (ds, f), where ds is the Grapple dataspace identifier, and f is partial function that maps a finite set of Grapple statement properties to variables and constants. A set of dataspace premises is defined as D.

Definition 2. [External Source Premise] An external source premise e is a 4-tuple (*uri, endpoint, namedGraph, T*), where *uri* is the informal identifier of the dataset, *endpoint* is the URI of SPARQL endpoint of the data source where the dataset is stored, *namedGraph* is the named graph that is used to store the dataset in the data source, and T is a basic graph pattern with at least one triple pattern. A set of external source premises is defined as E.

Definition 3. [Consequent] A consequent c is a dataspace premise (ds, f), where f is defined for at least *gc:subject*, *gc:predicate*, and *gc:object*, and at most also *gc:level*.

Definition 4. [A GDR Rule] A GDR rule r is a 3-tuple (M, A, c), with:

- M is a set of additional information of r, such as the name, description, and creator of the rule,
- A is the antecedent of the rule, which is a conjunction of premises written $p_1 \wedge \ldots \wedge p_n$, where $p_i \in (D \cup E)$ and $n > 0$,
- c is the consequent of the rule such that all variables in c appear in at least one premise of A.

In Section 4.3, we present the XML serialization format of GDR by an example. The next section introduces the engine that interprets and enforces given GDR rules.

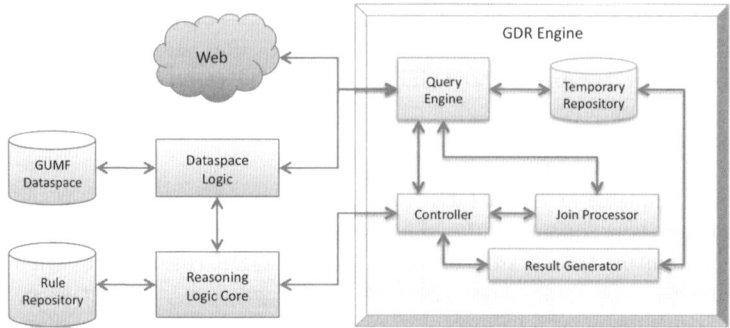

Fig. 4. The Architecture of GDR Engine

4.2   GDR Engine

The GDR Engine is responsible to derive new knowledge based on a "*recipe*" defined in a GDR rule that possibly effects the integration of data from different data sources. Fig. 4 depicts its architecture and interactions with other GUMF modules. The GDR Engine consists of five components: the *Controller*, the *Query Engine* (*QE*), the *Join Processor* (*JP*), the *Result Generator* (*RG*), and the *Temporary Repository* (*TR*).

The Controller manages the whole process happening inside the GDR Engine. It receives requests from the GUMF Reasoning Logic Core. It also utilizes the *QE* to fetch data and maintain intermediate data temporarily and the *JP* component to perform join operations. It exploits the *RG* to generate a set of newly derived Grapple statements. The *QE* inside the GDR Engine performs the following tasks: 1) by sending query requests to the GUMF Dataspace Logic, it fetches data from GUMF dataspaces; 2) it queries external data sources through SPARQL endpoints; 3) it reads and writes data that is temporarily maintained in the *TR*. The *TR* component is an RDF repository that is used to store the RDF triples of intermediate results (e.g. join results). The *JP* component is responsible in performing join operations. This component interacts with the *QE* whenever it wants to retrieve data from the *TR* and the external data sources as well as to put data into the *TR*. The *RG* analyzes the premises and consequent of the rule, generates a SPARQL query that will be issued against the GDR's temporary repository, and constructs a set of Grapple statements to be returned to the GUMF Reasoning Logic Core as the result.

When a client application issues a pattern-based query q to GUMF, the Dataspace Logic forwards query q to the Reasoning Logic. The Reasoning Logic Core module then checks if there are any GDR rules relevant for q. For each relevant rule r, the Reasoning Logic Core sends a request to the GDR Engine to process rule r.

The GDR Engine first evaluates all the dataspace premises of r and maintains the result of each dataspace premise evaluation in the *TR* by utilizing the *QE*. Based on the Grapple Query patterns specified in the dataspace premises, the *QE* sends requests to the GUMF Dataspace Logic to fetch data from dataspaces. If there is at least one dataspace premise evaluation that returns no result, then the GDR Engine stops processing r and returns *null* meaning that rule r derives no result. The intuition behind this is as following. The empty result of a dataspace premise d means that there is no Grapple statement that satisfies the pattern defined in premise d. Consequently, premise d does not hold, and thus rule r does not hold. In the case that all dataspace premise evaluations return results, the GDR Engine continues processing r.

Next, the GDR Engine exploits the *JP* to join the dataspace premises using the results stored in the *TR*. If two dataspace premises d_1 and d_2 share the same variables, then they can be joined. The join results are also temporarily stored in the *TR*. If two premises can be joined, but the join result is empty, then the GDR Engine stops processing r and returns *null*. Note that in the current implementation, the GDR Engine joins the dataspace premises based on the appearance order in r. Optimizing the join order is an interesting and non-trivial research problem. However, since the focus of this paper is to present a configurable method for integrating and enriching user data, the join optimization issue will be investigated in the future.

The next step is to process the external premises. The GDR Engine also processes the external source premises according to the appearance order in r. Given an external premise e, the *JP* checks if e can be joined with previously processed premises (both dataspace and external source premises). If e can be joined, then the *QE* is exploited to fetch the data of previously processed premises stored in the *TR*, to construct SPARQL queries based on the specified triple patterns and fetched data, and to send these queries to the SPARQL endpoint of e. The results of these queries are maintained by the TR. Note that the results can be used in processing other external source premises. If e can be joined with another external source premise e_j that is not processed yet, the *JP* will process e_j first before processing e. This process stops if all possible joins between premises are performed. If there is an external source premise e_k that cannot be joined with other premises, the *JP* requests the *QE* to constructs a SPARQL query only based on the specified triple patterns. In constructing the query, the *QE* takes into account whether or not premise e_k and the consequent of r share at least one variable. If they do not share any variables, then the *QE* rewrites the query to an `ASK` form to test whether or not it has a solution. The intuition is that if a premise cannot be joined with other premises and the data from this premise will not be used in the final result, then we only need to check whether this premise returns any results. The constructed query then is sent to the SPARQL endpoint of e_k, and the result is stored in the *TR*.

After all premises are processed, then the Controller sends request to the *RG* that generates a set of new Grapple statements. The *RG* analyzes rule r and generates a SPARQL query that is executed against the temporary data stored in the *TR*. The result of this SPARQL query is then modeled as Grapple statement and sent to the Controller, which subsequently sends it to the Reasoning Logic Core.

4.3 Extending GUMF with GDR

We implemented the GDR Engine in Java. For the Temporary Repository component, we choose to base our implementation on the open-source RDF framework Sesame[10]. Sesame offers a good level abstraction on connecting to and querying of RDF data, similar to JDBC. The GDR Engine is integrated into GUMF as a module inside the

---

[10] http://www.openrdf.org/

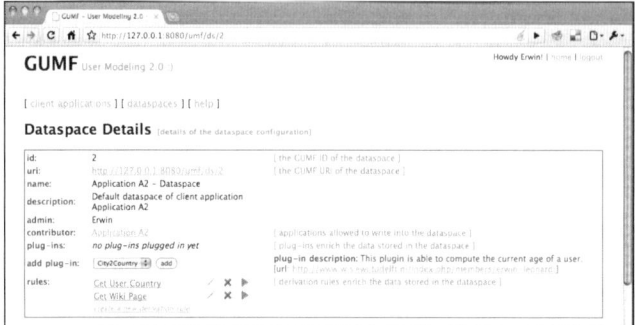

Fig. 5. GUMF Administrator Page

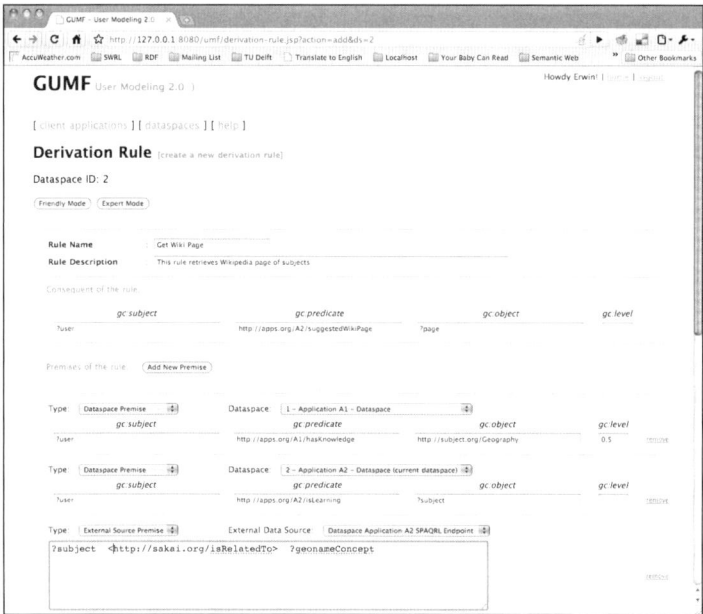

Fig. 6. GDR Rule Creation Page: Friendly Mode

Reasoning Logic. For this purpose, there are several components in GUMF that have to be extended. The "Plug-in & Rule Repository" of GUMF is extended to be able to store the GDR rules. We added a feature in the Reasoning Logic Core component to detect which GDR rules in the dataspace are relevant for a Grapple query sent by GUMF client. This can be done by analyzing the consequent of the rules. The Reasoning Logic Core has to be able to communicate with the GDR Engine. Furthermore, the GUMF administrator page is extended such that it shows the list of specified GDR and a hyperlink to the GDR rule creation page (Fig. 5). There are two ways of creating a GDR rule: *friendly mode* and *expert mode*. In the friendly mode (Fig. 6) the dataspace administrator specifies the rule by filling up provided form fields. In the expert mode, the administrator has to type the rule in XML format.

gc:subject	gc:predicate	gc:object	gc:level
user:anna	profile:origin	"Dubai"	1.0
user:anna	A1:hasKnowledge	subject:Geography	0.8
user:bob	profile:origin	"Delft"	1.0
user:cindy	profile:origin	"Johannesburg"	1.0
user:cindy	A1:hasKnowledge	subject:Geography	0.6
user:donald	profile:origin	"Beijing"	1.0
user:donald	A1:hasKnowledge	subject:Geography	0.4

(a) Grapple Statements in A₁ Dataspace

gc:subject	gc:predicate	gc:object	gc:level
user:anna	A2:isLearning	subject:Malaysia	0.0
user:bob	A2:isLearning	subject:Japan	0.0
user:cindy	A2:isLearning	subject:Delft	0.0
user:donald	A2:isLearning	subject:Bangkok	0.0
user:frans	A2:isLearning	subject:Japan	0.0

(b) Grapple Statements in A₂ Dataspace

subject	predicate	object
subject:Malaysia	A2:isRelatedTo	<http://sws.geonames.org/1733045/>
subject:Japan	A2:isRelatedTo	<http://sws.geonames.org/993960/>
subject:Delft	A2:isRelatedTo	<http://sws.geonames.org/2757345/>
subject:Bangkok	A2:isRelatedTo	<http://sws.geonames.org/2757345/>

(c) Derived Tripples in A₂ Dataspace

Fig. 7. The Snapshot of A1 and A2 Dataspaces (partial view)

5 Use Case

In this section, we showcase the extended GUMF in the e-learning domain in the context of the GRAPPLE project. GDR applied in GUMF allows for distributed user modeling across e-learning systems. Suppose there are two adaptive e-learning applications, namely, A1 and A2 that use GUMF. A1 that is a Moodle-based application is used for a basic Geography course, and A2 that is an AHA!-based application is used for an Urban Geography course. Fig. 7(a) depicts a set of Grapple statements in the A1 dataspace. Fig. 7(b) depicts a set of Grapple statements in the A2 dataspace. A set of triples derived by a semantic enhancement plug-ins that relates data in the dataspace to the GeoNames concepts is shown in Fig. 7(c).

The creator of A_2 would like to suggest Wikipedia pages about the subject that the students are currently taking for enhancing their knowledge about the subject if they have good basic knowledge about Geography. She knows that application A_1 provides the basic Geography course, and thus chooses to reuse data from A_1. She applies for a dataspace subscription to A_1 and the creator of A_1 approves this subscription request. Thus, A_2 is able to query data in the A_1 dataspace. Moreover, the creator of A_2 defines a GDR rule named "Get Wiki Page" as shown in Fig. 8 that can be used to integrate data from four distributed data source to get the URLs of Wikipedia pages.

There are two dataspace premises and three external source premises defined in the GDR rule. The first dataspace premise (Lines 06 – 11) is used to determine the students who have passed the Geography subject using application A_1 with at least a 50% score. The second one (Lines 12 – 16) retrieves a set of Grapple statements whose *gc:predicate* is *http://apps.org/A2/isLearning*. These dataspace premises are joined, and the result of join is as following.

user	subject
user:anna	subject:Malaysia
user:cindy	subject:Delft

Using this result, the external source premises are processed. For example, the bindings of variable *subject* that is one of the variables in the first external premise

```
01 <gdr:rule xmlns:gdr="http://www.grapple-project.org/grapple-derivation-rule/"
02          xmlns:gc="http://www.grapple-project.org/grapple-core/"
03          xmlns:rdf="http://www.w3.org/1999/02/22-rdf-syntax-ns#"
04      name="Get Wiki Page" creator="http://localhost:8080/umf/client/2"
05      description="This rule retrieves Wikipedia page of subjects">
06       <gdr:premise dataspace="1">
07          <gc:subject>?user</gc:subject>
08          <gc:predicate rdf:resource="http://apps.org/A1/hasKnowledge" />
09          <gc:object>http://subject.org/Geography</gc:object>
10          <gc:level>0.5</gc:level>
11       </gdr:premise>
12       <gdr:premise dataspace="2">
13          <gc:subject>?user</gc:subject>
14          <gc:predicate rdf:resource="http://apps.org/A2/isLearning" />
15          <gc:object>?subject</gc:object>
16       </gdr:premise>
17       <gdr:premise uri="http://localhost:8080/umf/ds/2" namedGraph=""
                  endpoint="http://localhost:8080/umf/rest/sparql/ds/2?client=2&amp;token=123">
18          <gdr:pattern>?subject
                  &lt;http://sakai.org/isRelatedTo&gt; ?geonameConcept</gdr:pattern>
19       </gdr:premise>
20       <gdr:premise uri="http://geonames.org/" namedGraph="http://geonames.org"
                  endpoint="http://localhost:8890/sparql">
21          <gdr:pattern>?geonameConcept
                  &lt;http://www.w3.org/2002/07/owl#sameAs&gt; ?dbpediaConcept</gdr:pattern>
22       </gdr:premise>
23       <gdr:premise uri="http://dbpedia.org/" namedGraph="http://dbpedia.org"
                  endpoint="http://dbpedia.org/sparql" >
24          <gdr:pattern>?dbpediaConcept
                  &lt;http://xmlns.com/foaf/0.1/page&gt; ?page</gdr:pattern>
25       </gdr:premise>
26       <gdr:consequent dataspace="2">
27          <gc:subject>?user</gc:subject>
28          <gc:predicate rdf:resource="http://apps.org/A2/suggestedWikiPage" />
29          <gc:object>?page</gc:object>
30       </gdr:consequent>
31 </gdr:rule>
```

Fig. 8. An Example of a GDR Rule in XML Syntax

(Lines 17 – 19) are available. Hence, the values of the bindings of variable subject and the triple pattern specified in this premise are used to construct SPARQL queries that will be sent to the SPARQL endpoint of the premise. For the first external source premise, the following SPARQL query is constructed.

```
SELECT ?geonameConcept ?subject
WHERE {
  {  ?subject <http://sakai.org/isRelatedTo> ?geonameConcept .
     FILTER (?subject = <http://subject.org/Delft> ) . }
  UNION
  {  ?subject <http://sakai.org/isRelatedTo> ?geonameConcept .
     FILTER (?subject = <http://subject.org/Malaysia> ) . }
}
```

The result of this query is stored in the Temporary Repository for further processes.

Basically, the external source premises specify the graph patterns across three different data sources (namely, dataspace A_2, GeoNames, and DBpedia) that must be matched in order to get the Wikipedia pages. For example, Fig. 9 depicts the path from resource *subject:Malaysia* to resource *http://en.wikipedia.org/wiki/Malaysia*. The GDR rule in Fig. 8 derives two Grapple statements as depicted in Fig. 10.

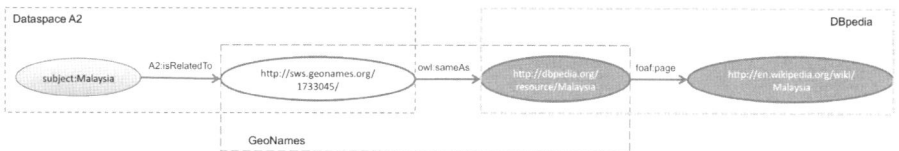

Fig. 9. Graph Patterns Across Three Different Data Sources

gc:subject	gc:predicate	gc:object
user:anna	A2:suggestedWikiPage	<http://en.wikipedia.org/wiki/Malaysia>
user:cindy	A2:suggestedWikiPage	<http://en.wikipedia.org/wiki/Delft>

Fig. 10. Derived Grapple Statements

6 Conclusion

In this paper, we have extended the Grapple User Modeling Framework (GUMF) with the Grapple Derivation Rule language (GDR), and thus the reasoning capability of GUMF is extended and enhanced by allowing Web applications to exchange, reuse, integrate, and enrich the user data using not only data in Grapple dataspaces, but also openly accessible data published on the Web as Linked Data in a flexible and configurable way. We have implemented and integrated our method into the GUMF and applied it in an e-learning setting where different e-learning systems (such as Moodle, AHA!, and CLIX) are connected. Our method successfully supports the integration and enrichment of user data as demonstrated by a representative use case.

As a continuation of our work, we plan to extend GDR specification such that we can derive not only Grapple statements, but also RDF graphs. We also plan to improve the join heuristic of GDR Engine as currently the join process only follows the order of appearance in the rule. We also would like to evaluate the GDR Engine in terms of performance and, especially, scalability to explore the limit of our approach as Semantic Web reasoning applications typically run into scalability issues.

Acknowledgments. This work was partially supported by the European 7th Framework Program project GRAPPLE ("Generic Responsive Adaptive Personalized Learning Environment"): http://www.grapple-project.org.

References

1. Abel, F., Henze, N., Herder, E., Krause, D.: Interweaving Public Profile Data on the Web, Technical Report, L3S Research Center, Hannover, Germany (2010)
2. Broeskstra, J., Kampman, A.: SeRQL: A Second Generation RDF Query Language. In: SWAD-Europe Workshop on Semantic Web Storage and Retrieval, Vrije Universiteit, Amsterdam, Netherlands (2003)
3. Firan, C.S., Nejdl, W., Paiu, R.: The Benefit of Using Tag-based Profiles. In: Proc. of LA-WEB 2007, Washington, DC, USA. IEEE Computer Society, Los Alamitos (2007)
4. Prud'hommeaux, E., Seaborne, A.: SPARQL Query Language for RDF. W3C Recommendation (January 2008),
   http://www.w3.org/TR/rdf-sparql-query/

5. Stewart, C., Celik, I., Cristea, A., Ashman, H.: Interoperability between aeh user models. In: Proc. of APS 2006 (2006)
6. Aroyo, L., Dolog, P., Houben, G., Kravcik, M., Naeve, A., Nilsson, M., Wild, F.: Interoperability in personalized adaptive learning. J. Educational Technology &Society 9(2), 4–18 (2006)
7. Berners-Lee, T.: Design Issues: Linked Data (2006), `http://www.w3.org/DesignIssues/LinkedData.html`
8. Quilitz, B., Leser, U.: Querying Distributed RDF Data Sources with SPARQL. In: Bechhofer, S., Hauswirth, M., Hoffmann, J., Koubarakis, M. (eds.) ESWC 2008. LNCS, vol. 5021, pp. 524–538. Springer, Heidelberg (2008)
9. Hartig, O., Bizer, C., Freytag, J.-C.: Executing SPARQL Queries over the Web of Linked Data. In: Bernstein, A., Karger, D.R., Heath, T., Feigenbaum, L., Maynard, D., Motta, E., Thirunarayan, K. (eds.) ISWC 2009. LNCS, vol. 5823, pp. 293–309. Springer, Heidelberg (2009)
10. Langegger, A., Wöß, W., Blöchl, M.: A semantic web middleware for virtual data integration on the web. In: Bechhofer, S., Hauswirth, M., Hoffmann, J., Koubarakis, M. (eds.) ESWC 2008. LNCS, vol. 5021, pp. 493–507. Springer, Heidelberg (2008)
11. Abel, F., Heckmann, D., Herder, E., Hidders, J., Houben, G.-J., Krause, D., Leonardi, E., van der Sluijs, K.: A Framework for Flexible User Profile Mashups. In: The Proc. of the APWEB 2.0 2009 Workshop in conjunction UMAP 2009 (2009)
12. Langegger, A.: Virtual data integration on the web: novel methods for accessing heterogeneous and distributed data with rich semantics. In: Proc. of iiWAS'08 (2008)
13. Schenk, S., Staab, S.: Networked graphs: a declarative mechanism for sparql rules, sparql views and rdf data integration on the web. In: Proc. of WWW '08 (2008)
14. Zemanek, J., Schenk, S., Svatek, V.: Optimizing sparql queriesover disparate rdf data sources through distributed semi-joins. In: ISWC 2008 Poster and Demo Session Proceedings. CEUR-WS (2008)
15. Horrocks, I., Patel-Schneider, P.F., Boley, H., Tabet, S., Grosof, B., Dean, M.: SWRL: A Semantic Web Rule Language Combining OWL and RuleML, `http://www.w3.org/Submission/SWRL/`
16. McGuinness, D.L., van Harmelen, F. (eds.): OWL Web Ontology Language Overview, W3C Recommendation (February 2004), `http://www.w3.org/TR/owl-features/`
17. OWL W3C: Working Group (eds.): OWL 2 Web Ontology Language Document Overview, W3C Recommendation (October 2009), `http://www.w3.org/TR/owl2-overview/`
18. Rule Markup Language Initiative. Rule Markup Language (RuleML), `http://ruleml.org/`
19. Kifer, M.: Rule Interchange Format: The Framework. In: Calvanese, D., Lausen, G. (eds.) RR 2008. LNCS, vol. 5341, pp. 1–11. Springer, Heidelberg (2008)
20. Heckmann, D., Schwartz, T., Brandherm, B., Schmitz, M., von Wilamowitz-Moellendorff, M.: GUMO - The General User Model Ontology. In: Ardissono, L., Brna, P., Mitrović, A. (eds.) UM 2005. LNCS (LNAI), vol. 3538, Springer, Heidelberg (2005)
21. Kuflik, T.: Semantically-Enhanced User Models Mediation: Research Agenda. In: Proc. of UbiqUM 2008 Workshop at IUI 2008, Gran Canaria, Spain (2008)
22. Finkel, J.R., Grenager, T., Manning, C.: Incorporating Non-local Information into Information Extraction Systems by Gibbs Sampling. In: Proc. of ACL 2005 (2005)

Ranking the Linked Data: The Case of DBpedia

Roberto Mirizzi[1], Azzurra Ragone[1,2],
Tommaso Di Noia[1], and Eugenio Di Sciascio[1]

[1] Politecnico di Bari – Via Orabona, 4, 70125 Bari, Italy
mirizzi@deemail.poliba.it, {ragone,dinoia,disciacio}@poliba.it
[2] University of Trento – Via Sommarive, 14, 38100 Povo (Trento), Italy
ragone@disi.unitn.it

Abstract. The recent proliferation of crowd computing initiatives on
the web calls for smarter methodologies and tools to annotate, query
and explore repositories. There is the need for scalable techniques able
to return also approximate results with respect to a given query as a
ranked set of promising alternatives. In this paper we concentrate on
annotation and retrieval of software components, exploiting semantic
tagging relying on Linked Open Data. We focus on DBpedia and propose
a new hybrid methodology to rank resources exploiting: (i) the graph-
based nature of the underlying RDF structure, (ii) context independent
semantic relations in the graph and (iii) external information sources such
as classical search engine results and social tagging systems. We compare
our approach with other RDF similarity measures, proving the validity
of our algorithm with an extensive evaluation involving real users.

1 Introduction

The emergence of the crowd computing initiative has brought on the web a new
wave of tools enabling collaboration and sharing of ideas and projects, ranging from simple blogs to social networks, sharing software platforms and even
mashups. However, when these web-based tools reach the "critical mass" one of
the problem that suddenly arises is how to retrieve content of interest from such
rich repositories. As a way of example, we can refer to a platform to share software components, like ProgrammableWeb[1], where programmers can share APIs
and mashups. When a user uploads a new piece of code, she tags it so that the
component will be later easily retrievable by other users. Components can be
retrieved through a keywords-based search or browsing accross *categories*, *most
popular* or *new updates*. The limits of such platforms, though very popular and
spread out on the entire web, are the usual ones related to keywords-based retrieval systems, e.g., if the user is looking for a resource tagged with either *Drupal*
or *Joomla!*[2], the resources tagged with *CMS* (*Content Management System*) will
not be retrieved. For example, in ProgrammableWeb, APIs as *ThemeForest* and

---

[1] http://www.programmableweb.com
[2] http://www.drupal.org, http://www.joomla.org

B. Benatallah et al. (Eds.): ICWE 2010, LNCS 6189, pp. 337–354, 2010.

Ecordia[3] are tagged with *CMS* but not with *Drupal* nor *Joomla*, even if in their abstracts it is explicitly written that they are available also for the two specific CMSs.

An effective system should be able to return also *approximate* results w.r.t. the user's query, results ranked based on the *similarity* of each software components to the user's request. Referring to the previous example, it means that the two mentioned APIs should be suggested as relevant even if the exact searched tag is not present in their description, due to their similarities with the query.

Another issue stricly coupled with the keyword-based nature of current tagging systems on the web is synonymy. Different tags having the same meaning can be used to annotate the same content. `Faviki`[4] is a tool for social bookmarking that helps users to tag documents using `DBpedia` [2] terms extracted from Wikipedia. Although it is a good starting point to cope with synonymy, it does not solve the problem of *ranking* tags w.r.t. a query. Moreover it does not provide the user with any suggestion of tags related to the ones selected during the annotation phase, e.g., if the user tags a page with *Drupal*, the tool does not suggest to tag the page with *CMS* too.

Partially inspired by `Faviki`, in this paper we propose a new hybrid approach to rank `RDF` resources within `Linked Data` [1], focusing in particular on `DBpedia`, which is part of the Linked Data Cloud. Given a query (tag), the system is able to retrieve a set of ranked resources (e.g., annotated software components) semantically related to the requested one. There are two main relevant aspects in our approach: (1) the system returns resources within a specific context, e.g., *IT, Business, Movies*; (2) the final ranking takes into account not only `DBpedia` links but it combines the `DBpedia` graph exploration with information coming from external textual information sources such as web search engines and social tagging systems.

A system able to compute a ranking among `DBpedia` nodes can be useful both during the annotation phase and during the retrieval one. On the one hand, while annotating a resource, the system will suggest new tags semantically related to the ones already elicited by the user. On the other hand, given a query formulated as a set of tags, the system will return also resources whose tags are semantically related to the ones representing the query. For instance, if a user is annotating an API for ProgrammableWeb with the tag *CMS* (which refers to `DBpedia` resource `http://dbpedia.org/resource/Content_management_system`), then the system will suggest related tags as *Drupal, Joomla* and *Magento* (each one related to their own `DBpedia` resource).

Main contributions of this work are:

– A novel *hybrid* approach to rank resources on `DBpedia` w.r.t. a given query. Our system combines the advantages of a *semantic-based* approach (relying on a `RDF` graph) with the benefits of *text-based* IR approaches as it also exploits the results coming from the most popular *search engines* (Google, Yahoo!, Bing) and from a popular *social bookmarking system* (Delicious). Moreover,

---

[3] `http://www.programmableweb.com/api/themeforest|ecordia`
[4] `http://www.faviki.com`

our ranking algorithm is enhanced by *textual* and *link analysis* (abstracts and wikilinks in `DBpedia` coming from Wikipedia).

- A *relative* ranking system: differently from PageRank-style algorithms, each node in the graph has not an importance value per se, but it is ranked w.r.t. its neighbourhood nodes. That is, each node has a different importance value depending on the performed query. In our system we want to rank resources w.r.t. a given query by retrieving a ranked list of resources. For this reason we compute a weight for each mutual relation between resources, instead of a weight for the single resource, as in PageRank-style algorithms.
- A back-end system for the semantic annotation of web resources, useful in both the *tagging* phase and in the *retrieval* one.
- An extensive evaluation of our algorithm with real users and comparison w.r.t. other four different ranking algorithms, which provides evidence of the quality of our approach.

The remainder of the paper is structured as follows: in Section 2 we introduce and detail our ranking algorithm *DBpediaRanker*. In Section 3 we present a prototype that highlight some characteristics of the approach. Then, in Section 4, we show and discuss the results of the experimental evaluation. In Section 5 we discuss relevant related works. Conclusion and future work close the paper.

2   DBpediaRanker: RDF Ranking in DBpedia

In a nutshell, *DBpediaRanker*[5] explores the `DBpedia` graph and queries external information sources in order to compute a *similarity value* for each pair of resources reached during the exploration. The operations are performed offline and, at the end, the result is a weighted graph where nodes are `DBpedia` resources and weights represent the similarity value between the two nodes. The graph so obtained will then be used at runtime, (i) in the annotation phase, to suggest *similar* tags to users and (ii) in the retrieval phase, to retrieve a list of resources, ranked w.r.t. a given query.

The exploration of the graph can be limited to a specific *context*. In our experimental setting we limited our exploration to the IT context and, specifically, to programming languages and database systems, as detailed in Section 2.3.

For each node in the graph a depth-first search is performed, stopped after a number of n hops, with n depending on the context. The exploration starts from a set of **seed nodes** and then goes recursively: in place of seed nodes, at each step the algorithm identifies a number of **representative nodes** of the context, i.e., popular nodes that, at that step, have been reached several times during the exploration. Representative nodes are used to determine if every node reached during the exploration is in the context and then should be further explored in the next step. This is done computing a similarity value between each node and representative ones – if this similarity value is under a certain threshold, the node will not be further explored in the subsequent step.

---

[5] For a more detailed description of the system the interested reader can refer to `http://sisinflab.poliba.it/publications/2010/MRDD10a/`

The similarity value is computed querying *external information sources* (search engines and social bookmarking systems) and thanks to *textual* and *link analysis* in DBpedia. For each pair of resource nodes in the explored graph, we perform a query to each external information source: we search for the number of returned web pages containing the labels of each nodes individually and then for the two labels together (as explained in Section 2.2). Moreover, we look at **abstracts** in Wikipedia and **wikilinks**, i.e., links between Wikipedia pages. Specifically, given two resource nodes a and b, we check if the label of node a is contained in the abstract of node b, and vice versa. The main assumption behind this check is that if a resource name appears in the *abstract* of another resource it is reasonable to think that the two resources are related with each other. For the same reason, we also check if the Wikipedia page of resource a has a *(wiki)link* to the Wikipedia page of resource b, and vice versa.

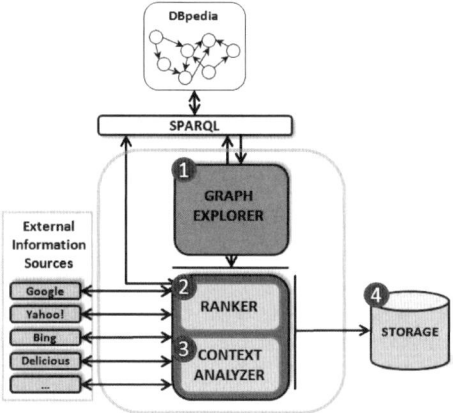

Fig. 1. The ranking system *DBpediaRanker*

In the following we will describe all the components of our system, whose architecture is sketched in Figure 1. The main data structure we use in the approach contains information about DBpedia resources[6] reached during the exploration. Hence, for each reached resource an associated data structure r is defined as:

$$r = \begin{cases} URI : \text{a DBpedia URI;} \\ hits : \text{how many times the URI has been reached during the exploration;} \\ ranked : \text{Boolean value representing if the URI has already been ranked with respect to its neighbors;} \\ in\_context : \text{Boolean value stating if the URI has to be considered within the context or not;} \end{cases}$$

As the exploration starts from the seed nodes, a global variable $\mathcal{R}$ is initialized with the set of *seed nodes* and then it is further populated with other nodes reached during the graph exploration (see Algorithm 1 in the Appendix). Seed nodes must belong to the context to explore and are selected by domain experts.

---

[6] From now on, we use the words *URI*, *resource* and *node* indistinctly.

The algorithm explores the DBpedia graph using a depth-first approach up to a depth of MAX_DEPTH (see Section 2.1).

2.1   Graph Explorer

This module queries DBpedia via its SPARQL endpoint[7]. Given a DBpedia resource, the explorer looks for other resources connected to it via a set of predefined properties. The properties of DBpedia to be explored can be set in the system before the exploration starts. In our initial setting, we decided to select only the SKOS[8] properties skos:subject and skos:broader[9]. Indeed, these two properties are very popular in the DBpedia dataset. Moreover, we observed that the majority of nodes reached by other properties were also reached by the selected properties, meaning that our choice of skos:subject and skos:broader properties does not disregard the effects of potentially domain-specific properties.

Given a node, this is explored up to a predefined distance, that can be configured in the initial settings. We found through a series of experiments that, for the context of programming languages and databases, setting $MAX\_DEPTH = 2$ is a good choice as resources within two hops are still highly correlated to the root one, while going to the third hop this correlation quickly decreases. Indeed, we noticed that if we set $MAX\_DEPTH = 1$ (this means considering just nodes directly linked) we lost many relevant relation between pairs of resources. On the other hand, if we set $MAX\_DEPTH > 2$ we have too many non relevant resources.

In order to find the optimal value for MAX_DEPTH, we initially explored 100 seed nodes up to a $MAX\_DEPTH = 4$. After this exploration was completed, we retrieved the top-10 (most similar) related resources for each node (see Section 2.2). The results showed that on the average the 85% of the top-10 related resources where within a distance of one or two hops. The resources two hops far from the seeds where considered as the most relevant the 43% of times ($\sigma = 0.52$). On the contrary the resources above two hops were rarely present among the first results (less than 15% of times). In figure 2 the average percentage of top-10 related resources w.r.t. to the distance from a seed (MAX_DEPTH) is shown.

The exploration starts from a node *root*. Given a DBpedia node *root* and a maximal depth to be reached, this module browses (using a depth-first approach) the graph from *root* up to a number of hops equal to MAX_DEPTH (see Algorithm 2 in the Appendix). For each node u discovered during the exploration, we check if u is relevant for the context and computes a similarity value between *root* and u. Such value is computed by the module *Ranker* as detailed in Section 2.2. As we said before, given a resource u, during the exploration of the RDF graph we analyze only the properties/links skos:subject and skos:broader for which u is either rdf:subject or rdf:object.

---

[7] http://www.w3.org/TR/rdf-sparql-query/

[8] http://www.w3.org/2004/02/skos/

[9] skos:subject has been recently deprecated in the SKOS vocabulary. Nevertheless, in DBpedia it has not been replaced by its corresponding dcterms:subject.

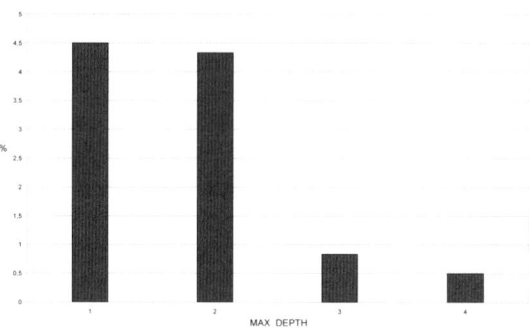

Fig. 2. Evaluation for MAX_DEPTH. It represents the average percentage (y axis) of the top-10 resources related to 100 seeds within a distance of 1 to 4 hops (x axis).

2.2 Ranker

This is the core component of the whole system. Given two resources u_1 and u_2 in the same graph-path, it compares how much they relate with each other exploiting information sources external to DBpedia such as search engines and social tagging systems (see Algorithm 3 in the Appendix).

The aim of this module is to evaluate how strong a semantic connection is between two DBpedia resources using information taken from external sources. In our current implementation we consider as external sources both web search engines (Google, Yahoo! and Bing) and social tagging systems (Delicious), plus Wikipedia-related information contained in DBpedia. Given two DBpedia resources u_1 and u_2, we verify how many web pages contain (or have been tagged by) the value of the rdfs:label associated to u_1 and u_2. Then we compare these values with the number of pages containing (or tagged by) both labels. We select more than one search engine because we do not want to bind the result to a specific algorithm of a single search engine. Moreover, we want to rank a resource not only with respect to the popularity of related web pages on the web, but also considering the popularity of such resources among users (e.g., in Delicious). In this way we are able to combine two different perspectives on the popularity of a resource: the one related to the words occurring within web documents, the other one exploiting the social nature of the current web. Through formula (1) we evaluate the related similarity of two resources u_1 and u_2 with respect to an external information source $info\_source$.

$$sim(u_1, u_2, info\_source) = \frac{p_{u_1, u_2}}{p_{u_1}} + \frac{p_{u_1, u_2}}{p_{u_2}} \qquad (1)$$

Given the information source $info\_source$, p_{u_1} and p_{u_2} represent the number of documents containing (or tagged by) the rdfs:label associated to u_1 and u_2 respectively, while p_{u_1, u_2} represents how many documents contain (or have been tagged by) both the label of u_1 and u_2. It is easy to see that the formula is symmetric and the returned value is in $[0, 2]$. *Ranker* does not use only external information sources but exploits also further information from DBpedia.

In fact, we also consider Wikipedia hypertextual links mapped in DBpedia by the property dbpprop:wikilink. Whenever in a Wikipedia document w_1 there is a hypertextual link to another Wikipedia document w_2, in DBpedia there is a dbpprop:wikilink from the corresponding resources u_1 and u_2. Hence, if there is a dbpprop:wikilink from u_1 to u_2 and/or vice versa, we assume a stronger relation between the two resources. More precisely, we assign a score equal to 0 if there are no dbpprop:wikilinks between the two resources, 1 if there is a dbpprop:wikilink just in one direction, 2 if both resources are linked by dbpprop:wikilink in both directions. Furthermore, given two resources u_1 and u_2, we check if the rdfs:label of u_1 is contained in the dbpprop:abstract of u_2 (and vice versa). Let n be the number of words composing the label of a resource and m the number of words composing the label which are also in the abstract, we also consider the ratio $\frac{m}{n}$ in the final score, with $\frac{m}{n}$ in [0,1] as $m \leq n$.

2.3   Context Analyzer

The purpose of *Context Analyzer* is to identify a subset of DBpedia nodes representing a context of interest. For instance, if the topics we are interested in are *databases* and *programming languages*, we are interested in the subgraph of DBpedia whose nodes are somehow related to *databases* and *programming languages* as well. This subgraph is what we call a **context**. Once we have a context $\mathcal{C}$, given a query represented by a DBpedia node u, first we look for the context u belongs to and then we rank nodes in $\mathcal{C}$ with respect to u. In order to identify and compute a context, we use *Graph Explorer* to browse the DBpedia graph, starting from an initial meaningful set of resources (**seed nodes**). In this preliminary step a domain expert selects a subset of resources that are representative of the context of interest. The set of seed nodes we selected for the context of databases and programming languages are *PHP, Java, MySQL, Oracle, Lisp, C#* and *SQLite*.

Since we do not want to explore the whole DBpedia graph to compute $\mathcal{C}$, once we reach nodes that we may consider out of the context of interest we need a criterion to stop the exploration. During the exploration we may find some special nodes that are more popular than others, i.e., we may find nodes in the context that are more interconnected to other nodes within $\mathcal{C}$. We call these resources **representative nodes of the context**. Intuitively, given a set of representative nodes of a context $\mathcal{C}$, we may check if a DBpedia resource u is within or outside the context of interest evaluating *how relevant u is* with respect to representative nodes of $\mathcal{C}$.

While exploring the graph, *Graph Explorer* populates at each iteration the set $\mathcal{R}$ with nodes representing resources reached starting from the initial seed nodes. Each node contains also information regarding how many times it has been reached during the exploration. The number of hits for a node is incremented every time the corresponding URI is found (see Algorithm 4 in the Appendix). This value is interpreted as *"how popular/important the node is"* within $\mathcal{R}$.

In DBpedia there are a special kind of resources called **categories**[10]. Since in Wikipedia they are used to classify and cluster sets of documents, in DBpedia they classify sets of resources. They might be seen as abstract concepts describing and clustering sets of resources. As a matter of fact, to stress this relation, every DBpedia category is also a rdf:type skos:Concept. Moreover, since DBpedia categories have their own labels we may think at these labels as names for clusters of resources. *Context Analyzer* uses these categories in order to find **representative nodes**. In other words, the representative nodes are the *most popular* DBpedia categories in $\mathcal{R}$.

Hence, for each new resource found during the exploration, in order to evaluate if it is within or outside the context, we compare it with the most popular DBpedia categories in $\mathcal{R}$. If the score is greater than a given threshold, we consider the new resource within the context. The value $THRESHOLD$ is set manually. After some tests, for the context of programming languages and database systems, we noticed that a good value for the context we analyzed is $THRESHOLD = 4.0$. Indeed, we noticed that many non-relevant resources were considered as in context if the threshold was lower. On the contrary, a greater value of the threshold was too strict and blocked many relevant resources.

2.4   Storage

For each pair of resources, we store the results computed by *Ranker*. We also keep the information on how many times a resource has been reached during the graph exploration. For each resource belonging to the extracted context, the *Storage* module stores the results returned by *Ranker*. For each resource *root* we store information related to it: $\langle root, hits, ranked, in\_context \rangle$, plus ranking results for each of its discovered nodes u_i: $\langle root, u_i, wikipedia, abstract, google, yahoo, bing, delicious \rangle$.

3   Not Only Tag

In this section we describe a concrete system that exploits the algorithms proposed in Section 2 in order to suggest semantically related tags.

We borrow directly from DBpedia the use case to "*classify documents, annotate them and exploit social bookmarking phenomena*"[11]. Terms from DBpedia can be used to annotate Web content. Following this idea, our system wants to offer a fully-semantic way of social tagging. Figure 3 shows a screenshot of the prototype of the system, *Not Only Tag*, available at http://sisinflab.poliba.it/not-only-tag.

The usage is very simple. The users starts by typing some characters (let us say "*Drup*") in the text input area (marked as *(1)* in Figure 3) and the system returns a list of DBpedia resources whose labels or abstracts contain

---

[10] http://en.wikipedia.org/wiki/Help:Category
[11] http://wiki.dbpedia.org/UseCases#h19-5

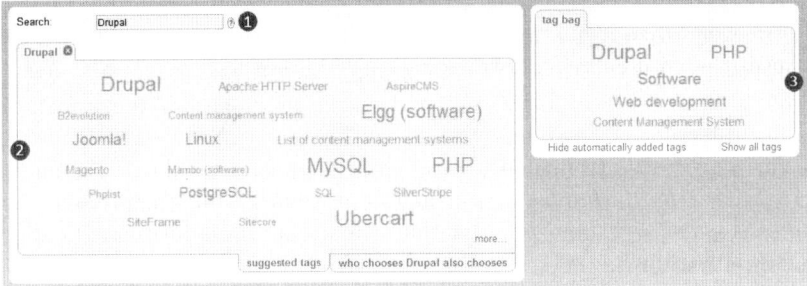

Fig. 3. Screenshot of *Not Only Tag* system

the typed string. Then the user may select one of the suggested items. Let us suppose that the choice is the tag *Drupal*. Then, the system populates a tag cloud (as shown by *(2)* in Figure 3), where the size of the tags reflects their relative relevance with respect to the chosen tag (*Drupal* in this case). The biggest tags are *Ubercart*, *PHP*, *MySQL*, *Elgg* and *Joomla!*. When the user clicks on a whatever tag, the corresponding cloud is created. Thanks to this feature the user can efficiently navigate the DBpedia subgraph just like she usually does when jumping from a web page to another one. The user can also drag a tag and drop it in her tag bag area (indicated by *(3)* in Figure 3) or just click on the plus icon next to each tag. Once the user selects a tag, the system enriches this area by populating it with concepts related to the dropped tag. For example, in the case of *Drupal*, its most similar concepts are *PHP*, *Software*, *Web Development*, *Content Management System* and so on. These ancestors correspond to Wikipedia Categories. As seen in Section 2.3 it is possible to discover them because they are the subject of a triple which has rdf:type as property and skos:Concept as object. Moreover skos:broader property links Categories with each other (specifically a subcategory to its category), whereas skos:subject relates a resource to a Category. By means of a recursive SPARQL query, filtered by the above mentioned properties, it is possible to check if a node is parent of another one.

4   Evaluation

In the experimental evaluation we compared our *DBpediaRanker* algorithm with other four different algorithms; some of them are just a variation of our algorithm but lack of some key features.

Algo2 is equivalent to our algorithm, but it does not take into account textual and link analysis in DBpedia.

Algo3 is equivalent to our algorithm, but it does not take into account external information sources, i.e., information coming from search engines and social bookmarking systems.

Algo4, differently from our algorithm, does not exploit textual and link analysis. Moreover, when it queries external information sources, instead of Formula (1), it uses the *co-occurrence* formula: $\frac{p_{u_1,u_2}}{p_{u_1}+p_{u_2}-p_{u_1,u_2}}$.

Algo5 is equivalent to *Algo4*, but it uses the *similarity distance* formula [3] instead of the co-occurrence one.

We did not choose to use either the co-occurrence formula or the similarity distance with *DBpediaRanker* since they do not work well when one of the two resources is extremely more popular than the other, while formula (1) allows to catch this situation.

In order to assess the quality of our proposal we conducted a study where we asked participants to rate the results returned by each algorithm. For each query, we presented five different rankings, each one corresponding to one of the ranking methods. The result lists consisted of the top ten results returned by the respective method. In Figure 4, results for the query *Drupal* are depicted. Looking at all the results obtained with our approach (column 3), we notice that they are really tightly in topic with *Drupal*. For example, if we focus on the first three results, we have *Ubercart*, that is the popular e-commerce module for *Drupal*, *PHP* which is the programming language used in *Drupal*, and *MySQL* the most used DBMS in combinance with *Drupal*. The other results are still very relevant, we have for example *Elgg* and *Joomla!*, that are the major concurrents of *Drupal*, and *Linux* which is the common platform used when developing with *Drupal*.

We point out that even if we use external information sources to perform substantially a textual search (for example checking that the word *Drupal* and the word *Ubercart* appear more often in the same Web pages with respect to the pair *Drupal* and *PHP*), this does not mean that we are discarding semantics in our search and that we are performing just a keyword-based search, as the inputs for the text-based search come from a semantic source. This is more evident if we consider the best results our system returns if the query is *PHP*. In fact, in this case no node having the word *PHP* in the label appears in the first results. On the contrary, the first results are *Zend Framework* and *Zend Engine*, that are respectively the most used web application framework when coding in PHP and the heart of PHP core. *PHP-GTK* is one of the first resources that contains the word *PHP* in its label and is ranked only after the previous ones.

During the evaluation phase, the volunteers were asked to rate the different ranking algorithms from 1 to 5 (as shown in Figure 4), according to which list they deemed represent the best results for each query. The order in which the different algorithms were presented varied for each query: e.g., in Figure 4 the results for *DBpediaRanker* algorithm appear in the third column, a new query would show the results for the same algorithm in a whatever column between the first and the last. This has been decided in order to prevent users to being influenced by previous results.

The area covered by this test was the *ICT* one and in particular *programming languages* and *databases*.

Please rate the following rankings:

1.	MySQL	Computer_Output_to_Laser_Disc	Ubercart	SilverStripe	Pennd
2.	List of content management systems	Magento	PHP	Joomla!	Slimweb
3.	PostgreSQL	CS_EMMS-Suite	MySQL	B2evolution	Tencent QQ
4.	PHP	Web software	Elgg (software)	AspireCMS	Molins
5.	Elgg (software)	CivicSpace	Joomla!	Magento	JSMS
6.	Linux	CityDesk	Linux	SiteFrame	ProjectWise
7.	Joomla!	Wiki software	List of content management systems	Ubercart	Soq
8.	Mambo (software)	Mambo (software)	Content management system	PHP	Invu PLC
9.	Net2ftp	SiteFrame	Web content management system	Sitecore	Powerfront CMS
10.	Apache HTTP Server	Mozilla Firefox	PostgreSQL	Phplist	Folding@home

Fig. 4. Screenshot of the evaluation system. The five columns show the results for, respectively, *Algo3*, *Algo4*, *DBpediaRanker*, *Algo2* and *Algo5*.

The test was performed by 50 volunteers during a period of two weeks, the data collected are available at `http://sisinflab.poliba.it/evaluation/data`. The users were Computer Science Engineering master students (last year), Ph.D. students and researchers belonging to the ICT scientific community. For this reason, the testers can be considered IT domain experts. During the testing period we collected 244 votes. It means that each user voted on average about 5 times. The system is still available at the website `http://sisinflab.poliba.it/evaluation`. The user can search for a keyword in the ICT domain by typing it in the text field, or she may directly select a keyword from a list below the text field that changes each time the page is refreshed. While typing the resource to be searched for, the system suggests a list of concepts obtained from `DBpedia`. This list is populated by querying the `DBpedia` URI lookup web service[12].

If the service does not return any result, it means that the typed characters do not have any corresponding resource in `DBpedia`, so the user can not vote on something that is not in the `DBpedia` graph. It may happen that after having chosen a valid keyword (i.e., an existing resource in `DBpedia`) from the suggestion list, the system says that there are no results for the selected keyword. This happens because we used the context analyser (see Section 2.3) to limit the exploration of the `RDF` graph to nodes belonging to *programming languages* and *databases* domain, while the URI lookup web service queries the whole `DBpedia`. In all other cases the user will see a screenshot similar to the one depicted in Figure 4. Hovering the mouse on a cell of a column, the cells in other columns having the same label will be highlighted. This allows to see immediately in which positions the same labels are in the five columns. Finally the user can start to rate the results of the five algorithms, according to the following scale: (i) one star: *very poor*; (ii) two stars: *not that bad*; (iii) three stars: *average*; (iv) four stars: *good*; (v) five stars: *perfect*. The user has to rate each algorithm before sending her vote to the server. Once rated the current resource, the user may vote for a new resource if she wants. For each voting we collected the time elapsed to

---

[12] `http://lookup.dbpedia.org/api/search.asmx`

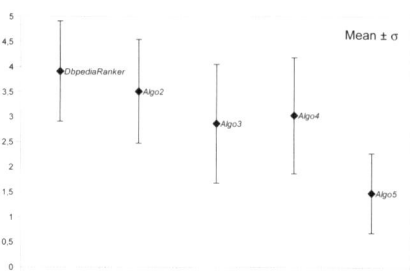

Fig. 5. Average ranks

rate the five algorithms: on the average it took about 1 minute and 40 seconds ($\sigma = 96.03$ s). The most voted resources were `C++`, `MySQL` and `Javascript` with 10 votings.

In Figure 5 we plotted the mean of the votes assigned to each method. Error bars represent standard deviation. *DBpediaRanker* has a mean of 3.91 ($\sigma = 1.0$). It means that, on the average, users rated it as *Good*. Examining its standard deviation, we see that the values are within the range of $\sim 3 \div 5$. In order to determine if the differences between our method and the others are statistically significant we use the Wilcoxon test [13] with $p < .001$. From the Wilcoxon test we can conclude that not only our algorithm performed always better than the others, but also that the (positive) differences between our ranking and the others are statistically significant. Indeed, the *z-ratio* obtained by comparing *DBpediaRanker* algorithm with *Algo2*, *Algo3*, *Algo4* and *Algo5* is respectively 4.93, 8.71, 7.66, 12.89, (with $p < 0.0001$). By comparing these values with the critical value of z^{13}, we can reject the null hypothesis (correlated rankings), and say that the differences between our algorithm and the others are statistically significant.

5   Related Work

Nowadays, a lot of websites expose their data as `RDF` documents; just to cite a few: the *DBPL* database, *RDF book mashup*, *DBtune*, *MusicBrainz*[14]. `SPARQL` is the *de-facto* standard to query `RDF` datasets. If the considered dataset has a huge dimension, `SPARQL` will return as result of the query a long list of *not-ranked* resources. It would be very useful to have some metrics able to define the relevance of nodes in the `RDF` graph, in order to give back to the user a *ranked* list of results, ranked w.r.t. the user's query. In order to overcome this limit several `PageRank`-like [11] ranking algorithms have been proposed [4,7,9,6]. They seem, in principle, to be good candidates to rank resources in a `RDF` knowledge base. Yet, there are some considerable differences, that cannot be disregard, between ranking web documents and ranking resources to which some semantics

---

[13] http://faculty.vassar.edu/lowry/ch12a.html

[14] http://www.informatik.uni-trier.de/~ley/db/,
http://www4.wiwiss.fu-berlin.de/bizer/bookmashup/,
http://dbtune.org/, http://musicbrainz.org/

is attached. Indeed, the only thing considered by the `PageRank` algorithm is the origin of the links, as all links between documents have the same relevance, they are just hyperlinks. For `RDF` resources this assumption is no more true: in a `RDF` graph there are several types of links, each one with different relevance and different semantics, therefore, differently from the previous case, a `RDF` graph is not just a graph, but a directed graph with labels on each edge. Moreover a `RDF` resource can have different origins and can be part of several different contexts and this information has be exploited in some way in the ranking process. *Swoogle* [4] is a semantic web search engine and a metadata search provider, which uses the *OntologyRank* algorithm, inspired by the `PageRank` algorithm. Differently from *Swoogle*, that ranks `RDF` documents which *refer* to the query, our main task is to rank `RDF` resources *similar* to the query. Nonetheless, we borrowed from Swoogle the idea of browsing only a predefined subset of the semantic links. Similarly to our approach also the *ReConRank* [7] algorithm explores just a specific subgraph: when a user performs a query the result is a topical subgraph, which contains all resources related to keywords specified by the user himself. In the subgraph it is possible to include only the nodes *directly* linked to the particular root node (the query) as well as specify the number n of desired hops, that is how far we want to go from the root node. The *ReConRank* algorithm uses a `PageRank`-like algorithm to compute the relevance of resources, called *ResourceRank*. However, like our approach, the *ReConRank* algorithm tries to take into account not only the relevance of resources, but also the "context" of a certain resource, applying the *ContextRank* algorithm [7]. Our approach differs from [7] due to the semantic richness of the `DBpedia` graph (in terms of number of links) the full topical graph for each resource would contain a huge number of resources. This is the reason why we only explore the links `skos:subject` and `skos:broader`. Hart et al. [6] exploit the notion of naming authority, introduced by [9], to rank data coming from different sources. To this aim they use an algorithm similar to `PageRank`, adapted to structured information such as the one contained in an `RDF` graph. However, as for `PageRank`, their ranking measure is absolute, i.e. it does not depend on the particular query. In our case, we are not interested in an absolute ranking and we do not take into account naming authority because we are referring to `DBpedia`: the naming authority approach as considered in [6] loses its meaning in the case of a single huge source such as `DBpedia`. Mukherjea et al. in [10] presented a system to rank `RDF` resources inspired by [9]. As in the classical `PageRank` approach the relevance of a resource is decreased when there are a lot of outcoming links from that, nevertheless such an assumption seems not to be right in this case, as if an `RDF` resource has a lot of outcoming links the relevance of such a resource should be increased not decreased. In our approach, in order to compute if a resource is within or outside the context, we consider as *authority* URIs the most popular `DBpedia` categories. Based on this observation, URIs within the context can be interpreted as *hub* URIs. *TripleRank* [5], by applying a decomposition of a 3-dimensional tensor that represents an `RDF` graph, extends the paradigm of two-dimensional graph representation, introduced by HITS, to obtain information on the resources and predicates of the analyzed graph. In the pre-processing phase they prune dominant predicates, such as `dbpprop:wikilink`, which, instead, have a fundamental role as shown in the experimental evaluation. Moreover in [5] they consider only objects of triples, while we look at both directions of statements. Finally, as for all the HITS-based algorithms, the ranking is just based on the graph structure. On the contrary we also use external information sources. *Sindice* [12], differently from the approaches already presented, does not provide a ranking based on any lexicographic or graph-based information. It ranks resources retrieved by `SPARQL` queries exploiting external ranking services (as Google popularity) and information related to hostnames, relevant statements, dimension of information sources. Differently from our approach, the main task of Sindice is to return `RDF` triples (data) related to a given query. Kasneci et al. [8] present a semantic search engine *NAGA*. It extracts information from several sources on the web and, then, finds relationships between the extracted entities. The system answers to queries about relationships already collected in it, which at the moment of the writing are around one hundred. Differently from our system, in order to query NAGA the user has to know all the relations that can possibly link two entities and has to learn a specific query language, other than know the exact name of the label she is looking for; while we do not require any technical knowledge to our users, just the ability to use tags. We do not collect information from the entire Web, but we rely on the `Linked Data` cloud, and in particular on `DBpedia` at the present moment.

6   Conclusion and Future Work

Motivated by the need of of annotating and retrieving software components in shared repositories, in this paper we presented a novel approach to rank `RDF` resources within the `DBpedia` dataset. The notion of context is introduced to reduce the search space and improve the search results. Semantic resources within a context are ranked according to the query exploiting the semantic structure of the `DBpedia` graph as well as looking for similarity information in web search engines and social tagging systems, and textual and link analysis. The approach has been implemented in a system for semantic tagging recommendation. Experimental results supported by extensive users evaluation show the validity of the approach. Currently, we are mainly investigating how to extract more fine grained contexts and how to enrich the context extracting not only relevant resources but also relevant properties. Moreover we are developing a wrapper for `ProgrammableWeb` using our tagging system as a backend for the annotation process. The aim is to facilitate the tagging process and the subsequently recommendation of software components in the retrieval phase.

Acknowledgment

We are very grateful to Joseph Wakeling for fruitful discussion and to the anonymous users who participated in the evaluation of the system. This research has been supported by Apulia Strategic projects PS_092, PS_121, PS_025.

References

1. Bizer, C., Heath, T., Berners-Lee, T.: Linked data - the story so far. International Journal on Semantic Web and Information Systems 5(3), 1–22 (2009)
2. Bizer, C., et al.: Dbpedia - a crystallization point for the web of data. In: Web Semantics: Science, Services and Agents on the World Wide Web (July 2009)
3. Cilibrasi, R., Vitányi, P.: The Google Similarity Distance. IEEE Transactions on Knowledge and Data Engineering 19(3), 370–383 (2007)
4. Ding, L., Finin, T., Joshi, A., Pan, R., Cost, S.R., Peng, Y., Reddivari, P., Doshi, V., Sachs, J.: Swoogle: a search and metadata engine for the semantic web. In: CIKM '04, pp. 652–659 (2004)
5. Franz, T., Schultz, A., Sizov, S., Staab, S.: TripleRank: Ranking semantic web data by tensor decomposition. In: Bernstein, A., Karger, D.R., Heath, T., Feigenbaum, L., Maynard, D., Motta, E., Thirunarayan, K. (eds.) ISWC 2009. LNCS, vol. 5823, pp. 213–228. Springer, Heidelberg (2009)
6. Harth, A., Kinsella, S., Decker, S.: Using naming authority to rank data and ontologies for web search. In: Bernstein, A., Karger, D.R., Heath, T., Feigenbaum, L., Maynard, D., Motta, E., Thirunarayan, K. (eds.) ISWC 2009. LNCS, vol. 5823, pp. 277–292. Springer, Heidelberg (2009)
7. Hogan, A., Harth, A., Decker, S.: ReConRank: A Scalable Ranking Method for Semantic Web Data with Context (2006)
8. Kasneci, G., Suchanek, F.M., Ifrim, G., Ramanath, M., Weikum, G.: Naga: Searching and ranking knowledge. In: ICDE 2008 (2008)
9. Kleinberg, J.M.: Authoritative sources in a hyperlinked environment. In: Proc. of the Ninth Annual ACM-SIAM Symposium on Discrete Algorithms (1998)
10. Mukherjea, S., Bamba, B., Kankar, P.: Information Retrieval and Knowledge Discovery utilizing a BioMedical Patent Semantic Web. IEEE Trans. Knowl. Data Eng. 17(8), 1099–1110 (2005)
11. Page, L., Brin, S., Motwani, R., Winograd, T.: The PageRank Citation Ranking: Bringing Order to the Web. Technical report (1998)
12. Tummarello, G., Delbru, R., Oren, E.: Sindice.com: Weaving the Open Linked Data. In: Aberer, K., Choi, K.-S., Noy, N., Allemang, D., Lee, K.-I., Nixon, L.J.B., Golbeck, J., Mika, P., Maynard, D., Mizoguchi, R., Schreiber, G., Cudré-Mauroux, P. (eds.) ASWC 2007 and ISWC 2007. LNCS, vol. 4825, pp. 552–565. Springer, Heidelberg (2007)
13. Wilcoxon, F.: Individual comparisons by ranking methods. Biometrics Bulletin 1(6), 80–83 (1945)

Appendix: Algorithms

Algorithm 1. *DBpediaRanker*

Input: a set $\mathcal{S} = \{u_i\}$ of seed nodes
Output: the context

1 $\mathcal{R} = \emptyset$;
   /* For each seed, we create the corresponding node. We impose each seed to be within
      the context.   */
2 **foreach** $u_i \in \mathcal{S}$ **do**
3 $\quad$ $create\_new\_node(r_i)$;
4 $\quad$ $r_i.URI = u_i$;
5 $\quad$ $r_i.hits = 1$;
6 $\quad$ $r_i.ranked = false$;
7 $\quad$ $r_i.in\_context = true$;
8 $\quad$ $\mathcal{R} = \mathcal{R} \cup \{r_i\}$;
9 **end**
10 $finished = false$;
11 **while** $finished == false$ **do**
   $\quad$ /* We expand only the DBpedia nodes whose corresponding URI is evaluated to be
      $\quad$ within the context.   */
12 $\quad$ **foreach** $r_i \in \mathcal{R}$ *such that both* $(r_i.in\_context == true)$ *and* $(r_i.ranked == false)$ **do**
13 $\quad\quad$ $explore(r_i.URI, r_i.URI, MAX\_DEPTH)$;
14 $\quad$ **end**
15 $\quad$ $finished = true$;
   $\quad$ /* After we updated $\mathcal{R}$ expanding nodes whose URI is within the context, we might
      $\quad$ have new representative nodes of the context. Hence, we check if nodes
      $\quad$ previously considered outside of the context can be reconsidered as part of it.
      $\quad$ */
16 $\quad$ **foreach** $r_i \in \mathcal{R}$ *such that* $(r_i.in\_context == false)$ **do**
17 $\quad\quad$ **if** $is\_in\_context(r_i.URI)$ **then**
18 $\quad\quad\quad$ $r_i.in\_content = true$;
19 $\quad\quad\quad$ $finished = false$;
20 $\quad\quad$ **end**
21 $\quad$ **end**
22 **end**

Algorithm 2. $explore(root, uri, depth)$. The main function implemented in *Graph Explorer*.

Input: a URI $root$; one of $root$'s neighbour URIs uri; $depth$: number of hops before the search stops

/* We perform a depth-first search starting from $root$ up to a depth of MAX_DEPTH. */

1 **if** $depth < MAX\_DEPTH$ **then**
2   **if** *there exists* $r_i \in \mathcal{R}$ *such that* $r_i.URI == uri$ **then**
        /* If the resource uri was reached in a previous recursive step we update its popularity. Moreover, if uri is evaluated to be within the context we compute how similar uri and $root$ are. */
3         $r_i.hits = r_i.hits + 1$;
4   **if** $is\_in\_context(uri)$ **then**
5 | $sim = similarity(root, uri)$;
6   **end**
7   **else**
        /* If the resource uri was not reached in a previous recursive step we create the corresponding node. Moreover, if uri is evaluated to be within the context we compute how similar uri and $root$ are, otherwise we mark uri as being outside of the context. */
8         $create\_new\_node(\overline{r_i})$;
9         $\overline{r_i}.URI = uri$;
10        $\overline{r_i}.hits = 1$;
11        $\overline{r_i}.ranked = false$;
12   **if** $is\_in\_context(uri)$ **then**
13           $sim = similarity(root, uri)$;
14           $\overline{r_i}.in\_context = true$;
15   **else**
16           $\overline{r_i}.in\_context = false$;
17   **end**
18   **end**
19 **end**
/* If we are not at MAX_DEPTH depth w.r.t. $root$, we create the set of all the resources reachable from uri via skos:subject and skos:broader. */
20 **if** $depth > 0$ **then**
21 | $\mathcal{N} = explode(uri)$;
22 **end**
/* We recursively analyze the resources reached in the previous step. */
23 **foreach** $n_i \in \mathcal{N}$ **do**
24 | $explore(root, n_i, depth - 1)$;
25 **end**
26 **save** $\langle root, uri, sim \rangle$;

Algorithm 3. $similarity(u_1, u_2)$. The main function implemented in *Ranker*.

Input: two DBpedia URIs
Output: a value representing their similarity

1 $wikipedia = wikiS(u_1, u_2)$;
2 $abstract = abstractS(u_1, u_2)$;
3 $google = engineS(u_1, u_2, google)$;
4 $yahoo = engineS(u_1, u_2, yahoo)$;
5 $bing = engineS(u_1, u_2, bing)$;
6 $delicious = socialS(u_1, u_2, delicious)$;
7 **return** $wikipedia + abstract + google + yahoo + bing + delicious$;

Algorithm 4. $is\_in\_context(uri, \mathcal{R})$. The main function implemented in *Context Analyzer*.

Input: a DBpedia URI uri
Output: $true$ if uri is considered part of the context, $false$ otherwise
1 $cont = 0$;
2 $r = 0$;
3 **foreach** $node\ r \in \mathcal{R}$ **do**

        `/*`
           `We consider the most popular DBpedia categories reached during the exploration`
           `as the representative URIs of the context.`
                                                      `*/`
4   **if** $r.URI$ *is one of the ten most popular DBpedia categories reached so far during the search* **then**
5          $s = s + similarity(uri, r.URI)$
6   **end**

        `/*`
           `If the similarity value computed between `uri` and the representative URIs of the`
           `context is greater than a threshold we consider `uri` as part of the context.`
                                                `*/`
7   **if** $s \geq THRESHOLD$ **then**
8   **return** $true$
9   **end**
10 **end**
11 **return** $false$

Linkator: Enriching Web Pages by Automatically Adding Dereferenceable Semantic Annotations

Samur Araujo[1], Geert-Jan Houben[1], and Daniel Schwabe[2]

[1] Delft University of Technology, P.O. Box 5031, 2600 GA Delft, The Netherlands
[2] Informatics Department, PUC-Rio Rua Marques de Sao Vicente, 225, Rio de Janeiro, Brazil
{s.f.cardosodearaujo,g.j.p.m.houben}@tudelft.nl,
dschwabe@inf.puc-rio.br

Abstract. In this paper, we introduce Linkator, an application architecture that exploits semantic annotations for automatically adding links to previously generated web pages. Linkator provides a mechanism for dereferencing these semantic annotations with what it calls semantic links. Automatically adding links to web pages improves the users' navigation. It connects the visited page with external sources of information that the user can be interested in, but that were not identified as such during the web page design phase. The process of auto-linking encompasses: finding the terms to be linked and finding the destination of the link. Linkator delegates the first stage to external semantic annotation tools and it concentrates on the process of finding a relevant resource to link to. In this paper, a use case is presented that shows how this mechanism can support knowledge workers in finding publications during their navigation on the web.

Keywords: auto-linking, semantic annotation, semantic link, dereferencing, dynamic links, navigation.

1 Introduction

The links between HTML pages offer the main mechanism for users to navigate on the web. It allows a user, with a simple mouse click, to go from one page to another. However, most websites only contain links that are considered essential for the functioning of the web site, therefore leaving some potentially relevant terms on the web pages unlinked. This can be observed specially in the long tail of the web, where pages are created manually, and where adding extra links can be a laborious and time-consuming task. Because of this situation, a recurring scenario has risen when users browse web pages: they very often select a piece of text (e.g., a telephone, an address, or a name) from a web page and copy and paste it into search engine forms, as the only reasonable procedure available for accessing relevant resources related to those terms. However, this procedure may be significantly reduced, and the navigation improved, by automatically adding links semantically related to those terms.

The problem of automatically adding links to an existing web page can be divided in two main tasks. First, identifying candidate terms (anchors) for adding links – typically they denote concepts in which the user is interested. Second, identifying a

B. Benatallah et al. (Eds.): ICWE 2010, LNCS 6189, pp. 355–369, 2010.

web resource to be the link target. The first task can be solved by using information extraction techniques for identifying candidate terms to be linked. Two major sub-problems here are to determine whether a term should be linked, for avoiding "overlinking", and to disambiguate candidate terms to the appropriate concepts, since the terms can have different meanings and consequently demand different links. The second task, that is the focus of this paper, demands using an external source of knowledge in order to discover links related to the terms – actually, the concepts – that were found in the first task. The major sub-problem here is to select a source of data for finding the destination to the link. It can be achieved by querying a pre-defined knowledge base (e.g. Wikipedia) or by querying a distributed and wider data space such as the Semantic Web. The latter approach requires also a pre-selection of the sources to be consulted, since only a few sources in the whole data space can contain relevant knowledge about the concept being exploited.

Although several solutions are being proposed to automatically add links to web pages, they focus on the first part of the problem, i.e. to disambiguate keywords in the text of the page, and typically select the target of the link to be in a single website (e.g., Wikipedia or DBpedia). However, the full problem is only solved after finding the most significant destination for the link. For instance, [3, 4, 5, 6] are focused on linking keywords on web pages to Wikipedia articles, which reduces to disambiguating terms using the Wikipedia as a knowledge base. In spite of their relevance, Wikipedia articles are not always the most adequate objects to link to. These approaches support relatively well users that are interested in, for example, encyclopedic knowledge, however they do not adequately support for example users that are shopping and need to find more information about products, or knowledge workers that are interested in finding bibliographic references for their research.

This paper introduces *Linkator*[1], a framework that uses Semantic Web technology to build dereferenceable semantic annotations, which in this paper are called *semantic links*. It shows how information extraction technology can be composed with semantic technology for automatically and semantically annotating terms in web pages, and subsequently exploiting these semantic annotations to define a target of the link. Linkator expects the extractor component to have the intelligence to disambiguate the terms and annotate them properly, and focuses on determining the appropriate target of the link.

The paper shows the use of the Linked Data cloud[2] as an external source of knowledge for discovering link destinations for the recognized terms. With Linked Data the destination of a link can be semantically determined, instead of restricting it to just one source on the web. The source in the Linked Data to be queried and the query itself are defined based on the semantics of the annotation of the semantic link. By combining these two processes, given an input document, the Linkator system has the ability to identify, on the fly, the important concepts in a text, and then link these concepts to semantically related resources on the web in a contextually relevant way. Later in this paper we will present an example of how Linkator can be used to support knowledge workers.

---

[1] Linkator prototype is available at:
  http://www.wis.ewi.tudelft.nl/index.php/linkator
[2] Linked Data - http://linkeddata.org/

2   Related Work

2.1   Auto-linking

Augmentation of text with links is not novel. Hughes and Carr [6] discussed the Smart Tag system, a Microsoft agent for automatically enriching documents with semantic actions. Google's AutoLink is another tool for augmenting web pages with links. It uses a pattern-based approach to detect numbers related to entities on web pages, such as street addresses, ISBN numbers, and telephone numbers, and to add links to them. In this tool, users can select an action to be performed after a click action, however it is limited in the number of recognizable entities and the number of sources to be linked to.

Some authors proposed to augment web pages by adding links to Wikipedia articles, therefore having an impartial (e.g., not commercially oriented) target for the link and still adding value to the user navigation. In [3, 4, 5, 6] the same approach for the problem is used. They integrate a term extraction algorithm for automatically identifying the important terms in the input document, and a word sense disambiguation algorithm that assigns each term with the correct link to a Wikipedia article. NNexus [14] proposes an approach that works for general sources of data. Basically, it builds a concept graph of terms extracted from a specific knowledge base and indexes it with the destination document. Then that index is exploited for linking a concept with a document.

Another related research area is that of link recommendation systems. Whereas the idea is similar, these systems are typically oriented towards recommending additional links to entire web pages, as opposed to specific items within a page. Whereas some of the algorithms may be adapted for Linkator's purposes, we have not focused on this aspect here.

2.2   Semantic Annotations

Annotating existing and new documents with semantics is a requirement for the realization of the Semantic Web. A semantic annotation tags an entity on the web with a term defined in an ontology. The annotation process can be done manually or automatically, and the latter demands a specialized system to accomplish the task. The automatic process of annotating is composed basically of finding terms in documents, mapping them against an ontology, and disambiguating common terms. The systems that solve this problem differ in architecture, information extraction tools and methods, initial ontology, amount of manual work required to perform annotation, and performance [7]. The result of the annotation process is a document that is marked-up semantically. For that concern, some markup strategies were proposed. Microformats[3] is an approach to semantic markup for XHTML and HTML documents, that re-uses existing tags to convey metadata. This approach is limited to a few set of published Microformats standards. Moreover, it is not possible to validate Microformats annotations since they do not use a proper grammar for defining it. An

---

[3] http://microformats.org/

evolution of Microformats is eRDF (embedded RDF)[4], an approach for annotating HTML pages using RDF[5], however it faces the same criticism than Microformats, since they use the same strategy for annotating pages. Another approach for semantically annotating pages is *RDFa*[6] (Resource Description Framework - in - attributes). RDFa is a W3C Recommendation that adds a set of attribute level extensions to XHTML for embedding RDF metadata within web documents.

2.3 SPARQL Endpoint Discovery

Querying the Semantic Web implies querying a distributed collection of data. Federated querying over linked data has been addressed in [8, 9, 10, 11, 12, 13]. Among others, a part of this problem is related to selecting the proper sources of data to be queried. For this concern, two approaches stand out. The first approach requires the designer of the query to specify, declaratively, in the body of the query, which sources of data should be queried [9, 11]. In those cases, the endpoints to be queried are fixed and pre-determined by the user. The second approach tries to select the sources automatically, by finding correspondences between the terms mentioned in the query and the sources available in the Semantic Web. For instance, [8, 12] exploit the fact that recursively dereferencing resources mentioned in the query provides the data that then can be used for solving the query. The authors point out that this approach works well in situations where incomplete results are acceptable, since the dereferencing process does not reach all available graphs that match with the query pattern. Also, this approach is limited to a subset of SPARQL queries, since some elements must be present in the query in order to trigger this mechanism.

Another approach that tries to exploit such a correspondence uses statistics about the data in the SPARQL endpoints. For instance, [10] summarizes the endpoint data into an index containing statistics about the data space. In order to determine relevant sources, it tries to locate, for each triple pattern in the query, which entry in the index matches it. The selection of the endpoint is determined if the pattern matches with an entry in the index. In [13] middleware is used that catalogs the instances and classes in each endpoint. This middleware selects the right endpoint by matching the resources used in the query with the resources registered in the catalog.

3 Semantic Link – Definition

In an HTML page, an HTML link is denoted by the HTML tag *A*. Normally, it addresses a specific URL which when clicked triggers an HTTP request that retrieves an HTML document that a browser can render in a human-readable representation of this URL. Fig. 1 shows an example of a conventional HTML link. Once clicked, this link redirects the user to the URL described in the *href* attribute: *www.st.ewi.tudelft.nl/~leonardi/*

In these cases, HTML links are defined through an explicit intervention of an author, at the time the page is created or programmed.

---

[4] http://research.talis.com/2005/erdf/wiki/Main/RdfInHtml
[5] http://www.w3.org/TR/rdf-primer/
[6] http://www.w3.org/TR/xhtml-rdfa-primer/

```
1   <html>
2   <body>
3   <a href="www.st.ewi.tudelft.nl/~leonardi/">Erwin Leonardi</a>
4   </body>
5   </html>
```

Fig. 1. Example of a conventional HTML link

Definition 1: A *semantic link* is an HTML tag *A* that is semantically annotated with *RDFa*, which implies that *RDF* triples are associated to the link. The *semantic link* must contain the attribute *property* or *rel,* which is defined in the RDFa specification; it semantically relates the link to another resource or content. The triples associated to the *semantic link* are determined by the semantics defined in the RDFa specification[7]. Linkator uses the semantics of these triples to compute, dynamically, the URL of the link. Based on the semantics of these triples, it selects sources in the Linked Data cloud to search for a URL for the link. The next example (in Fig. 2) shows how two links with the same anchor (e,g., "Erwin Leonardi") can take the user to distinct pages based on the semantics in the link: Erwin Leonardi's Facebook page and Erwin Leonardi's DBLP[8] publications page.

```
1    <?xml version="1.0" encoding="utf-8"?>
2    <!DOCTYPE html PUBLIC "-//W3C//DTD XHTML+RDFa 1.0//EN"
3    "http://www.w3.org/MarkUp/DTD/xhtml-rdfa-1.dtd">
4    <html version="XHTML+RDFa 1.0" xmlns="http://www.w3.org/1999/xhtml"
5    xmlns:foaf="http://xmlns.com/foaf/0.1/" xmlns:dc="http://purl.org/dc/elements/1.1/">
6    <body>
7    <a href="" typeof="foaf:Person"
8    about="http://www.theleonardi.com/foaf.rdf" property="foaf:name" >Erwin Leonardi</a>
9
10   <a href="" typeof="swrc:Author"
11   about="http://www.theleonardi.com/foaf.rdf" property="dc:creator" >Erwin Leonardi</a>
12   </body>
13   </html>
```

Fig. 2. Examples of semantic links

Note that conceptually, based on the semantics of the RDFa annotations, there are two triples associated to the link in line 7 of Fig. 2 (expressed here in the notation N3[9]). Fig. 3 represents these triples.

```
1    @prefix rdf: <http://www.w3.org/1999/02/22-rdf-syntax-ns#> .
2    @prefix foaf: <http://xmlns.com/foaf/0.1/>  .
3    @prefix dc: <http://purl.org/dc/elements/1.1/> .
4
5    <http://www.theleonardi.com/foaf.rdf> rdf:type foaf:Person ;
6    foaf:name "Erwin Leonardi" .
```

Fig. 3. Example of an HTML link

---

[7] http://www.w3.org/TR/rdfa-syntax/
[8] http://www.informatik.uni-trier.de/~ley/db/indices/a-tree/l/
   Leonardi:Erwin.html
[9] http://www.w3.org/2000/10/swap/Primer

Those triples add the following semantics to the link: it is about a person, named "Erwin Leonardi" that is described in the resource *http://www.theleonardi.com/ foaf.rdf*. One possible result for this link is to show Erwin Leonardi's Facebook page, since this homepage is a relevant resource semantically related to those triples. Such information could be obtained in Linked Data that describes people or in the resource *http://www.theleonardi.com/foaf.rdf* that describes the person mentioned. On the other hand, for the link in line 10 of Fig. 2, it would be more relevant to show *Erwin Leonardi*'s DBLP publications page, since *author* and *creator* are concepts (see the triples in Fig. 4) that are semantically more related to DBLP[10] Linked Data than the previous mentioned sources.

```
1  @prefix dc: <http://purl.org/dc/elements/1.1/> .
2  @prefix swrc: <http://ontoware.org/swrc/swrc_v0.3.owl#> .
3
4  <http://www.theleonardi.com/foaf.rdf> rdf:type swrc:Author;
5  dc:creator "Erwin Leonardi" .
```

Fig. 4. Example of an HTML link

The main benefit of semantic links here is that Linkator can use this extra information for finding a significant value for the *href* attribute, i.e. the link URL, automatically. Therefore, based on vocabularies used for annotating the semantic links, Linkator can select a source on Semantic Web to look for a human-readable representation related to the concept behind the link. This process is further explained in detail in the rest of this paper.

4 Adding Semantic Links

Adding semantic links to a web page means to detect the relevant entities on this page and add anchors that will lead the user to a related document on the web. In this section we will describe these two processes. Fig. 5 illustrates the full process described in this paper.

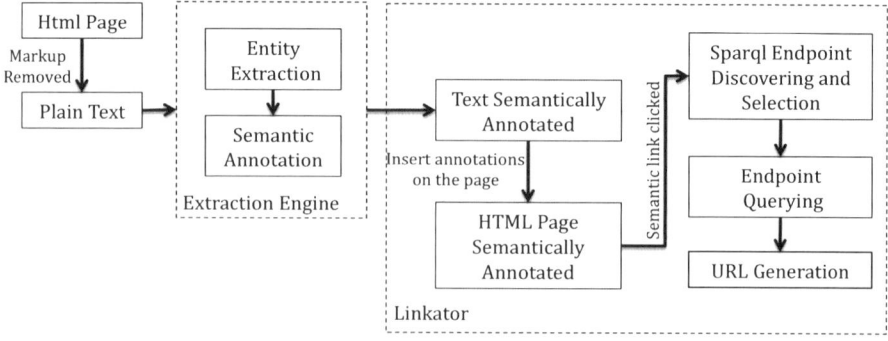

Fig. 5. Semantic link processing flow

---

[10] http://dblp.l3s.de/d2r/

The first stage in the whole process is to detect the relevant entities to be linked. For this, *Linkator* can use any available information extraction engine, due to its flexible and plug & play architecture. The main advantage of this approach is that *Linkator* delegates the intelligence of entity extraction to the extractor tools. However, as most of the extractors do not produce annotations in *RDF* or *RDFa*, a major additional step here is to convert the annotation made by the extractor into *RDFa* annotations. This step is achieved by mapping a specific domain ontology to the schema used by the extractor, and later using that mapping to convert the extractor annotations into *RDFa*. As the result of that entire process, the entities extracted are annotated in a domain specific ontology representing the domain that the extractor was trained for.

Let us give an example. Suppose that we decided to plug the *Freecite*[11] extractor into *Linkator*. *Freecite* is an entity extraction engine for bibliographic citations. It is able to detect citations in text documents and retrieve a structured XML file containing *authors, title, year of publication, location, number of pages,* and *volume* of the citation.

```
1   <?xml version="1.0" encoding="UTF-8"?>
2   <html>
3     <head>
4       <title>Erwin's Home Page</title>
5     </head>
6   <body>
7   This is one publication of Erwin Leonardi <br>
8   Erwin Leonardi, Sri L. Budiman, Sourav S Bhowmick.
9   Detecting Changes to Hybrid XML Documents Using Relational Databases.
10  In the Proceedings of the 16th International Conference on Database and Expert
11  Systems Applications (DEXA 2005). Springer Verlag, Copenhagen, Denmark, August, 2005
12  </body>
13  </html>
```

Fig. 6. Sample HTML page with a bibliographic reference

Fig. 6 shows an example of a web page with a bibliographic reference. The result of the parsing and extraction of this page in *Freecite* is shown in Fig. 7.

```
1   <?xml version="1.0" encoding="utf-8"?>
2   <citations>
3       <citation valid='true'>
4           <authors>
5               <author>Erwin Leonardi</author>
6               <author>Sri L Budiman</author>
7               <author>Sourav S Bhowmick</author>
8           </authors>
9           <title>Detecting Changes to Hybrid XML Documents Using Relational Databases</title>
10          <booktitle>In the Proceedings of the 16th International Conference on Database and
11          Expert Systems Applications (DEXA 2005)
12          </booktitle>
13          <publisher>Springer Verlag</publisher>
14          <location>Copenhagen, Denmark</location>
15          <year>2005</year>
16      </citation>
17  </citations>
```

Fig. 7. Output extracted with Freecite in XML

---

[11] FreeCite - Open Source Citation Parser - freecite.library.brown.edu

Notice that, in particular, Freecite does not directly mark the source page with the corresponding extracted semantic annotation. As was mentioned before, in order to get the original document semantically annotated as output from the extractor, an extended version of *Freecite* was developed in Linkator for directly annotating the original document with *RDFa*. For this purpose, the DBLP ontology[12] (the set of vocabularies used in this dataset) was used as the underlying ontology for the annotation, although any other similar ontology or vocabulary can be used in this process. This Freecite extension is able to transform the Freecite XML output to RDF format and embed the annotations in the original web page. Basically, it transforms the XML file to RDF using an XSL transformation[13]. The resulting file is then automatically injected into the original web page with RDFa annotations. Linkator places the annotation into the original page by matching the literal objects of the RDF triples against the text in the page. The result of this transformation is shown in Fig. 8.

Like this, the first step of the process has been completed as the web page has been extended with semantic links. Yet the computation of the target URL has to be done, and in the next section we choose to do this by exploiting the triples that are associated to the semantic link, which represent the semantics of the link in the page.

```
1   <?xml version="1.0" encoding="utf-8"?>
2   <!DOCTYPE html PUBLIC "-//W3C//DTD XHTML+RDFa 1.0//EN"
3   "http://www.w3.org/MarkUp/DTD/xhtml-rdfa-1.dtd">
4   <html version="XHTML+RDFa 1.0" xmlns="http://www.w3.org/1999/xhtml"
5   xmlns:swrc="http://ontoware.org/swrc/swrc_v0.3.owl#"
6   xmlns:dcterms="http://purl.org/dc/terms/"
7   xmlns:dc="http://purl.org/dc/elements/1.1/">
8   <head>
9       <title>Erwin's Home Page</title>
10  </head>
11  <body>
12  <p>
13  One publication of <a property="dc:creator" href="">Erwin Leonardi</a>.
14  </p>
15  <div  about="[_:cite01]" typeof="swrc:Article" >
16  <a property="dc:creator" href="">Erwin Leonardi</a>,
17  <a property="dc:creator" href="">Sri L Budiman</a>,
18  <a href="" property="dc:creator"> Sourav S Bhowmick</a>,
19  <a property="dc:title" href="">Detecting Changes to Hybrid XML Documents Using Relational
20  Databases</a>,
21  <a property="swrc:jornal" href="">In the Proceedings of the 16th International
22  Conference on Database and Expert Systems Applications (DEXA 2005).</a>
23  <a href="" property="dc:publisher">Spring Verlag</a>,
24  <a href="" property="dc:location">Copenhagen, Denmark</a>,
25  August, <a href="" property="dcterms:issued">2005</a>
26  </div>
27  </body>
28  </html>
```

Fig. 8. Original page with semantic links expressed as RDFa annotations

5   Dereferencing Semantic Links

The next step in the process is to define a destination for the added link. Regardless of the destination of the anchor being defined during the annotation process, as

---

[12] http://dblp.l3s.de/d2r/
[13] http://www.w3.org/TR/xslt

illustrated in [7, 8, 9, 10], in Linkator, it is computed at the moment the user clicks on the *semantic link*. Therefore, it can select the most significant source to find the destination of the anchor, based on the current context where the anchor occurs. To improve performance Linkator uses the Linked Data cloud for discovering destinations for the semantic link as opposed to querying search engines or a fixed knowledge base. For that reason, the destination is computed at the user click, since it would be very slow to query the linked data for find all destination beforehand.

Explained in brief, in order to dereference the semantic link, Linkator queries the Linked Data cloud or an RDF source that exposes a *SPARQL endpoint*, looking for a human-readable representation of the triples related to the semantic link. Thus, the dereferencing process reduces to defining the endpoints to be queried and defining the query itself.

5.1 Endpoint Resolution

SPARQL endpoint resolution is a problem that has recently attracted attention of researchers in the field of federated queries [15, 16, 17, 18, 19, 20]. In spite of the fact that many approaches have been proposed as (partial) solution for this problem there is not yet a consensus or standard. Due to this fact, the Linkator architecture implements its own solution.

```
1    <http://dblp.l3s.de/d2r/resource/void>
2        a void:Dataset;
3        foaf:homepage <http://dblp.l3s.de/>;
4        rdfs:label "dblp.l3s.de Linked Data Repository";
5        dcterms:title "dblp.l3s.de Linked Data Repository";
6        dcterms:description "This repository contains data from the DBLP Bibliography.";
7        dcterms:publisher <http://dblp.l3s.de>;
8        dcterms:date "2010-01-03T06:40:15"^^xsd:date;
9        void:uriRegexPattern "^http://dblp\\.l3s\\.de/d2r/resource/.+";
10       void:sparqlEndpoint <http://dblp.l3s.de/d2r/sparql/>;
11       void:exampleResource <http://dblp.l3s.de/d2r/resource/publications/conf/vldb/VhanHY93>;
12       void:vocabulary <http://swrc.ontoware.org/ontology#>
13       void:vocabulary <http://purl.org/dc/terms/>
14       void:vocabulary <http://purl.org/dc/elements/1.1/>
15       void:vocabulary <http://www.w3.org/2000/01/rdf-schema#>
16       void:vocabulary <http://xmlns.com/foaf/0.1/>
17       dcterms:subject <http://dbpedia.org/resource/Person>;
18       dcterms:subject <http://dbpedia.org/resource/Category:Computer_scientists>;
19       dcterms:subject <http://dbpedia.org/resource/Publication>;
20       dcterms:subject <http://dbpedia.org/resource/Organization>;
21       dcterms:subject <http://dbpedia.org/resource/Computer_science>;
```

Fig. 9. Excerpt from the DBLP voID descriptor

In *Linkator*, the *SPARQL endpoint* to be queried is determined based on three things:

1. the triple associated to the semantic link itself,
2. the *RDF* graph of the annotations that exist on the web page,
3. the *voID* (Vocabulary of Interlinked Datasets) [5] descriptors of the available endpoints.

Linkator selects available endpoints based on the vocabularies that they use. The vocabulary that an endpoint uses defines the semantics of the data that it contains to a great extent. Therefore, by matching the vocabulary used on the semantic link with the endpoint vocabularies, *Linkator* can resolve which endpoint may contain significant information about the resources associated with the semantic link. The vocabularies used by the endpoints can be obtained from their *voID* descriptors. *VoID* is an *RDF* vocabulary used to describe linked datasets in the Semantic Web. Fig. 9 shows a fragment of a *voID* descriptor for the *DBLP*[14] endpoint.

When *Linkator* receives a dereferencing request, it executes the algorithm illustrated in Fig. 10 to choose the most relevant endpoint:

```
1    FUNCTION SelectEndpoint
2        E := Array
3        R : = select all rdf:type objects associated to the semantic link
4        T := ExtractVocabulary(R)
5
6        FOR EACH vocabulary in T DO
7        {
8          E.add (select endpoints that contain this vocabulary)
9        }
10       IF E = Empty
11       {
12         R := select all predicates associated to the semantic link
13         T := ExtractVocabulary(R)
14
15         FOR EACH vocabulary in T DO
16         {
17                E.add (select endpoints that contain this vocabulary)
18         }
19       }
20       RETURN E
21
22   FUNCTION ExtractVocabulary (R)
23       V := Array
24       FOR EACH resource in R DO
25       {
26         V.add (extract the vocabulary from the resource)
27       }
28       RETURN V
```

Fig. 10. Algorithm of the *SelectEndpoint* function that resolves which endpoint will be queried based on the matching between vocabularies.

Let us illustrate how this mechanism works. Suppose that a user clicks on the semantic link *Erwin Leonardi* illustrated in line 16 of Fig. 8. In addition to the triple of this link shown in Fig. 11, all triples annotated and associated to the semantic link can be used during this process.

```
_:cite01 <http://purl.org/dc/elements/1.1/creator> 'Erwin Leonardi'
```

Fig. 11. Triple associated to the semantic link *Erwin Leonardi* in line 16 of Fig. 8

The *SelectEndpoint* function (see Fig. 10, line 3) looks in those triples for an *rdf:type* object. In this example, it matches the resource *swrc:Article* (that represents

---

[14] The DBLP SPARQL server can be accessed at: http://dblp.l3s.de/d2r/

the expanded URL *http://ontoware.org/swrc/swrc_v0.3.owl#Article*). In the next step (Fig. 10, line 4), it extracts the vocabulary associated to this resource. This is done by the function *ExtractVocabulary* (Fig. 10, line 22) that retrieves the vocabulary *http://ontoware.org/swrc/swrc_v0.3.owl,* in this example. The last step in this process is to find the endpoints that use this vocabulary. In line 8, the procedure queries the *voID* descriptor of the available *SPARQL* endpoints, looking for such a vocabulary. The endpoints that contain it will be used during the querying process, as described in the next subsection.

Note that the *SelectEndpoint* function can retrieve more than one endpoint, and in that case, all of them will be used during the querying process. Also, note that in this example, the algorithm found an *rdf:type* object in the annotations. However, in the case where *rdf:type* is not available the engine uses the *predicate* associated to the semantic link, which, in this last example, would be the resource *http://purl.org/dc/elements/1.1/creator.*

In spite of the existence of endpoints covering a broad range of topics (e.g., DBpedia) and using hundreds of vocabularies, most of them are domain-specific and use a small set of vocabularies, which justifies this as a reasonable approach for detecting the endpoint semantically associated to the link being dereferenced.

Nevertheless, several other heuristics can be used to semantically disambiguate the endpoint, for instance, it could be based on the conceptual description of the content of the endpoint. In *voID*, the *dcterms:subject* property should be used to describe the topics covered by the datasets, and in a future version we intent to exploit such information to implement an approach based on concept mapping between the content in endpoint and the semantic links. We also intend to use the voID Store[15] service that aims to aggregate voID descriptors of public endpoints available in the Semantic Web.

5.2 Query Formulation

As mentioned earlier, *Linkator* queries an *RDF* source in order to find a URL for the semantic link. The query itself is created based on the object of the triples associated to the semantic link. The resulting query varies depending on whether the object is a URL (i.e., an *ObjectProperty*) or an *RDF literal* (i.e., a *DatatypeProperty*). In this approach, triples where the object is an *RDF blank node* are discarded. In the case where the object is a literal, the query is *Select ?s where {?s rdfs:label literal}*. For example, for the triple in Fig. 11, one of the *SPARQL* queries generated is shown in Fig. 12.

```
1   Select ?s where {?s rdfs:label 'Erwin Leonardi'.}
```

Fig. 12. Sample query to find URLs for links

---

[15] http://void.rkbexplorer.com/

In fact, the query exemplified in Fig. 12 is just one of the queries computed. *Linkator* executes a set of queries in order to overcome the limitations posed by the endpoints. For the previous example, the entire set of queries generated is shown in Fig. 13:

```
1  Select ?s where {?s rdfs:label 'Erwin Leonardi'.}
2  Select ?s where {?s rdfs:label 'Erwin Leonardi'^^xsd:string.}
3  Select ?s where {?s rdfs:label 'Erwin Leonardi'@en.}
4  Select ?s where {?s rdfs:label ?o. Filter (regex (?o, 'Erwin Leonardi', 'i')).}
```

Fig. 13. Set of queries generated

Note however, that the query in line 4 is only executed if the endpoint supports keyword search, otherwise, in practice, such a query has a large probability of timing out or even returning an error. The support for keyword search can be obtained from the *voID* descriptor of the endpoint.

The final step in the dereferencing process is to find a URL to be inserted in the generated XHTML. Since a resource is retrieved in the query, *Linkator* tries to find a URL that contains a human-readable representation associated to the resource. Otherwise, it dereferences the resource URL directly. This same process also applies in the case where the object of the triple is a URI and not a literal. In order to find a human-readable representation for that resource, *Linkator* searches for predicates in the target resource that contain the string 'seealso', 'homepage', 'web' or 'site'. This means that it is able to match predicates such as: *foaf:homepage, akt:has-web-address, rdfs:seeAlso*, that in general, contain a human-readable representation for RDF resources, meaning, a URL for a website. The whole dereferencing process ends with Linkator redirecting the user request to the URL found.

6 Proof of Concept

Linkator has been implemented as an extension to the Firefox browser, together with a backend service used by the extension. In this section, it is illustrated how Linkator can be used for supporting users in their searches where they try to locate PDF files and author's pages for references that they encounter in web pages with citations. In this scenario, the main problem is that some researchers or research groups mention their works on their homepages but they often do not add a link to the referred documents. Therefore, other researchers spend a considerable time looking for copies of the documents on the web: Linkator is able to automatically do this job without any intervention of the user or author. The Linkator prototype used to exemplify this mechanism can be found at: http://www.wis.ewi.tudelft.nl/index.php/linkator

To exemplify this particular scenario, consider Erwin Leonardi's personal homepage: http://www.theleonardi.com/. That page contains a list of Erwin Leonardi's publications, as exhibited in Fig. 14.

Publications
(**C** Conference paper; **J** Journal paper; **W** Workshop paper; **B** Book chapter)

1. **B** Kees van der Sluijs, Geert-Jan Houben, Erwin Leonardi, Jan Hidders. **Hera: Engineering Web Applications Using Semanti** Information Management: A Model-Based Perspective, De Virgilio, Roberto; Giunchiglia, Fausto; Tanca, Letizia (Eds.), Chapter 1!

2. **C** Erwin Leonardi, Sourav S Bhowmick, Mizuho Iwaihara. **Efficient Database-Driven Evaluation of Security Clearance for Fe** Proceedings of 15thInternational Conference on Database Systems for Advanced Applications (DASFAA 2010). Springer Verlag,

3. **C** Sourav S Bhowmick, Erwin Leonardi, Zhifeng Ng, Curtis Dyreson. **Towards Non-Directional XPath Evaluation in a RDBMS** on Information and Knowledge Management (ACM CIKM 2009). ACM Press, Hong Kong, China, November, 2009. (Short paper)

4. **W** Fabian Abel, Dominikus Heckmann, Eelco Herder, Jan Hidders, Geert-Jan Houben, Daniel Krause, Erwin Leonardi, Kees van **Modeling Framework**. In the Proceedings of the 17th Workshop on Adaptivity and User Modeling in Interactive Systems (FGABI

Fig. 14. Excerpt from Erwin Leonardi's publications listed in http://www.theleonardi.com

The same page, after it has been (automatically) processed by Linkator, is shown in Fig. 15. The semantic links added by Linkator are shown in blue.

Publications
(**C** Conference paper; **J** Journal paper; **W** Workshop paper; **B** Book chapter)

1. **B** Kees van der Sluijs, Geert-Jan Houben, Erwin Leonardi, Jan Hidders. **Hera: Engineering Web Applications Using Semanti** Information Management: A Model-Based Perspective, De Virgilio, Roberto; Giunchiglia, Fausto; Tanca, Letizia (Eds.), Chapter 1!

2. **C** <u>Erwin Leonardi</u>, <u>Sourav S Bhowmick</u>, <u>Mizuho Iwaihara</u>. **Efficient Database-Driven Evaluation of Security Clearance for Fe** Proceedings of 15thInternational Conference on Database Systems for Advanced Applications (DASFAA 2010). <u>Springer Verlag</u>,

3. **C** <u>Sourav S Bhowmick</u>, <u>Erwin Leonardi</u>, Zhifeng Ng, <u>Curtis Dyreson</u>. **Towards Non-Directional XPath Evaluation in a RDBMS** on Information and Knowledge Management (ACM CIKM 2009). <u>ACM Press</u>, <u>Hong Kong, China, November</u>, 2009. (Short paper)

4. **W** <u>Fabian Abel</u>, <u>Dominikus Heckmann</u>, Eelco Herder, <u>Jan Hidders</u>, <u>Geert-Jan Houben</u>, Daniel Krause, <u>Erwin Leonardi</u>, <u>Kees</u> van **Modeling Framework**. In the Proceedings of the 17th Workshop on Adaptivity and User Modeling in Interactive Systems (FGABI!

5. **W** Kees van der Sluijs, Jan Hidders, Erwin Leonardi, Geert-Jan Houben. **GAL: A Generic Adaptation Language for Describin**

Fig. 15. Excerpt from Erwin Leonardi's publications enriched with semantic links

After Linkator processes the page, clicking on the name of an author (annotated with *dc:creator*) makes Linkator search for a human-readable representation of the author in the DBLP SPARQL endpoint, as described in the previous section. As a result, it retrieves a homepage of this author, as registered in the DBLP records. For instance, by clicking on "Jan Hidders" the homepage http://www.wis.ewi.tudelft.nl/ index.php/personal-home-page-hidders is retrieved, while by clicking on "Geert-Jan Houben" the homepage http://wwwis.win.tue.nl/~houben/ is retrieved, since both URLs are stored in the DBLP endpoint. The same applies to the title of the citations that were annotated by Linkator with semantic links. By clicking on a title (annotated with *dc:title*), Linkator searches in DBLP for a human-readable representation of this title. For instance, in Erwin Leonardi's page, by clicking in the title "XANADUE: A System for Detecting Changes to XML Data in Tree-Unaware Relational Databases", it retrieves the ACM page of this article (http://portal.acm.org/citation.cfm?doid= 1247480.1247633). In this page the user can find the PDF file for the article.

This proof of concept scenario shows how Linkator operates. Although the example solves a trivial task, the full approach exemplifies how the problem of auto-linking page can be decomposed to support a domain-independent and an adaptive approach.

7   Conclusion and Future Work

Linkator is an architecture that brings the benefits of semantic annotation closer to end-users. It automatically adds links to web pages, which increases the connectivity of the web page with related external resources, consequently improving user navigation and access to web resources. This paper shows how this auto-linking problem can be solved by incorporating elements of the Semantic Web into a solution based on semantic links. Also, it focuses on the process of finding a URL for the semantic link, using the Semantic Web as underlying knowledge base. Linkator is able to select a source to be queried based on the semantics of the annotations on the page. Therefore, it can determine the URL of the link, by exploiting a knowledge base that is semantically related to the concept that is being linked. The Linkator prototype exemplifies how knowledge workers can benefit from such a mechanism to find documents related to bibliographic citations mentioned in web pages.

As future work, we intend to extend and measure Linkator's discovery and query model for improving the dereferencing mechanism; investigate the use of more general adaptation and recommendation approaches that can be enriched with Linkator's added semantics; improve the Linkator resolution mechanism when the endpoint retrieves more than one human-representation for an semantic link; and investigate how the link generation can benefit from integrating with classical search engines as well.

References

1. Alexander, K., Cyganiak, R., Hausenblas, M., Zhao, J.: Describing Linked Datasets - On the Design and Usage of voiD, the 'Vocabulary of Interlinked Datasets'. In: Linked Data on the Web Workshop (LDOW '09), in conjunction with 18th International World Wide Web Conference, WWW '09 (2009)
2. Hughes, G., Carr, L.: Microsoft Smart Tags: Support, ignore or condemn them? In: Proceedings of the ACM Hypertext 2002 Conference, Maryland, USA, pp. 80–81 (2002)
3. Medelyan, O., Witten, I.H., Milne, D.: Topic Indexing with Wikipedia. In: Proceedings of the AAAI 2008 Workshop on Wikipedia and Artificial Intelligence (WIKIAI 2008), Chicago, IL (2008)
4. Mihalcea, R., Csomai, A.: Wikify!: linking documents to encyclopedic knowledge. In: Proceedings of the 16th ACM Conference on Information and Knowledge management (CIKM '07), Lisbon, Portugal, pp. 233–242 (2007)
5. Milne, D., Witten, I.H.: Learning to link with wikipedia. In: Proceeding of the 17th ACM conference on Information and knowledge management, Napa Valley, California, USA, October 26-30 (2008)
6. Gardner, J., Xiong, L.: Automatic Link Detection: A Sequence Labeling Approach. In: International Conference on Information and Knowledge Management, CIKM '09 (2009)
7. Reeve, L., Han, H.: Survey of semantic annotation platforms. In: Proceedings of the 2005 ACM Symposium on Applied computing, Santa Fe, New Mexico, March 13-17 (2005)
8. Hartig, O., Bizer, C., Freytag, J.: Executing SPARQL Queries over the Web of Linked Data. In: Bernstein, A., Karger, D.R., Heath, T., Feigenbaum, L., Maynard, D., Motta, E., Thirunarayan, K. (eds.) ISWC 2009. LNCS, vol. 5823, pp. 293–309. Springer, Heidelberg (2009)

9. Schenk, S., Staab, S.: Networked graphs: a declarative mechanism for SPARQL rules, SPARQL views and RDF data integration on the web. In: Proceeding of the 17th international conference on World Wide Web (WWW '08), New York, NY, USA, pp. 585–594 (2008)

10. Harth, A., Hose, K., Karnstedt, M., et al.: On Lightweight Data Summaries for Optimised Query Processing over Linked Data (2009)

11. Zemanek, J., Schenk, S., Svatek, V.: Optimizing SPARQL Queries over Disparate RDF Data Sources through Distributed Semi-Joins. In: ISWC 2008 Poster and Demo Session Proceedings. CEUR-WS (2008)

12. Bouquet, P., Ghidini, C., Serafini, L.: Querying the Web of Data: A Formal Approach. In: The Semantic Web, pp. 291–305 (2008)

13. Langegger, A., Wöß, W., Blöchl, M.: A Semantic Web Middleware for Virtual Data Integration on the Web. In: Bechhofer, S., Hauswirth, M., Hoffmann, J., Koubarakis, M. (eds.) ESWC 2008. LNCS, vol. 5021, pp. 493–507. Springer, Heidelberg (2008)

14. Gardner, J.J., Krowne, A., Xiong, L.: NNexus: An Automatic Linker for Collaborative Web-Based Corpora. IEEE Trans. Knowl. Data Eng. 21(6), 829–839 (2009)

A Generic Proxy for Secure
Smart Card-Enabled Web Applications

Guenther Starnberger, Lorenz Froihofer, and Karl M. Goeschka

Vienna University of Technology
Institute of Information Systems
Argentinierstrasse 8/184-1
1040 Vienna, Austria
{guenther.starnberger,lorenz.froihofer,karl.goeschka}@tuwien.ac.at

Abstract. Smart cards are commonly used for tasks with high security
requirements such as digital signatures or online banking. However, systems that Web-enable smart cards often reduce the security and usability
characteristics of the original application, e.g., by forcing users to execute privileged code on the local terminal (computer) or by insufficient
protection against malware. In this paper we contribute with techniques
to generally Web-enable smart cards and to address the risks of malicious
attacks. In particular, our contributions are: (i) A single generic proxy to
allow a multitude of authorized Web applications to communicate with
existing smart cards and (ii) two security extensions to mitigate the effects of malware. Overall, we can mitigate the security risks of Web-based
smart card transactions and—at the same time—increase the usability
for users.

Keywords: Smart cards, Web applications, Digital signatures, Security.

1 Introduction

Despite ongoing efforts to Web-enable smart cards [1] there is still a *media dis-
continuity* when using smart cards in combination with Web applications, as
smart cards typically require a *native* helper application as proxy to communicate with the Web browser. One reason is that the Web security model is
fundamentally different from the smart card security model, leading to potential
security issues even for simple questions such as: "Is a particular Web application
allowed to access a particular smart card?".

Ongoing research to Web-enable smart cards typically either requires computational capabilities at smart cards higher than the capabilities provided by
today's smart cards or requires users to install software customized to particular
types of Web applications [2]. In contrast, our generic mapping proxy enables
access from arbitrary Web applications to arbitrary smart cards, while using
access control to protect smart cards from malicious Web applications, without
requiring any on-card software modifications.

However, guarding only against malicious Web applications is not sufficient,
if the local computer is potentially controlled by malware. Consequently, we

B. Benatallah et al. (Eds.): ICWE 2010, LNCS 6189, pp. 370–384, 2010.
© Springer-Verlag Berlin Heidelberg 2010

enhance our mapping approach to provide end-to-end security between a user and a smart card, but this enhancement requires the possibility to adapt on-card software. In particular, we allow the user to (i) either use the TPM (Trusted Platform Module) inside her computer or, (ii) alternatively, use a trusted secure device to secure communication with the smart card.

Summarized, the contributions of our paper are:

- A smart card Web communication protocol that provides a secure way for Web applications to interact with existing smart cards. Unlike state-of-the-art technologies, our approach allows any Web application to interact with any given smart card where communication is allowed based on our authorization and access control mechanisms.
- A first extension to our protocol that uses the Trusted Computing facilities part of recent PC (Personal Computer) hardware. This allows us to mitigate the effects of malware on the local computer, but requires modification of the on-card software.
- A second extension to our protocol that uses QR-TANs (Quick Response – Transaction Authentication Number) [3] instead of a TPM. Thus, the security is provided by an external security device instead of a PC.

Section 2 discusses related work before Sect. 3 presents the architecture and trust model of our application. Sect. 4 introduces our generic Web mapping. Sect. 5 extends our mapping approach with TPM-based attestation, while Sect. 6 provides alternative security measures based on QR-TAN authentication. Finally, Sect. 7 concludes the paper and provides an outlook on future work.

2   Related Work

In this section we discuss related work to Web-enable smart cards as well as to improve client-side security.

Web-enabling smart cards. Itoi et al. [4] describe an approach for secure Internet smart cards that allows users to access remote smart cards over the Internet. In contrast, we provide the client-side part of a Web application running in a Web browser with access to smart cards at the local computer. Thus, the security assumptions and implementation details differ fundamentally. An expired IETF Internet draft for SmartTP by P. Urien [5] specifies a unique software stack applicable to different types of smart cards, but—unlike our approach—requires software support from the smart card. Hence, it is not applicable to legacy cards. The TLS-Tandem approach [6] seems to use smart cards for access control to a Web server, while we aim at Web-enabling smart cards to mitigate man-in-the-middle attacks. Further details on approaches to provide smart cards with network access can be found in [1].

Improving client-side security. Lu et al. [7] increase security with respect to online identity theft by placing confidential information inside the smart card from where it can be transferred to a remote authenticated server. This reduces the risk of confidential information being captured by malware at a user's computer. However, it does not guarantee that data entered on the computer are not changed on the way to the server, which is the focus of our two security extensions. Bottoni and Dini [8] use a secure device to secure transactions between a user and a merchant. This is conceptually similar to our QR-TAN approach [3]. However, in this work we secure transactions between the user and the smart card itself. While our techniques rely on a trusted device or execution platform, Aussel et. al [9] include security-hardened monitors into applications running on untrusted platforms and use USB (Universal Serial Bus) smart cards to verify the log data provided by the monitors. Consequently, this approach could complement our techniques if no trusted execution environment is available, providing less security than a dedicated secure hardware device, of course.

Conclusion. Related work and existing implementations prove that connectivity of smart cards is a well researched topic. However, *real* Internet smart cards [10] able to communicate directly using the IP protocol are not yet widely available on the market. Furthermore, all existing evaluated solutions require custom on-card software for communication with the terminal. In comparison to existing work, our approach strives to partition trust requirements between Web applications and different smart cards and, additionally, features advanced security capabilities that allow to mitigate attacks due to insecure terminals. While Internet smart cards do not require our proxy application for Internet access, they do not provide equivalent capabilities for request filtering on the terminal. However, combining Internet smart cards with our two security mechanisms discussed in Sections 5 and 6 would allow to improve these cards' security in regard to man-in-the-middle attacks.

3   Architecture and Trust Model

This section presents our overall system architecture and the different security constraints. Due to the different trust requirements of the different entities, the problem we solve can be seen as a type of *multilateral security* [11] problem. For example, the user and the Web server both trust the smart card, but neither does the user trust executable code provided by the Web server, nor does the Web server trust executable code provided by the user. And while the user may place considerable trust into her own hardware, this hardware may not be trustworthy enough for the Web application in regard to non-repudiability requirements.

An overview of our architecture is given in Figure 1, which illustrates the major components and communication paths, but not the sequence of interactions detailed later. Figure 1(a) shows the architecture when used in combination with TPM and Figure 1(b) shows the architecture when used in combination with QR-TAN. The black arrows indicate direct communication paths between two entities while the highlighted broader lines in the background depict the secure channels in our system. If a secure channel spans several black arrows, this means that the intermediate entities are untrusted and data are passed through that entities by means of encryption or digital signatures. The trust relations

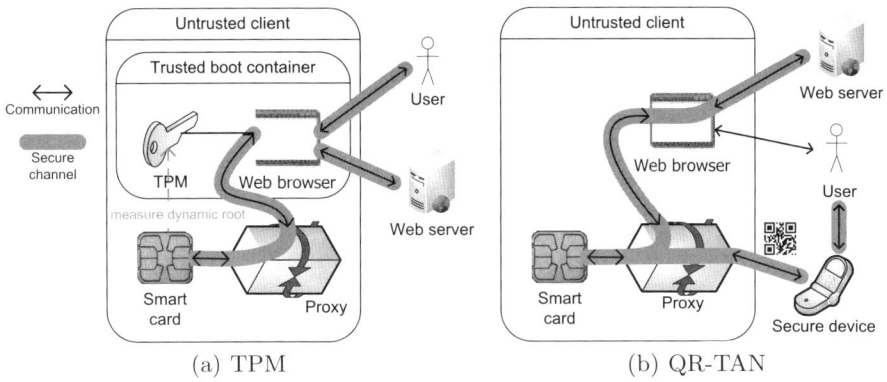

(a) TPM   (b) QR-TAN

Fig. 1. System components

between the constituents described in the following paragraphs are depicted in Figure 2. Arrows labelled "high" indicate that a component is highly trusted, while "partial" indicates a lower trust relationship.

Web application and Web server. The Web application is an entity that wants to interact with the smart card; for example a banking site that requires a digital signature before conducting a transaction. The user trusts the Web application for communication with the smart card. However, the user does not trust the Web application with unrestricted access to her computer—e.g., to execute binary code obtained from the Web application. The server-side part of the Web application is running on the Web server, while the client-side part of the Web application is implemented in JavaScript and running on the Web browser. The term "Web application" refers to the combination of these two components.

Web browser. The Web browser is the entity used to interact with the smart card. It hosts the client-side part of the Web application and interacts with the server-side part of the Web application and the smart card. The user needs to trust the Web browser for the type of executed transaction. For low security transactions such as reading a stored-value counter, a normal Web browser can be used. For high security transactions, the trust in the Web browser can either be increased by executing the Web browser inside a *trusted environment* (see Sect. 5), or the trust requirements in the Web browser can be decreased by outsourcing part of the transaction to a trusted secure device (see Sect. 6).

Proxy. The proxy is our application responsible for mapping requests from a Web browser to a smart card. Combined with our generic mapping approach (Sect. 4), only a single generic proxy provided by a trusted vendor is required to be installed in order to allow access to smart cards from a multitude of authorized Web applications using state-of-the-art Web technologies. The trust requirements in the proxy are two-fold: From a user's perspective, the proxy is running on a semi-trusted platform as the proxy is started on her local operating system. Thus, some security features—such as controlling which type of APDUs

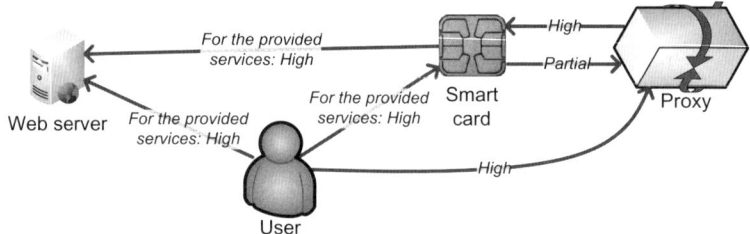

Fig. 2. Trust relations

(Application Protocol Data Unit) can be transmitted to smart cards are taken care of by the proxy. However, from a smart card issuers perspective the proxy is not necessarily trusted, as malware could control the computer. Thus, the smart card issuer can mandate additional security measures such as the authentication over a TPM (Sect. 5) or QR-TANs [3] (Sect. 6).

Smart card. A smart card is issued by an entity such as a local bank and is responsible for signing sensible data and/or for executing sensitive transactions. User and Web application have trust in the smart card's correctness. However, the user does not have a direct input and output path to the smart card. Thus, malware can manipulate the user's communication with the smart card.

4   Our Generic Proxy and Mapping Approach

Our generic proxy Web-enables smart cards without installation of custom on-card software and hence is applicable to a large range of existing smart cards. In sections 5 and 6, we enhance our approach to provide even better security in particular against malware for cases where the on-card software can be adapted.

The developed mapping technique uses a proxy to provide Web applications with access to smart cards and—in addition—protect the smart card from malicious Web applications. In contrast to existing techniques, our *generic mapping* allows a single executable to be used for a diverse set of Web applications and smart cards, not requiring the user to trust and execute code obtained from different Web sites. We validated our concepts with prototype implementations, showing that (i) the proxy concept is feasible in practice, (ii) correctly serves as a filter between Web applications and smart cards, and (iii) allows Web access to smart cards using state-of-the-art Web technologies.

Our system uses a mapping configuration to map abstract method calls to APDUs; an example is shown in Figure 3. This configuration defines how invoked methods with their arguments are mapped to APDUs and how results are mapped back to a structure. Furthermore, it includes a list of trusted origins, defining Web sites that may use the mapping, and a list of ATRs (*Answer To Reset*—an ATR *identifies* a smart card), identifying accessible cards. An AID (*Application Identifier*) identifies the respective smart card application.

```
<mapping>
  <smartcard atr="3b134028351180" aid="A00000006203010C0202" />
  <method name="login">
    <request>
      <args><arg name="pin" type="STRING" /></args>
      <apdu—mapping is="D4"><argument name="pin" /></apdu—mapping>
    </request>
    <response />
  </method>
</mapping>
```

Fig. 3. Request mapping example

The mapping is cryptographically signed by the card issuer with a key certified by a Privacy CA (Privacy Certificate Authority).

Mapping procedure. The following steps detail how a Web application can call a particular method defined in the mapping file. The interaction between the different components is shown in Figure 4.

1. The Web browser obtains a mapping definition and transmits the mapping to the proxy via an RPC (Remote Procedure Call) call by using JavaScript.
2. The proxy verifies that the origin of the client-side Web application part running in the Web browser matches the origin defined in the mapping file—e.g., by providing a secret to the Web application over a callback—and verifies the signature of the mapping.
3. The proxy verifies that there is a card in the reader and that the ATR of the card matches the ATR of the mapping. If the ATR is different, or if during the remaining process the ATR changes (e.g., because the card is replaced), the proxy will reset.
4. The proxy either asks the smart card if the public key used to sign the mapping should be trusted, or—alternatively—only verifies if the public key has been signed by a trusted certificate authority. If one of the options succeeds, the process is continued. Otherwise, the process is aborted.
5. The proxy receives RPC requests from the Web application's JavaScript code running inside the browser and converts them to APDUs according to the mapping. These APDUs are subsequently transmitted to the smart card. After the response APDU is received from the smart card it is converted and sent back to the application running in the Web browser.

Summarized, to protect smart cards from malicious Web applications we first identify the type of smart card connected to the PC. We proceed by verifying if the smart card provider has authorized the mapping file[1] provided by the Web application for this particular type of smart card. If this verification succeeds, this mapping is then used to restrict which Web applications can access the smart card and to restrict the type of APDUs that the Web application may transmit to the smart card.

---

[1] The authorization of a Web application's mapping file is an administrative task in contrast to the development and installation of on-card software, which would require re-distribution of smart cards.

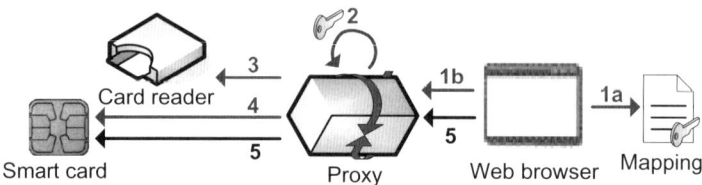

Fig. 4. Mapping procedure

5   Smart Card-Based TPM Attestation

This and the next section show extensions for establishment of a secure channel between the smart card and either (i) the Web browser running in a secure environment, or (ii) a mobile device trusted by the user. Both extensions require the support of on-card software.

The approach of this section uses an end-to-end security protocol between smart card and Web browser that provides authentication, integrity and confidentiality between the two endpoints. Our protocol works on a layer between APDU transmission and APDU interpretation. Conceptually, it can be compared to TLS. However, instead of desktop computers it targets smart cards and uses the remote attestation features of TPM that allow a remote party to verify if a computer is running a particular software configuration. The main requirements of our protocol are: (i) *Authentication*: Each party must be able to securely authenticate the other party. (ii) *Integrity*: Each party must be able to verify that the transmitted data has not been manipulated. (iii) *Confidentiality* (optionally): End-to-end encryption between the parties must be possible.

The first two items are required to prevent manipulation of transmitted data: Without authentication of the remote party, an attacker could directly establish a connection to one of the parties. Furthermore, without integrity of individual data items, an attacker could use a man-in-the-middle attack to manipulate these data items while in transit. The third item is optional: Without encryption, a man-in-the-middle is able to read transmitted information, but she is not able to manipulate this information. By using encryption, we can prevent an attacker from learning information about ongoing transactions.

In the following sections we first introduce the TPM functionality we use for remote attestation. Afterwards, we continue with a description of our secure channel that provides authentication, integrity and confidentiality. While a secure channel is already part of the Global Platform specification (`http://www.globalplatform.org/`), the specification assumes that there is a shared secret key between smart card and accessor. However, as we want to enable access to smart cards from different Web sites, a shared secret between the smart card and each individual Web site is infeasible.

5.1   Secure Computer Model

For the endpoint of our end-to-end security protocol on the local computer (see Figure 1(a)) we assume a computer model that allows to create a *secure runtime partition* in which software is executed that cannot be accessed or manipulated by the user's (default) operating system. This secure partition hosts a browser instance used for communication with the smart card. Furthermore, the secure partition allows for remote attestation—allowing a remote entity to securely identify the executed software. To provide compatibility across different types of trusted environments, our model does not assume any further features. In particular, we do *not* assume that it is possible to open a secure channel to I/O devices such as smart card readers. This is in accordance with the current state of trusted environments, where applications can open secure channels only to some types of I/O devices such as monitors and keyboards [12, 13].

There are different technologies that allow for the creation of such a secure partition. One technology is Intel's *Trusted Execution Technology* (Intel TXT) that complements the functionality of a TPM by allowing a secure hypervisor to provide *virtual environments* that are protected from access by malicious applications. For remote attestation, a TPM can be used. The Xen hypervisor provides a *vTPM* [14] implementation that provides *virtual* TPM chips [15] to the executed instances. These virtual TPMs use features of the host's hardware TPM for the secure implementation of their different functions.

5.2   Establishing a Shared Secret for HMAC and Encryption

As basis for encryption and authentication we use a shared secret between smart card and Web browser running in a trusted environment. To establish this secret we use an authenticated Diffie-Hellman (DH) key exchange [16] as depicted in Figure 5. The variables g, p, A, B in the figure are Diffie-Hellman parameters. For authentication, we sign the parameter set sent by each of the parties with digital signatures that prevent man-in-the-middle attacks. On the smart card we use an asymmetric key that is certified by the smart card manufacturer, while on the TPM we use the AIK (Attestation Identity Key) for authentication.

Instead of using Diffie-Hellman, it would also be possible to generate the symmetric key on one of the endpoints and use asymmetric encryption to transfer this key to the other endpoint. However, with such a method the long-term security of the communication would depend on the security of the particular asymmetric key. If the asymmetric key would be broken, each symmetric key encrypted with this key in the past would be compromised. With Diffie-Hellman on the other hand, an attacker needs to crack each session key individually.

5.3   Mutual Authentication and Integrity

Mutual authentication allows each endpoint of a conversation to authenticate the identity of the opposing endpoint. Authentication and integrity are intertwined concepts: When endpoint authentication is used without data integrity, an adversary can exchange data while in transit. Likewise, if data integrity is used without endpoint authentication, an endpoint knows that the data have not been modified, but does not know the identity of the remote endpoint.

In this section we provide our approach for authentication and integrity. There are two endpoints (see Figure 1(a)): The smart card and the Web browserrunning

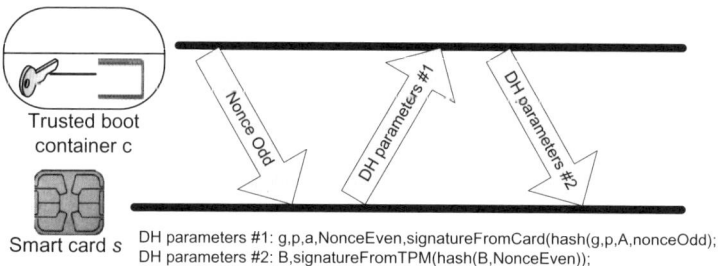

Fig. 5. Exchange of Diffie-Hellman parameters for secure channel. A nonce is used to prevent replay attacks. Each endpoint transfers cryptographically signed DH parameters to the other endpoint. These parameters are then used to establish the key for the secure channel.

in a trusted environment. Authentication uses authenticated Diffie-Hellman key exchange, while integrity uses HMACs (Keyed-Hash Message Authentication Codes).

Each smart card stores a custom key pair generated on initialization and digitally signed by the smart card issuer. When transmitting Diffie-Hellman parameters, the smart card signs them with its private key and appends the public key together with the certificate of the issuer to the data structure as shown in Figure 5.

On the PC, the `TPM_Quote` command of the TPM is used to sign the content of particular PCRs (Platform Configuration Registers) that contain measurements of the executed software used to identify a particular software configuration. When PCRs are updated, the TPM combines the existing value with the new value, thus software running on the computer is not able to set the registers to arbitrary values. By cryptographically signing the values within the registers, the TPM chip can attest the state of the system to a remote system. This functionality is also called *remote attestation*.

The certification of the TPM's key is more complex than in the case of the smart card. Figure 6 shows the certification and attestation procedure, illustrating which entity certifies which other entity to build up a chain of trust that allows the smart card to verify a correct software execution environment. The TPM contains two different types of keys: The unmodifiable EK (Endorsement Key) generated during production and certified by the TPM's manufacturer and a modifiable AIK used for attestation. A Privacy CA (Privacy Certificate Authority) with knowledge about EK and AIK is responsible for certifying the AIK [12,13]. During attestation, the AIK signs a set of values that identifies the executed software. The smart card compares these values with a set of reference values identifying a particular browser appliance and signed by a trusted party (e.g., the smart card issuer).

5.4   APDU Encryption and Authentication

For the encryption and authentication of APDUs we use a protocol similar to smart card *secure messaging* defined in ISO/IEC 7816-4. The main reason why

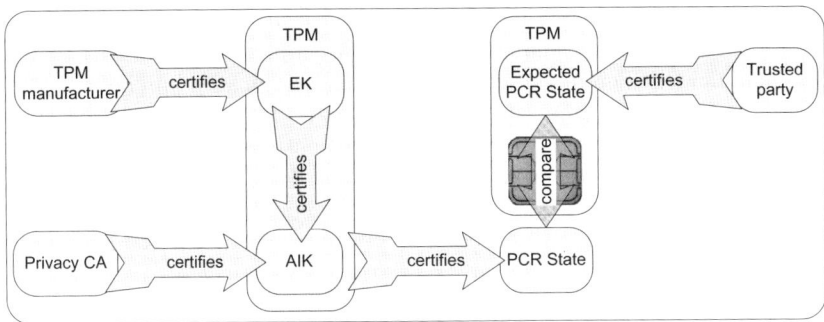

Fig. 6. Certification and attestation procedure

we cannot directly use secure messaging is that secure messaging requires a shared key between smart card and terminal. However, in our application scenario, this is not feasible as we want to enable secure communication with the smart card from a wide range of Web applications.

In theory, it would be possible to use the secret we established in Section 5.2 as the basis of a ISO/IEC 7816-4 compliant communication established directly by the smart card's operating system. However, as common smart card operating systems do not allow for access to the particular layer by client applications, this option is not possible. Instead, we re-implement comparable functionality inside the application layer, transmitting secured data as payload in APDUs. Therefore, APDU encryption and authentication allows us to establish a secure channel between smart card and Web browser.

APDUs are encrypted and authenticated by calculating *encrypt(session_key, hmac(session_key, orig_apdu + counter) + orig_apdu + counter)* using a symmetric encryption protocol such as AES (Advanced Encryption Standard). If no encryption is required, using an HMAC without encryption is possible. The HMAC serves to detect manipulation attempts of the APDU. The counter allows to detect replay attacks: In the beginning, a counter value derived from the shared key is used. As the session proceeds, each endpoint increases the counter value by one for each request and each response. Furthermore, each endpoint can detect if the received counter value matches the expected counter value. As a side effect, the counter also acts as a type of *initialization vector* (IV), as two equal APDUs encrypt to two different cipher texts.

5.5 Security Discussion

One issue with the certification of the PCR state by the smart card is that if the Privacy CA only certifies that the AIK belongs to *any* valid TPM implementation, it would be sufficient for an adversary to obtain the private key of *any* certified TPM to apply signatures. To cause a security problem the adversary would need to (i) break the user's environment so that the browser does not run inside a secure environment with the effect that malware has access to the software and (ii) to sign the TPM's side of the transaction with a certified key from another broken TPM implementation.

One option for mitigation of such an issue would be for the Privacy CA to not only certify that the key belongs to any TPM implementation, but to further include a user identifier in the certificate. This certificate can then be used by the smart card to verify that the TPM belongs to an authorized user. Another option is to store the first TPM key used for authentication and to only allow this particular key for future transactions. While this does not help against malware on a freshly installed PC, it protects the user against later attacks. However, to use another PC, a user would first need to obtain a certificate from the card issuer that instructs the smart card to reset the stored key.

When a secure channel is used, there is an important difference in the behavior of the intermediate proxy: In unencrypted communication, the proxy is responsible for filtering requests, i.e., to only allow requests *whitelisted* in the mapping to pass. However, with encryption, such a filtering is not possible as the proxy does not have access to the plain text. Thus, the proxy cannot verify if a particular encrypted APDU is allowed by the mapping file. As the proxy cannot filter requests sent to smart cards in that case, smart cards need to be developed with the assumption that potentially *any* Web site can send requests. While a majority of smart cards is already designed to withstand external attacks, the optimal mitigation strategies in our scenario are different. Traditionally, a smart card does, e.g., deactivate itself, if large amounts of failed authentication attempts are detected. However, if any Web application can communicate with the smart card, a Web application could abuse such a behavior for a DoS (Denial of Service) attack: By deliberately causing failed authentication attempts, any Web application could disable the smart card.

As mitigation strategy, smart cards should not take any *destructive* actions in case of failed authentication attempts or other types of security alerts that can be caused by external Web applications. For example, instead of deactivating the smart card in case of multiple failed authentication attempts the smart card could just increase the minimum interval required between each authentication attempt. As an alternative, the on-card application responsible for the secure connection can filter requests according to the information in the mapping—and thereby accomplish the filtering task of the proxy.

6   Authentication with QR-TAN

This section presents the second option to increase the security of the proxy, by extending our system with QR-TANs [3]. QR-TANs are a transaction authentication technique that uses a secure device to allow users to securely confirm electronic transactions on remote systems. A user scans a two dimensional barcode containing information about a transaction with a secure device. The secure device allows the user to verify the transaction. To approve the transaction, the user transmits a TAN (Transaction Authentication Number) dynamically generated by the secure device to the remote system. In comparison to our original QR-TAN approach [3], our modifications allow the use of QR-TANs without the interaction of a server.

Conceptually, it is sufficient to replace the RTC (Remote Trusted Computer) in the original approach with a smart card. However, due to the different capabilities of smart cards and servers, modifications to the original approach allow for better integration. In particular, our modifications address the following issues:

1. On smart cards, the generation of textual authentication requests intended for humans is more complicated than on servers. Especially as the smart card's memory restricts the amount of stored localizations and as it is rather complex to update the messages once the card has been issued.
2. The smart card should have the capability to decide if external transaction authentication is required. For example, in banking applications a smart card may allow daily transactions of up to a particular total value without authentication, only requiring authentication above that value.
3. Usage of QR-TAN should be transparent to applications using the proxy. Thus, only the smart card, the proxy, and the secure device should contain QR-TAN specific code.

To enable these properties, we extend our mapping description to contain information about QR-TAN authentication. In particular, we introduce a new <auth /> section that describes which status words in the APDU response indicate that QR-TAN authentication is required and how human readable text is generated from the structure returned by the smart card. For authentication the following steps depicted in Figure 7 are used:

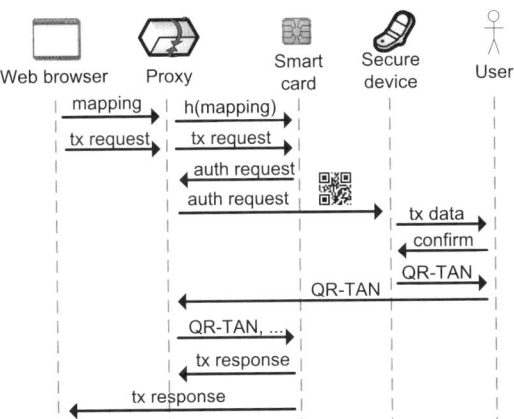

Fig. 7. QR-TAN authentication steps

1. On initialization the browser provides the mapping to the proxy. The proxy transmits a hash and a certificate of the used mapping file signed by a trusted third party to the smart card. The smart card verifies that the certificate allows use of the given mapping file.
2. The Web browser sends its transaction request to the smart card.
3. The smart card responds with a particular status word indicating that QR-TAN authentication is required for this type of transaction. The data of the response contains a structure with information about the transaction, a nonce, and the secure hash of the used mapping file.

4. The proxy reads the response by the smart card and converts it to a QR code that is subsequently displayed to and scanned by the user's secure device.
5. The secure device first checks if it has stored the mapping file indicated in the QR code. If not, it prompts the user to install the mapping file—e.g., by scanning a compressed QR code containing the mapping file. Otherwise, it generates a human readable text according to the information in the mapping file and asks the user for confirmation.
6. If the user confirms, the secure device generates a QR-TAN over the data structure of the original request (in step 2), the nonce, and the hash of the mapping file and presents this QR-TAN to the user.
7. When the user provides this QR-TAN to the proxy, it generates an APDU with this information and sends it together with the nonce and the hash of the mapping file to the smart card. The smart card validates if the QR-TAN request for the particular nonce matches the hash of the information within the APDU.
8. In case of success, the smart card returns the response of the original APDU request issued in step 2.

The conversion of the transaction data to a human readable text can be done on either one of the two endpoints of the QR-TAN authentication: Inside the smart card or inside the trusted secure device. It is not possible to do this conversion on any device between these two endpoints as this would prevent from successful end-to-end authentication. In our approach we perform this conversion inside the secure device. While generating the text directly within the smart card would be conceptually simpler, it would require the smart card to store potentially large amounts of textual data—e.g., if multiple localizations are required. Furthermore, it is not easily possible to adapt the text once the smart card has been issued.

As the secure device uses a mapping file to convert the structure with information about the transaction to a textual format, it must ensure that the mapping file is also authenticated. Otherwise, an attacker would be able to send a manipulated mapping file to the secure device, causing the device to show incorrect transaction information to the user. It is not sufficient for the secure device to only validate if the mapping file has been signed by a trusted party: As the secure device cannot securely obtain the ATR (Answer To Reset) of the smart card, it cannot ensure that the mapping file belongs to a particular smart card. Instead, we include a hash of the mapping file in the structure that is hashed by the secure device, allowing the smart card to ensure that the QR-TAN belongs to a particular mapping. While the smart card does not need to know the content of the mapping file, it needs to know the digital signature to decide if it can trust the mapping. Thus, the proxy can send the signature of a mapping to the smart card via APDUs.

By integrating our QR-TAN approach directly with the proxy, the proxy is responsible for displaying the QR code and for forwarding the QR-TAN back to the smart card. Thus, the whole process can be transparent to applications using the mapping, as the only difference between authentication and non-authentication is the additional delay of the authentication process.

7   Conclusion and Outlook

We presented a secure approach for Web-based smart card communication. Our overall contributions are: (i) a secure technique using a single generic proxy to allow a multitude of authorized Web applications to communicate with *existing* smart cards and (ii) techniques for *new* smart cards that allow for secure end-to-end communication between a user and a smart card. In particular, our security extensions cover the usage of (i) a TPM and (ii) QR-TANs to secure communication with smart cards.

Especially with citizen cards recently introduced in several countries and with high security requirements in online banking, a secure solution for Web-enabled smart cards is required. Compared to related approaches, our system works with existing smart cards without requiring changes to on-card software. Thereby, we can increase the security of the user's system, by not requiring the user to install privileged software distributed by Web sites that require access to a smart card. Furthermore, in cases where it is feasible to adapt on-card software, we can increase the security over the state-of-the-art even further, as we can use the TPM or QR-TANs to secure transactions that would otherwise be affected by malware on the terminal.

Overall, we see that Web to smart card communication techniques are an area where further research is required. In particular, researching the possibilities to use state-of-the-art Web protocols for secure mashups [17] may provide viable results, allowing smart cards to use standard Web protocols to identify and authorize Web applications. However, with increasing usage of Web technologies in smart cards also new kinds of attacks against smart cards are viable, as malicious applications can now target the Web browser to gain access to the smart card. Therefore, end-to-end security techniques are required to allow smart cards to mitigate the risk of such attacks. Additionally, new approaches [18] in automatic generation of network protocol gateways can allow for the more efficient generation of Web to smart card mapping files.

Concluding, our research can serve as basis for a newer, more secure generation of smart card to Web communication. By combining the security features of smart cards with the features provided by TPMs and QR-TANs, we can mitigate the effects of the terminal problem [19] as the smart card is able to assert that the transaction data has not been manipulated. While future smart card generations may require modifications to the specific techniques introduced in this paper, the overall approach will still be applicable. Furthermore, recent developments such as the Trusted Execution Module (TEM) [20] allow for the implementation of more powerful request mapping approaches on smart cards.

Acknowledgments. The authors would like to thank Markus Wilthaner for the proof-of-concept prototype implementation of the work described in this paper. This work has been partially funded by the Austrian Federal Ministry of Transport, Innovation and Technology under the FIT-IT project TRADE (Trustworthy Adaptive Quality Balancing through Temporal Decoupling, contract 816143, http://www.dedisys.org/trade/).

References

1. Lu, H.K.: Network smart card review and analysis. Computer Networks 51(9), 2234–2248 (2007)
2. Leitold, H., Hollosi, A., Posch, R.: Security architecture of the austrian citizen card concept. In: ACSAC, pp. 391–402. IEEE Computer Society, Los Alamitos (2002)
3. Starnberger, G., Froihofer, L., Goeschka, K.M.: QR-TAN: Secure mobile transaction authentication. In: International Conference on Availability, Reliability and Security. ARES '09, Fukuoka, pp. 578–583 (March 2009)
4. Itoi, N., Fukuzawa, T., Honeyman, P.: Secure internet smartcards. In: Attali, I., Jensen, T.P. (eds.) JavaCard 2000. LNCS, vol. 2041, pp. 73–89. Springer, Heidelberg (2001)
5. Urien, P.: Smarttp smart transfer protocol. Internet Draft (June 2001)
6. Urien, P.: TLS-tandem: A smart card for WEB applications. In: 6th IEEE Consumer Communications and Networking Conf. CCNC 2009, pp. 1–2 (January 2009)
7. Lu, H.K., Ali, A.: Prevent online identity theft - using network smart cards for secure online transactions. In: Zhang, K., Zheng, Y. (eds.) ISC 2004. LNCS, vol. 3225, pp. 342–353. Springer, Heidelberg (2004)
8. Bottoni, A., Dini, G.: Improving authentication of remote card transactions with mobile personal trusted devices. Computer Communications 30(8), 1697–1712 (2007)
9. Aussel, J.D., d'Annoville, J., Castillo, L., Durand, S., Fabre, T., Lu, K., Ali, A.: Smart cards and remote entrusting. In: Future of Trust in Computing, pp. 38–45. Vieweg/Teubner (2009)
10. Márquez, J.T., Izquierdo, A., Sierra, J.M.: Advances in network smart cards authentication. Computer Networks 51(9), 2249–2261 (2007)
11. Rannenberg, K.: Multilateral security a concept and examples for balanced security. In: NSPW '00: Proceedings of the 2000 workshop on New security paradigms, pp. 151–162. ACM, New York (2000)
12. Müller, T.: Trusted Computing Systeme. Xpert.press/Springer (2008)
13. Challener, D., Yoder, K., Catherman, R., Safford, D., Van Doorn, L.: A practical guide to trusted computing. IBM Press (2007)
14. Berger, S., Caceres, R., Goldman, K.A., Perez, R., Sailer, R., van Doorn, L.: vTPM: Virtualizing the Trusted Platform Module. In: Proceedings of the 15th USENIX Security Symposium, USENIX, pp. 305–320 (August 2006)
15. England, P., Löser, J.: Para-virtualized tpm sharing. In: Lipp, P., Sadeghi, A.-R., Koch, K.-M. (eds.) Trust 2008. LNCS, vol. 4968, pp. 119–132. Springer, Heidelberg (2008)
16. Diffie, W., Hellman, M.E.: New Directions in Cryptography. IEEE Transactions on Information Theory IT-22(6), 644–654 (1976)
17. Hammer-Lahav, E., Cook, B.: The oauth core protocol. Internet Draft draft-hammer-oauth-02 (March 2009)
18. Bromberg, Y.D., Réveillàre, L., Lawall, J.L., Muller, G.: Automatic generation of network protocol gateways. In: Bacon, J.M., Cooper, B.F. (eds.) Middleware 2009. LNCS, vol. 5896, pp. 21–41. Springer, Heidelberg (2009)
19. Gobioff, H., Smith, S., Tygar, J.D., Yee, B.: Smart cards in hostile environments. In: WOEC'96: Proc. of the 2nd USENIX Workshop on Electronic Commerce, Berkeley, CA, USA, USENIX Association, p. 3 (1996)
20. Costan, V., Sarmenta, L.F.G., van Dijk, M., Devadas, S.: The trusted execution module: Commodity general-purpose trusted computing. In: Grimaud, G., Standaert, F.-X. (eds.) CARDIS 2008. LNCS, vol. 5189, pp. 133–148. Springer, Heidelberg (2008)

Efficient Term Cloud Generation for Streaming Web Content

Odysseas Papapetrou, George Papadakis,
Ekaterini Ioannou, and Dimitrios Skoutas

L3S Research Center, Hannover, Germany
{papapetrou,papadakis,ioannou,skoutas}@L3S.de

Abstract. Large amounts of information are posted daily on the Web, such as articles published online by traditional news agencies or blog posts referring to and commenting on various events. Although the users sometimes rely on a small set of trusted sources from which to get their information, they often also want to get a wider overview and glimpse of what is being reported and discussed in the news and the blogosphere. In this paper, we present an approach for supporting this discovery and exploration process by exploiting term clouds. In particular, we provide an efficient method for dynamically computing the most frequently appearing terms in the posts of monitored online sources, for time intervals specified at query time, without the need to archive the actual published content. An experimental evaluation on a large-scale real-world set of blogs demonstrates the accuracy and the efficiency of the proposed method in terms of computational time and memory requirements.

1 Introduction

The popularity of online news sources has experienced a rapid increase recently, as more and more people use them every day as a complement to or replacement of traditional news media. Therefore, all major news agencies make nowadays their content available on the Web. In addition, there exist several services, such as Google News or Yahoo! News, that aggregate news from various providers. At the same time, more and more people maintain Web logs (blogs) in a regular basis as a means to express their thoughts, present their ideas and opinions, and share their knowledge with other people around the globe. Studies about the evolution of the Blogosphere [3,19] report around $175,000$ new blogs and 1.6 million new blog posts every day. Micro-blogging has also emerged as a special form of such social communication, where users post frequent and brief text messages, with Twitter constituting the most popular example. This results in an extremely large and valuable source of information that can be mined to detect themes and topics of interest, extract related discussions and opinions, and identify trends. Perhaps the most interesting aspect that distinguishes these sources from other information available on the Web is the temporal dimension. Several recent research activities have focused on analyzing and mining news and

B. Benatallah et al. (Eds.): ICWE 2010, LNCS 6189, pp. 385–399, 2010.

blogs for detecting events or stories (e.g., [2,1,15,19]), identifying trends (e.g., [5,11]), and extracting opinions and sentiments (e.g., [8,18,22]).

In the Web, users very often need to explore large collections of documents or pages, without having a-priori knowledge about the schema and the content of the contained information. Various techniques, such as clustering, faceted browsing, or summarization, are used in such scenarios to help the user more effectively navigate through the volume of the available information, to obtain an overview, and to drill down to the items of interest. One simple visualization technique that has become very popular, especially in Web 2.0 applications, is presenting to the user a visual overview of the underlying information in the form of a *tag cloud*. This is typically a set comprising the most popular (i.e., frequent) tags contained in or associated to the underlying data collection, where most frequent tags are visualized in a more prominent way (i.e., using larger font size). In many applications, the cloud is constructed from the tags assigned to the data explicitly by users. For example, Flickr provides tag clouds of the most popular tags entered by the users in the last 24 hours or over the last week[1]. In other cases, where no such tags or keywords are available, the cloud can be constructed from terms (or other semantically richer items, such as entities) automatically extracted from the text, as for example in Blogscope [2]. The advantage of tag clouds lies mainly in the fact that they are very simple, intuitive, and visually appealing to the user, while serving two purposes; first, they provide an easy and quick way to the user to get hints about what is contained in the underlying data, e.g., what is being reported and discussed in the news. Second, they allow the user to navigate and explore the available information, by selecting an interesting term in the cloud and viewing the corresponding documents.

In this work, our goal is to exploit these benefits of tag clouds to facilitate users in exploring and obtaining an overview of Web content made available in a streaming fashion, such as news-related information provided by online media and the blogosphere. Once the candidate terms are available, either explicitly provided or automatically extracted, the construction of the cloud comprises just a single task: computing the frequencies of these terms, and selecting the most frequent ones to visualize in the cloud. This computation is straightforward and can be done on the fly in most cases, when dealing with a relatively small set of documents. However, this is not scalable to very large collections of items, as in our scenario, which should scale, for example, to millions of news and blog posts, and over potentially very long time periods. Moreover, we do not want to restrict the interests of the users to pre-defined time periods, e.g., for the current day, as typically happens in existing applications. Instead, the users should be able to request the generation of a term cloud for arbitrarily small or large, as well as for arbitrarily more or less recent, time intervals, such as for the last hour(s), day(s) or week(s), for the last year, for the summer of 2008, etc.

To deal with these challenges, we cast the above problem as one of finding frequent items in incoming streams of data. Specifically, we assume a stream consisting of the terms extracted from the newly published posts that are made

---

[1] http://www.flickr.com/photos/tags/

available to the system. Our goal is to maintain space-efficient data structures that allow the computation of the top-k frequent terms in the stream, for arbitrary time intervals and as accurately as possible. This formulation allows us to exploit the advantages of efficient algorithms that have been proposed for mining frequent items in streaming data. In particular, our main contributions are as follows.

- We exploit term clouds to facilitate the navigation and exploration of streaming Web content, such as incoming news-related posts from online media and the blogosphere.
- We present an efficient method for generating term clouds dynamically for arbitrary time periods, based on techniques for mining frequent items in data streams.
- We validate experimentally the efficiency and the effectiveness of our method, using a large-scale real-world collection of blog posts.

The remainder of the paper is organized as follows. The next section discusses related work. Section 3 presents our approach in detail. In Section 4, we present our experimental evaluation. Section 5 concludes the paper.

2   Related Work

As discussed above, our purpose is to help users to get an overview and to explore and navigate large streams of news-related information, by identifying and visualizing in the form of a cloud the most popular terms appearing in the posts. Such clouds have become a very popular visualization method, especially in Web 2.0 applications, for presenting frequent tags or keywords for exploration and navigation purposes. Blogscope [2], for instance, presents in its front page a cloud of keywords extracted from the posts of the current day. Similarly, Grapevine [1], which builds on Blogscope, displays also a cloud of the main entities related to a selected topic or story, using the font size to denote the popularity of the entity, and the font color to denote its relatedness to the topic or story. In a different domain, PubCloud [14] is an application that allows querying the PubMed database of biomedical literature, and employs term clouds as a visualization technique to summarize search results. In particular, PubCloud extracts the words from the abstracts retrieved for the given query, performs stopword filtering and stemming, and visualizes the resulting terms as a cloud. It uses different font sizes to denote term frequency and different font colors to denote the recency of the corresponding publications in which the term appears. Through a user study, the authors show that this summarization via the term cloud is more helpful in providing descriptive information about the query results compared to the standard ranked list view. TopicRank [4] generates a term cloud from documents returned as a result to a query, where the positioning of the terms in the cloud represents their semantic closeness. Page History Explorer [9] is a system that uses term clouds to visually explore and summarize the contents of a Web page over time. Data clouds are proposed in [13] for combining the advantages of keyword search over structured data with those of tag clouds for summarization and navigation. Complex objects are retrieved from a database as responses to a keyword query; terms found in these objects are ranked and visualized as a cloud. Other tools also exist, such as ManyEyes[2] or Wordle[3], for generating word clouds from user provided text.

In all these applications, the term cloud is generated from a corpus that has a relatively limited size, e.g., blog posts of a particular day or story, query results, or a document given by the user. Hence, efficiently computing the most frequent terms for visualization in these scenarios does not arise as a challenging problem per se, and, naturally, it does not constitute the focus of these works. Instead, in our work, frequent terms need to be computed on demand from very long streams of text (e.g., news articles and blog posts spanning several months), and therefore efficiency of the computation becomes a crucial issue.

To meet these requirements, we formulate the problem as finding frequent items in streaming data. Several space-efficient algorithms have been proposed that, given a stream of items, identify the items whose frequency exceeds a given threshold or estimate the frequency of items on demand. A survey and comparison of such algorithms can be found in [6,16]. In our case, we are interested in finding the top-k frequent items. Hence, these algorithms are not directly applicable, since there is no given frequency threshold nor is it efficient to maintain a set of *all* possible items and then calculate their frequencies on demand to select the top-k ones. The problem of finding the top-k frequent items in a stream is addressed in [21]. The authors propose two algorithms, one based on the Chernoff bound, and one that builds on the Lossy Counting algorithm from [17]. However, these solutions consider either the entire data stream or a sliding window that captures only the most recent items. Instead, we want to allow the users to construct the term clouds on arbitrary time intervals of the stream. The TiTi-Count+ algorithm has been proposed in [20] for finding frequent items in ad hoc windows. This algorithm, though, finds the items with frequency above a given threshold and not the top-k frequent items. Thus, the method we present in this paper builds on the ideas and advantages of the solutions presented in [20,21] in order to meet the requirements of our case.

3   Term Clouds for Streaming Text

In this section, we present our approach for constructing term clouds from streaming text. A term cloud is composed of the top-k frequent terms in the data collection, which in our case is a substream of text covering a specified time interval, defined by the query. First, we describe the main components of the system and we formally define the problem. Then, in Section 3.2, we describe the data structures used to support the efficient computation of the term cloud. Using these data structures, we describe in Section 3.3 how the term cloud for a desired time interval is generated dynamically upon request.

---

[2] http://services.alphaworks.ibm.com
[3] http://www.wordle.net

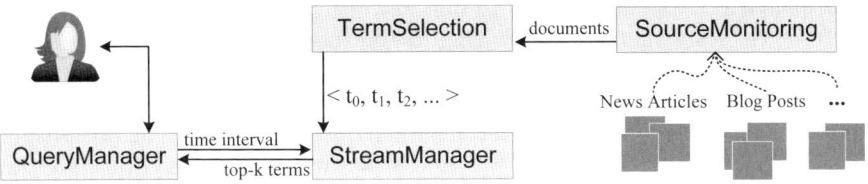

Fig. 1. System overview

3.1 System Overview

The overall system architecture is depicted in Figure 1. The *SourceMonitoring* component monitors online sources for new posts (e.g., online news agencies and blogs) and updates the system when new information is published. These incoming posts are provided as input to the *TermSelection* component, which is responsible for extracting a set of terms from each post. These are the candidate terms from which the term clouds are generated. These terms consist the input to the *StreamManager* component, which maintains a set of data structures for efficiently generating term clouds. Finally, the *QueryManager* component is responsible for receiving and evaluating user queries for term clouds, using the data structures provided by the *StreamManager*.

Each component poses a set of challenges. For example, regarding the *SourceMonitoring* component, the main issues that arise are how to select or dynamically discover sources, and how to monitor and extract the useful text from sources that do not provide feeds. A main challenge for the *TermSelection* component is how to tokenize the incoming text and to select potentially useful and meaningful terms. Solutions for this may vary from simply filtering out some terms using a stopword list to applying more sophisticated NLP techniques for extracting keywords and identifying named entities. In this paper, we focus on the *StreamManager* and *QueryManager* components, which we describe in detail in the following sections. This modular architecture for the processing pipeline allows us to easily plug in and try different solutions for the other components.

3.2 Data Structures

We now describe the data structures maintained by *StreamManager* for supporting the efficient computation of term clouds.

As described above, the *TermSelection* component extracts a set of candidate terms from each newly published post that is detected by the *SourceMonitoring* component, and provides the resulting terms as an input to the *StreamManager* component. Hence, the input of *StreamManager* is a stream of terms $T = <t_0, t_1, t_2, \ldots >$. For each term t_i, we also maintain a pointer to the document from which it was extracted. As will be described later, only a subset of the incoming terms are maintained by the system. Accordingly, only the pointers to the documents containing these terms are finally stored. Maintaining these pointers is required to enable the navigation functionality in the term cloud. Also note that the actual contents of the posts are not stored in main memory;

they are fetched (from the Web or from a local cache in secondary storage) only if the user requests them by navigating in the term cloud. The order in which the terms are appended to the stream is according to the publishing time of the corresponding posts. This implies that the *SourceMonitoring* component fetches the posts in batches, and sorts them by their publishing time, before pushing them further down the processing pipeline. Posts with the same publishing time are ordered arbitrarily, while terms extracted from the same post are ordered according to their appearance in the post. Hence, the first (last) term in the stream is the first (last) term of the least (most) recent post that has been processed by the system.

Efficiently computing the top-k frequent terms in the stream involves two main issues: (a) a method to estimate the frequency of a given term, and (b) a method to distinguish which terms have high probability to be in the top-k list, so that only these terms are maintained by the system to reduce memory requirements. In the following, we describe how these issues are addressed.

Instead of storing the whole stream, we maintain a compact summary of the observed terms and their frequencies (i.e., number of occurrences) using an *hCount structure* [10], as briefly described in the following. hCount is composed of a set of counters, which allow us to estimate the frequency of a term with high accuracy. Specifically, this structure is a 2-dimensional array of counters comprising h rows and m columns. Each row is associated with a hashing function, which maps a given term to an integer in the range $[0, m - 1]$. When the next term t in the stream arrives, its hashing values for all the h hashing functions are computed. Then, for each row i, the counter of the column $h_i(t)$ is incremented by 1. The frequency of a term t can then be estimated as follows: the term is hashed as before, to identify the corresponding counters, and the minimum value of these counters is used as an estimation of the term's frequency. The benefit is that this operation now requires only $h \times m$ counters instead of $|\mathcal{T}_d|$ which would be required for storing the exact frequencies of all the distinct terms $\mathcal{T}_d$. hCount also provides probabilistic guarantees for the frequency estimation error of a term t in a given stream $\mathcal{T}$. According to [10], if $\frac{e}{\epsilon} \times \ln\left(-\frac{|\mathcal{T}_d|}{\ln \rho}\right)$ counters are stored, the frequency estimation error for each term is not more than $\epsilon \times |\mathcal{T}|$ with probability ρ. In our experiments in Section 4, a sufficiently accurate estimation of the term frequencies was achieved using only a very small number of counters, i.e., $(h \times m) << |\mathcal{T}_d|$, and thereby with negligible memory requirements.

The problem that arises next is how to determine for which terms to probe the *hCount* structure for their frequencies, in order to get the top-k frequent terms. A straightforward solution, albeit prohibitively expensive, is to maintain a list of all the distinct terms, and look them up in the *hCount* structure to get their estimated frequencies. To significantly reduce the memory requirements and execution time, we need to prune this list, maintaining instead only a subset of terms, such that all the top-k frequent terms are contained in this subset with high probability. To determine this subset, we employ a method based on Chernoff bounds, as proposed in [21]. The method considers the terms in the stream in batches of fixed length l. For each batch B, the exact frequencies of all the contained terms are calculated, and the frequency of the k-th most frequent term in the batch, denoted by $freq_k(B)$, is found. The *support* of this term in terms of B is computed as $s_k(B) = freq_k(B)/l$. Then, using Chernoff bounds, we filter out all terms in the batch that do not belong in the top-k terms of the stream with probability higher than a predefined value δ. In particular, only terms that have observed support:

$$s(B) \geq s_k(B) - 2\sqrt{\frac{2 \times s_k(B) \times \ln(2/\delta)}{l}} \qquad (1)$$

need to be maintained in a term pool $\mathcal{P}$, with $\mathcal{P} \subseteq \mathcal{T}_d$, and it is sufficient to use only these terms for probing the hCount structure. The final size of $\mathcal{P}$ is typically substantially smaller than the total number of distinct terms $|\mathcal{T}_d|$ observed from the beginning of the stream. In our experiments presented in Section 4, we were able to obtain highly accurate results by maintaining no more than $2 \times k$ terms in $\mathcal{P}$ (which was at least 3 orders of magnitude less than the total number of distinct terms occurring in the stream). At query time, to extract the top-k terms from $\mathcal{T}_d$, the system estimates the frequency of each term $t \in \mathcal{T}_d$ from the hCount structure, it sorts all the terms on their estimated frequency, and it returns the top-k ones to the user.

The method described so far allows us to efficiently compute the top-k frequent terms for the *whole* stream. However, different users may be interested in different time periods. Therefore, we want to generate term clouds for arbitrary time intervals specified by the query. A query is a tuple $Q = <i, j>$, defining a substream comprising all the terms between the positions i and j ($i < j$).

To allow users to view term clouds for arbitrary time intervals, we divide the stream into time windows, as proposed in the TiTiCount+ algorithm [20]. Each window covers a different time interval, i.e., a different substream, as illustrated in Figure 2. Specifically, the first window, i.e. w_0, has size b, which means that it contains summary information for the b most recently received terms, where b is a predefined system parameter. Each subsequent window w_i, $i > 0$, has size $2^{i-1}b$. This particular organization of the windows is motivated by the assumption that users will more often be interested in more recent time periods; hence, the more recent parts of the stream are covered by windows of higher resolution (i.e., contain less terms) than the least recent ones. In the presence of these windows, the method described above is adapted by maintaining a separate hCount structure and a term pool of potentially frequent terms for each window, instead of the whole stream. The same number of counters and pool sizes are used for all the windows; hence, all windows have the same memory requirements. Consequently, the maintained summary in windows that cover less recent parts of the stream, and therefore have larger size, has typically lower accuracy.

Initially, the hCount structures in all the windows are empty. Each new incoming term is received by the first window. When a batch of b terms has been received, this window becomes full. Then, the contents of this window are shifted to the second one, so that new terms can be received. When the second window reaches its maximum size, its contents are shifted to the third window, and this continues recursively. If the window that receives the new content already

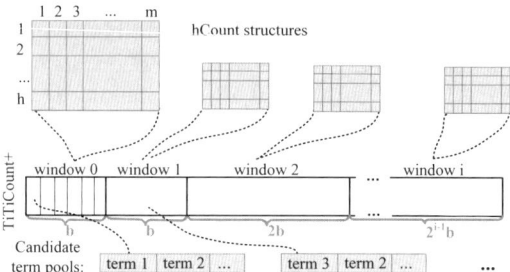

Fig. 2. An illustration of the StreamManager

contains some information, i.e., the counters in its hCount structure are not 0, the old contents need to be merged with the new one. Merging two subsequent windows involves: (a) merging the two hCount structures, as proposed in [20], and, (b) computing the new set of candidate top-k terms for the result of the merge, which is to replace the contents of the least recent of the two windows. Merging of the hCount structures is straight-forward; each counter of the resulting merged window is set to the sum of the corresponding hCount counters from the initial windows. Computing the new set of candidate top-k terms proceeds as follows. Let $\mathcal{P}_i$ and $\mathcal{P}_{i+1}$ denote, respectively, the corresponding pools of candidate terms of the original windows w_i and w_{i+1}. The new candidate terms $\mathcal{P}'_{i+1}$ are initially set to $\mathcal{P}_i \cup \mathcal{P}_{i+1}$. Then, the size of $\mathcal{P}'_{i+1}$ is further reduced by re-applying the Chernoff bound to the merged pool, with probability again δ, but now using the new window size as the length of the batch, i.e., $2^i \times b$. As before, the estimation for the Chernoff bound requires computing the support of the k most frequent term in the pool $\mathcal{P}'_{i+1}$. For this, the hCount structure of the new window is used to estimate the frequencies of all terms in $\mathcal{P}_i \cup \mathcal{P}_{i+1}$ with high accuracy. Since the tightness of the Chernoff bound increases with the length of the batch (see Equation 1), the final number of terms in $\mathcal{P}'_{i+1}$ is typically substantially smaller than the number of terms in $\mathcal{P}_i \cup \mathcal{P}_{i+1}$.

3.3   Query Execution

The *QueryManager* component is responsible for receiving the user query and evaluating it using the data structures described above. Mainly, this involves the following operations: (a) identifying the windows corresponding to the time interval specified in the query; (b) collecting the candidate top-k terms from each window; and (c) estimating their frequencies using the corresponding hCount structures, and identifying the top-k terms.

The first operation is rather straightforward. Given a query $Q = <i, j>$, the windows that are involved in the query execution are those covering a time interval that overlaps with the interval requested in the query. Once these relevant windows W_Q have been identified, the *QueryManager* constructs the set of the candidate top-k terms for Q, denoted by $\mathcal{P}_Q$. This set is the union of all the candidate top-k terms in the relevant windows, i.e., $\mathcal{P}_Q = \bigcup_{w_i \in W_Q} \mathcal{P}_i$. For each of the terms in $\mathcal{P}_Q$, the term frequency is estimated using the respective hCount structures. The terms are sorted descending on their estimated frequencies, and the top-k terms are returned to the user.

Note that, the starting and ending position specified in the query will typically not coincide with window bounds. Hence, some windows will be involved only partially in query execution. For example, consider a window of size ℓ that has an overlap of size ℓ' with the query. Assume also a term t in the pool of this window, with estimated frequency f_t provided by the hCount structure of this window. Then, this estimation is adjusted as $f'_t = (\ell'/\ell) \times f_t$. Even though this relies on a uniform distribution of the occurrences of the term in the duration of this window, our results show that the introduced error from non-uniform term distributions, i.e., bursty terms, is usually small.

4   Experimental Evaluation

4.1   Experimental Setup

To evaluate our approach, we have implemented a prototype as described in Section 3. Recall that in this paper we focus on the StreamManager and Query-Manager components (see Figure 1); hence, for the TermSelection component we have only implemented standard stemming and stopword filtering, and we have omitted the SourceMonitoring component, using instead the publicly available ICWSM 2009 dataset[4], which is a crawled collection of blogs. In particular, this is a large, real-world dataset comprising 44 million blog posts from the period between August 1st and October 1st, 2008. The posts are ordered by posting time, enabling us to simulate them as a stream. They are also separated by language. In our experiments, we have used only the English posts, which yields a number of 18.5 million posts. After the stemming and stopword filtering, the total number of distinct terms in the stream was 5 million, while the total number of terms was 1.68 billion. The statistics of this data collection used in our experiments are summarized in Table 1.

In the conducted experiments, we measure the performance of our system with respect to memory usage and for different query types. The examined parameters and their value ranges are summarized in Table 2. In our implementation, we set the size of the first window to be $1,000,000$ terms, which corresponds roughly to half an hour. Moreover, we set the error parameter for the Chernoff-based filtering to $\epsilon = 0.0001$, which resulted to maintaining approximately $2 \times k$ terms per window. The results of our evaluation are presented below.

4.2   Accuracy Versus Memory

First, we investigate how the amount of memory that is available to the system affects the quality of the results. Recall from Section 3 that the amount of memory used is related to: (a) the number of counters used by the hCount

---

[4] http://www.icwsm.org/2009/data/

Table 1. Statistics for the ICWSM 2009 Blog dataset (English posts)

Days	47
Blog posts	18, 520, 668
Terms	1, 680, 248, 084
Distinct terms	5, 028, 825
Average blog posts per day	394, 056
Average terms per post	90.72

Table 2. Parameters for the experiment. The default values are emphasized.

Parameter	Values
Counters per window and total memory required	10,000 (0.45Mb), 25,000 (1.14Mb), **40,000 (1.83Mb)**, 70,000 (3.20Mb), 100,000 (4.57Mb), 130,000 (5.95Mb)
Top-k terms to retrieve	25, **50**, 75, 100, 125
Query length	$[1, \ 1.68 \times 10^9]$, $\mathbf{10^7}$

structure in each window (i.e., the accuracy when estimating the frequency of a term), and (b) the number of candidate terms maintained in each window (i.e., the probability to falsely prune a very frequent term). To measure the quality of our results, we compare the list $\mathcal{L}$ of top-k frequent terms computed by our method to the exact list $\mathcal{L}_0$ of top-k terms computed by the brute-force method, i.e., by counting the exact occurrences of all the terms in the query interval. In particular, we evaluate accuracy using two standard measures, recall[5] and Spearman's footrule distance [12], which are briefly explained in the following.

The recall measure expresses the percentage of the actual top-k frequent terms that have been correctly retrieved by our system, i.e., the overlap between the two lists $\mathcal{L}$ and $\mathcal{L}_0$:

$$recall(\mathcal{L}, \mathcal{L}_0) = \frac{|\mathcal{L} \cap \mathcal{L}_0|}{k} \qquad (2)$$

However, simply identifying the actual top-k frequent terms is not sufficient. Since the contents of a cloud are typically visualized with different emphasis based on their frequency, identifying also the correct ordering of the terms is important. Hence, the ordering of the terms in $\mathcal{L}$ should be as close as possible to that in $\mathcal{L}_0$. To measure this, we use Spearman's footrule distance [12], a popular distance measure for comparing rankings. In particular, we use the extended version proposed by Fagin et al. [7], henceforth referred to as *Spearman's distance* (F^*), which also handles the case where the overlap of the two compared rankings is not complete. This measure is computed as follows:

$$F^*(\mathcal{L}, \mathcal{L}_0) = \sum_{i \in \mathcal{D}} |pos(i, \mathcal{L}) - pos(i, \mathcal{L}_0)| / maxF^* \qquad (3)$$

where $\mathcal{D}$ denotes the set of terms in $\mathcal{L} \cup \mathcal{L}_0$, and function $pos(i, \mathcal{L})$ gives the position of i in the list $\mathcal{L}$ if $i \in \mathcal{L}$, or $(|\mathcal{L}| + 1)$ otherwise. $maxF^*$ denotes the

---

[5] Since we are retrieving a fixed number of terms k, recall and precision have always the same value.

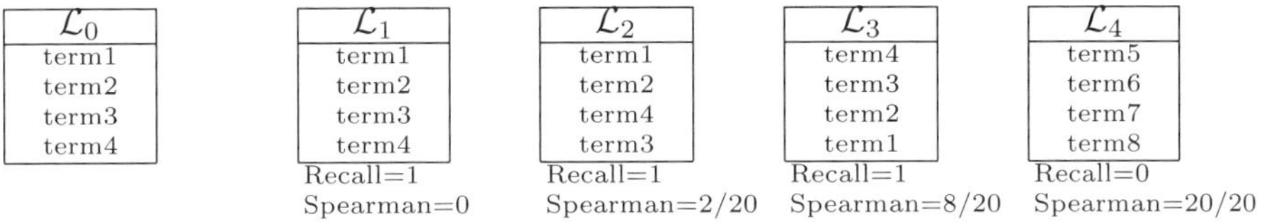

$\mathcal{L}_0$
term1
term2
term3
term4

$\mathcal{L}_1$
term1
term2
term3
term4

Recall=1
Spearman=0

$\mathcal{L}_2$
term1
term2
term4
term3

Recall=1
Spearman=2/20

$\mathcal{L}_3$
term4
term3
term2
term1

Recall=1
Spearman=8/20

$\mathcal{L}_4$
term5
term6
term7
term8

Recall=0
Spearman=20/20

Fig. 3. Comparing various rankings, $\mathcal{L}_1$, $\mathcal{L}_2$, $\mathcal{L}_3$, $\mathcal{L}_4$, to a correct ranking $\mathcal{L}_0$

Table 3. Sample queries (the most recent window is 0 and the oldest is 11)

Query	Start time	End time	Length	Involved windows
Q1	1	10000000	10000000	11
Q2	1600000001	1610000001	10000000	6, 7
Q3	1672000004	1680248081	8248078	0 - 4

maximum possible value that F^* can take, which equals to $(k+1) \times k$. Some examples are illustrated in Figure 3.

Figure 4 displays the Spearman's distance and the recall for the three sample queries depicted in Table 3, with respect to the system's memory. We see that both quality measures increase by increasing the available system memory. This happens because with more memory available, TiTiCount+ is automatically configured with more counters per window, thereby substantially reducing the probability of collisions in the hCount structure and increasing the accuracy. The memory increase is beneficial especially for the queries that involve many windows (e.g., Q3). A near-maximum accuracy of the system is achieved by using only 3Mb memory, even for the most difficult queries that involve small parts of very old windows (e.g., Q1). Increasing the available memory beyond 3 Mb still contributes to the accuracy, though at a lower rate.

4.3 Accuracy Versus Query Characteristics

We now examine how the accuracy of the system varies with different types of queries, and in particular with respect to (a) the query starting point, (b) the query length, and (c) the number k of most frequent terms to be retrieved. Throughout this section, we present the experimental results for three different memory configurations.

For the first set of experiments, we fix the query length to 10 million terms, and we vary the query starting point. Figure 5 presents the values for Spearman's distance and recall for the experiments with $k = 50$. As expected, the most recent queries are more accurate, since they are evaluated against windows of higher resolution. Nevertheless, even the queries that involve the least recent windows are still accurate, giving a minimum recall of 0.83 and a maximum Spearman's distance of 0.15. As shown by the previous experiments, we can further increase the accuracy for all queries by increasing the available system memory.

We also notice that both quality measures have some small peaks, e.g., at the queries starting at ca. 500 million terms. These peaks are due to the existence or

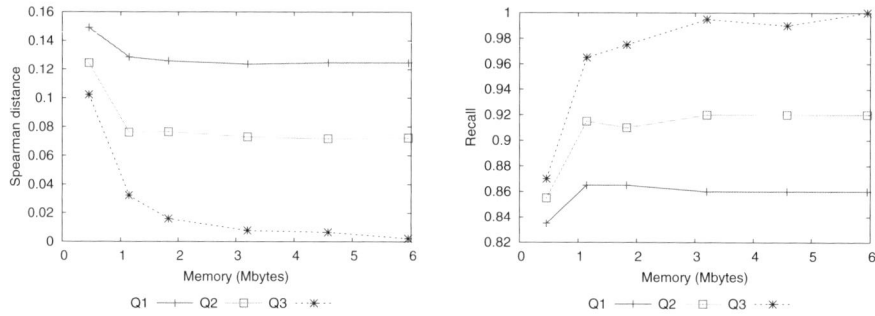

Fig. 4. Varying the allowed system memory: (a) Spearman's distance (b) Recall

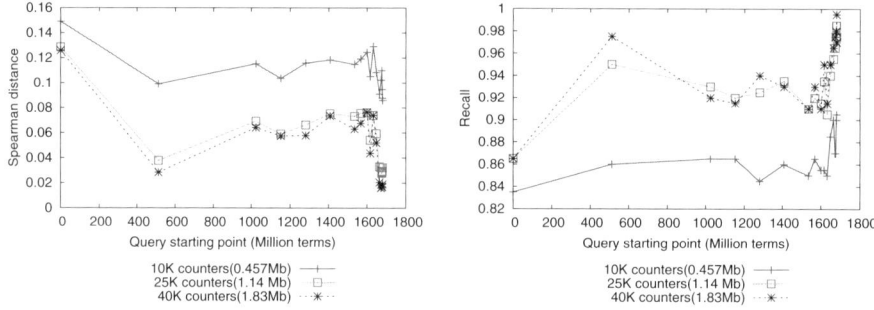

Fig. 5. Varying the query starting point: (a) Spearman's distance (b) Recall

absence of bursty terms in the particular query range: As explained in Section 3, when a query does not fully overlap with a window but involves only a part of it, then the computation of the term frequencies for the overlapping part are computed by multiplying the frequency given by hCount with the ratio of the overlap with the window length. This assumes a uniform distribution of the occurrences of the term in the duration of this window; therefore, if bursty terms are included in this overlap period, then a small error is introduced. In future work, we plan to deal with this issue by identifying these bursty terms and handling them differently, thereby providing even more accurate results.

We also conducted experiments varying the length of the query. For this, we fix the ending point of the query to the most recent term read from the stream, and we increase the query length with exponential steps. Figure 6 presents the two quality measures related to these experiments, for three example memory configurations. We see that the query length has a small effect on the system's accuracy. In particular, the accuracy is higher for the queries that involve the most recent windows, since these have higher resolution. Nevertheless, also the queries involving older windows are still answered with high accuracy.

Finally, we performed experiments varying the desired number of k most frequent terms to retrieve, keeping the rest of the system parameters fixed. Figure 7 presents the results for the three queries described in Table 3, for different values

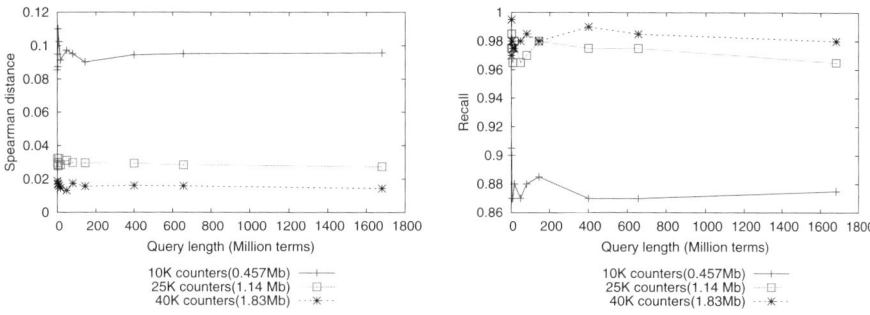

Fig. 6. Varying the query length: (a) Spearman's distance (b) Recall

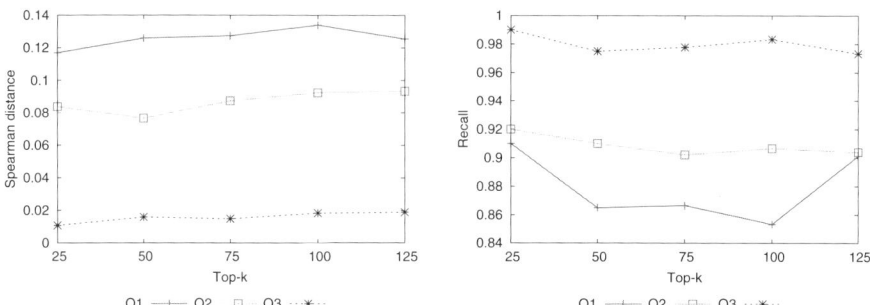

Fig. 7. Varying the desired number of top-k terms: (a) Spearman's distance (b) Recall

of k. These results were obtained using 1.83 Mb memory (40000 counters per window). We see that the value of k has no noticeable effect on the system's accuracy. This is expected, because as explained in Section 3, the system adapts to the value of top-k.

4.4   Execution Time Versus Query Characteristics

We also investigated how query execution time changes with system memory and with query characteristics. Figure 8(a) plots the execution time per query while varying the query length, for three example memory configurations. The results are averaged over 100 query executions, on a single 2.7 GHz processor. We see that query execution time scales linearly with query length. This is expected, since the query length determines the number of the involved windows, thereby also the number of number of hCount probes that need to be executed. Nevertheless, even for the query involving the whole stream range, the required execution time is negligible, below 200 msec. The execution time is also slightly affected by the system configuration, i.e., the number of counters, but this effect is also negligible.

Figure 8(b) plots the query execution time in correlation to query length, for five values of k. The execution time scales sublinearly to the value of k. This happens because with a higher k, the number of candidate terms that need to

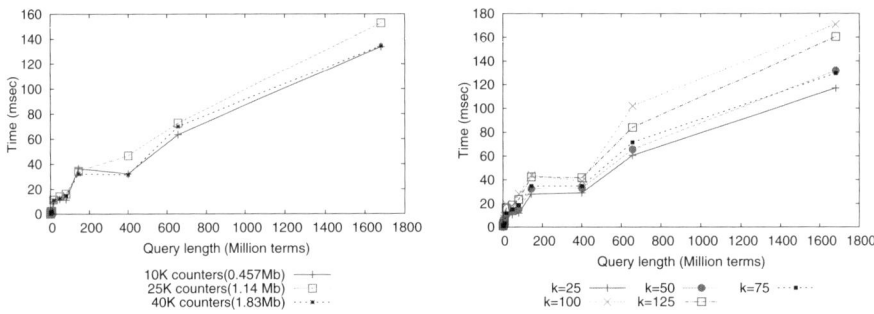

Fig. 8. Execution time: (a) varying the query length and the setup configuration, and (b) varying the query length and the number of top-k terms

be probed in the windows increases slightly, requiring more time. Nevertheless, all queries are executed at less than 200 msec, even for $k = 125$.

5 Conclusions

We have presented a system for generating term clouds from streaming text. The proposed method can be used to answer queries involving arbitrary time intervals, and it is very efficient with respect both to execution time and memory requirements. Our approach works by summarizing the term frequencies in compact in-memory structures, which are used to efficiently detect and maintain the top-k frequent terms with high accuracy. We have also conducted a large-scale experimental evaluation of the approach, using a real-world dataset of blog posts. The experimental results show that the proposed system offers high accuracy for identifying and correctly ordering the top-k terms in arbitrary time intervals, measured both with recall and Spearman's distance. Owing to the small memory footprint and the small query execution time, the system can be used for summarizing huge streams, and for efficiently answering top-k queries.

Our current and future work focuses mainly on two directions. First, we are working on increasing the resolution of the windows with respect to terms with uneven distribution, i.e., bursty terms. As shown in our experiments, when such terms occur in the query range, they introduce a small error in the accuracy of the results. This error can be avoided by identifying and handling such uneven term distributions differently. Furthermore, we will also focus on the other two components, the *TermSelection* and the *SourceMonitoring* component, to address the additional challenges mentioned in Section 3.1.

Acknowledgments

This work was partially supported by the FP7 EU Projects GLOCAL (contract no. 248984) and SYNC3 (contract no. 231854).

References

1. Angel, A., Koudas, N., Sarkas, N., Srivastava, D.: What's on the grapevine? In: SIGMOD, pp. 1047–1050 (2009)
2. Bansal, N., Koudas, N.: Blogscope: spatio-temporal analysis of the blogosphere. In: WWW, pp. 1269–1270 (2007)
3. Bansal, N., Koudas, N.: Searching the blogosphere. In: WebDB (2007)
4. Berlocher, I., Lee, K.-I., Kim, K.: TopicRank: bringing insight to users. In: SIGIR, pp. 703–704 (2008)
5. Chi, Y., Tseng, B.L., Tatemura, J.: Eigen-trend: trend analysis in the blogosphere based on singular value decompositions. In: CIKM, pp. 68–77 (2006)
6. Cormode, G., Hadjieleftheriou, M.: Finding frequent items in data streams. In: PVLDB, pp. 1530–1541 (2008)
7. Fagin, R., Kumar, R., Sivakumar, D.: Comparing top k lists. In: SODA, pp. 28–36 (2003)
8. He, B., Macdonald, C., He, J., Ounis, I.: An effective statistical approach to blog post opinion retrieval. In: CIKM, pp. 1063–1072 (2008)
9. Jatowt, A., Kawai, Y., Tanaka, K.: Visualizing historical content of web pages. In: WWW, pp. 1221–1222 (2008)
10. Jin, C., Qian, W., Sha, C., Yu, J.X., Zhou, A.: Dynamically maintaining frequent items over a data stream. In: CIKM, pp. 287–294 (2003)
11. Juffinger, A., Lex, E.: Crosslanguage blog mining and trend visualisation. In: WWW, pp. 1149–1150 (2009)
12. Kendall, M., Gibbons, J.D.: Rank Correlation Methods. Edward Arnold, London (1990)
13. Koutrika, G., Zadeh, Z.M., Garcia-Molina, H.: Data clouds: summarizing keyword search results over structured data. In: EDBT, pp. 391–402 (2009)
14. Kuo, B.Y.-L., Hentrich, T., Good, B.M., Wilkinson, M.D.: Tag clouds for summarizing web search results. In: WWW, pp. 1203–1204 (2007)
15. Leskovec, J., Backstrom, L., Kleinberg, J.M.: Meme-tracking and the dynamics of the news cycle. In: KDD, pp. 497–506 (2009)
16. Manerikar, N., Palpanas, T.: Frequent items in streaming data: An experimental evaluation of the state-of-the-art. Data Knowl. Eng. 68(4), 415–430 (2009)
17. Manku, G.S., Motwani, R.: Approximate frequency counts over data streams. In: VLDB, pp. 346–357 (2002)
18. Melville, P., Gryc, W., Lawrence, R.D.: Sentiment analysis of blogs by combining lexical knowledge with text classification. In: KDD, pp. 1275–1284 (2009)
19. Platakis, M., Kotsakos, D., Gunopulos, D.: Searching for events in the blogosphere. In: WWW, pp. 1225–1226 (2009)
20. Tantono, F.I., Manerikar, N., Palpanas, T.: Efficiently discovering recent frequent items in data streams. In: SSDBM, pp. 222–239 (2008)
21. Wong, R.C.-W., Fu, A.W.-C.: Mining top-k frequent itemsets from data streams. Data Mining and Knowledge Discovery 13, 193–217
22. Zhang, W., Yu, C.T., Meng, W.: Opinion retrieval from blogs. In: CIKM, pp. 831–840 (2007)

Experiences in Building a RESTful Mixed Reality Web Service Platform

Petri Selonen, Petros Belimpasakis, and Yu You

Nokia Research Center, P.O. Box 1000, FI-33721 Tampere, Finland
{petri.selonen,petros.belimpasakis,yu.you}@nokia.com

Abstract. This paper reports the development of a RESTful Web service platform at Nokia Research Center for building Mixed Reality services. The platform serves geo-spatially oriented multimedia content, geo-data like street-view panoramas, building outlines, 3D objects and point cloud models. It further provides support for identity management and social networks, as well as for aggregating content from third party content repositories. The implemented system is evaluated on architecture qualities like support for evolution and mobile clients. The paper outlines our approach for developing RESTful Web services from requirements to an implemented service, and presents the experiences and insights gained during the platform development, including the benefits and challenges identified from adhering to the Resource Oriented Architecture style.

Keywords: Web Services, REST, Mixed Reality, Web Engineering.

1 Introduction

Mixed Reality (MR) refers to the merging of real and virtual worlds to produce new environments and visualizations where physical and digital objects co-exist and interact in real time, linked to each other [8]. As a concept, it encompasses both Augmented Reality (AR) and Augmented Virtuality (AV). As smart phones become more potent through graphical co-processors, cameras and a rich set of sensors like GPS, magnetometer and accelerometers, they turn out to be the perfect devices for AR/MR applications: interactive in real time and registering content in 3D [2] [6].

MR related research has been active for many years at Nokia Research Center (NRC). Most AR/MR applications developed at NRC (e.g. [5][9]) have focused on studying user interface and interaction, and therefore been either stand alone clients or clients linked to project specific service back-ends. The lack of having open Application Programming Interfaces (APIs) have kept these applications isolated, resulting in limited amount of content, prohibited sharing it with other similar systems, and not providing support for third party mash-ups. Some AR/MR prototypes have utilized content from Internet services like Flickr that have not been built for AR/MR applications, and therefore lack proper support for rich media types and geographical searching. In practice, each new AR/MR application has implemented another backend to host their content.

B. Benatallah et al. (Eds.): ICWE 2010, LNCS 6189, pp. 400–414, 2010.

This paper describes an effort to move from stand-alone service back-ends to a common Mixed Reality Web services platform, referred to as MRS-WS. An essential aim was to build a common backend for easily creating Mixed Reality services and applications for both mobile clients and Web browsers. Taking care much of the complexity at the backend infrastructure side, the creation of applications on top of this service platform would be faster, simpler and open also to 3rd parties.

As we essentially serve content, we have chosen REST [4] as the architectural style for the platform. Up to date, there does not exist a systematic method for proceeding from requirements to an implemented RESTful API, so we briefly outline the approach emerged during the development. In what follows, we present the platform architecture, explore its architectural qualities and discuss some of the experiences and insights gained during building the platform.

2 Requirements for the Service Platform

The MRS-WS platform is developed within the context of a larger NRC Mixed Reality Solutions program building a multitude of MR solutions, each integrating with the platform. At the time of writing, a first client solution has been deployed and is currently being maintained, and a second one is under development. In addition to the functional features required by the dedicated program clients, there are other more generic requirements to the platform architecture, outlined next.

2.1 Overview to Functional Requirements

The first of the two clients allows the user to explore photos at the location of the image spatially arranged with its original orientation and camera pose. Once a photo is taken using a mobile phone, the photo is uploaded to a content repository with metadata acquired from the GPS, digital compass (magnetometer) and accelerometers of the device. The user and his friends can later add and edit comments, tags and descriptions related to the photos. There are both a Web browser based client for exploring the user's content rendered on map and 3D space (Augmented Virtuality) and a mobile client that can show the photo on location, overlaid on top of a view-through camera image *in situ* (Augmented Reality). Screenshots of this solution, called Nokia Image Space[1], is shown in Fig. 1. To enable access to masses of pre-existing content, the platform supports aggregation of content and social relationships from 3rd party content repositories, starting with Flickr.

The second client fetches street-view panoramas, building outlines and points of interest based on a location, and visualizes annotations and related comments belonging to the user overlaid on top the panorama. The client enables the users to annotate particular buildings and other landmarks through touch and share these annotations with other users. The geo-data is post-processed from commercial Navteq content using advanced computer vision algorithms for e.g. recognizing building outlines from panoramic images.

---

[1] http://betalabs.nokia.com/apps/nokia-image-space

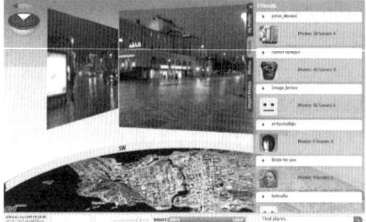 

Fig. 1. Image Space Web and mobile AR clients

Generalizing from the above client requirements and general MR domain knowledge, we can identify high-level functional requirements for the platform as support for

- managing MR metadata—geographical location, orientation and accelerometer data —for storing and visualizing content in 3D space;
- storing content of different media types, such as photos, videos, audio and 3D objects, as well as non-multimedia like point clouds and micro-blogging entries;
- managing the typical social media extensions for every content item, including comments, tags and ratings;
- managing application specific metadata for content items, allowing clients with specific needs to store pieces of information without modifications at the platform side;
- managing user identities, social connections and access control lists; and
- creating, modifying, retrieving, searching and removing the above content.

In addition, there are requirements related to content hosted on external repositories:

- aggregating content and social connections from, and synchronizing with, third party repositories to provide a uniform API to all content available to the clients, making the clients agnostic about the interfaces and existence of external repositories; and
- serving commercial geo-data and content post-processed from it, including 360-degree street view panoramic images, POIs and building outlines.

The platform is supporting both aggregating content from other content providers and exposing it in a uniform manner, and enabling other services to link our platform to them. The clients are able to discover, browse, create, edit and publish geo-content useful for MR applications.

2.2   Overview to Non-functional Requirements

In addition to the functional requirements collected above, we consider a few of the most important non-functional requirements to the platform, identified as

- usability of the platform to ensure that its users—end developers—find it easy to use and to integrate with;
- modifiability and extensibility of the platform to ensure that new requirements and clients can be rapidly integrated;

- support for mobile devices—the main devices through which augmented reality is realized—translating to efficient bandwidth utilization and fine-grained content control for the clients;
- security and privacy to ensure user generated content is safely stored and transferred; and
- scalability and performance to make sure the system architecture will scale up to be used in global scale with potentially millions of users.

While all of the above qualities are important, we place special emphasis in realizing the three first ones. The main contribution expected from the platform is to provide an easy-to-use API for developers to build MR applications and Web mash-ups. In the case of mash-ups, external services should be able to link their unique content to the MRS-WS platform to be seamlessly visualized in MR view by our platform clients. As a simple example, consider a restaurant business "pushing" their menu offering to our platform which would then be visible in AR view to a potential customer passing by the restaurant and exploring the area via his mobile phone, as a "magic lens" to reality. The key in the quest for killer applications is proper support for open innovation.

3   Developing Resource Oriented Architectures

REST architecture style provides a set of constraints driving design decisions towards such properties as interoperability, evolvability and scalability. To this end, RESTful APIs should exhibit qualities like addressability, statelessness, connectedness and adhering to a uniform interface. The MRS-WS platform serves what essentially is content: the core value of the service is in storing, retrieving and managing interlinked, MR enabled content through a unified interface. Therefore, REST and Resource Oriented Architecture [10] were chosen as the architecture style for the platform. We have also previously explored using REST with mobile clients, and the lessons learned were taken into use in the platform building exercise.

However, so far there does not exist a commonly agreed, systematic and well-defined way to proceed from a set of requirements to an implemented RESTful API. The best known formulation of designing RESTful Web services has been presented by Richardson and Ruby [10] which can be summarized as finding resources and their relationships, selecting uniform operations for each resource, and defining data formats for them. This formulation is too abstract to be followed as a method and further, it does not facilitate communication between service architects and other stakeholders. In our previous work [7] we have explored how to devise a process for developing RESTful service APIs when the set of service requirements are not content oriented but arbitrary functional features. We took the learnings of this work and constructed a more lightweight and agile approach, suited for the Mixed Reality program and its a priori content oriented domain. In a way, the result is something resembling a lightweight architecture style for developing Resource Oriented Architectures, in a way a ROA meta-architecture.

3.1 Resource Model, Information Model and Implementation

In the heart of the approach is the concept of a *resource model*. A resource model, adapted from Laitkorpi et al [7], organizes the elements of a domain model to addressable entities that can then be mapped to elements of a RESTful API, service implementation and database schema, while still being compact and understandable by the domain experts. The resource model divides resources into *items*, which represent individual resources having a state that can be created, retrieved, modified and deleted, *containers* that can be used for retrieving collections of items and creating them, and *projections* that are filtered views to containers. Resources can have subresources and references to other resources.

A natural way to map the service requirements into a resource mode is to first collect them to a domain model called the *information model*—expressed for example as a UML class diagram—which is essentially a structural model describing the domain concepts, their properties and relationships, and annotated with information about required searching, filtering and other retrieval functionality. The concepts of the information model are mapped to the resource model as resources; depending on their relationship multiplicities, they become either items or containers containing items. Composition relationships form resource–subresource hierarchies while normal relationships become references and collections of references between resources. Attributes are used to generate resource representations with candidates for using standard MIME singled out when possible. Each search or filtering feature defined by an annotation—e.g. UML note—is represented by a projection container. There are obviously fine-grained details in the mapping, but they are left out as they are not relevant for the high-level architectural picture.

The concepts of a resource model, i.e. containers, items and projections, are then mapped to implementation level concepts in service specifications (e.g. WADL), database schema and source code. For example, containers can manifest themselves as tables in relational databases, items as rows in a table, item representation as columns, and subresources and references as links to other tables. Similarly, containers and items can be mapped to HTTP resource handlers in the target REST framework (say, Ruby on Rails, Django or Restlet).

3.2 MRS-WS Implementation Binding and Architecture

The concrete binding between the MRS-WS domain model, resource model and the implementation is done by mapping resource model concepts above to the concepts of the selected technology stack: Restlet, Hibernate, Java EE and MySQL. The overall approach is illustrated in Fig. 2. The top-left corner shows a simple information model fragment, mapped to a resource model shown at bottom-left corner and further to an implementation shown on the right-hand side. Based on the information model fragment, the resource model contains Annotations container, Annotation item and Annotations-By-Category projection that are in turn mapped into implementation model elements: AnnotationsResource, AnnotationResource, AnnotationsView, AnnotationView, Annotation model and AnnotationDAO. The binding is further explained in Table 1.

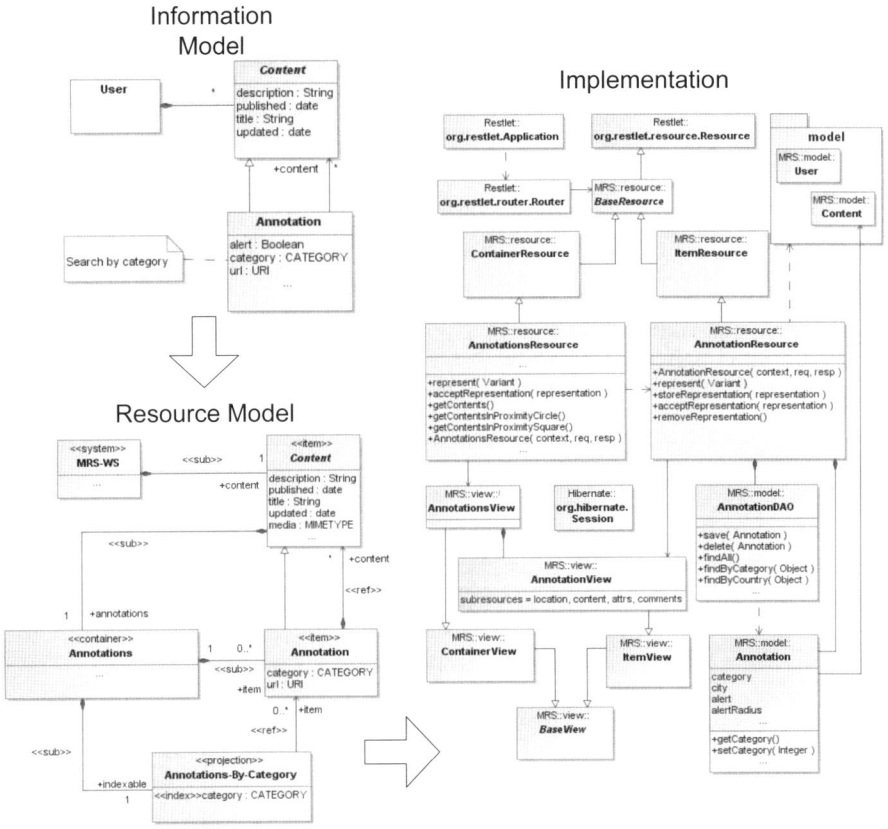

Fig. 2. Overview of the ROA architecture

Fig. 3. shows the architectural layers of the platform. The design of the architecture is to provide a uniformed REST API set for MR data plus mashed-up contents from third party content providers available to the service clients. Each resource handler is identified by a URI and represents the unique interface for one function and dataset. The operation function delegates the request to concrete business logic code which works closely in the persistence layer. After returning from the business logic code, the operation function transforms the returned internal data to the corresponding data format through the representation classes along the output pipe to the web tier and the client eventually.

Table 1. MRS-WS implementation binding

	API (Restlet)	Representation (XML/JSON)	Model (Hibernate, Java EE)	Persistence (MySQL)
Item	Restlet resource bound to the URI. Supported default operations are GET, PUT and DELETE.	Representation parsing/generation based on the item attributes. Subresources inlined per request basis.	A native Java object (POJO) generated for each item with a Hibernate Data Access Object and binding to database elements.	Items are rows in respective database table with columns specified by item attributes. References map to foreign keys.
Container	Restlet resource bound to the URI. Supported default operations are GET and POST.	Representation parsing/generation delegated to Items.	Basic retrievals to database, using item mappings.	Containers are database tables.
Projection	Implemented on top of respective Containers.	Representation generation delegated to Container.	Extended retrievals to database, using item mappings.	Stored procedures for more advanced database queries. Tables implied by Container.

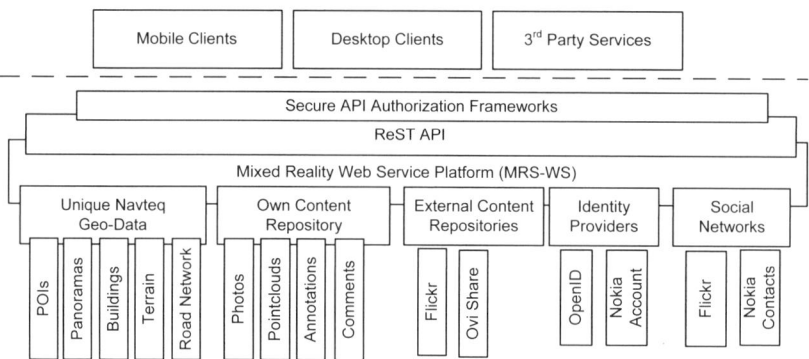

Fig. 3. Overview of the MRS Platform Architecture

To federate and provide aggregated contents, the platform defines a generic interface and a set of standard data models for general geo-referenced data. The platform aggregates the contents returned from the interfaces and transform them into the federated representations requested by the clients. The returned data is then transformed and merged by the aggregator as same as the interface approach. The advantage of this approach is to scale up the platform easily. Unlike the interface approach, which the business logic must be implemented by each every connector and integrated closely to the aggregator component inside the platform, the callback approach enables the implementation hosting outside the platform, typically residing on the third party controlled servers. This gives much flexibility and freedom to the third parties and eases the platform maintenance.

4   Examples of MRS-WS API

Following the discussion above, we give some concrete examples of MRS-WS APIs, starting with the Annotation API. MR annotations can be attached to a particular location or a building, and they can comprise a title and a textual description, as well as link to other content elements. Essentially an Annotation is a way of the users to annotate physical entities with digital information and share those with other users.

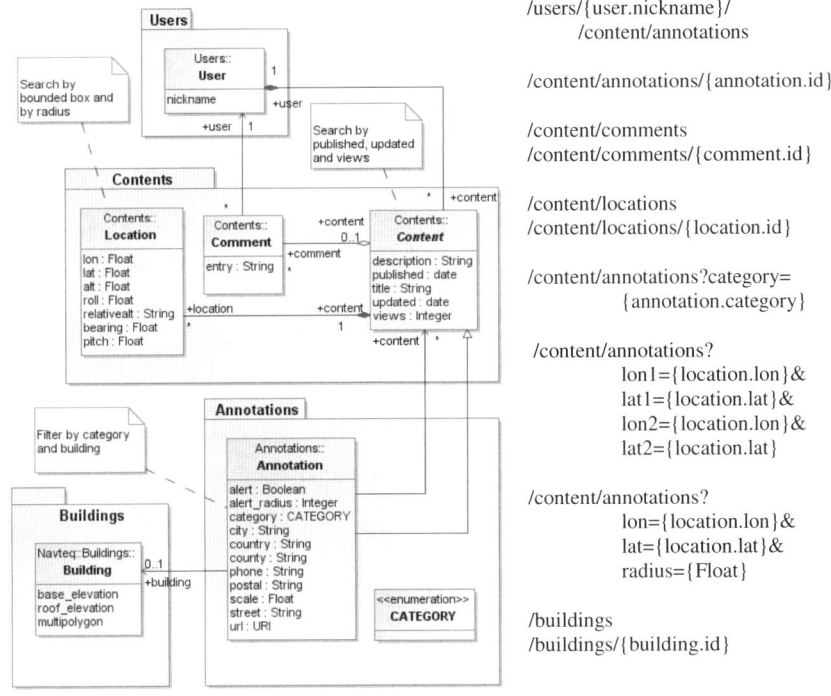

Fig. 4. Annotation Information Model and Implied Resource Hierarchy

Fig. 4 shows an Annotation information model, related resources and generated resource hierarchy. The default operations are as stated earlier: GET and POST for containers, GET, PUT, POST and DELETE for items and GET for projections. A sample representation subset for an annotation resource is given below, showing attributes inherited from Content and Annotation classes. Location subresources and Building reference are inlined to the representation, as shown in Fig. 5.

```
<annotation href="...">
    <id>4195042682</id>
    <updated>2009-12-18 04:01:13.0</updated>
    <published>2009-12-18 04:01:02.0</published>
    <title>I have an apartment for rent here!</title>
    <alert>true</alert>
```

```
    ...
    <locations>
      <location href="...">
        <lat>61.44921</lat>
        <lon>23.86039</lon>
          ...
        <pitch></pitch>
      </location>
    </locations>
    <building href="..">...</building>
  </annotation>
```

Fig. 5. Example of content type application/vnd.research.nokia.annotation

4.1 Other MRS-WS APIs

The photo, video, audio and point cloud data objects can be accessed and modified via similar operations and representations. The API also offers access to other read-only geo-content originating from Navteq, such as

— building (/buildings) footprints and 3D models (Fig. 6. a & b);
— terrain (/terrain) tiles of earth's morphology (Fig. 6. c & d);
— street-view 360 panorama (/panoramas) photos (Fig. 6. e); and
— point-of-interest (/pois), with details about businesses and attractions.

To all those resources and containers, user-generated content and static geo-data, a uniform set of operations can be applied. Examples include:

— **geo-searching** for performing searches in a bounded box (e.g. /photos/?lat1=41.95&lon1=-87.7&lat2=41.96&lon2=-87.6) or in proximity (e.g. /photos /?lat=41.8&lon=-87.6&radius=1);
— **pagination** [x-mrs-api: page(), pagesize()] for controlling the number of objects fetched from a container per request by setting the page size and number;
— **verbosity** [x-mrs-api: verbosity()] for controlling the representation subset, ie.e selected attributes for the retrieved items; and
— **inlining** [x-mrs-deco: inline()] for selecting which subresources and referenced items are included per resources in the response to decide when to reduce the number of requests to the server and when to reduce the amount of data transferred.

Inlining is heavily used in MRS-WS for exposing advanced geo-data associations. For example, we have pre-calculated from the raw Navteq data the specific buildings that are visible in a given street view panorama, along with their associated POIs. A client requesting a specific panorama image can also request building data (e.g. links to 3D building models) to be inlined. This combined information can allow very advanced applications, such as touching and highlighting buildings, in an augmented reality view, for interacting with the real world (e.g. touch a building to find the businesses it hosts). Finally, when it comes to user-generated content, the following operations can be additionally applied:

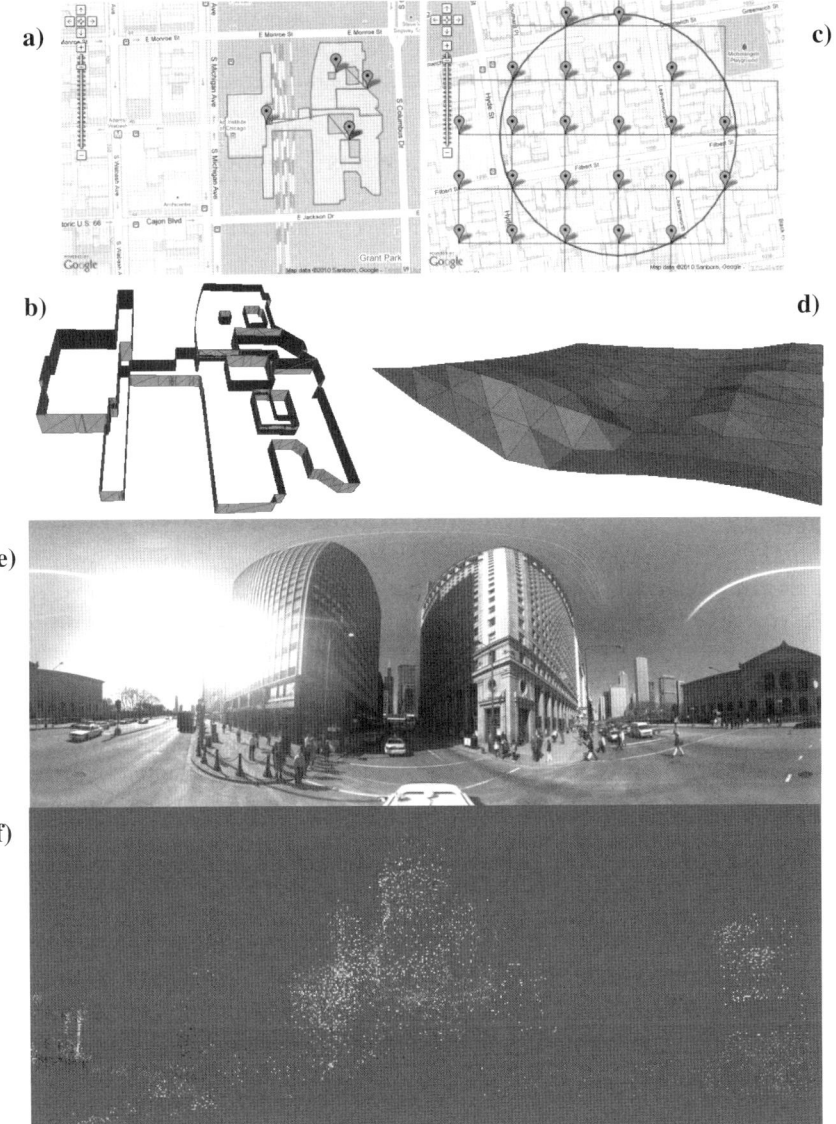

Fig. 6. Visualization of different data offered by our platform a) 2D footprint of a building, b) 3D model of a building, c) Searching terrain tiles in a given radius, d) 3D model of a terrain tile, e) Street-view 360 degree panorama, f) Point cloud of a church in Helsinki, created by multiple user photos

- **Temporal searching.** It is possible to limit the scope of the search within a specific time window. In the search query the time window (since- until) needs to be specified, in the form of UNIX timestamps. The matching is done against an item's "Updated time" or "Taken time", or combination of both (e.g. /photos/ lat=41.8&lon=-87.6&radius=1&updated_since=1262325600)

– **Mixed Reality enabled.** Client can request results that explicitly include geo-spatial metadata, having in addition to geo-location at least associated bearing (yaw) metadata. This flag is set via the URI parameter mr_only=true.

5 Architectural Evaluation and Lessons Learned

Next, we revisit the non-functional requirements of the platform and review some of its other architectural quality attributes.

5.1 Non-functional Requirements Revisited

Usability defined as the ease of use and training of the service developers who build their clients against the platform. So far we have experience with several client teams building their applications against the platform and the feedback has been positive. We have not yet performed formal reviews of the usability, but we are currently preparing to adopt a 12-step evaluation process from Nokia Services as a medium for collecting feedback.

Modifiability interpreted as the ease with which a software system can accommodate changes to its software. Modifiability, extensibility and evolvability have proved to be the key architectural characteristics achieved through using our architecture. Integrating new requirements from the program clients has been very easy: the new requirements are integrated with the domain model, mapped to the resource model concepts, and to database schema and implementation level concepts. Based on our experiences, adding new concepts now takes around a few days. This is tightly related to the Resource Oriented Architecture style: mapping of the new concepts to the system is straightforward and supporting infrastructure for implementing functional requirements pre-exists.

Security and privacy as the ability to resist unauthorized attempts at usage, and ensuring privacy and security for user generated content. The platform supports a variant of Amazon S3 and OAuth for authorizing each REST request in isolation, instead of storing user credentials as cookies or with unprotected but obfuscated URIs. Our system has been audited for security by an independent body.

Scalability and performance in terms of the response time, availability, utilization and throughput behavior of the system. In the beginning of the program, we outlined goals for the scalability and performance in terms of number of requests, number of users, amount of data transferred, amount of content hosted and so forth. The main goal is to have an API and platform architecture that has built-in support for scalability and performance once the platform is productized and transferred to be hosted at business unit infrastructure. After initial performance tests with hosting a few million users and content elements, the system has so far been able to meet its requirements. While we have reason to believe that proper REST API, resource design and statelessness should be able to support scalability, for example through resource sharding, most performance characteristics like e.g. availability and utilization are not related to REST or Resource Oriented Architecture *per se* and are thus outside the scope of this paper.

Support for mobile devices is not one of the more common architectural quality attributes but is in the heart of the MRS-WS platform. By using the results of our previous work with REST and mobile devices, we support advanced content control for the clients through verbosity and inlining, and more effective bandwidth utilization through JSON and on-demand compression. Basically the clients can control the amount of data they want to receive and thus try to optimize the amount of requests and the amount of data being transmitted. It should be noted that optimizing for mobile clients includes a variety of topics, ranging from HTTP pipelining and multipart messages to intelligent device-side caching. Such techniques are outside the scope of this paper. We do claim, however, that the architecture we have chosen provides substantial support for implementing efficient mobile clients, and this is even more important in the domain of Mixed Reality services.

There are other system qualities, like conformance with standard Internet technologies and Nokia Services business unit technology stack, and still others that with a suitable interpretation are quite naturally supported by Resource Oriented Architecture like interoperability, one of the key promises of REST, but they are outside the scope of this paper. However, there is one additional quality that was not originally considered but that proved to be harder to achieve than initially expected: testability.

Testability interpreted as the ease with which platform can be tested. Testing complex platforms is *a priori* a hard task. It became obvious quite early in the development that as most resources share similar common functionality, changes to one part in the implementation can typically have effects throughout the API, making automated API level regression testing very desirable to ensure rapid refactoring. However, this is not currently properly supported by existing tools and frameworks. We have therefore built our own test framework for in-browser API level testing using JavaScript and XHR. The tests can be run individually, or using JSUnit as a way for running several test pages and reporting the results. The additional benefit is that whenever a new version of the platform is deployed, either locally, at the development server or at the alpha servers, the tests are deployed as well. Even though our test framework takes care of e.g. user authentication and basic content control through HTTP headers, in practice the browser environment still makes writing test cases more laborious than would be good. Therefore we unfortunately often fall back to using curl command line tool for doing initial tests.

5.2 Insights on Developing RESTful APIs and Resource Oriented Architecture

One of the main benefits of having ROA based architecture is that the RESTful API exposes concepts that are very close to the domain model, thus making understanding of the API in itself relatively easy. The uniform interface and uniformity of the searching and filtering operations across the platform makes the learning curve for different API subsets smooth. However, implementing ROA based services can be difficult for software architects lacking previous experience. Therefore having a lightweight process of building ROA services that is not tied to a particular implementation technology would be very useful for a project to be successful. We also found out that while a REST API is in principle simple, programming against it requires a change in the programmer mindset from arbitrary APIs to a uniform interface with resource manipulations, which sometimes can be more challenging than expected.

Interoperability and its serendipitous form, ad-hoc interoperability, are hard to achieve without commonly agreed content types. One of the few ways of making interoperable systems is to use Atom, but without MR specific features it reduces the system into yet another content repository. Another topic for further research is how to evaluate a RESTful API against principled design characteristics like addressability, connectedness, statelessness and adherence to uniform interface. Issues like proper resource granularity, supporting idempotency, having one URI for one resource, and balancing between number of requests and amount of data are not commonly laid out. There is a clear lack of REST patterns and idioms, although the upcoming RESTful Web Services Cookbook [1] is a step towards the right direction.

5.3   Linking to Multiple Service Providers and Support for Multiple Sign-Ins

Generally users have multiple accounts on different services coming with different identity providers. Our service requires mashing up authorized external data from different services on behalf of the users. Instead of storing raw user credentials, our service saves the access tokens or opaque identifiers from external SPs to authenticate the requests on his behalf. Users can revoke the rights at the SP side easily to invalidate the access token, i.e. the authorized connection between our service and the SP. Additional benefit for the users is the ability to sign in with to their favorite identify provides without a need to create new passwords. The benefit of having multiple sign-in is to minimize the risk of identity service breakdown. The downside of the multiple sign-in is that our service has difficulties to maintain a least common denominator of user profiles.

5.4   Legal and Terms-of-Service Issues for a Service Platform

Our initial design goal was to create a generic platform on top of which our first lead service could be publically deployed. The subsequent services could later be deployed on top of the very same platform instance, allowing the users to be already registered and to potentially have pre-existing content, seamlessly accessible by the different clients, as simply different views or ways to explore the same data. However, it turned out that this approach, while very attractive engineering-wise, was legally considered very problematic. The Terms of Service (ToS) and the related Privacy Policy have to be very specific to the users on what they are registering for, and how the service is using their content. Asking the users to register for a single service now and later on automatically deploying more services, having their content automatically available, is against the policies for simple and clear ToS. One solution to the problem would be to explicitly ask the users to opt-in for the new services and get their permission to link their existing content, before making it available to the new service, which would also have its own ToS.

5.5 Related Work

While a relatively new topic, a number of books on the development of RESTful Web services have come about during the last year or are in the pipeline after the original [10] book came out. The upcoming RESTful Web Services cookbook [1] outlines recipes on the different aspects of building REST Web services. However, most of its recipes lean more towards identifying a design concern than giving a solution or stating the different forces related to using it. There are also books on how to implement RESTful Web services on particular frameworks like Jersey and Apache Tomcat and .NET. As one example, [3] gives a good if short overview on how to concretely implement RESTful Web services using Java and JAX-RS. The resource modeling approach described in this paper is an effort to outline the approach from REST and ROA point of view without any particular implementation platform in mind—in fact, it was originally developed and applied with Ruby on Rails.

6 Concluding Remarks

This paper described the development of a RESTful Web service platform for Mixed Reality services. We outlined our approach for developing RESTful Web services from requirements, summarized the platform architecture, evaluated some of its quality attributes and described insights gained during the development. We also presented some of the benefits identified from adhering to the Resource Oriented Architecture style and RESTful Web services, including having a close mapping from domain model to architecture in the case of content oriented systems, decoupling of clients and services, improved modifiability of the platform, and ability to support mobile clients.

We also encountered some challenges. While REST has become something of a buzzword in developing Internet Web services, there is a surprisingly small amount of real-life experience reports available concerning the development of services conforming to the Resource Oriented Architecture. In particular, there is a lack of modeling notations and methods for systematically developing RESTful services. The team also had to develop its own testing tools as no suitable ones existed.

As future work, the team will continue developing the platform for future clients to be released during spring 2010. From a Web services point of view, the team plans to continue research on the RESTful Web services engineering while from a Mixed Reality point of view, the focus will be on supporting more advanced contextuality, providing related and relevant information through associative browsing, and orchestrating content and metadata coming from different sources.

Acknowledgments. The authors wish to thank Arto Nikupaavola, Markku Laitkorpi, the former NRC Service Software Platform and the NRC Mixed Reality Solutions program for their valuable contribution.

References

1. Allamaraju, S., Amudsen, M.: RESTful Web Services Cookbook. O'Reilly, Sebastopol (2010)
2. Azuma, R.: A Survey of Augmented Reality. Presence: Teleoperators and Virtual Environments 6(4), 355–385 (1997)
3. Burke, B.: RESTful Java with JAX-RS. O'Reilly, Sebastopol (2009)
4. Fielding, R.: Architectural Styles and the Design of Network-based Software Architectures. Doctoral Thesis, University of California, Irvine (2000)
5. Föckler, P., Zeidler, T., Brombach, B., Bruns, E., Bimber, O.: PhoneGuide: museum guidance supported by on-device object recognition on mobile phones. In: 4th international Conference on Mobile and Ubiquitous Multimedia MUM '05, vol. 154, pp. 3–10. ACM, New York (2005)
6. Höllerer, T., Wither, J., DiVerdi, S.: Anywhere Augmentation: Towards Mobile Augmented Reality in Unprepared Environments. In: Location Based Services and TeleCartography. Lecture Notes in Geoinformation and Cartography. Springer, Heidelberg (2007)
7. Laitkorpi, M., Selonen, P., Systä, T.: Towards a Model Driven Process for Designing RESTful Web Services. In: International Conference on Web Services ICWS '09. IEEE Computer Society, Los Alamitos (2009)
8. Milgram, P., Kishino, F.: A Taxonomy of Mixed Reality Visual Displays. IEICE Transactions on Information Systems E77-D (12), 1321–1329 (1994)
9. Reitmayr, G., Schmalstieg, D.: Location based applications for mobile augmented reality. In: Biddle, R., Thomas, B. (eds.) Fourth Australasian User interface Conference on User interfaces 2003 - Volume 18, Adelaide, Australia, vol. 18. ACM International Conference Proceeding Series, vol. 16, pp. 65–73. Australian Computer Society, Darlinghurst (2003)
10. Richardson, L., Ruby, S.: RESTful Web Services. O'Reilly, Sebastopol (2007)

WebRatio BPM: A Tool for Designing and Deploying Business Processes on the Web

Marco Brambilla[1], Stefano Butti[2], and Piero Fraternali[1]

[1] Politecnico di Milano, Dipartimento di Elettronica e Informazione
P.za L. Da Vinci, 32. I-20133 Milano - Italy
{marco.brambilla,piero.fraternali}@polimi.it
[2] Web Models S.r.l., I-22100 Como - Italy
stefano.butti@webratio.com

Abstract. This paper presents WebRatio BPM, an Eclipse-based tool that supports the design and deployment of business processes as Web applications. The tool applies Model Driven Engineering techniques to complex, multi-actor business processes, mixing tasks executed by humans and by machines, and produces a Web application running prototype that implements the specified process. Business processes are described through the standard BPMN notation, extended with information on task assignment, escalation policies, activity semantics, and typed dataflows, to enable a two-step generative approach: first the Process Model is automatically transformed into a Web Application Model in the WebML notation, which seamlessly expresses both human- and machine-executable tasks; secondly, the Application Model is fed to an automatic transformation capable of producing the running code. The tool provides various features that increase the productivity and the quality of the resulting application: one-click generation of a running prototype of the process from the BPMN model; fine-grained refinement of the resulting application; support of continuous evolution of the application design after requirements changes (both at business process and at application levels).

1 Introduction

Business process languages, such as BPMN (Business Process Management Notation) [13], have become the *de facto* standard for enterprise-wide application specification, as they enable the implementation of complex, multi-party business processes, possibly spanning several users, roles, and heterogeneous distributed systems. Indeed, business process languages and execution environments ease the definition and enactment of the business constraints, by orchestrating the activities of the employees and the service executions.

This paper presents an approach and a supporting toolsuite to the specification, design and implementation of complex, multi-party business processes, based on a Model-Driven Engineering (MDE) methodology and on code generation techniques capable of producing dynamic Web applications from platform independent models.

B. Benatallah et al. (Eds.): ICWE 2010, LNCS 6189, pp. 415–429, 2010.

The proposed approach is a top down one: the (multi-actor, multi-site) business process is initially designed in an abstract manner, using the standard BPMN notation for schematizing the process actors, tasks, and business constraints. The resulting BPMN model is an abstract representation of the business process and cannot be used directly for producing an executable application, because it lacks information on essential aspects of process enactment such as: task assignment to humans or to Web Services, data flows among tasks, service invocation and user interface logics. Therefore, the standard BPMN specification is manually annotated with the missing information, to obtain a *detailed process model* amenable to a two-step transformation:

- A first model-to-model transformation (*Process to Application*) translates the detailed process model into: 1) a platform-independent model of the Web user interface and of the Web Service orchestrations needed for enacting the process, expressed in a Domain Specific Language called WebML [4]; 2) a Process Metadata Model, representing the business constraints (e.g., BPMN precedence constraints, gateways, etc).
- A second model-to-text transformation (*Application to Code*) maps the Application Model and the Process Metadata Model into the running code of the application. The resulting application is runtime-free and runs on any standard Java Enterprise Edition platforms.

The original contributions of the paper are: (i) a two-step generative framework comprising a first model transformation from a detailed Business Process Model to an Application Model and a second transformation for producing the executable code from the Application Model; and (ii) an extended version of the WebRatio toolsuite [17], called WebRatio BPM, that fully implements the proposed transformation steps. The tool is currently in beta version and will be released in the second quarter of 2010. However, a major European banking customer is already adopting WebRatio BPM for the development of a large, multi-country and multi-user business process based portal. Therefore, the validation of the approach is already ongoing and several lessons learned have been collected.

The advantages of having a two steps modeling process are multifold: (i) the BP model (BPMN) and the web model (WebML) allows the designer to separate the different concerns in the design, keeping the process issues separate from the hypertext and interface issues; (ii) the transformation of the BP model to a Web-specific model allows fine-grained description of the interfaces, while remaining at a modeling level; (iii) the Web modeling level allows seamless integration of resulting applications within non-BP-based application models (e.g., web portals, B2C e-commerce sites, and so on); (iv) having distinct models allows different user roles (i.e., business analysts at the BP level and Web engineers at the Web modeling level) to work together and independently at the same application design. These advantages, together with the one-click deployment option, make our proposal unique also considering the plethora of BPM tools existing on the market.

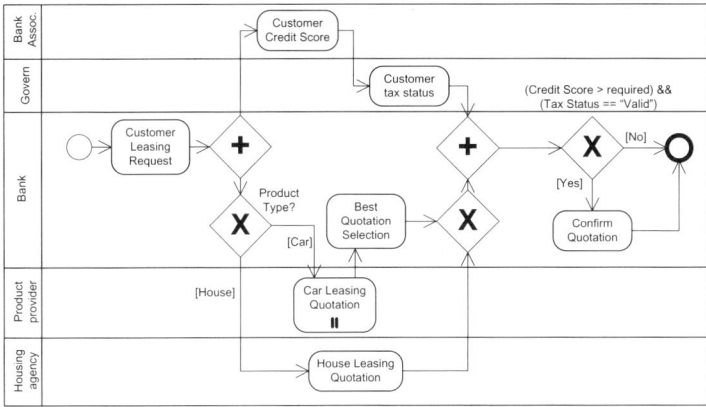

Fig. 1. Business process model of the leasing running example

The paper is organized as follows: Section 2 discusses the background technologies and notations; Section 3 discusses the approach to application development; Section 4 and Section 5 illustrate the extended process model and the application model, respectively; Section 6 describes the implementation of the WebRatio BPM tool; Section 7 discusses the related work; and Section 8 draws the conclusions.

2   Background: BPMN, WebML, and WebRatio

This work builds upon existing methods and tools to cover the different design phases.

BPMN [13] supports the specification of business processes, allowing one to visually specify actors, tasks, and constraints involved. Precedence constraints are specified by arrows, representing the control flow of the application, and gateways, representing branching and merging points of execution paths. Parallel executions, alternative branches, conditional executions, events, and message exchanges can be specified. BPMN allows analysts to describe complex orchestrations of activities, performed by both humans and machines. Figure 1 shows an example of BPMN, describing a simplified leasing process for houses and cars.

WebML [4] is a Domain Specific Language for data-, service-, and process-centric Web applications [3]. It allows specifying the conceptual model of Web applications built on top of a data schema and composed of one or more hypertexts used to publish or manipulate data. The data model can be specified through standard E-R or UML Class diagrams. Upon the same data model, different hypertext models (*site views*) can be defined (e.g., for different types of users or devices). A site view is a graph of *pages*, consisting of connected *units*, representing data publishing components. Units are related to each other through *links*, representing navigational paths and carrying parameters. WebML

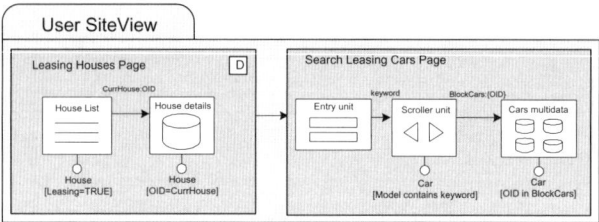

Fig. 2. WebML hypertext model example

allows specifying also update *operations* on the underlying data (e.g., the creation, modification and deletion of instances of entities or relationships) or operations performing arbitrary actions (e.g. sending an e-mail, invoking a remote service [9], and so on). Figure 2 shows a simple site view containing two pages, respectively showing the list of houses and a form for searching cars available for leasing. Page *Search Leasing Cars* contains an entry unit for inputting the car model to be searched, a scroller unit, extracting the set of cars meeting the search condition and displaying a sequence of result blocks, and a multidata unit displaying the cars pertaining to a block of search results.

WebML is supported by the WebRatio CASE tool [17], which allows the visual specification of data models and site views and the automatic generation of J2EE code. The tool consists of a set of Eclipse plug-ins and takes advantage of all the features of this IDE framework. It also supports customized extensions to the models and code generators, model checking, testing support, project documentation, and requirements specifications. The main features of WebRatio are the following: it provides an integrated MDE approach for the development of Web applications and Web services, empowered by model transformations able to produce the complete running code; it unifies all the design and development activities through a common interface based on Eclipse, which includes the visual editing of models, the definition of presentation aspects, and the extension of the IDE with new business components and code generation rules; it includes a unified workspace for projects and a version control and collaborative work environment.

3   Development Process

The development process supported by WebRatio BPM is structured in five main steps, represented in Figure 3 according to the SPEM notation [12].

Initially, business requirements are conceptualized in a *coarse Business Process Model* by the business analyst. Figure 1 is an example of BPM that can be obtained as a requirement specification of the leasing application. Subsequently, the BPMN schema is refined by a BPMN designer, who annotates it with parameters on the activities and data flows.

The resulting *refined Process Model* is subject to a first model transformation, which produces the WebML *Application Model* and *Process Metadata Model*.

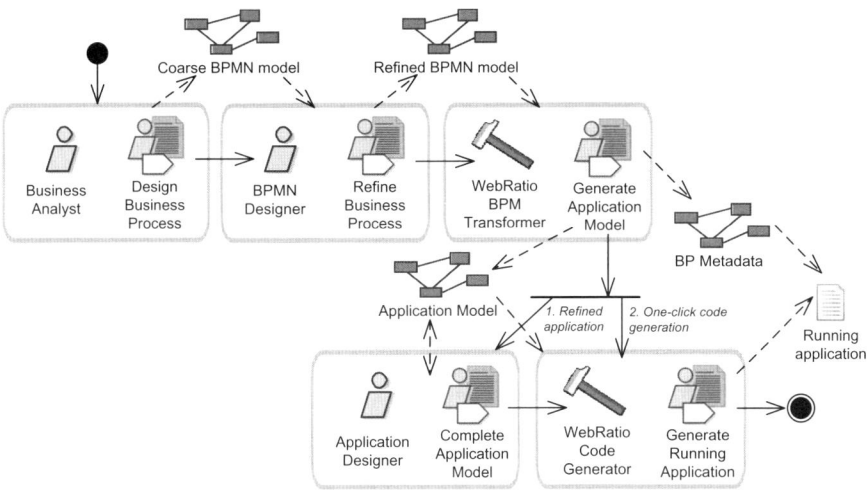

Fig. 3. Development process overview (SPEM notation)

The Application Model (discussed in Section 5.2) specifies the details of the executable application according to the WebML notation, representing the hypertext interface for human-directed activities. The Process Metadata Model (discussed in Section 5.1) consists of relational data describing of the activities of the process and of the associated constraints, useful for encapsulating the process control logic. This transformation extends and refines the technique for model-driven design of Web applications from business process specification initially proposed in [3]. Subsequently, the generated Application Model can be either used as is by a one-click prototype generation transformation to get a first flavour of the application execution (option 2 in the branching point), or it can be refined manually by the application designer, to add domain-dependent information on the execution of activities (option 1 in the branching).

Finally, the Application Model is the input of a second transformation, which produces the code of the application for a specific technological platform (in our case, J2EE); this step is completely automated thanks to the code generation facilities included in WebRatio.

4   Refined Process Model

The high-level BPMN process model designed in the requirement specification phase is not detailed enough to allow the generation of the application code. Its refinement is done using an extended BPMN notation, which enables a more precise model transformation into a WebML Application Model and then into the implementation code. In particular, the process model is enriched with information about the data produced, updated and consumed by activities, which is expressed by typed activity parameters and typed data flows among activities.

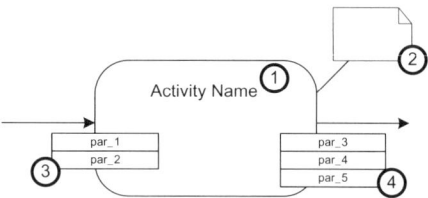

Fig. 4. Extended activity notation

Furthermore, activities are annotated to express their implicit semantics, and gateways (i.e., choice points) that require human decisions are distinguished.

Figure 4 shows the graphical notation of the extended BPMN activity. An activity is associated with a *Name* (1), which is a textual description of its semantics, and possibly an *Annotation* (2), which describes the activity behaviour using an informal textual description. An activity is parametric, and has a (possibly empty) set of input (3) and output (4) parameters. The actual values of input parameters can be assigned from preceding activities; the output parameters are produced or modified by the activity. Analogous extensions are defined for gateways; these are further refined by specifying whether they are implemented as manual or as automatic branching/merging points. *Manual gateways* (tagged as Type "M") involve user interaction in the choice, while *Automatic gateways* (tagged as Type "A") automatically evaluate some condition and decide how to proceed with the process flow without human intervention. The output flow of gateways can be associated to a guard condition, which is an OCL Boolean expression over the values of the parameters of the gateway; the semantics is that the activity target of the flow link with the guard condition can be executed only if the condition evaluates to true.

5   Application Model

Starting from the Detailed Process Model presented above, an automatic transformation produces: (1) Process Metadata Model, describing the process constraints in a declarative way as a set of relations; (2) the Domain Model, specifying the application-specific entities; (3) and the Application Model, including both the site views for the user interaction and the service views for Web service orchestration.

Hence, the transformation consists of two sub-transformations:

- Detailed Process Model to Process Metadata: the BPMN precedence constraints and gateways are transformed into instances of a relational representation compliant to the Process Metamodel shown in Figure 5, for enabling runtime control of the process advancement;
- Detailed Process Model to Application Model: the BPMN process model is mapped into a first-cut Application Model, which can be automatically

transformed into a prototype of the process enactment application or sub-sequently refined by the designer to incorporate further domain specific aspects.

Thanks to the former transformation, the BPMN constraints, stored in the Process Metadata Model, are exploited by the components of the Application Model for executing the service invocation chains and enacting the precedences among human-executed tasks.

5.1 Process Metadata Generation

Figure 5 shows, as a UML class diagram, the schema of the metadata needed for executing an BPMN process at runtime.

A *Process* represents a whole BPMN diagram, and includes a list of *Activities*, each described by a set of input and output *ParameterTypes*. A *Case* is the instantiation of a process, and is related to the executed *Activity Instances*, with the respective actual *Parameter Instances*. The evolution of the status history is registered through *CaseLogEntry* and *ActivityLogEntry*. *Users* are the actors that perform a task and are clustered into *Groups*, representing their roles.

Notice that the diagram spans two modeling levels in the same data schema, namely process model and process instance information. The BPMN part is confined to the entities in the upper part of the figure, while the lower part regards execution data.

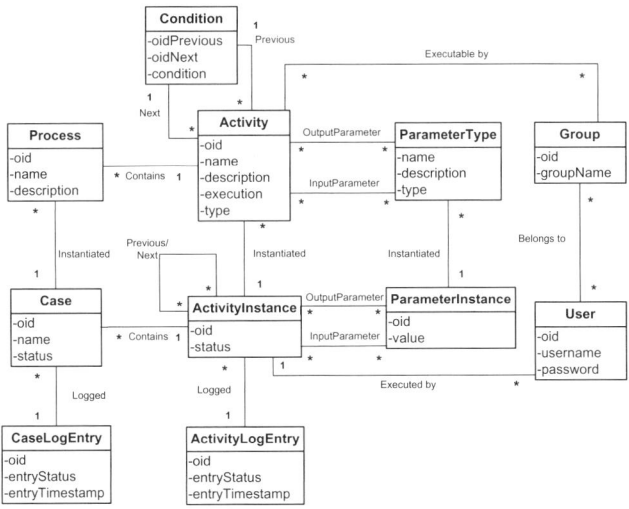

Fig. 5. Process Metadata describing the BPMN constraints

The transformation from the extended BPMN to the Process Metadata produces a relational encoding of the BPMN concepts: each process model is transformed to a *Process* instance; each activity is transformed into an *Activity* instance; each flow arrow is transformed into a *nextActivity/previousActivity* relationship instance; each guard condition is transformed into a *Condition* instance.

Process Metadata generation has been formalized as an ATL transformation from the BPDM metamodel to the Process Model of Figure 5.

5.2   Application Model Generation

The transformation from Refined Process Models to WebML coarse models of services and hypertext interfaces considers the type (human or automatic) of the gateways and the information on the data flows. The application models produced by the transformation still need manual refinement, to add domain-specific elements that cannot be expressed even in the enriched BPMN notation. However, by exploiting information about the activity type, a first-cut application model can be generated, which needs reduced effort for manual refinement.

The computation of the next enabled activities given the current state of the workflow is encapsulated within a specific WebML component, called *Next* unit, which factors out the process control logic from the site view or service orchestration diagram: the *Next* unit exploits the information stored in the Process Metadata to determine the current process status and the enabled state transitions. It needs the following input parameters: *caseID* (the currently executed process instance ID), *activityInstanceID* (the current activity instance ID), and the *conditionParameters* (the values required by the conditions to be evaluated). Given the *activityInstanceID* of the last concluded activity, the *Next* unit queries the Process Metadata objects to find all the process constraints that determine the next activity instances that are ready for activation. Based on the conditions that hold, the unit determines which of its output links to navigate, which triggers the start of the proper subsequent activities.

The *Process to Application Model Transformation* from BPMN to WebML consists of two main rules: the *Process transformation rule*, addressing the structure of the process in-the-large; and the *Activity transformation rule*, managing the aspects of individual activities: parameter passing, starting and closing, and behavior. For modularity and reusability, the piece of WebML specification generated for each activity is enclosed into a WebML *module*, a container construct analogous to UML packages.

Figure 6 shows an overview of the outcome of the *Process transformation rule*: the hypertext page for manually selecting the process to be started and for accessing the objects resulting from process termination. This WebML fragment models the process wrapping logic, generated from the *Start Process* and *End Process* BPMN events.

The generated WebML model further comprises: (1) the process orchestration site view, that contains the logic for the process execution; (2) a site view or service view for each BPMN pool; (3) a set of hypertext pages for each human-driven BPMN activity; (4) one service invocation (triggering the suitable actions for application data updates) for each automatic activity.

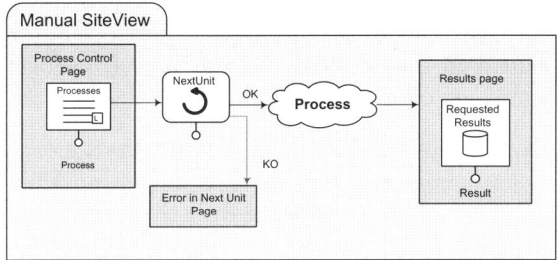

Fig. 6. Excerpt of a WebML application model generated from a BPMN model

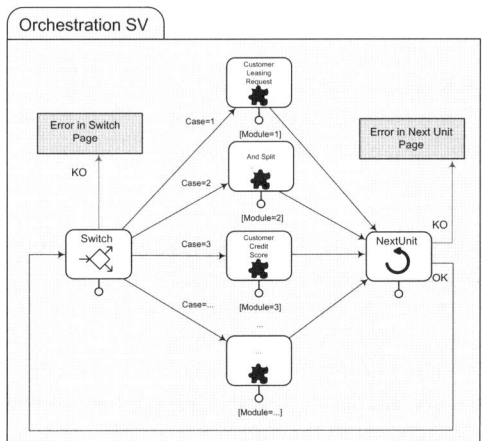

Fig. 7. WebML Orchestration Siteview

Figure 7 shows the model of the *orchestration site view*. The enactment of the process is performed through a loop of *WebML module* invocations, each representing the implementation of one of the activities. At the beginning, the initiation logic (shown in Figure 6) invokes the first module in the loop. The invoked module, be it a Web service call or a Web interface for the user to perform the activity, upon termination returns the control to the Next unit, which determines the modules to be executed next.

The *Activity transformation* rule is based on the BPMN activity and gateway specifications, taking into account aspects like the actor enacting the activity (e.g., a human user or the system). For each BPMN activity and gateway, a WebML module implementing the required behavior is generated. Each generated module has a standard structure: an input collector gathers the parameters coming from previous activities; the activity business logic part comprises a form with fields corresponding to the output of the activity and a Create unit that stores the information produced by the activity persistently, for subsequent use. For gateways, the transformation rule behaves according to the BPMN semantics and to the kind of executor assigned to the gateway (human or automatic): if the gateway is tagged as human-driven, a hypertext is generated for allowing the user to choose how to proceed; if the gateway is tagged as automatic, the choice condition is embedded in the application logic. The transformation of BPMN gateways is conducted as follows:

- *AND-splits* allow a single thread to split into two or more parallel threads, which proceed autonomously. The WebML model for AND-split automatic execution generates a set of separate threads that launch the respective subsequent activity modules in parallel, while manual execution allows the user to select and activate all the possible branches.
- *XOR-splits* represent a decision point among several mutually exclusive branches. Automatic XOR-splits comprise a condition that is automatically evaluated for activating one branch, while manual XOR-splits allow the user to choose one and only one branch.
- *OR-splits* represent a decision for executing one or more branches. Automatic OR-splits comprise a condition that is automatically evaluated for activating one or more branches, while the manual version allows the user to choose the branches to activate.
- *AND-joins* specify that an activity can start if and only if all the incoming branches are completed. This behavior is usually implemented as automatic.
- *XOR-joins* specify that the execution of a subsequent activity can start as soon as one activity among the incoming branches has been terminated. This behavior is usually implemented as automatic.
- *OR-joins* specify that the execution of the subsequent activity can start as soon as all the started incoming branches have been terminated. This behavior is usually implemented as automatic, possibly through custom conditions on the outcome of the incoming branches.

Figure 8 shows two simplified examples of generated modules: the *XOR (ProductType)* module (Figure 8.a) implements the automatic evaluation of the XOR

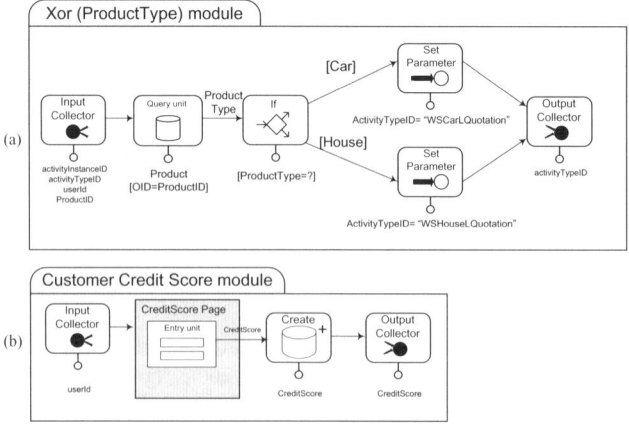

Fig. 8. WebML Modules for XOR gateway and Customer Credit Score

gateway in the BPMN model of Figure 1: given the ProductID, it extracts its type and checks whether it is a car or a house. The next activity to be performed is set accordingly, and this information is passed to the Next unit in the orchestration site view. The *Customer Credit Score* module in Figure 8.b shows the generated hypertext page that allows the user to enter and store the credit score value for the customer, which is the output parameter of the *Customer Credit Score* activity of Figure 1.

The whole approach is specified by an ATL transformation organized into the three above specified rules: a *Process transformation rule* generates the process actions and then invokes the *Activity rule* that manages untyped activities. A set of *type-specific Activity rules* inherit from the general transformation and refine it.

6   Implementation of WebRatio BPM

The illustrated method has been implemented as a new major release of WebRatio, called WebRatio BPM. To achieve this result, all three major components of the tool suite have been extended: the model editing GUI, the code generator, and the runtime libraries. The *model editing GUI* has been extended by: 1) creating an Eclipse-based workflow editor supporting the definition of the refined BPMN Process Model; and 2) adding the Next unit as a new component available in the WebML Application Model editor. The *code generator* has been extended in two directions: 1) the BPMN to WebML transformation has been integrated within the toolsuite, thus allowing automatic generation of the WebML Application Models and of the Process Metadata. 2) the code generation from WebML has been augmented to produce the instances of the Process Metadata and to integrate the novel components (e.g., the Next unit) into the existing J2EE code generation rules.

Moreover, a one-click publishing function has been added to the BPMN editor, thus allowing the immediate generation of a rapid prototype of the BPMN process. The prototype is a J2EE dynamic, multi-actor application with a default look & feel, produced from the WebML Application Models automatically derived from the BPMN diagrams, according to the previously described techniques. The process prototype comprises a few exemplary users for each BPMN actor, and allows the analyst to impersonate each role in the process, start a process and make it progress by enacting activities and both manual and automatic gateways. Figure 9 shows a snapshot of the user interface of the WebRatio BPMN editor.

The WebRatio BPM tool is being tested in a real industrial scenario of a major European bank, that needs to reshape its entire software architecture according to a BPM paradigm with a viable and sustainable design approach. The first set of developed applications addressed the leasing department. The running case presented in this paper is inspired by the leasing application that is under development. The real application involves more than 80 business processes, which orchestrate more that 500 different activities.

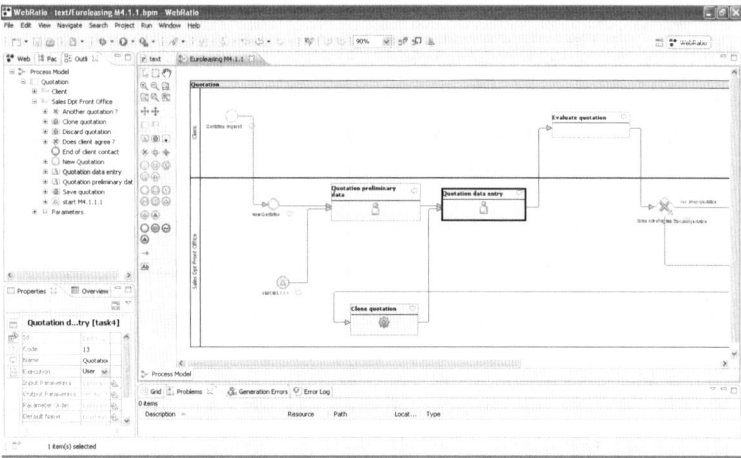

Fig. 9. WebRatio BPM user interface

7   Related Work

A plethora of tools exist for business process modeling and execution, produced by major software vendors, open source projects, or small companies. In our review of existing tools, we identified more than fifty relevant tools in the field. A report from Gartner [6] describes the magic quadrant of the field and selects the most promising solutions. Among them, we can mention Lombardi Teamworks, Intalio, webMethods BPMS, Tibco iProcess, Savvion BusinessManager, Adobe Livecycle ES, Oracle BPM Suite, IBM WebSphere Dynamic Process Edition. Most of them rely on Eclipse as IDE environment and include a visual designer of business models and a generator of configurations for associated workflow engines. In our analysis, we considered more than 50 tools: each of them exposes different strengths and weaknesses. Some 50% of them adopt BPMN as modeling notation; a few provide quick prototyping features (e.g., Oracle, Tibco, BizAgi), while only one provides fine grained simulation capabilities, i.e., the possibility of visualizing hypothetical executions over the visual model, considering stochastic properties of the executions themselves (IBM); some of them are also very strong on BAM features (business analysis and monitoring), such as Oracle and BizAgi; owever, the ones that provide a good level of personalization of the user interfaces allow to do so only at the code level. The main innovations of our approach with respect to the competitors are: (1) quick generation of a running prototype with no coding required; (2) possibility of refinement of the prototype at the web modeling level; (3) clear separation of concerns and assignment to design tasks to roles; (4) model transformation and code generation of the final Web application through MDD.

In the scientific community, some other works have addressed the challenge of binding the business processing modeling techniques with MDD approaches addressing Web applications development.

In this work we focus on process- and data-centric application design, a field where several MDD-based approaches has proven valid. The challenge, though, is to define methods for effectively binding the business processing modeling techniques with MDD approaches addressing, for instance, Web applications development. The Process Modeling language (PML) [11], for instance, is an early proposal for the automatic generation of simple Web-based applications starting from imperative syntax, that allows users to enact their participation to the process .

Koch et al. [7] approach the integration of process and navigation modeling in the context of UWE and OO-H. The convergence between the two models is limited to the requirement analysis phase, while the design of the application model, is separated. In our work, both aspects are considered: like in UWE, we preserve the *process model* as an additional domain model in the application data; as in OO-H, we provide semi-automatic generation of WebML navigational model skeletons directly from the process model. Among the other existing models, we can mention Araneus [10], that has been extended with a workflow conceptual model, allowing the interaction between the hypertext and an underlying workflow management system. In OOHDM [14], the content and navigation models are extended with activity entities and activity nodes respectively, represented by UML primitives. In WSDM [16], the process design is driven by the user requirements and is based on the ConcurTaskTrees notation. An alternative approach is proposed by Torres and Pelechano [15], where BPM and OOWS [5] combines to model process-centric applications; model-to-model transformations are used to generate the Navigational Model from the BPM definition and model-to-text transformations can produce an executable process definition in WS-BPEL. Liew at al. [8] presents a set of transformations for automatically generating a set of UML artifacts from BPM.

With respect to our previous work, this paper extends and refines the technique initially proposed in [3] with several major aspects. The main innovation point is that the BP model and the application model are now treated at the same level and can be evolved separately, thanks to the topology of the generated application models, which insulates the process control logic from the interface and navigation logic of the front-end. Specifically, while in our previous proposal the BPM constructs were transformed into control flows in the application model (c.g., as links in the WebML hypertexts), practical use demonstrated that this approach led to severe difficulties during process and application maintenance and evolution; therefore, we had to better encapsulate process with the help of the process metadata.

Another related research area is the specification and deployment of service orchestrations [1] (e.g., as WS-BPEL specifications). These approaches lack management of the user interactions and of results presentation.

8   Conclusion

This paper presented a methodology and a tool called WebRatio BPM for supporting top-down, model-driven design of business-process based Web applications. The tool is now available for testing purposes and will be commercially distributed starting from October 2009. Our approach is based on model transformations and allows designers to produce applications (both as early prototypes and refined products) without coding. Thanks to the two different modeling levels (business model and Web hypertext model), the needs and skills of different design roles can be accommodated, thus allowing easy joint work between business analysts and web engineers. The high-level BPM perspective provides easy specifications of business models and quick generation of running prototypes, while the hypertext model covers the need of refined design of the final application, thus provide separation of concerns.

The tool, albeit still in a beta status, is have been used in a large banking application for 6 months now. In this period, we collected useful user feedbacks and new requirements, that were considered in the refinement of the system. The experiment was applied on large scale on a real industrial application:

- three different user roles worked together on the same large project: 3 business analysts, 6 application developers, and a few key customers interacted in the definition of the problems and of the solutions;
- the users were spread across Europe and across 4 different companies (the business consultancy company, the application development company, the banking customer, and WebRatio itself);
- the size and volume of the subprojects was so big that it challenged the code generation performances, bringing to applications that included more than 100,000 software components and XML configuration descriptors.

Although the size and the complexity of the project was so large, the need raised only for refinements and small fixes to the approach, which therefore proved valid. Two big requirements were collected on the field: the need for a BAM (Business Analysis and Monitoring) console associated to the approach and for a refined support to requirements and process changes. The BAM console design task will be simplified by the possibility of building the console itself with the model-driven WebML approach;being specified through models, the console will be configurable and customizable at will depending on the customer needs. The support to requirements and process changes is crucial in process-based applications, since the evolution of processes must be supported even when the application is in use, and therefore several process instances can be ongoing while the process change is applied. This requires to preserve all the versions of process models (and of associated applications) at the same time, to grant correct execution of both ongoing and new processes.

Further tasks will include quantitative evaluation of productivity of the developers and of quality of the implemented applications, and coverage of further aspects of BPMN semantics (i.e., customized events and exceptions).

References

1. Benatallah, B., Sheng, Q.Z.: Facilitating the Rapid Development and Scalable Orchestration of Composite Web Services. Distrib. Parallel Databases 17(1), 5–37 (2005)
2. Brambilla, M., Ceri, S., Fraternali, P., Manolescu, I.: Process Modeling in Web Applications. ACM TOSEM 15(4), 360–409 (2006)
3. Brambilla, M., Ceri, S., Fraternali, P., Manolescu, I.: Process Modeling in Web Applications. ACM TOSEM 15(4), 360–409 (2006)
4. Ceri, S., Fraternali, P., Bongio, A., Brambilla, M., Comai, S., Matera, M.: Designing Data-Intensive Web Applications. Morgan Kaufmann Publishers Inc., San Francisco (2002)
5. Fons, J., Pelechano, V., Albert, M., Pastor, O.: Development of web applications from web enhanced conceptual schemas. In: Song, I.-Y., Liddle, S.W., Ling, T.-W., Scheuermann, P. (eds.) ER 2003. LNCS, vol. 2813, pp. 232–245. Springer, Heidelberg (2003)
6. Gartner. Magic quadrant for business process management suites. Technical report, Gartner (February 2009)
7. Koch, N., Kraus, A., Cachero, C., Meliá, S.: Integration of business processes in web application models. J. Web Eng. 3(1), 22–49 (2004)
8. Liew, P., Kontogiannis, K., Tong, T.: A framework for business model driven development. In: STEP '04: Software Tech. and Engineering Practice, pp. 47–56. IEEE, Los Alamitos (2004)
9. Manolescu, I., Brambilla, M., Ceri, S., Comai, S., Fraternali, P.: Model-Driven Design and Deployment of Service-Enabled Web Applications. ACM Transactions on Internet Technologies (TOIT) 5(3), 439–479 (2005)
10. Merialdo, P., Atzeni, P., Mecca, G.: Design and development of data-intensive web sites: The Araneus approach. ACM Trans. Internet Techn. 3(1), 49–92 (2003)
11. Noll, J., Scacchi, W.: Specifying process-oriented hypertext for organizational computing. J. Netw. Comput. Appl. 24(1), 39–61 (2001)
12. OMG. Spem - software process engineering meta-model, version 2.0. Technical report (2008), http://www.omg.org/technology/documents/formal/spem.htm
13. OMG, BPMI. BPMN 1.2: Final Specification. Technical report (2009), http://www.bpmn.org/
14. Schmid, H.A., Rossi, G.: Modeling and designing processes in e-commerce applications. IEEE Internet Computing 8(1), 19–27 (2004)
15. Torres, V., Pelechano, V.: Building business process driven web applications. In: Dustdar, S., Fiadeiro, J.L., Sheth, A.P. (eds.) BPM 2006. LNCS, vol. 4102, pp. 322–337. Springer, Heidelberg (2006)
16. De Troyer, O., Casteleyn, S.: Modeling complex processes for web applications using wsdm. In: Ws. on Web Oriented Software Technology (IWWOST), pp. 1–12. Oviedo (2003)
17. Webratio, http://www.webratio.com

A Visual Tool for Rapid Integration of Enterprise Software Applications

Inbal Tadeski[1], Eli Mordechai[2], Claudio Bartolini[3],
Ruth Bergman[1], Oren Ariel[2], and Christopher Peltz[4]

[1] HP Labs, Haifa, Israel
[2] HP Software, Yahud, Israel
[3] HP Labs, Palo Alto, California, USA
[4] HP Software, Fort Collins, Colorado, USA

Abstract. Integrating software applications is a challenging, but often very necessary, activity for businesses to perform. Even when applications are designed to fit together, creating an integrated solution often requires a significant effort in terms of configuration, fine tuning or resolving deployment conflicts. This is often the case when the original applications have been designed in isolation. This paper presents a visual method allowing an application designer to quickly integrate two products, taking the output of a sequence of steps on the first product and using that as input of a sequence of steps on the second product. The tool achieves this by: (1) copying UI components from the underlying applications user interface; (2) capturing user interaction using recording technology, rather than by relying on the underlying data sources; and (3) exposing the important business transactions that the existing application enables as macros which can then be used to integrate products together.

1 Introduction

Integrating software applications is hard. Even when such applications are designed to fit together, creating an integrated solution often requires a significant effort in terms of configuration, fine tuning or resolving deployment conflicts. If the original applications had been designed in isolation and not originally intended to provide an integrated solution, then it might easily turn into a nightmare. In fact, this scenario is very common. If you want to scare off an IT executive, tell them that "it's just a simple application integration effort". In reality, there is no simple and cheap application integration effort. Difficult as it is, software integration application is often necessary. Due to the dynamic nature of most enterprises, many disparate software applications live in the enterprise space. The creation of new business operations, for example, as the result of mergers and acquisitions, brings about the need to use multiple applications. The enterprise staff has to juggle between them to complete new business processes. Indeed, such processes sometimes require operators to have multiple screens open at the same time and manually move data from an application to the other. So, the IT executive has to choose between an error prone and time consuming

B. Benatallah et al. (Eds.): ICWE 2010, LNCS 6189, pp. 430–444, 2010.

process, or an application integration effort. The visual integration tool presents a way out for the IT executive, a method for rapid, programming-free software integration.

Software applications integrations comes in various flavors, the most prevalent of which are Enterprise Application Integration (EAI) and, more recently, web 2.0 mashups. The business relevance of this problem is exemplified by that the total available market for EAI alone, defined as the use of software and computer systems architectural principles to integrate a set of enterprise computer applications, is expected to reach 2.6 billion by 2009 (Wintergreen research). From a technology point of view, what make the problem hard are issues with data representation, data semantics, connector semantics, error control, location and non-functional requirements.

In this paper, we introduce a novel approach to software application integration, and a visual tool that embodies it, called **visual product integrator**. The main guiding principle for our approach is that software application integration should be made *faster* and *cheaper*. Our approach aims to make the *integration process* better, since our tool will enable quick proof-of-concept integrations that can go from inception to use in a very short turn-around time. We do not expect that solutions developed using it will be as robust and as complete as full standard application integration solutions. Having said this, our approach and tool, when used in parallel with standard techniques, provide the advantages of shortening developing cycles and enabling the quick and cheap creation of successive throw-away versions of the final integrated solution. Last but not least, our approach demands very little in terms of knowledge about the internal data representation and semantics of the original applications.

In a nutshell, the visual product integrator allows an application designer to quickly realize use cases for integrations of two products that can generally be defined as taking the output of a sequence of steps on the first product and using that as input of a sequence of steps on the second product. Visual product integrator achieves this by

- copying UI components from the user interface of the underlying applications.
- capturing user interaction using recording technology, rather than by relying on underlying data sources or on programming.
- exposing the important business transactions that the existing application enables as *macros*.

We refer to a recorded sequence of interactions as a macro. Once macros of one or more existing applications have been created they may be re-used in any new application. The main contributions of our approach and visual product integrator tool are:

- blend together the roles of the designer and the user of the integrated solution.
- require no programming of scripting for creating the graphical user interface of the integrated solution.

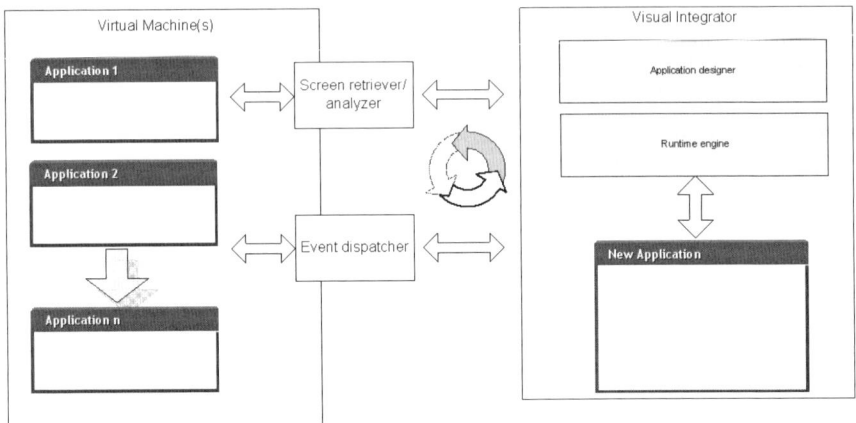

Fig. 1. Architecture Diagram for the Visual Product Integrator

- record the actions of the application user/designer similarly to what is done when recording a macro in Microsoft Excel - only this happens when juggling between different software applications rather than spreadsheets.
- force the designer/user to think about business transactions rather than specific business objects: visual product integrator captures the sequence of steps of a business process and its interactions with all the related business objects.

2   Architecture

The visual product integrator has two main functionalities. First, it allows to user to design a new application with a new user interface. Second, it provides a runtime environment for the new application. The architecture of the visual product integrator, shown in Figure 1, supports both functionalities.

At design time, the visual integrator behaves like a WYSIWYG Graphical User Interface builder. It has the usual GUI builder capabilities for placing UI components into the application interface, and assigning actions to these UI components. The visual integrator goes beyond standard GUI builders, because it can copy UI components from an existing application. To copy UI components the user is prompted to select an application, which will be started within the visual integrator. The Application Screen Retriever/Analyzer module of the visual product integrator initiates and displays the application. It also connects to the object hierarchy and captures events. The user can then select the UI components he wishes to copy into the new application: the visual integrator has the ability to duplicate the selected components in the new application. Copied components are tagged, so that at run time they can be synchronized with the original application. The user can continue to copy objects from any number of applications. An additional enhancement in the visual product integrator is that the actions assigned to UI components may include *macros*. At design time, macros are recorded using the record/replay engine. It is also possible to import macros that were recorded or manually written elsewhere.

At run time, the visual product integrator must do all of the following

1. Run the new application
2. Refresh copied UI components from the original application
3. Run macros on original application

Referring to the architecture diagram in Figure 1, the Runtime Engine is responsible for running the new application. The critical new piece of this technology is the Application Event Dispatcher, which handles the events of the new application that refer to any of the original application. It dispatches events from copied UI components of the new application to the original application. Similarly, it refreshes the state of copied UI components based on the state of the respective component of the original application. The Application Event Dispatcher also handles macros using the record/replay engine.

Notice that we describe the visual product integrator independently of the technology used for the underlying applications. Our architecture is rich enough to enable combining Windows applications with HTML applications, Java applications, and so on. All is required is that the Application Screen Retriever/Analyzer be able to manipulate and analyze the object hierarchy of the application. It must be able to convert UI components from the original technology to the technology used for the new application. For example, if the technology for the new application is HTML and the underlying application is .NET, the Screen Retriever/Analyzer must recognize that a selected UI component is, e.g., a .NET button, so that it can tell the Application Designer module to create an HTML button. Likewise, the Application Event Dispatcher must be able to replicate events on all the technologies of the underlying applications and query the objects for their current state. These capabilities require a deep model of the object hierarchy for each supported technology, similar to what is used for automated software testing, e.g., in HP Quick Test Pro [3], for example. By leveraging this technology, the visual integrator can support standard Windows, Visual Basic, ActiveX controls, Java, Oracle, Web Services, PowerBuilder, Delphi, Stingray, Web (HTML) and .NET.

The integrator is a GUI builder application, among other things. As such it builds new applications in some particular technology. This technology can be Java Swing, .NET, HTML or any other GUI technology. It is important to note that every application built by the visual product integrator will use this same GUI technology.

To achieve a good runtime experience, we want to present a single application to the user. Although there are several underlying applications and all these applications are running while the new application runs, we want to hide the presence of these applications from the user. In the architecture, therefore, the original applications run using a remote protocol, either on a remote machine or on a Virtual Machine (VM), rather than on the local machine. At design time, the Screen Retriever/Analyzer module displays the underlying application to the user and captures the UI objects that the user selects. At run time, the Application Event Dispatcher passes events to and from the VM, and refreshes the components of the new application.

3   Implementation

The prototype implementation of the visual product integrator utilizes the architecture described in the previous section. We made several simplifying assumptions that enable us to rapidly prototype and demonstrate feasibility of the visual product integrator. The technology used to develop the new application is HTML. In our first embodiment of the visual product integrator, we are also restricting the underlying applications to a single technology, again HTML. With this restriction, the Application Screen Retriever/Analyzer and the Applications Event Dispatcher need only have a model of a single technology. Much of the functionality of the Applications Event Dispatcher is the macro record and replay. This capability is available in the software testing application Quick Test Pro (QTP) [3]. For the prototype we use QTP's record/replay engine to dispatch events on the underlying applications. QTP itself runs on the VM in order to give the user the expected interaction with the application.

Figures 2(a) and 2(b) show screen shots of the Visual Application Integrator prototype. To design a new application, the user does the following:

Select Application. Figure 2(a) shows the design time view in the integrator. The user chooses one or more applications, which the integrator initiates and displays.

Copy UI Components. The user can select UI objects for duplication in the new application. Selected objects appear in the design view of the integrator, as shown in Figure 2(b).

Add New UI Components. The user may add additional UI components from the toolbar.

Set Actions. For each object the user assigns an action. Figure 2(b) demonstrates the set of possible actions.

  1. If the element was copied from one of the underlying applications, the user selects the "keep original behavior" mode, then the object will execute the original action in the original application.

  2. New objects may be assigned actions by writing a java script function. These functions will be executed locally.

  3. Another type of action for a new object is a "workflow action". The workflow is set up by connecting the action with a QTP script. The QTP script will run on the original application in the VM. In the current prototype we do not enable recording the QTP script from the integrator. Instead, we assume that the QTP script is pre-recorded. This limitation is related to the implementation of the record/replay engine in QTP. Future implementations using a record/replay engine built specifically for the integrator will not have this limitation.

(a) A view of capturing elements from the underlying application

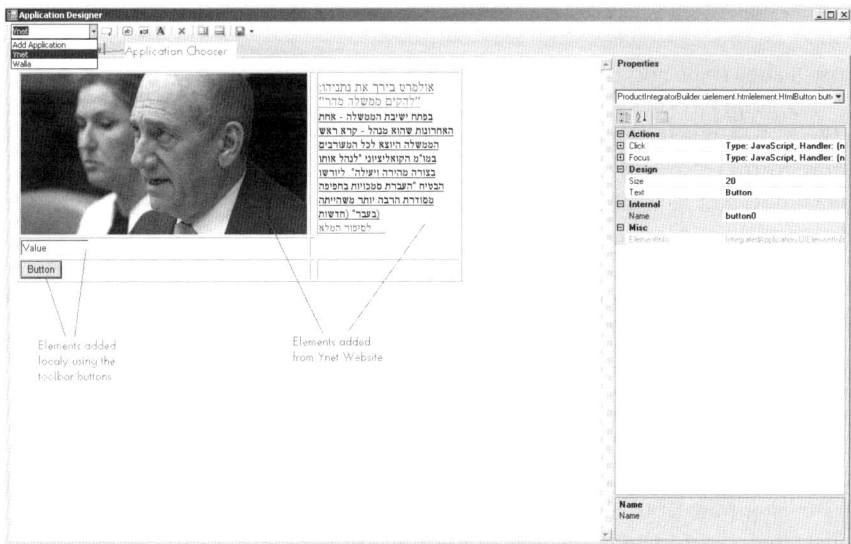

(b) The user can arrange UI elements, and assign actions to each element

Fig. 2. Visual Product Integrator screen shots depicting design capability

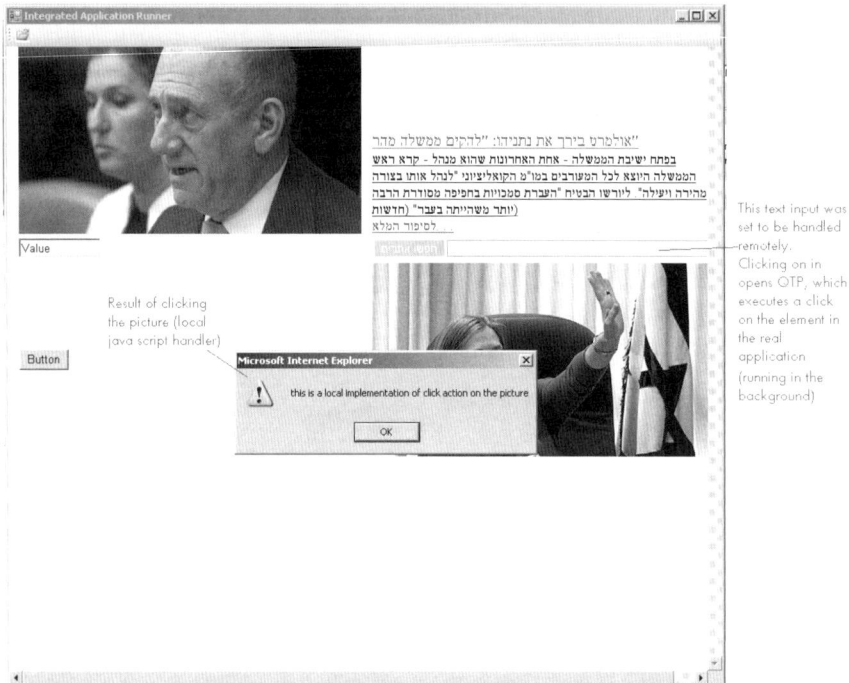

Fig. 3. Visual Product Integrator screen shot. A view of the application at run time.

Figure 3 shows the run time view in the integrator. The integrator is now running a HTML application in a browser object. But this application must run inside the integrator because of the connection with the underlying applications. First, the integrator refreshes copied objects, as necessary. Second, actions on copied UI objects and QTP scripts are executed on the original applications. The results of these actions are shown in the new application. The integrator uses QTP to start and drive the application. QTP gets its instruction via its remote automation API, thus hiding the presence of these applications from the user. Screen data is captured and reflected in the host machine via new remote API.

4   Validation

In order to validate our tool, we used it to build a **Search-Calculator** application. This application uses functionality from two Web applications, a calculator application [1] and the Google search application [2].

To build the new application the user first opens each application in a viewer. Figure 4 shows each application in the viewer. Next the user selects UI elements for copying. Figure 4(a) shows the calculator application, in which all the elements making up the calculator interface have been selected for copying to the

(a) Object capture from the calculator application

(b) Object capture from the Google search application

Fig. 4. Building the Search-Calculator application. A view of the underlying applications inside the viewer.

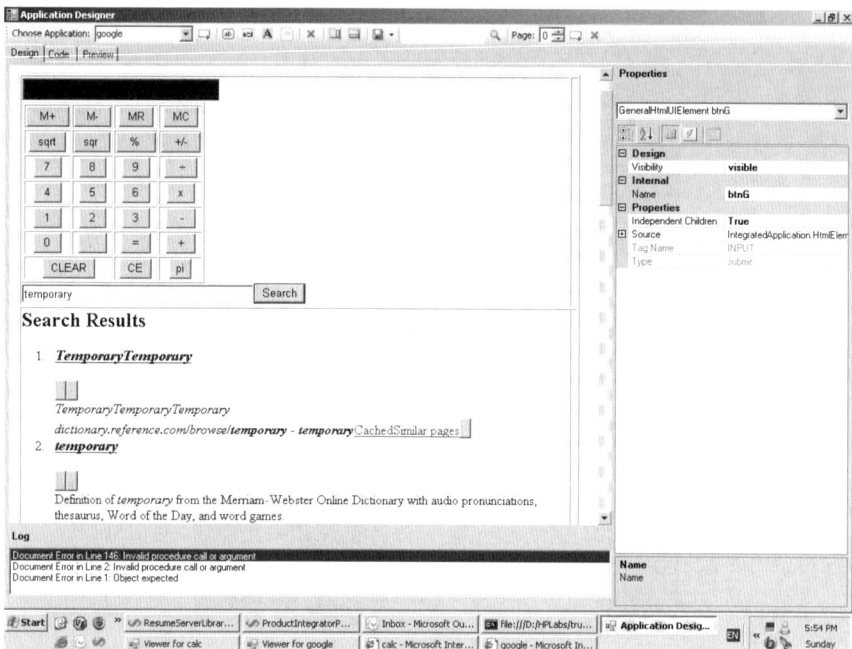

Fig. 5. Building the Search-Calculator application. The user arranges UI elements and sets actions in the designer.

new application's UI. Figure 4(b) shows the selection of the elements on the Google page that are needed for search, i.e., the edit box and the search button.

The next step of building the new application includes arranging UI elements, and assigning actions to each element. The designer supports these activites, as shown in Figure 5. For the **Search-Calculator** application, the actions of the calculator components are set to remote, meaning that actions on these buttons will be passed through to the original calculator application. Likewise, the calculator result text box is remote, so the value calculated in the original application will be automatically updated in the new application. Similarly, remote actions are set on the elements copied from the Google application. Finally, the workflow action of searching for the calculator result is set using a QTP script. That completes the application.

At run time, the **Search-Calculator** application will search the Web for the value resulting from any computation. Figure 6 demonstrates this behavior. The user pressed the PI button on the new application's interface. The action was passed to the calculator application, which displays the value 3.14159. The new application refreshes the value on the calculator display due to the change in the underlying application. Based on the workflow set up earlier, the value 3.14159 is copied to the search edit box and the search is started. When the search results are available in the Google search application, the new application interface is refreshed to show the results.

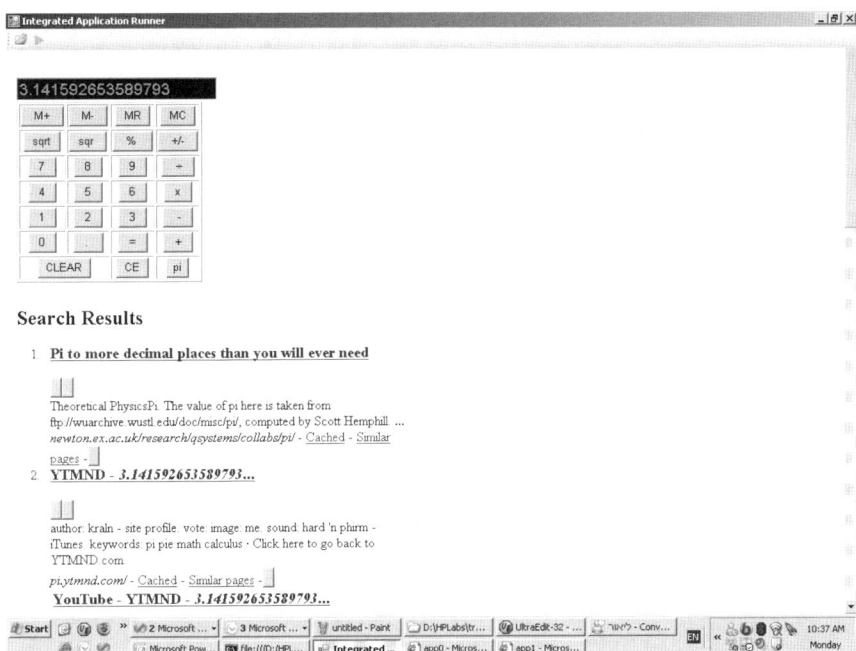

Fig. 6. Visual Product Integrator screen shot. A view of the **search-calculator** application at run time.

5 An Enterprise Application Integration Use Case

The benefit of this tool for enterprise application integration is illustrated by the following a common scenario. In this scenario, an operator has to juggle between two applications to carry out their daily task. The operator opens the Customer Relationship Management (CRM) applications and finds a list of activities. Among them, he sees a new "Contract job by partner" activity. He clicks on the *detail* tab of the activity. The operator then can navigate through the details of the activity, shown in Figure 7(a). This screen shows the activity creator *camptocamp*, a partner company.

At this point - as is often the case with enterprise applications - the operator will need to pull up an Enterprise Resource Planning (ERP) application, to retrieve collateral information about the partner, in the partner events panel. He will have to enter in the basic search form (possibly copy-and-paste) the name of the partner as it appears in the CRM application. On obtaining the search results, the operator has to add a new event for this partner and fill the details manually with information based on the activity he found in the CRM application (Figure 7(b)).

Consider using the visual product integrator to expose the operator to a single interface, and to automate this error-prone, manual task. To build the new application, the application designer selects and copies a UI element from the first

(a) A Customer Relationship Management (CRM) application. View of a new
"Contract job by partner" activity.

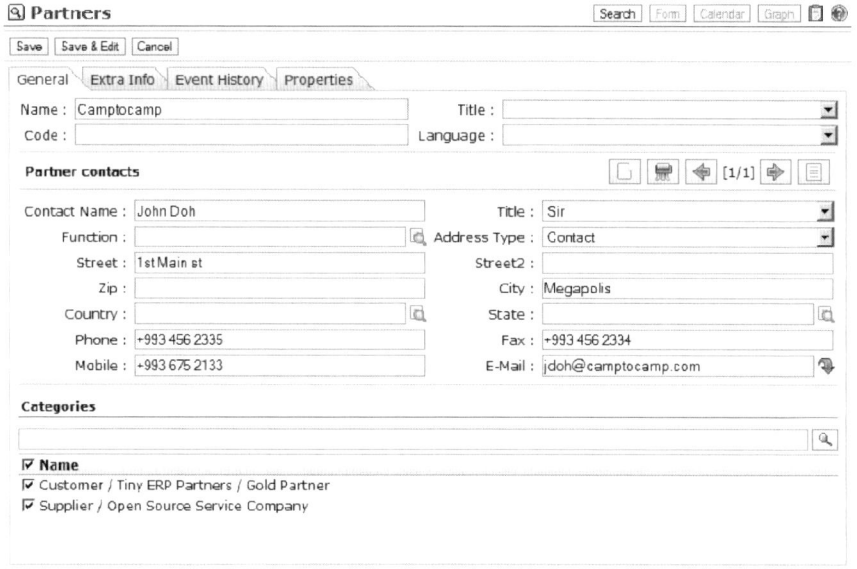

(b) An Enterprise Resource Planning (ERP) application. Entering new partner
activity by copying from the CRM application.

Fig. 7. An enterprise application integration example

original application, i.e., the CRM application shown in Figure 7(a). He then
adds a new action button, labeled "update ERP". At this point, the tool makes
use of recording technology to capture the business flow that the operators has
to execute, and attaches a description of that to the button. The tool records
the sequence of copy and paste operations between the selected fields from the
first original application (CRM) to the second original application (ERP). As
the designer performs a function on the values from the selected fields in the first

original application to fill in the value of the selected field in the second original application, visual product integrator records that function and sets this macro as the action on the button in the new application. Thus, this button represents the business flow of creating a new event for a partner and copying the details from the activity.

6   Related Work

The Visual Product Integrator tackles the problem space covered by enterprise application integration (EAI), but aiming at making it significantly cheaper and faster, getting inspiration from UI integration concepts such as mashups and Web 2.0, which today are aimed at web application integration.

6.1   Enterprise Application Integration

EAI applications usually implement one of two patterns [14]:

Mediation: The EAI system acts as the go-between or broker between the underlying applications. Whenever an interesting event occurs in an application an integration module in the EAI system is notified. The module then propagates the changes to other relevant applications.

Federation: The EAI system acts as the front end for the underlying applications. The EAI system handles all user interaction. It exposes only the relevant information and interfaces of the underlying applications to the user, and performs all the user's actions in the underlying applications.

Most EAI solutions, whether they implement the mediation or federation approach, focus on middleware technology [15], including: Message-oriented middleware (MOM) and remote procedure calls (RPCs); Distributed objects, such as at CORBA and COM; Java middleware standards; and Message brokers The visual product integrator, by contrast, uses the federation pattern.

A long list of EAI solutions may be found in [18]. We mention some leading solutions in this area. Microsoft BizTalk Server [16] is used to connect systems both inside and across organizations. This includes exchanging data and orchestrating business processes which require multiple systems.

Microsoft Host Integration Server allows enterprise organizations to integrate their mission-critical host applications and data with new solutions developed using the Microsoft Windows platform. All these middleware solutions require the application integrator to go deep into the original application's source code and write wrappers that bring the original application into the fold of the new technology.

In contrast to most EAI solutions, our approach eliminates the need to access the application's source code, substituting that with a representation of the user-interaction that does not depend on the original applications information model.

6.2   Integration at the Presentation Layer (Web Applications)

The web engineering community has so far typically focused on model-driven design approaches. Among the most notable and advanced model-driven web engineering tools we find, for instance, WebRatio [11] and VisualWade [10]. However these approaches tend to be heavy on modeling methods, thereby being more similar to traditional EAI in spirit than the approach we take here.

A very good account of the of the problem of integration at the presentation layer for web applications is found in [13]. On thoroughly reviewing academic literature, the authors drew the conclusion and concluded that there are no real UI composition approaches readily usable in end-users composition environments. There are some proprietary Desktop UI component technologies such as .NET CAB [9] or Eclipse RCP [8], and their being proprietary in nature limits their possible reuse for software product integration in general. On the other hand, browser plug-ins such as Java applets, Microsoft Silverlight, or Macromedia Flash are easily embedded into HTML pages, but provide limited and cumbersome interoperability through ad-hoc plumbing.

Mashup approaches, based on Web 2.0 and SOA technologies have been gaining broad acceptance and adoption [13]. Existing mashup platforms typically provide easy-to-use graphical user interfaces and extensible sets of components for development composite applications. Yahoo! Pipes [12] integrate data services through RSS or Atom feeds, and provide a simple data-flow composition language. UI integration is not supported. Microsoft Popfly [7] uses a similar approach and does provide a graphical user interface for the composition of data, application, and UI components, however does not support operation composition through different components. IBM Lotus Mashups [4] provides a wiki-based (collaborative) mechanism to glue together JavaScript or PHP-based widgets. Intel MashMaker [5] takes a different approach to mashup of web application in that it makes "mashup creation part of the normal browsing process". As a user navigates the web, MashMaker gives them suggestions for useful mashups to be applied to the page that they are looking at. As a result, mashups are composed "on-the-go", but the main use case for MashMaker is still "Pipes"-like. It is not oriented to producing a new stand-alone application like our visual integrator. JackBe Presto [6] has an approach similar to Pipes for data mashups and provides mashlets, or UI widgets used for visualizing the output. Throughout all these example, we observe that mashup development is a process requiring advanced programming skills.

6.3   Other Hybrid Approaches

One closely related solution for product integration is the OpenSpan platform [17]. This platform may be used to build new composite applications by integrating any legacy applications and/or business workflows in a rapid manner. By dynamically hooking into applications at run-time,

OpenSpan can create integration points with an application without requiring modification or access to the application's source code, APIs or connectors. At design time, the user can set up the workflow of a business process, by hooking into UI objects and specifying the data flow between the UI objects. We observe two main differences between the OpenSpan platform and our Visual Product Integrator:

1. Our integrator captures the workflow or business process as macros, using record/replay technology. This technology spares the designer the manual task of setting up the flows.
2. Our integrator allows the creation of a new graphical user interface. The user of the new application only sees this interface. While the underlying applications are running, they are not visible. Thus, the user experience is more satisfying.

7   Conclusions

We presented a visual tool for rapid integration of software products to create a new application. The main advantages of this tool over prior solutions are

- It goes beyond simple juxtaposition and syndication of data feeds well into the realm of visual application design.
- It eliminates the need to access the application's source code, substituting that with a representation of the user-interaction that does not depend on the original application's information model.
- It captures the workflow or business process as macros, using record/replay technology. This technology spares the designer the manual task of setting up the flows.
- It allows the creation of a new graphical user interface. The user of the new application only sees this interface. While the underlying applications are running, they are not visible. Thus, the user experience is more satisfying.

The current prototype implementation has several limitations. The most severe limitation is that it uses Quick Test Pro (QTP) as the record/replay engine. The advantage of QTP is that it is a mature product. The disadvantage is that it is cumbersome. We make it transparent to the user by running it on a VM. In future development, we would replace QTP with a lighter weight record/replay engine. A second limitation of the prototype is that, at present, it is restricted to HTML applications. Future work will add support for additional GUI technologies, such as .NET and Java. Longer term work is aimed to eliminate the need to model specific GUI technologies by modeling the object hierarchy directly from the application's visual interface, i.e., the screen images.

References

1. CalculateForFree, http://www.calculateforfree.com/
2. Google search, http://www.google.com/

3. HP Quick Test Pro,
   https://h10078.www1.hp.com/cda/hpms/display/main/hpms_content.jsp?
   zn=bto&cp=1-11-127-241352_4000_100__
4. IBM Lotus Mashups, http://www-01.ibm.com/software/info/mashup-center/
5. Intel Mash Maker, http://mashmaker.intel.com/web/
6. JackBe Presto, http://www.jackbe.com/
7. Microsoft popfly, http://en.wikipedia.org/wiki/Microsoft_Popfly
8. Rich Client Platform. Technical report, The Eclipse Foundation
9. Smart Client - Composite UI Application Block. Technical report, Microsoft Corporation
10. VisualWade, http://www.visualwade.com/
11. WebRatio, http://www.webratio.com
12. Yahoo! Pipes, http://pipes.yahoo.com/pipes/
13. Daniel, F., Yu, J., Benatallah, B., Casati, F., Matera, M., Saint-Paul, R.: Understanding ui integration: A survey of problems, technologies, and opportunities. IEEE Internet Computing 11, 59–66 (2007)
14. Hohpe, G., Bobby, W.: Enterprise Integration Patterns: Designing, Building, and Deploying Messaging Solutions. Addison-Wesley, Reading (2003)
15. Linthicum, D.S.: Enterprise Application Integration. Addison-Wesley, Reading (2000)
16. Microsoft Biztalk Server,
    http://www.microsoft.com/biztalk/en/us/default.aspx
17. OpenSpan, http://www.openspan.com/index.php/software_platform.html
18. Wikipedia,
    http://en.wikipedia.org/wiki/Enterprise_application_integration

Customization Realization in Multi-tenant Web Applications: Case Studies from the Library Sector

Slinger Jansen[1], Geert-Jan Houben[2], and Sjaak Brinkkemper[1]

[1] Utrecht University, P.O. Box 80.089, 3508TB Utrecht, The Netherlands
[2] Delft University of Technology, P.O. Box 5031, 2600 GA Delft, The Netherlands

Abstract. There are insufficient examples available of how customization is realized in multi-tenant web applications, whereas developers are looking for examples and patterns as inspiration for their own web applications. This paper presents an overview of how variability realization techniques from the software product line world can be applied to realize customization when building multi-tenant web applications. The paper addresses this issue by providing a catalogue of customization realization techniques, which are illustrated using occurrences of customization in two practical innovative cases from the library sector. The catalogue and its examples assist developers in evaluating and selecting customization realization techniques for their multi-tenant web application.

1 Introduction

Web applications profit greatly from customization, as they make them applicable in a broader context, enable component reusability, and make them more user-specific. Especially when taking the software as a service model into account, the benefits of one centralized software application with multiple tenants, over the alternative of multiple deployments that have to be separately maintained, become apparent. Customization and multi-tenancy, however, do not come for free. In customized web applications code becomes more complex, performance problems impact all tenants of the system, and security and robust design become much more important. A multi-tenant web application is an application that enables different customer organizations ('tenants') to use the same instantiation of a system, without necessarily sharing data or functionality with other tenants. These tenants have one or more users who use the web application to further the tenant's goals. Software developers implementing these systems have all at some point wondered what the available Customization Realization Techniques (CRTs) are. This leads to the following research question:

How are customization and configurability realized in Multi-tenant web applications?

There are three research areas that are directly related to CRTs in multi-tenant web applications: variability in software product lines, end-user personalization in web applications, and multi-tenancy architectures.

From the area of software product lines variability models can assist in determining techniques for customization as is already determined by Mietzner [1]. Svahnberg, van Gurp, and Bosch [2] present a taxonomy of variability realization techniques, some of which were previously encountered in a case study [3]. These variability realization techniques, however, are only partially relevant for two reasons: (1) web applications are generally one-instance systems and (2) the variability realization techniques

B. Benatallah et al. (Eds.): ICWE 2010, LNCS 6189, pp. 445–459, 2010.

of Svahnberg have different binding times, whereas variability in web applications is generally bound at run-time only.

Rossi et al. [4] introduce the following scenarios for web application customization: static customization, link customization, node structure customization, node content customization, and context customization. These scenarios are interesting, but do not include customizations that impact the structure of the web application, such as the integration with another (varying) application or the adaptation of the object model on which the system is built. Chong and Carraro [5] suggest there are four ways to customize web applications in a multi-tenant environment, being user interface and branding, workflow and business rules, data model extensions, and access control. The practical view of Chong does include data model extensions, but also does not take into account integration with varying other applications. Another specific problem that is not covered by the customizations of Chong is user interface adaptations based on user specific properties, such as experience level.

WUML [6] is a modeling language that enables customization by splitting up functionality in a generic and a customized part. Unfortunately, this technique only enables a priori customization and end-user customization is not possible at present. A generic framework for web site customization is introduced by Lei, Motta and Domingue [7], that provides customization capabilities based on any logic condition for web-based applications. The framework is based on the mature OntoWeaver web application modeling and development framework. The capabilities of the customization framework appear to be geared towards the views and controller of applications. The framework does not cover, however, customization by linking to other applications, accessing different databases with one code base, or customer-specific model adjustments.

Mietzner et al. [1,8] propose a variability modeling technique for multi-tenant systems. The variability modeling technique is also rooted in the software product line research area, and defines internal (developer view) and external (customer view) variability. Their variability modeling technique enables developers to speculate and calculate costs for unique instances within a multi-tenant deployment. Furthermore, Mietzner also addresses different deployment scenarios for different degrees of tenant variability. Unfortunately, they do not address the CRTs. Guo et al. [9] provide a framework for multi-tenant systems. Their framework makes a relevant distinction with regards to security, performance, availability, and administration isolation, where issues are addressed that would not occur in a single-tenancy environment. The framework also aims to provide as much flexibility for end-users, even to the extent that end-users create their own customizations on the fly. Again, the issue of how such customizations are implemented is not addressed.

In regards to customization and its relationship to end-user personalization, much can be learned about the reasons for changing web application functionality based on a user's context. Goy et al. [10] state that customization is based on information about the user (knowledge, interests and preferences, needs, and goals), the device used by the user to interact with the system, and information about the context of use (physical context, such as location, and social context). Schilit, Adams, and Want define context as the 'where you are', 'who you are with', and 'what resources are nearby' [11]. Such information is generally stored in a user model [12,13]. Ardissono et al. [13]

address the problem of web application customization using an advanced evaluation model that questions the user on his interests and that automatically derives what parts of the application the user is interested in. Fiala and Houben present a tool that adapts a web page after it has been generated but before it is presented to the user, using a set of transcoding rules [14]. The models used for these techniques are, however, not re-usable for multi-tenant systems, since the meta-configuration data stored per tenant is dissimilar from the information that is stored per user in these user models. It must be noted that most literature that refers to such user models [12,11] addresses dynamic adaptation of web applications, generally to improve the system behaviour for specific end-users, such as showing or removing options based on the level of the end-user [12]. The dynamic adaptations, however, conflict directly with the topic of this paper, i.e., the description of static customizations per tenant. The work on these user models and customization is therefore considered out of scope.

In this paper we present a catalogue of CRTs for web applications and show how these are implemented in practice using two case studies. The catalogue assists web application developers and architects in selecting the right techniques when implementing multi-tenant web applications. Without such a catalogue, developers of multi-tenant web applications need to reinvent the wheel and are unable to profit from existing architectural patterns used for implementing multi-tenancy.

This research was initiated because a lack was noticed in current literature regarding customization techniques in web applications, specifically with regards to multi-tenant web applications. The aim was set to create an overview of such techniques for developers, such that they can reuse these techniques and their associated implementation details. The research consisted of three steps. First, the two case studies were performed, using the theory-building multi-case study approach [15,16]. The case study results were discussed with and confirmed by lead developers from both cases. Secondly, the CRTs overview was created. Thirdly, this overview was evaluated by five developers of multi-tenant web applications, who were not part of the core development teams of the case studies. In both cases the first author played a major part as lead designer. The developers who were interviewed were all at some point involved in the development of multi-tenant systems. They were interviewed and in-depth discussions were had about the proposed CRTs.

Section 2 continues with a discussion of the different types of customization. Sections 3 and 4 describe the two case studies of native multi-tenant library web applications and the customization techniques encountered. Finally, in sections 5 and 6 the findings and conclusions are presented.

2   Definition of Core Concepts: Multi-tenancy and Customization

In a multi-tenant web application there are tenants (customer companies) and the tenant's users. Users are provided with features that may or may not be customized. A feature is defined as a distinguishing characteristic of a software item (e.g., performance, portability, or functionality). A customized feature is a feature that has been tailored to fit the needs of a specific user, based on the tenant's properties, the user's context and properties, or both. Customized features are customized using customization techniques, which are in turn implemented by CRTs. These are implemented by variability realization techniques.

Variability realization techniques are defined by Svahnberg, van Gurp, and Bosch [2] as the way in which one can implement a variation point, which is a particular place in a software system where choices are made as to which variant to use. Svahberg's variability realization techniques have specific binding times, i.e., times in the software lifecycle at which the variability must be fixed. Because the systems under study are multi-tenant systems, it is impossible to apply all variability realization techniques into account that have a binding time that is not at run-time, since there is just one running instance of the system. This constraints to the following applicable variability realization techniques: (a) infrastructure-centered architecture, (b) run-time variant component specializations, (c) variant component implementations, (d) condition on a variable, and (e) code fragment super-imposition. Typically, in a multi-tenant environment a decisions must be made whether to store a tenant data in a separate databases or in one functionally divided database [5]. By following the variability techniques of Svahnberg this major decision is classified as 'condition on a variable'. The same is true, however, if the tenant requires a change in the view by adding the tenant's logo to the application. The difference between these two examples is the scale on which the change affects the architecture of the web application. Svahnberg's techniques provide some insight into the customization of web applications, but a richer vocabulary is required to further specify how customizations are implemented in practice. The richer vocabulary is required to identify patterns in the creative solutions that are employed to solve different functional customization problems. To bring variation realization techniques and customized features (subject to developer creativity) closer together, the CRTs are introduced.

Before we continue providing the CRTs, the descriptions of the variability realization techniques of Svahnberg et al. are repeated here in short. The (a) infrastructure-centered architecture promotes component connectors to a first class entity, by which, amongst other things, it enables to replace components on-the-fly. (b) Run-time variant component specializations enable different behaviors in the same component for different tenants. (c) Variant component implementations support several concurrent and coexisting implementations of one architectural component, to enable for instance several different database connectors. (d) Condition on a variable is a more fine-grained version of a variant component specializations, where the variant is not large enough to be a class in its own right. Finally, (e) code fragment super-imposition introduces new considerations into a system without directly affecting the source code, for instance by using an aspect weaver or an event catching mechanism.

Five CRTs can be identified from two types of customization: *Model View Controller (MVC) customization* and *system customization*. By MVC customization we mean any customization that depends on pre-defined behaviour in the system itself, such as showing a tenant-specific logo. We base it on the prevalent model-view-controller architectural pattern [17], that is applied in most web applications as the fundamental architectural pattern. In regards to the category of MVC customization we distinguish three changes: *model changes*, *view changes*, and *controller changes*. By system customization we mean any customization that depends on other systems, such as a tenant-specific database or another web application. For the category of system cus-tomization two types of changes are distinguished: *connector changes* and *component changes*. Please note that each of the changes can be further specialized, but presently we aim to at least be able to categorize any CRT.

The first customization is that of **model change**, where the underlying (data-)model is adjusted to fit a tenant's needs. Typical changes vary from the addition of an at-tribute to an object, to the complete addition of entities and relationships to an existing model. Depending on the degree of flexibility, the ability to make model changes can be introduced at architecture design time or detailed design time. Binding times can be anywhere between compilation and run-time, again depending on the degree of freedom for the tenants and end-users. An example of an industrial application is SalesForce[1], in which one can add entities to models, such as domain specific entities. The variability for SalesForce was introduced at architecture design time and can be bound at run-time. The model change assumes there is still a shared models between tenants. If that is not the case, i.e., tenants get a fully customized model, the customization is considered a full component change (being the model itself).

The second type of customization is that of **view change**, where the view is changed on a per-tenant basis. Typical changes vary from a tenant-specific look and feel, to complete varying interfaces and templates. Again, depending on the degree of flexi-bility, variations can be introduced between design time and run-time. Binding times can be anywhere between compilation and runtime, also depending on the degree of freedom for the end-users. An example of an industrial application is the content man-agement system Wordpress, in which different templates can be created at runtime to show tenant-specific views of the content.

Controller change is the third type of customization, where the controller responds differently for different tenants and guides them, based on the same code, through the application in different ways. The simplest example is that of tenant specific data, which can be viewed by one tenant only. More extreme examples exist, such as different li-cense types, where one tenant pays for and uses advanced features, opposed to a light version with simple features for a non-paying tenant. Binding times are anywhere be-tween design time and runtime and again depend on how much freedom the tenants are given. An example of an industrial application is the online multi-tenant version of Microsoft CRM, which enables the tenant to create specific workflows for end-users.

The system changes are of a different magnitude: they concern the system on an ar-chitectural level and use component replacement and interfaces to provide architectural variability. The fourth type of customization is **system connector change**, where an extension that connects to another system is made variable, to enable connecting to dif-ferent applications that provide similar functionality. An example might be that of two tenants that want to authenticate their users without having them enter their credentials a second time (single-sign-on), based on two different user administration systems. The introduction time for any system connector change is during architecture design time. The binding time can be anywhere after that up to and including during runtime. A practical example is that of photo printing of Google's Picasa, a photo management ap-plication. Photo printing can be done by many different companies and depending on

---

[1] http://www.force.com

Table 1. Customization techniques, their realization techniques, and their latest introduction times

Customization Realization Technique	Latest introduction time	a	b	c	d	e
Model change	Design		✓		✓	
View change	Detailed design		✓		✓	✓
Controller change	Detailed design				✓	✓
System connector change	Architecture design		✓	✓	✓	
System component change	Architecture design	✓		✓	✓	
(a) Infrastructure centered architecture, (b) Run-time variant component specialization, (c) Variant component implementations, (d) Condition on variable, (e) Code fragment super-imposition						

which company suits the needs of the Picasa user, he or she can decide which company (and thus connector) will be most suitable for his or her picture printing needs.

Finally, the fifth type of customization is **system component change**, where similar feature sets are provided by different components, which are selected based on the tenants' requirements. An example is that of a tenant that already developed a solution for part of the solution provided by a multi-tenant web application that the tenant wants to keep using. The component in the web application is, in this example, replaced by the component of the tenant completely. Depending on the level of complexity, further integration might be needed to have the component communicate with the web application, when necessary. Introduction time for system component changes is during architectural design. Binding time can be anywhere between architectural design and runtime. A practical use of the system component change is that of Facebook, a social networking site, which enables an end-user to install components from both Facebook and third parties to gain more functionality. System component changes also include the provision of a tenant-specific database or a tenant-specific model.

Table 1 lists the customization techniques of our model and their latest time of introduction. These customizations have their latest binding time at run-time, since all tenants use the same running instance of the system. The introduction time, however, is during either architecture design or detailed design of the web application. For the model change technique, for instance, it must be taken into account during architecture design that end-users might be able to add to the object model of a web application.

The CRTs presented here provide deeper insight into how tenant-specific customizations can be created in multi-tenant web applications. The list assists developers in that it provides an overview of ways in which customization can be implemented for multitenant systems. These techniques are elaborated and embellished with examples from the two case studies in sections 3 and 4, to further help developers decide what practical problems they will encounter when implementing such customization.

3   Case Study 1: Collaborative Reading for Youngsters

In 2009 a program was launched in a Dutch library that aims to encourage reading in public schools for youngsters between the ages of eight and twelve. The aim within the

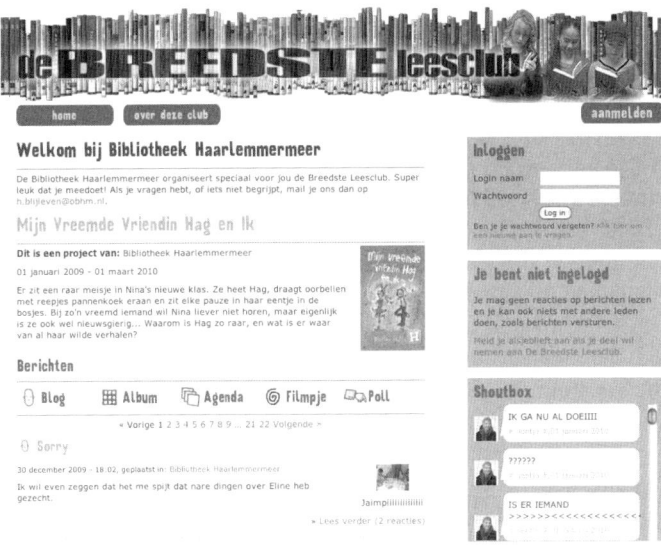

Fig. 1. Screenshot Collaborative Reading Support Information System

program is to have groups of students read one specific book over a short period of time in a so-called reading club. In that time, several events are organized at local libraries, museums, and schools, to further encourage the student to actively participate in the reading process. These events consist of reading contests (who reads the most excitingly and beautifully out loud?), interviews with writers, airings of screenings of the book, poetry contests, and other related activities. The children involved are informed using a custom built content management system. A screenshot is shown in figure 1. The program has been running for approximately one year. The program was launched by one library, which means that other libraries and schools pay the initiating library a fee per participant. Projects can run simultaneously, but also on different moments in time. Tenants are libraries wishing to run a series of reading projects. End-users are members of the library in the age group of 10-12 years old.

3.1 Tenant-Specific Customization

No **model** or **view changes** were made for the Reading Support Information System (RSIS). All end-user and tenant-specific customizations are realized in the RSIS using the **controller change technique**. The controller customizations are complex, however, due to the unique views on the content for each user. For example, the RSIS provides children with a child-specific view of the project in which they participate. The view includes an activity agenda, a content management system on which both participants and organizers can post, a shoutbox (a place where short unfiltered messages can be posted by members on the frontpage), a reading progress indicator, and a project information page. The content management system enables organizers and participants to post polls, pictures, movies, and basic textual blog items. The RSIS also has some basic social network facilities, which enables users to befriend each other and enables users to communicate. The system is not fully localized to one tenant. Participants from two different tenants can see each other's profiles, view each other's blog posts, and become friends using the social networking facilities, to enable two participants from different areas in the country to connect.

Content is also only in part limited to the tenant. Most of the content, such as the local organizer's content items and that of the participants can only be seen (by default) in the local scope. Some of the content, however, should be published nationally, when two projects are running in different locations simultaneously and the content item spans multiple projects. On the other hand, when a local organizer chooses to edit a national item, a new instance of that content item is created that becomes leading within the tenant's scope. The tenant can choose to revert to the previous version, deleting the tenant-specific version.

The changes to the controller are complex and required a lot of testing. During the first pilot projects it was discovered that members could sometimes still see content from other tenants, such as in the shoutbox, where different projects were run. Also, a downside was discovered when doing performance testing: the home page requires several complex queries for every refresh, thereby introducing performance problems. The main upside of multi-tenancy, which is found in having to maintain one deployment across different customers, was almost removed completely by the performance issue. Several solutions were implemented, such as deploying the shoutbox on a dedicated server, and some query optimization. In all encounters of customization, controller changes were implemented using the tenant variable as the configuring condition, i.e., if the end-user belongs to a library, the end-user stays in that particular scope.

No other customizations were necessary: the **model** does not require extensions, the **views** must look uniform for different tenants, there are no **connectors to external components**, and none of the **components** need to be replaced. One could argue that some changes to the view are made since local organizers (libraries) can customize the view offered to participants by adding a tenant specific logo. However, this logo is stored in the database, and the controller determines which logo to show, which by our classification makes it a controller change.

4   Case Study 2: Homework Support System for Schools

In 2007 the proposal for a Homework Support System (HSS) for secondary schools was approved by an innovation fund for public libraries. The idea was to set up a new web service for high school students, where help is provided to perform tasks like show-and-tell speeches, essays and exam works for which bibliographic investigation is required. The platform now forms the linking pin between public libraries, teachers, students, and external bodies such as schoolbook publishers, parents, and homework assistance institutes. The goal of the HSS is to increase the adoption of the public library for homework assignments by the target group of high school students, to provide education institutions with a tool to help them organise their work and save time, and for libraries to remain an attractive place for students to meet and work, both online and on-premise. The role of the public library as a knowledge broker would be, if the homework support system is successful, reaffirmed in the age group of 12–20. This age group is well-known for its disregard for the public library, whereas on the other hand they experience lots of difficulty when searching on the web with commercial search engines [18]. The tenants are libraries and the users are members of the library in the age group 12–20.

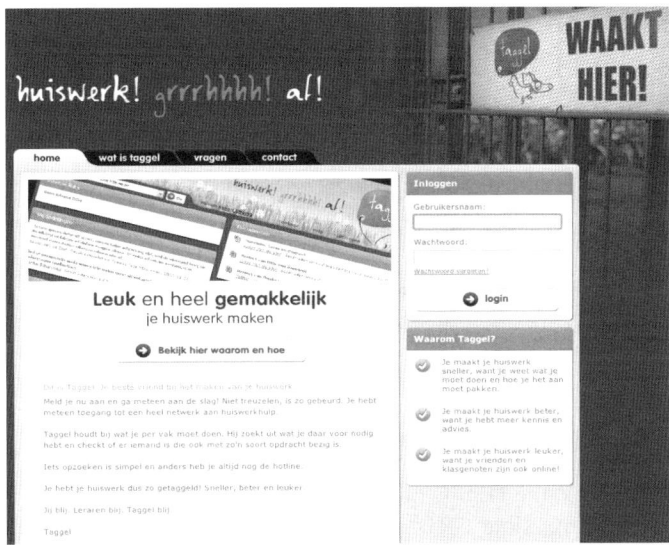

Fig. 2. Screenshot Collaborative Homework Support Information System

The HSS (aptly named Taggel.nl, screenshot in figure 2) that was developed is a web application that helps students find reliable sources of information while doing research for homework projects. The HSS consists of a number of main components being groups (management), a search component, a collaboration component, a knowledge link and rating component, and a portfolio component. Presently, the HSS is running a pilot with three schools and five libraries in one Dutch region. Results have been promising, but many technical challenges are still to be solved.

The HSS consists mostly of commercial and off-the-shelf components. These components are modeled in figure 3. Central to the HSS architecture is the HSS controller, a relatively light-weight navigation component that coordinates each user to the right information. The controller produces views for students and fills these views with data from the database, which is based on the object model of the application. The object model stores students, libraries, and schools as the main entities. When a student logs in through the controller, the associated school and library are determined. Data is synchronized regularly with the Student Administration System (SAS). Different SASs can be used, based on a configuration option that is set by the developer or administrator of the system. The HSS also communicates with the national library administration system. When the system has confirmed that the user is a member of the library, the user gains several extra features, such as automatic book suggestions in the e-learning environment and book reservation.

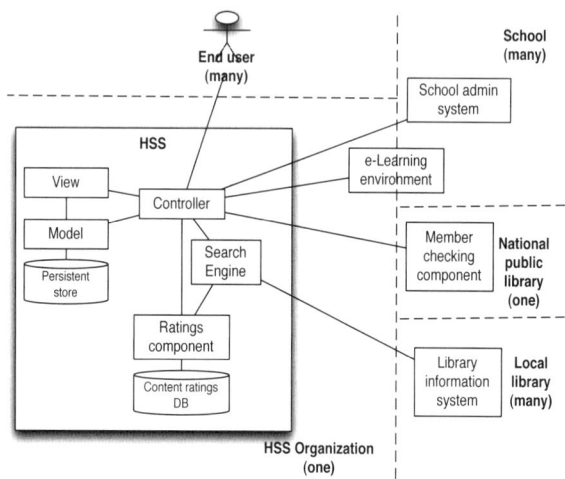

Fig. 3. Component and Component Location Overview of HSS

4.1   Tenant-Specific Customization

With regards to CRTs, the HSS is more interesting than the RSIS, since the HSS employs almost all the possible CRTs. Customizations are triggered by several tenant-
and user-specific properties. Each of the different customization types will be discussed
here, including a rationale and the advantages and disadvantages of using the CRT.

There are no changes to the **models** or **views**, since no custom model extensions
are required (the domain was known at design time) and the system must look uniform
across different tenants.

In regards to the **controller changes**, several customizations are specific to a tenant.
First, end-users have specific scopes in which they are active, such as their school, class-
room, or library. The scope determines what content they can see, in which discussion
groups they can take part, and to which library a homework question is directed. The
scope is determined by either the library or the school of which the student is a member.
This scoping is a typical instrument used in multi-tenancy systems, as was already seen
in the RSIS case study, the main advantage being that different tenants hide data from
each other. Much like in the RSIS case, scoping was complex compared to traditional
multi-tenant web applications, because some content is shared between tenants.

The controller component controls access to Fronter, a commercial e-learning platform. The e-learning platform presently provides a lot of the functionality of the HSS,
and its Application Programming Interface (API) is used to develop other applications
that interact with the e-learning platform. The platform provides the portfolio component, the collaboration component, and the groups functionalities. Furthermore, it
provides a real-time chat functionality, enabling librarians to provide real-time support
to students. If the school already uses Fronter or wants to use another e-learning platform, it can replace the multi-tenant version of Fronter used for HSS. Simple html page
encapsulation is used to capture Fronter pages in the HSS, though more integration can
be and has been implemented using the other CRTs.

In regards to the **interface change**, several CRTs are used for the HSS. First, students have to be able to search in the databases of their local libraries. Secondly, in case the schools already have single sign-on solutions, these solutions can be used to sign on to the HSS, and different component interfaces were built to support single sign-on. The rationale behind both is simple: per tenant different systems need to be approached with different interfaces. The development of the interfaces was relatively simple, since in both cases previous solutions could be reused.

In regards to **component change**, customization is required for schools that already have an e-learning environment to replace the HSS supported environment. The Fronter component is in this case replaced by an already deployed solution on the school's servers. Presently, this tenant specific customization is well supported by the architecture, since it only requires some links to change. However, the true integration between systems in this case is lost. For example, it is presently possible to add any source found in the search engine, such as a book or an online newspaper article, to any content, such as a homework assignment or discussion, in the e-learning environment with one click. When the component is replaced and no new integration is applied, this ceases to work. In the development of the HSS was found that component change introduces complexity on an organizational level, where suddenly schools are confronted with adjustments to their own systems. To enable the schools in solving these problems, an API has been constructed, which enables schools with the same student information system products to exchange and independently maintain proprietary SAS API code.

At present, the system interface changes are built into the HSS rather haphazardly. Customization code is built into HSS (catching specific events) in different places. Mixing system customization and tenant customization code is generally considered bad practice, and will soon be separated into several separate code files.

4.2   Combining Mechanisms for Advanced Search

An example of the combination of two CRTs is found in the search engine. The HSS was designed to provide results ordered by quality and by relevance for a student, depending on the student's context (student level, school programme, source quality, locality). Furthermore, the HSS was designed to obtain these search results from the HSS content database and from a library specific database that contains all books and materials that can be taken out of that local library. It has not been visualized in figure 3, but there are in fact two sources that feed the search results, which are in turn ordered by yet another source with content ratings from librarians and school teachers.

The customization of these search results is taken care of by the search component, which can be considered part of the controller component of the HSS. customization happens on three levels, being (1) which library database to pick, (2) which content to show, and (3) in which order to show that content. The library database is chosen based on the library of which this student is a member, or, if that library does not participate in the HSS project, the library that is closest to the school that participates (it is not uncommon for Dutch schools to have a steady relationship with the nearest local library). The content that is shown from the HSS database is based on the school the student goes to, although most content in the local HSS database presently is public.

The customization mechanisms used are system connector change and controller change. The controller (i.e., search component) does most of the work here, since it has the responsibility to determine how queries are sent to which library catalogue instance (system connector change), how search results are ordered (controller change), and which results are shown (controller change). The variability mechanisms used are conditional variables that are stored in the database, i.e., not all variants were made explicit when the system was started up. Please note that the order is determined by the ratings component, which contains content quality and level ratings from teachers and librarians.

5   Case Study Findings

Table 2 lists the encountered CRTs in the two cases studies. The table also shows which variability realization technique has been employed for realizing the CRT. The 'controller change' CRT is encountered most frequently. This is not surprising, since it is easy to implement (compared to for instance model changes), has little architectural overhead, can be used when the number of variations is implicit or explicit, and the mechanism can be reused over the complete web application. Also, in both cases the model was pre-defined and no further changes were needed to the views (on the contrary, both systems have to present a uniform look and feel). One thing must be noted about controller changes, and that is that controller change code must be tested extensively. This is even more so when the controller shares some content across different tenants, which can introduce privacy problems when one tenant erroneously sees sensitive data from another tenant.

System customizations introduce challenging architectural challenges. In regards to architecture, depending on the direction of the data flow (inbound or outbound to the system) different architectural patterns can be used to arrange communication between the systems. An example of such an architectural pattern is the use of the API of the other system, such as in the case of the single sign-on feature of the HSS, through varying component implementations. The use of such an API requires the developers to know the number of variations a priori. Another example is that of an infrastructure centered architecture that only starts to look for available components at runtime, such as the example that the e-learning component can be replaced dynamically, based on one variable. Since the e-learning component runs in a frame of the HSS, it can easily be replaced. Of course, integration with other functionality of the e-learning system has to be built in again.

Even larger *organizational* challenges are introduced when doing system customizations. Whereas the developers and architects generally have a 'it can be built' attitude, customer organizations are reluctant to add new functionality to their existing systems and these co-development projects (both the school and HSS team, for instance) take far more time than necessary when looking at the actual code that is built in. For example, the e-learning environments of schools are generally hard to adjust, if possible at all. Also, because the interface code that is built needs to be maintained, this introduces considerable increase in total cost of ownership for the tenants. The introduction of generic

Table 2. Customization Realization Techniques used in the Two Cases

Case	Functionality	Customization parameters	Customization realization	Variability realization technique
1	Social networking	Tenant	Controller change	(d)
1	Content views	Tenant	Controller change	(d)
1	Content add	Tenant	Controller change	(d)
1	Calendar	Tenant	Controller change	(d)
1	Shoutbox	Tenant	Controller change	(d)
1	Projects	Tenant, payment, date	Controller change	(d)
2	Discuss with peers	School	Controller change	(d)
2	Access school content	School	Controller change	(b)
2	Search in book and KB	Library, Student level	Controller change	(d)
2	Discuss with librarian	Library	Controller change	(d)
2	Be active in group	Student context	Controller change	(d)
2	Access library database	Library	System interface ch.	(b), (d)
2	Discuss with teacher	School, SAS	System interface ch.	(c), (d)
2	Access SAS for single sign-on	School	System interface ch.	(c), (d)
2	Reuse functionality in e-learning environment	School	System component change	(a), (d)
(a) Infrastructure centered architecture, (b) Run-time variant component specialization, (c) Variant component implementations, (d) Condition on variable, (e) Code fragment super-imposition. SAS = School Administration System, KB = Knowledge Base					

APIs could help here, but in the case of e-learning environments such generic APIs will not be available in the near future. These APIs can be seen as enablers for an implicit number of variations using system interface change, whereas without a generic API one can only have a limited number of components.

Another finding from these cases is that the binding time for each of the customizations determines how much effort needs to be spent on the customization. Generally, the later the binding time, the larger the effort to build the multi-tenant customization in the system. For example, in both cases the model was known a priori and no adjustments had to be made at run-time by end-users. If such functionality had to be built in, both systems would need an end-user interface and a technical solution that enables run-time model changes. In the two cases these were not required and developers could simply make changes to the models at design time. The same holds for the whether there were an implicit or explicit number of variations. When the number of variations is explicit, such as in the case of the single sign-on solution, the variation points are easy to build in. However, when the number of variations is implicit, such as the number of projects that can be run in the RSIS, the code becomes much more complex because it needs to deal with projects running at differing times that share content and also differing projects running at differing times that do not share content.

The evaluation with developers provided relevant observations, which are summarized as follows. First, they indicated that the customization overview works well on a high level when selecting techniques, but when implementing a technique, several combinations of customization and variability realization techniques will be required.

Also, the second case study was mentioned specifically as being a much better example than the first, since the first case study merely provides insight into some of the simplest mechanisms proposed in the framework. The discussions with developers during the evaluation led to changes to the model: (a) an extra variability realization technique ("code fragment imposition") was added for the CRT "controller change" when one of the developers mentioned that he had implemented such a mechanism. The main advantage of using code fragment imposition, using an event catching mechanism, was that the API would function similar to the web application itself. The developers knew of several examples where the 'wrong' variability realization technique had been used to implement a customization technique. The question was posed whether advice can be given whether a certain type of customization is best applied to a specific situation. The developers listed several negative effects of using the wrong customization technique, such as untraceable customization code over several components, unclear behavior of an application when using code fragment super-imposition, and database switches that were spread out over the whole codebase.

6   Conclusions

It is concluded that current variability realization techniques insufficiently describe multi-tenant customization. This paper provides an overview of CRTs in multi-tenant web applications. These CRTs are further elaborated using examples from two innovative web applications from the library sector. The findings assist developers in choosing their own CRTs and helps them avoid problems and complexities found in the two cases. The provided list of customizations is just the beginning: we consider annotating the CRTs with customization experiences as future work. Typical examples of this knowledge are technical implementation details, code fragments, architectural patterns used, and non-functional requirements implications, such as performance and reliability problems.

References

1. Mietzner, R., Metzger, A., Leymann, F., Pohl, K.: Variability modeling to support customization and deployment of multi-tenant-aware software as a service applications. In: ICSE Workshop on Principles of Engineering Service Oriented Systems, pp. 18–25. IEEE Computer Society, Los Alamitos (2009)
2. Svahnberg, M., van Gurp, J., Bosch, J.: A taxonomy of variability realization techniques: Research articles. Software Practice and Experience 35(8), 705–754 (2005)
3. Jaring, M., Bosch, J.: Representing variability in software product lines: A case study. In: Chastek, G.J. (ed.) SPLC 2002. LNCS, vol. 2379, pp. 15–36. Springer, Heidelberg (2002)
4. Rossi, G., Schwabe, D., Guimaraes, R.: Designing personalized web applications. In: WWW '01: Proceedings of the 10th international conference on World Wide Web, pp. 275–284. ACM, New York (2001)
5. Chong, F., Carraro, G.: Architecture strategies for catching the long tail, Microsoft white paper (2006),
   http://msdn.microsoft.com/en-us/architecture/aa479069.aspx

6. Kappel, G., Pröll, B., Retschitzegger, W., Schwinger, W.: Modelling ubiquitous web applications - the wuml approach. In: Arisawa, H., Kambayashi, Y., Kumar, V., Mayr, H.C., Hunt, I. (eds.) ER Workshops 2001. LNCS, vol. 2465, pp. 183–197. Springer, Heidelberg (2002)

7. Lei, Y., Motta, E., Domingue, J.: Design of customized web applications with ontoweaver. In: K-CAP '03: Proceedings of the 2nd international conference on Knowledge capture, pp. 54–61. ACM, New York (2003)

8. Mietzner, R., Unger, T., Titze, R., Leymann, F.: Combining different multi-tenancy patterns in service-oriented applications. In: Enterprise Distributed Object Computing Conference, IEEE International, pp. 131–140. IEEE Computer Society Press, Los Alamitos (2009)

9. Guo, C.J., Sun, W., Huang, Y., Wang, Z.H., Gao, B.: A framework for native multi-tenancy application development and management. In: E-Commerce Technology and the 4th IEEE International Conference on Enterprise Computing, E-Commerce, and E-Services, CEC/EEE 2007, pp. 551–558 (2007)

10. Goy, A., Ardissono, L., Petrone, G.: Personalization in e-commerce applications. In: Brusilovsky, P., Kobsa, A., Nejdl, W. (eds.) Adaptive Web 2007. LNCS, vol. 4321, pp. 485–520. Springer, Heidelberg (2007)

11. Schilit, B., Adams, N., Want, R.: Context-aware computing applications. In: Proceedings of the Workshop on Mobile Computing Systems and Applications, pp. 85–90. IEEE Computer Society, Los Alamitos (1994)

12. Ardissono, L., Goy, A.: Tailoring the interaction with users in web stores. User Modeling and User-Adapted Interaction 10(4), 251–303 (2000)

13. Ardissono, L., Felfernig, A., Friedrich, G., Goy, A., Jannach, D., Petrone, G., Schäfer, R., Zanker, M.: A framework for the development of personalized, distributed web-based configuration systems. AI Mag. 24(3), 93–108 (2003)

14. Fiala, Z., Houben, G.J.: A generic transcoding tool for making web applications adaptive. In: CAiSE Short Paper Proceedings (2005)

15. Jansen, S., Brinkkemper, S.: Applied Multi-Case Research in a Mixed-Method Research Project: Customer Configuration Updating Improvement. In: Steel, A.C., Hakim, L.A. (eds.) Information Systems Research Methods, Epistemology and Applications (2008)

16. Yin, R.K.: Case Study Research - Design and Methods, 3rd edn. SAGE Publications, Thousand Oaks (2003)

17. Reenskaug, T.: Models, views, controllers, Xerox PARC technical note (December 1979)

18. Fidel, R., Davies, R.K., Douglas, M.H., Holder, J., Hopkins, C.J., Kushner, E.: A visit to the information mall: Web searching behavior of high school students. Journal of the American Society for Information Science (1), 24–37 (1999)

Challenges and Experiences in Deploying Enterprise Crowdsourcing Service

Maja Vukovic[1], Jim Laredo[1], and Sriram Rajagopal[2]

[1] IBM T.J. Watson Reserach Center, 19 Skyline Dr, Hawthorne, NY 10532, USA
{maja,laredoj}@us.ibm.com
[2] IBM India, Chennai, India
srirraja@in.ibm.com

Abstract. The value of crowdsourcing, arising from an instant access to a scalable expert network on-line, has been demonstrated by many success stories, such as GoldCorp, Netflix, and TopCoder. For enterprises, crowdsourcing promises significant cost-savings, quicker task completion times, and formation of expert communities (both within and outside the enterprise). Many aspects of the vision of enterprise crowdsourcing are under vigorous refinement. The reasons for this lack of progress, beyond the isolated and purpose-specific crowdsourcing efforts, are manifold. In this paper, we present our experience in deploying an enterprise crowdsourcing service in the IT Inventory Management domain. We focus on the technical and sociological challenges of creating enterprise crowdsourcing service that are general-purpose, and that extend beyond mere specific-purpose, run-once prototypes. Such systems are deployed to the extent that they become an integrated part of business processes. Only when such degree of integration is achieved, the enteprises can fully adopt crowdsourcing and reap its benefits. We discuss the challenges in creating and deploying the enterprise crowdsourcing platform, and articulate current technical, governance and sociological issues towards defining a research agenda.

Keywords: Enterprise crowdsourcing, Governance, Crowdsourcing Process.

1 Introduction

With the realization of Web 2.0, the trend of harnessing large crowds of users for mass data collection [1] and problem-solving [2], has become a growth industry employing over 2 million knowledge workers, contributing over half a billion dollars to the digital economy. Crowdsourcing is nowadays being employed in many domains, ranging from advanced pharmaceutical research [3] to T-shirt design [4].

Two types of methods have been employed to catalyze the crowd participation. Firstly, the more traditional ones, rely on motivating the participants to share information and thereby either gain creditability or get reciprocal information [1, 5]. Secondly, many systems provide tangible incentives to crowd participants for their contributions (e.g., monetary prizes) [3,4,5,6].

B. Benatallah et al. (Eds.): ICWE 2010, LNCS 6189, pp. 460–467, 2010.
© Springer-Verlag Berlin Heidelberg 2010

Tapscott and Williams [7] discuss how businesses can harness collective capability of outside experts to facilitate innovation, growth, and success. In contrast, our research investigates applicability of crowdsourcing methodology within the enterprise, thereby engaging internal networks of knowledge experts.

Section 2 describes the IT Inventory Management problem to which we have applied enterprise crowdsourcing service. Section 3 discusses our experience in developing and deploying an enterprise crowdsourcing service. We present challenges in building a general-purpose service, which integrates with the existing business processes, and the necessary development support to facilitate shorter turn-arounds in customizing new use cases. Finally, we discuss incentives and governance issues. Section 4 outlines a set of challenges for the true realization of enterprise crowdsourcing.

2   Use Case: Crowdsourcing for IT Inventory Management

IT inventory management captures and manages the enterprises' IT assets in numerous repositories, which are often outdated and incomplete. Thus they often fail to provide consolidated, global, views of the physical infrastructure and actionable data (e.g. which business applications would be affected if a specific data center is consolidated). However, this information can be found in the core of the organization - the knowledge workers that understand and drive the business and the IT itself. Yet, there is little transparency on who knows what within an enterprise. Locating such critical business information becomes intractable, especially as the desired knowledge is transferred between experts when they transition within the organization.

To address this IT inventory management challenge within a large enterprise, we used crowdsourcing ("wisdom or crowds") approach to discover, integrate and manage the knowledge about the physical infrastructure that hosts *business applications*. An example of such a business application would be a company's support website, which resides on a number of servers. As such, business application is differentiated from the actual middleware that may support such as a web server or messaging queue, as those can be discovered by utilizing advanced scripts.

The goal of our crowdsourcing service was to find out and verify the information about 4500 business applications. We have idenitified initial enterprise crowd from the existing business application registry. Before sending of the e-mails with task requests, we verified that the targeted crowd was reachable and still in the same role within the enterprise. Where possible, we used 'delegate' application owners. Additional feature of our crowdsourcing service was the capability to re-assign tasks to other team members.This was useful in the scenarios where a) different team members possessed partial knowledge about the task and b) business application owner was no longer responsible for the given application.

For each business application contributors were asked to:

1) Verify the application ownership
2) Provide compliance information (e.g. is the application subject to ITAR?)
3) Identify servers that are hosting it (e.g. enter fully qualified hostname) and their type (e.g. production, development, testing, etc.).

There were three exceptional scenarios that were also considered: 1) application has already been sunset (it's no longer running on the infrastructure), 2) application is hosted on a 3rd-party server and 3) application is not hosted on any server (e.g. application could be a spreadsheet file on a destkop).

Using the enterprise crowdsourcing service, we harnessed an expert network of 2500 application owners to execute the IT Inventory Management exercise - gathering information on the mapping of 4500 business applications to more than 14,000 of IT Systems (servers). Crowdsourcing service has achieved 30X improvement in process efficiency, in contrast to the traditional approach of employing two full time experts to manually reach out to the application owners and gather the information. This process is rather time consuming, and typically would result in 100-200 applications being captured over the two months time. Furthermore, the powerful knowledge network, generated as a result of crowdsourcing run, can be situationally engaged for other large-scale business and IT transformation initiatives, such as cloud transformation.

3   Experience in Enteprise Crowdsourcing Service

Not all the crowdsourcing use cases within the enterprise are the same. Inherently they impose a set of requirements on the actual crowdsourcing process. For example, is the task to be initially sent to one or more experts? Can the task be separated in a number of (concurrent) subtasks? What support for task sequencing needs to be supported? Can the user reassign the task or parts of the task to other users? How do we create tasks, what sort of existing business repositories are available and can serve to initiate the crowdsourcing tasks? Finally, how does one go about validating the collected data? In this section, we discuss the crowdsourcing process, its elements, and key requirements for the development support to enable efficient deployment and customization of new use cases.

3.1 Task Management and Crowdsourcing Process

When designing our crowdsourcing service we have been striving towards an approach that allows drawing best practices and eventually leading us to build a self service system where we can reuse our key elements. The use cases we have identified are of a knowledge seeking nature, such as the IT Inventory Management one that we describe in this work. We build a business process that facilitates the capture of knowledge, usually around a business object (e.g. a business application or a server).

We look at the business process as a sequence of steps where people either contribute knowledge to the business object or make decisions upon the knowledge that has been captured, such as validation, request for more information, or simply complete the step in the process. A step in the process we define as a task.

In its purest form, crowdsourcing tasks should be available for everyone to apply and attempt to complete them. In an enterprise environment we have the benefit of other intrinsic knowledge as part of the organization that allows us to pre-assign tasks to initiate the process. We call this the seeding process. It uses some information about the business object, such as prior, current ownership, or lead process owner, for example. Once the tasks are assigned, task owners may provide requested knowledge, segment the work and refer segments to different parties, or simply forward to someone else to take care of the task. One concept under consideration is the use of optional fields, if the capture process allows it, we may want to capture information as quick as possible, yet not delay key steps in the process when prior ones have been completed. By using optional fields we can expose and attempt to gather other information, yet if it is not captured by the time the mandatory information is captured, we may proceed to the next step.

Task ownership allows for other support services to accelerate the completion of the tasks, such as the use of reminders and escalations. Once a task has been referred to, there is the possibility of a delay due to lack of attention or misdirection. It is important to raise awareness and force an action as soon as possible. In our studies we were able to complete 50% of the tasks in 4 days [8], and with use of reminders and escalations we were able to keep the process alive and close almost 90% of the tasks in the following 3 weeks.

The lessons learned have helped us improve the task management process, and we are driving to a design that allows us to define any business object and apply the task management on top of it.

As tasks are completed, our system captures who is modifying each field of the business object and completing tasks. This information allows us to build an audit trail of all changes. This information creates user community around the business artifact as a by product of the process. This user community has many dimensions, It could be around a particular business object, or around the skills required to complete the task. The community can be invited or a member can be referred to help on an open task. The community usually outlives the process, and can be invoked at a later stage when rerunning the business case or for new extensions of the process.

Finally, crowdsourcing process requires certain governence support. Various use cases, may call for the admistrative and mediator roles, to provide capabilities such as task cancellations, task management or resolution of conflicts (e.g. esp. on the marketplaces where task requestor offers monetary rewards).

3.2 Implementation of the Enterprise Crowdsourcing Service

We have build our enterprise crowdsourcing service as a Web-application, designed to support knowledge acquisition activity. It enables task creation based on existing business artifacts, such as a task for the existing application in the business application registry (e.g. task for the "Support Website"). Tasks are either assigned to single or multiple users (based on ownership information available in the business artifacts). The task list for each user is displayed in the home page upon logging into the crowdsourcing service. When tasks can be worked upon by multiple users concurrently, an application level lock is introduced to enable concurrency. Users are notified whenever the task they are working on is concurrently modified. This helps multiple users collaborate on a single task in a transparent manner. Users are also presented with the modification history of artifacts enabling them to know who changed information and what changes were made.

Our crowdsourcing service relies on the Enterprise Directory (that is, a list of employeed, their contact details, and organizational structure) for user authentication and for obtaining the user profile information, such as such as first name, last name, location and manager's details. It further utilizes Enterprise Directory to enable quicker and easier lookup of employee names and e-mail IDs when re-assigning tasks. Business artifacts are sourced from existing internal repositories, such as server and business application registries. Enterprise crowdsourcing service relies heavily on the recency of data in such applications as tasks are assigned to users based on the same. Wherever required, it validates such data against the enterprise dricetory to confirm validity.

Enterprise crwodsourcing service also features reminder capability, to send out e-mails to users with tasks pending for more than a predetermined number of days in their task list. The reminder schedule can be customized by the administrator.

The application also enables users refer part of or the entire task to other users. Preferred fields of the task can be selected before sending a referral and these are highlighted when the referred user works on the task. Auto complete of referral emails in the email selection box is enabled for easy and quick user lookup against the enterprise directory.

Administrative users are provided with the capability to manage tasks by enabling them to cancel, complete or reassign open tasks; reopen canceled and completed tasks and editing task information without changing ownership. Administrators can also view reports and logs of the modification history of the artifacts.

3.3   Development Support for Building Crowdsourcing Solutions

As a crowdsourcing application needs to manage tasks in a context different from that of traditional workflow systems – multiple users collaborate on a single task with the ability to forward part of or the whole task to one or more users. Hence, the persistence layer requires a framework or tool that provides transaction isolation and transaction transparency – each user working on a task needs to know what information has been contributed/ modified about the task by other users and who did the same. Transaction transparency can be achieved on the persistence layer by alerting the user carrying out modifications about any intermediate transactions. Since tasks go through various stages of its lifecycle, development would be easier if the persistence layer also supports state-modeled artifacts.

The data layer should provide efficient techniques to cache and retrieve look-up and drop down list data that are used frequently. This is especially critical when the crowdsourcing platform caters to capturing information about physical assets in the organization which typically number hundreds of thousands in big organizations. Also critical is the efficiency of generating reports based on various criteria as the data size can run up considerably. The servers hosting the application should be clustered to achieve load balancing and failover.

The application also requires access to the enterprise's employee look up database to validate if tasks are forwarded to valid users and for ease of lookup of employee details such as email id, name and location. Access to details such as the project of the task owner would be helpful in cases where no response is obtained from the owner so that other members in the project may be assigned the task. Integration with other organizational systems also facilitates in developing a good crowd sourcing system by providing data on assets, language preferences of users, whether the user is still active within the organization or is on short/long leave during task creation etc.

The application also requires a robust email service which can handle the load required by the system. The email service should provide templating capabilities to send out personalized and customized email messages. Some organizations have restrictions on sending mass mailers to employees within the organization. The support of management is essential as crowdsourcing systems usually involve sending out high numbers of emails to employees – by way of initial information seeking emails, referral emails and reminders for pending tasks.

3.4   Incentives and Governance

The success of any collaborative endeavor, both within an organization or when an outside community is engaged, heavily depends on the incentive mechanisms put in place to achieve the desired outcome.   Numerous incentive mechanisms exist and make collaborative production challenging [9]. Traditional award schemas are presently employed by enterprises, e.g. salary, performance bonuses. Incentives for crowdsourcing raise new legal challenges (e.g. compliance and taxation).

Types of Incentives
When building a crowdsourcing service, there are two types of incentives that designers need to consider. Firstly, the challenges is how to attract new members? Secondly, and more importantly, the question is how to encourage the contributions and sustain the community of the time?

Effective integration of new members is critical factor to successfully building an online community. Challenges with attracting and enabling contributions of new members can be grouped in the following categories [10,11,12]:

1. New members tend to be less committed
2. They need to learn the norms of the behavior
3. Need to understand the structure of content and engagement.

To encourage contribution in a crowdsourcing exercise, one needs to reduce any entry barriers, and provide clearly identified goals. A critical factor for participants, aside from the clearly identifiable goal and skillset, is that benefits outweught the cost of participating. Cosely et al. [13], present an intelligent task routing mechanism to increase contribution by targeting contributors'efforts to where they are more needed, and based on their expertise.

Incentives can further be grouped into material and non-material (social) ones. Examples of material incentives, include monetary, such as the ones employed by Mechanical Turk and Innocentive. In the same group, are (material) prizes, which are utilized by TopCoder and Netflix for example, to award the best performing members of the crowd.

Our experience

The deployed crowdsourcing application included capabilities that allowed participants to collect virtual points for crowdsourcing task, as well as for any successful referrals. The points, however, at this stage were not exchangeable for a tangible, material award. The users had the ability to see their rating, compared to other participants. Finally, many users considered access to this consolidated repository, which was a result of the crowdsourcing, to be an incentive on its own.

When introducing an incentive for a crowdsourcing task within an enterprise, a number of questions are immediatelly raised: how would incentives affect employees of different status (contractor vs. full-time), or similarly how does one engage a service-type employee (with a billable utilization rate), as opposed to the traditional employee with a flat-rate compensation. Following is a set of questions that form a research topic within the enterprise crowdsourcing domain:

1. How are compliance, taxation and labor laws addressed as crowdsourcing is being adopted?
2. How do all of the above apply to a global enterprise that coexists in many national/regional jurisdictions?
3. How do we give incentives in a global company across borders?
4. How does a company view employees doing work (e.g. crowdsourcing tasks) that is not their "day job"?

4   Challenges

Deploying enterprise crowdsourcing service provides an insight into its design and use, but deploying systems beyond the limited existing, specific- prototypes will require significant progress toward integration with the existing business processes. We consider major challenges that must overcome before achieving this goal. This list is by no means exhaustive; it provides a set of open questions that define the requirements on components of enteprise crowdsourcing services.

1. Process: How does crowdsourcing become an extension of the existing business process?
2. Technical: How can we rapidly develop and deploy new use cases for enterprise crowdsourcing?
3. Sociological and governance: What are the effective incentives that can be deployed within the global enterprise, and what implications do they have on the existing compliance, tax and labor laws?

References

1. Olleros, F.X.: Learning to Trust the Crowd: Some Lessons from Wikipedia. In: Proceedings of the 2008 International MCETECH Conference on E-Technologies (2008)
2. Brabham, D.C.: Crowdsourcing As A Model For Problem Solving: An Introduction And Cases. Convergence: The International Journal of Research into New Media Technologies 14(1), 75–90 (2008)

 3. http://www.innocentive.com
 4. http://www.threadless.com
 5. Bobrow, D.G., Whalen, J.: Community Knowledge Sharing in Practice: The Eureka Story. Journal of the Society of Organizational Learning and MIT Press 4(2) (Winter 2002)
 6. Kittur, A., Chi, E.H., Suh, B.: Crowdsourcing user studies with Mechanical Turk. In: CHI '08 (2008)
 7. Don, T., Wikinomics, W.A.D.: How Mass Collaboration Changes Everything. Portfolio Hardcover (December 2006)
 8. Vukovic, M., Lopez, M., Laredo, J.: People cloud for globally integrated enterprise. In: The 7th ICSOC. 1st International Workshop on SOA, Globalization, People, & Work, SG-PAW 2009 (2009)
 9. Bartlett Christopher, A., Ghoshal, S.: Transnational management: text, cases, and readings in cross-border management, 3rd edn. Irwin/McGraw Hill, Boston (2000)
10. Kraut, Burke, Riedl, van Mosh: Dealing with newcomers. Working paper 12/7/07 (2007)
11. Arguello, J., Butler, B.S., Joyce, L., Kraut, R., Ling, K.S., Ros, C.P., et al.: Talk to me: Foundations for successful individual-group interactions in online communities. In: CHI 2006: Proceedings of the ACM Conference on Human-Factors in Computing Systems, pp. 959–968. ACM Press, New York (2006)
12. Bryant, S.L., Forte, A., Bruckman, A.: Becoming Wikipedian:Transformation of a Participation in a Collaborative Online Encyclopedia. In: Proceedings, GROUP '05, Sanibel Island, Florida, November 6-9 (2005)
13. Cosley, D., Frankowski, D., Terveen, L., Riedl, J.: Suggestbot: Using intelligent task routing to help people find work in wikipedia. In: Proceedings of the 12th ACM international conference on intelligent user interfaces, ACM Press, New York (2007)

Business Conversation Manager: Facilitating People Interactions in Outsourcing Service Engagements

Hamid R. Motahari-Nezhad, Sven Graupner, and Sharad Singhal

Hewlett Packard Labs
Palo Alto, USA
{hamid.motahari,sven.graupner,sharad.singhal}@hp.com

Abstract. People involved in outsourcing services work through *collaboration, conversations and ad-hoc* activities and often follow guidelines that are described in best practice frameworks. There are two main issues hindering the efficient support of best practice frameworks in outsourcing services: lack of *visibility* into how the work is done that prevents *repeatability*, and conducting best practice processes that are *ad-hoc* and *dynamically defined and refined*. In this paper, we present Business Conversation Manager (BCM) that enables and drives business conversations among people around best practice processes. It supports the dynamic definition and refinement of a process in a collaborative and flexible manner. The ad-hoc processes are backed with a semi-formal process model that maintains the model of interactions and an execution engine. We present the implementation of a prototype BCM and its application in outsourcing services. It supports making processes from best practices among people more transparent, repeatable and traceable.

Keywords: Ad-hoc Business Processes, Collaborations, Outsourcing Services.

1 Introduction

Outsourcing services are offered through organizations of people [1]. The service design and delivery entails several lifecycle phases, in which business artifacts are generated and transferred from one phase to another. People involved in service delivery often work through collaboration, conversations and ad-hoc activities. Currently, it remains hard to provide *visibility* on how a service is created and delivered across all lifecycles and *track* it as information about work results is scattered across many systems such as document repositories, project management systems, and emails. This leads to efficiency issues in the process, timing and economics of service delivery.

There is a push towards facilitating and streamlining processes followed in delivering outsourcing services so that they are delivered in a repeatable, traceable and cost efficient manner. While ad-hoc, the interactions among people in service delivery are not random: they often follow some high-level process flow described in process frameworks (such as ITIL [2]) or in best practice documents in repositories. For instance, the Transformation Services Catalog (TSC) from HP Enterprise Services provides collateral templates (e.g., sales brochure, deployment guide, precedence

B. Benatallah et al. (Eds.): ICWE 2010, LNCS 6189, pp. 468–481, 2010.

diagram, etc) for different lifecycle stages of several services. The templates are used as guidelines for people engaged in advertising, designing and delivering services.

While the main activities conducted in delivering services are often given in templates or defined by project participants, not all of these activities are known in advance nor are they fixed: new activities may be introduced or existing ones in the template not needed in a given engagement. Let us refer to the interaction among a group of people to discuss and work on a business problem as a *business conversation* (or conversation for short). The workplan of a given conversation should be able to be *adapted* in an *ad-hoc* and *flexible* manner, often through brainstorming and *collaboration* between people engaged in the service delivery. The activities within a conversation are often inter-related, so that some may need to be done before others. In addition, different conversations in an engagement are also related. Therefore, there is a need for capturing *relationships* and *dependencies* of activities and conversations. Addressing these problems could make outsourcing service delivery significantly more cost efficient and scalable.

Current workflow management systems do not support such ad-hoc interactions among people as often workflow systems need a well-structured and rigid definition of processes ahead of execution time [15, 16, 21]. Document management systems [18] such as Microsoft Sharepoint are passive repositories of documents and tasks and do not drive interactions among people. Collaboration tools simplify the communication between people and creating and sharing content in a collaborative manner, however, they are unaware of the work context.

This paper presents a system called *business conversation manager* that supports the *guided* interaction of people in a business context in a *flexible, adaptive* and *collaborative* manner. It is capable of establishing business conversations among a number of people (and from a template), drive work between people, and enable them to conduct and adapt the workplan collaboratively (*as they do the work*). The system does not necessarily need a starting template and can be used to start and drive ad-hoc business conversations among an agile-formed group of people. It builds on top of existing document management systems and collaboration systems. We envision that users would use the business conversation manager besides and in integration with productivity tools such as email, MS SharePoint and communication tools.

The conversation manager is backed with an engine that builds and maintains a formal model of the workplan in the form of a task dependency graph that allows nested task modeling, and enables automatic work allocation, progress monitoring and dependency checking through the analysis of the underlying graph model. The conversation manager is based on a minimal number of concepts. The core concept is *business conversation* which is a conceptual container for the interactions among people in order to achieve a business goal. It includes participants (*person* and *role*), *documents* that are consumed (input) or produced (output) in the conversation as well as a *workplan* for achieving the goal. A workplan consists of a set of *tasks* and their *dependency* relationships. Participants of a business conversation can use a number of communication channels (e.g., chat, web-based dialog mechanisms and email) to interact with each other and update the system on the progress of tasks.

The rest of paper is structured as follows. Section 2 presents the definition of concepts introduced in business conversation manager. Section 3 presents the functionalities of business conversation manager from a system point of view. In

Section 4, we present the architecture and implementation of the business conversation manager. We discuss related work in Section 5. Finally, we conclude and present areas of future work in Section 6.

2   Business Conversation Manager: Concepts and Design

2.1   Characteristics of Business Processes for People Services

People services are offered through collaborative work of a group of people. These services have a lifecycle that includes several phases such as inception, design, delivery and operation. Several business processes are involved in each lifecycle phase. These business processes are often described in best practice documents provided either by the vendor corporation or coming from standards such as ITIL (IT Infrastructure Library) [2]. By their very nature, these business processes do not include precise workflows and a strict definition of what activities and in which specific order they have to be executed. Rather, the processes define high level descriptions of activities, some of which are optional in service engagements, as well as a coarse-grained description of constraints on the ordering of activities.

The main characteristics of these processes include: they are (i) *non-structured,* i.e., no strict or formal definition of the process exists, (ii) *non-deterministic*, i.e., the execution order of the activities is not well specified, and the order may change in different engagements, (iii) *adaptive*, i.e., the identified activities may be updated in engagements and at runtime, some may be skipped and new ones added. In general, there is no separation between definition and execution phases of the process, (iv) *templated or ad-hoc*, i.e., there may exist templates for such processes that suggest an initial set of activities, however, such processes may be defined at runtime by the people in an ad-hoc fashion, and (v) *collaborative*, i.e., both the definition as well as execution of the process may be performed in a collaborative manner between involved people.

There are studies showing that capturing definition of processes in a complete and accurate way is often not practical [3] due to sometimes incorrect and incomplete information from various sources. Having adaptive and flexible mechanisms to define processes enables updating the process models at runtime to account for such incompleteness as well as in-accuracy. We refer to such semi-structured processes as *people processes* in the context of designing and delivering people services.

2.2   People Processes: Basic Concepts

We define a minimal number of concepts for people processes informally described in the following. The core concept is *business conversation* (conversation, for short) which is a conceptual container for the interactions among people in order to achieve a business goal. A conversation includes a number of participants that are either real persons or roles, and a set of documents that are consumed (input) or produced (output) in the conversation. Finally, a conversation has a workplan for achieving the goal.

A workplan consists of a set of *tasks*. A task is defined by the set of its input documents and output documents. A task can have a state of "new", "assigned", "pooled" (can be picked up by one of participants), "enabled" (ready to be performed), "started", "completed" and "in-active". Tasks can be composed of other tasks in a hierarchical manner. Therefore, tasks can be either "composite" (having subtasks) or atomic. Atomic tasks are executable, i.e., a participant may perform it and it is completed, however, composite tasks are abstract and their completion requires the completion of all its abstract and composite subtasks.

Tasks may have dependencies on one another. We define the *dependency relationship* between two tasks as their data dependency. That is an input document of dependent tasks may be produced by the depended task. The data dependency between tasks is used to draw implicit control flow dependencies between tasks. This frees users from explicit identification of control flows. Indeed, this is an important feature that minimizes the amount of information needed from users to specify the process. It comes from the lesson learned from the observation that most users involved in a project in outsourcing deals do not like the burden of specifying a control flow model (in other words a process model), but rather they are more concerned about their own role and function in the process.

There are two types of task dependencies: "start" and "completed". In the "start" type, the dependent task cannot start until the depended task is started, as well. The "completed" type specifies that the dependent task can start only if the depended task is completed. We define the concept of communication channel to represent a mechanism (email, chat, web-based dialog systems) through which participants of a business conversation interact with each other and the system to perform work and report on the progress of tasks. Similar to tasks, conversations may have dependencies of the same types.

Finally, a participant (a role or a person) can take one of four levels of involvement in a task: "Responsible", "Accountable", "Consulted" and "Informed" (similar to RACI chart [4] for assigning roles and responsibilities in a project plan). The role/person that is responsible should perform the task, while the accountable role/person is ultimately required to make sure of good performance of the task. People with consulted role are those who can be approached for brainstorming or information, and finally, people with informed role have an interest to be informed of the progress and the result of performing the task. Note that not all these roles have to be assigned for a given task, but any task should have a role/person assigned as responsible (which in this case is by accountable as well).

2.3 Towards a Formal Model for People Processes

In this section, we formally define the notion of people processes starting by the concept of "business conversation".

Definition 1 (Business Conversation).   A business conversation c is a triple $c =< P, D, W >$ in which P is the set of participants, D is the set of documents manipulated in the conversation and W is the workplan.

The participants $P = \{p \mid p \in R \vee p \in M\}$ where R is the set of roles and M is the set of people in the enterprise. D is the set of documents that are either consumed or generated in the conversation, and typically stored in document repositories such as MS SharePoint. We define a workplan W as follows:

Definition 2 (Workplan). A workplan W is a hierarchical directed graph represented with tuple $W = \langle T, X \rangle$ where T is the set of tasks (nodes in W), and $X \subseteq T \times T$ is the set of transitions. A task $t \in T$ is defined with the tuple $t = \langle I, O, s \rangle$ in which $I, O \subseteq D$ is the set of input (outputs) respectively, and $s \in \{new, assigned, pooled, enabled, started, completed, in\text{-}active\}$ is its status. A transition x is represented as tuple $\langle t_1, t_2, q \rangle \in X$ meaning that the execution of task t_2 depends on that of t_1 with the dependency type $q \in \{start, completion\}$. If $q = "start"$ then t_2 is not enabled unless t_1 is started, and if $q = "completion"$ then t_2 is not enabled until t_1 is completed. A task can be "composite" or "atomic". To a composite task t a child workplan W' is associated.

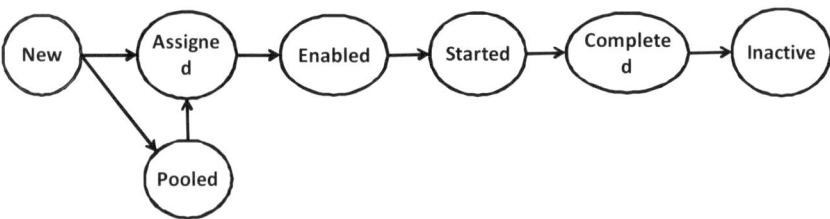

Fig. 1. The lifecycle of an atomic task

Figure 1 shows the lifecycle of an atomic task, i.e., a task that is performed by a participant. A composite task observes only some of these states, i.e., "new", "enabled", "started", "completed" and "in-active". A composite task is "enabled" when all its dependencies are resolved (started or completed respectively to the type of each dependency). It is "started" when at least one of its atomic sub-tasks (in any level) is started, and becomes "completed" when all its sub-tasks are completed. Note that the progression (the status) of a task is updated and for composite tasks can be over-written by participants.

2.4   Execution Semantics of People Processes

Conceptually, we map the workplan model of a business conversation into a colored, hierarchical Petri net [5] and therefore we adopt the execution semantics of such a model for a workplan model. Mapping to the concepts of Petri-net, tokens could be used to represent documents flowing in a business conversation. In a colored Petri net, tokens can take values that are defined by a simple or complex type (historically the value of a token is referred as its color). Workplans allow hierarchical definition to enable having multiple levels of abstraction in the process model. This simplifies the work for different participants such as managers, workers, etc and allows gradual development of the plan with more details concerning specific higher-level tasks. In order to model this aspect of workplans, the semantics of hierarchical Petri nets (HP-Net) [6] are adopted.

Note that this mapping is conceptual in the sense that we incorporate these semantics in our execution engine. Adopting this model allows capturing and reacting to events related to the status update or progression of tasks and document exchanges by humans in a business conversation. Note that the dependency graph of a workplan may form a set of disconnected sub-graphs. Each connected component in this model is related to a set of dependent tasks. When evaluating the dependency model from the execution semantic perspective, we can form a single HCP-net by creating a fake initial place and transitions through which all sub-graphs are connected to form the overall HCP-net representing the whole workplan.

2.5 Templates

In many domains and particularly in services outsourcing context, there are corporate repositories offering templates of documents that are used in the engagements as well as activities that have to be followed in each stage. In some areas such as IT service management, there are standards such as ITIL [2] that describe such processes and activities at a high level. In our previous work (reported in [7]), we have taken an approach to formalize part of these processes as templates that are used to initiate working processes among people based on ITIL processes. Process templates are described as RDF graphs that are refined: more details are added to them as the work progresses among people.

In this paper, we build on our previous work for encoding and formalizing processes in the context of outsourcing engagements as templates. In particular, we have taken collateral templates (high level description) for different lifecycle stages of service delivery from HP TSC catalog. We formalize them as templates and capture the knowledge in those processes as RDF graphs. These templates are made available in the system to be used by participants (e.g., managers) in a business conversation as the initial workplan that could be tailored for a specific engagement.

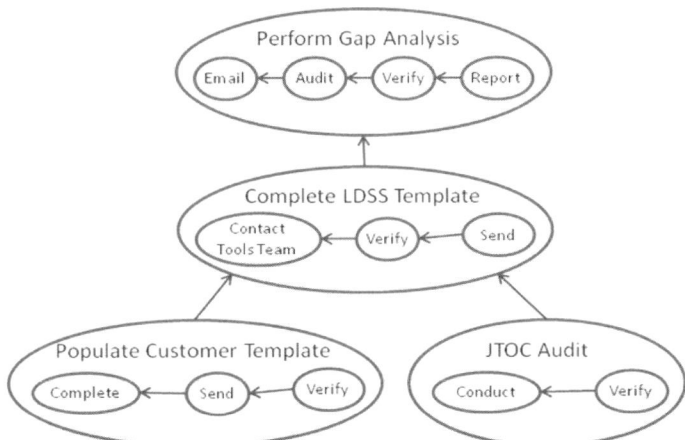

Fig. 2. Part of the hierarchical task dependency model for a workplan

For example, Figure 2 shows part of the (high-level) hierarchical dependency graph for the workplan of an "assessment" business conversation in the context of an incident management process. The ovals represent tasks and the links between them the dependencies between them. For details on how this information is encoded in RDF, please refer to [7].

3   Business Conversation Manager

In this section, we describe our system called "business conversation manager" (BCM) for the establishment, management as well as adaptation of business conversations among people.

3.1   Establishing Conversations and Implicit Dependency Model Management

While business conversations in BCM are supported with formal models, it is very important to note that these models are not exposed to users in BCM. Indeed, BCM creates and maintains the task dependency model of the workplan in the backend automatically. Therefore, participants do not work with the workplan model explicitly. Participants are concerned with the definition and progression of individual tasks and if a given task maintains dependency relationships with others. This is consistent with the nature of the job that they are doing. The intention is to introduce the least amount of overhead of the automation tool for participants.

   One of the main features of BCM is that it allows gradual and level-wise definition of the model by different participants with different expertise and levels of knowledge about the process. Therefore, there is no need for apriori definition of the whole process. The participants can start with a high-level and incomplete definition of the tasks. The hierarchical feature of the model allows the participants to refine abstract tasks into more finer grained ones in a level-wise fashion by people who are responsible for the next level of detail. Another important feature of this system is that there is no separation between the definition and the execution of the model. The process is considered executing right from the time that the highest level tasks are defined. Therefore, the runtime and design of the process are inter-leaved.

   In identifying the dependencies between the tasks, BCM does not mandate identifying which specific input document(s) of the dependent task depends on that (those) of a depended task. This is to provide more flexibility for participants. Instead, BCM establishes this relationship after the participants of the dependent tasks pull and use documents produced (or manipulated) during the performance of the depended task. Internally, this is managed in BCM by tagging documents as "input", "manipulated" or "output" for documents that are used or produced in the context of a task. However, the dependency between two tasks must be explicitly specified. It should be noted that dependency is specifiable between tasks at the same level in the hierarchy.

3.2 Management and Adaptation of Conversation Workplans

People processes are often ad-hoc and require adaptation, especially if the workplan of a conversation is defined according to a template. BCM provides the following facilities for adapting definition of processes.

Adding and updating a task: This method allows the definition of a task in a workplan. For each task, properties such as *start-date*, *due-date*, *end-date*, *status*, *actorIDs*, *documents*, *dependsOn*, *parent, type* and *method* can be specified. ActorIDs provides the list of the participants involved with their role(s) in accomplishing the task. A task can be one of atomic or composite types, and can be accomplished using one of two methods: "human" or "automated". By default tasks are human-offered. If the method of a task is "automated", the API interface details are needed and an adapter to call the Web service at the runtime is generated. "dependsOn" takes the list of other tasks that the current task depends on. A task may not have any dependencies.

The parent property takes as value the workplan or another composite task if it is its immediate child. A composite task is a place-holder for a set of other tasks that are not yet known. The update method enables updating various properties of a task at runtime.

Remove a task: This method introduces a "consensus-based removal" approach for removing tasks in a collaborative manner between conversation participants. When the removal request for a task is made by a participant, BCM triggers an event so that a notification message is sent to all participants with "accountable" and "responsible" roles from the list of actorIDs of this task and all the tasks that depend on it directly to inform them of the request for its removal. The message asks them to react if they object to the removal of the task. If nobody objects to its removal within a pre-specified timeline, it is removed from the workplan. Upon removal, the dependency list of tasks that depends on the removed task is updated. The status of the removed task is set to "in-active" meaning that it is not part of on-going conversation. It is not shown in the list of tasks of the workplan but is maintained in the back-end repository for historical reasons, as well, it can be restored to its last state before removal if needed by the participant who removed it.

Join a business conversation: This method allows a person in the enterprise to request to join a team involved in a business conversation. The membership request is sent to the participants with "accountable" role in the business conversation for approval (approval of one suffices). BCM creates a workspace for the new participants in which he can review the history of the business conversation progression.

Add/remove a participant: This method allows the participants in a business conversation to invite new people to join the business conversation, and allows participants with "accountable" role to remove a participant from the conversation. In case the person is removed, the workspace of the person gets the status of "in-active" and maintained in the repository for historical reasons.

4   Architecture and Implementation

We describe the architecture and implementation of prototype business conversation manager in the following.

4.1   Architecture

The business conversation manager is offered as a service that exposes a set of APIs. There are three main categories of components in BCM: those supporting the definition and execution of workplans, and those related to communication channels, and the client-side portal. The system's architecture is shown in Figure 3. It has the following components:

The service APIs: The APIs expose the functionality of business conversation manager as a Web service.

Business activity portal: This is a Web-based software component that implements the user interface and supports

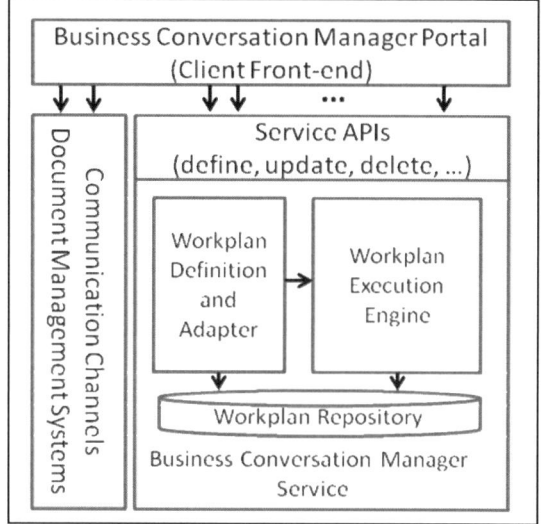

Fig. 3. The architecture of business conversation manager as a service

ad-hoc user interactions to define, view and update the workplan/activity details.

Workplan repository: This is the component that stores information about workplan templates from best practices for process frameworks such as ITIL as well as people service catalogs such as TSC. It also stores the information of on-going conversations.

Workplan definition and adapter: This component enables definition of workplans and updating their definitions through methods such as adding/removing tasks, as well as updating the workplans definition in an ongoing conversation (at runtime) initiated by events triggered from the portal.

Workplan execution engine: This component supports the execution of the workplan of a business conversation and the execution semantics introduced in Section 2.4. This component coordinates the flow of tasks among people based on task dependencies.

Communication channels: The communication channels are those that enable the communications between participants such as chat and email, and document management systems such as Microsoft SharePoint that enable storing information used in the context of BCM.

4.2   Implementation and Use

We have implemented the components for the prototype business conversation manager service in Java. The client-side portal has been implemented using Google Web Toolkit (GWT) (code.google.com/webtoolkit/).

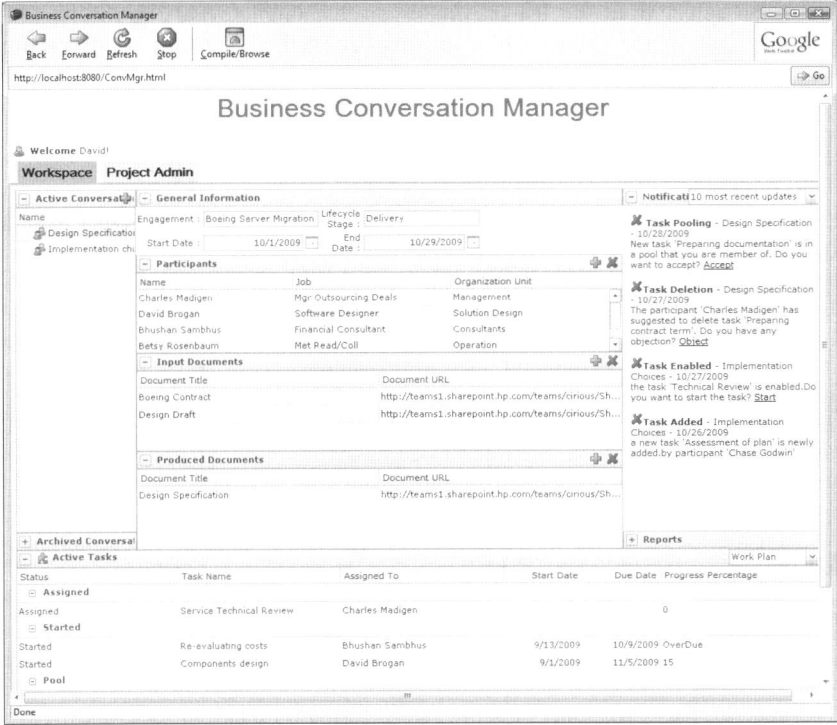

Fig. 4. The screenshot of the frontend of Business Conversation Manager prototype

In the prototype BCM tool, the participants of a business conversation are provided with a workspace within which they can manage the conversations that they are participating in. The workspace enables participants to monitor the status and progress of tasks as well as update the workplan. The workspace also provides highlights such as tasks that are (over-) due.

Figure 4 shows the screenshot of the workspace for a user called "David". In the left panel it shows the list of conversations (active and archived). The middle panel shows the information about a given conversation including which service engagement and lifecycle stage it belongs to, as well as the start and end date of the conversation. The list of participants and the input and output documents are also part of the middle panel. The bottom panel shows the list of tasks in the conversation. David has the option of limiting the task view to the ones allocated to him or all tasks in the workplan of the conversation. The right panel shows updates from all conversations in which David is participating (e.g., enablement of a particular task because its dependent tasks are completed, and new tasks that are in the pool and David is one of the nominated people to take it). The notification messages are visible in the workspace and can be optionally sent by email to participants, as well.

When an outsourcing manager comes to the business activity portal, he/she can create a new business conversation using a wizard. The system asks the manager whether the new conversation is defined based on a pre-existing template, in which case she can select from the list of already stored templates. Using the wizard the people involved in the conversation are also invited. A business conversation is considered active from the moment that it is defined.

5   Related Work

A novel contribution of the proposed business conversation manager is that is builds a bridge among structured and rigid process-centric systems, completely ad-hoc and unstructured conversations between people, and the use of productivity tools in work environments to facilitate the efficiency of conducting best practice processes such as ITIL. Therefore, related work spans across the areas of best practice frameworks; business processes; knowledge and document management systems; and collaboration environments.

Best practice frameworks. Processes from best practice frameworks such as ITIL [2] and eTOM [9] have been often described as textual descriptions. There have been efforts to support people in formalizing and following best practice frameworks. For example, some approaches propose using semantic Wiki [8] and also ontologies [10] to represent processes. However, these efforts only look at this representation as a knowledge base rather than actionable processes.

Business processes. Business process modeling and management tools such ARIS, SAP or standards such as BPEL4People [11] allow definition of well-defined and structured business processes. However, many processes in the enterprise, especially in the context of outsourcing services, involve human interactions that are semi-structured and ad-hoc. In the same line of work, [13] formalizes ITIL processes as precise business process models expressed in process modeling languages such as BPMN.

Definition of ad-hoc and flexible processes has also gained attention recently. For instance, Caramba [14] enables definition of ad-hoc processes for virtual teams. In this work, the process needs to be explicitly defined by the team members using graphical process modeling tools prior to its execution. However, in the context of best practices: (i) processes are not well-specified to enable formal definitions directly, (ii) process users are knowledge workers that are only familiar with productivity tools; they find it difficult to work with formal process models, And (iii) in our approach the process models are used in the backend to support the user but not explicitly exposed to users.

Change management for adaptive and dynamic workflows is studied in the literature [15, 16, 21]. Adaptable workflows address changes that affect the workflow definition (structure, type, etc), while dynamic workflows are concerned with changes to runtime instances of workflows. ADEPTflex [15] enables operators to (manually) change the running instances of a statically defined workflow, while ensuring its correctness. In [16] a high level definition of a workflow is assumed and the concept of *worklet* is introduced to assign concrete activities from a library to realize tasks in the predefined, high level workflow. We do not assume availability of a library of tasks that could be used to realize high level tasks, as most often changes to tasks in process templates are specific to the project context. Our proposed model enabled both static and dynamic changes. The changes to the ad-hoc processes and instances are made as the participants work.

Knowledge and document management systems. Many existing requirement capture and management tools and business process analysis tools such as ProVison simplify the tasks of gathering, documenting, tracking and managing requirements and process definitions in an enterprise [17]. Typically these tools help document requirements and processes, and in some instances simulate the impact of changes. They are geared towards implementing and executing projects and processes in IT systems not among people.

Collaboration approaches and tools. The proposed system differs from wiki-based collaboration systems (e.g. Semantic Media Wiki [8]) as Wikis provide a passive knowledge base. DOMINO [23] and OpenWater [22] are examples of early efforts to support more flexible and cooperative processes in organizations. Unified activity management [12] is another thread of work which aim at providing an integrated work environment for all activities of a person across various productivity tools and organizing them and supporting the collaborations of people around activities. Business Conversation Manager takes a step forward by combining informal interactions about the process with the semi-structured definition of the processes while supports ad-hoc (best practice) processes which was not the focus in earlier works. Recently, there has been a rapid growth in social collaboration tools and techniques such as Google Wave (wave.google.com) as well as Web 2.0 types of collaboration techniques. These tools and techniques are complementary to our work, and in our platform they play a role as communication channels between conversation participants. There has been also recently some works that allow collaborative definition of processes, e.g., based on Google Wave platform (e.g. Gravity [19] from SAP Research that allows collaborative process modeling) or Workflow-on-Wave (WoW) [20]. These tools aim at defining the business process model prior to their execution (in a collaborative manner), however, we do not assume the existence of a business process model (other than templates) ahead of execution time and the process definition emerges and becomes updated on the fly (while people work) in a flexible and collaborative manner among people.

6 Conclusion and Future Work

In this paper, we have presented an enterprise-grade system for establishing, managing and conducting business conversations to support ad-hoc people processes from best practice frameworks such as ITIL. We have implemented a prototype system in the context of supporting people processes for delivering outsourcing services. Our main aim has been to reduce the burden of using the system as well as its overhead for knowledge workers in terms of amount of process-related information that they need to learn. At the same time, we have designed the system so that it is backed with formal modeling and execution semantics of processes and makes uses of them in a transparent manner for users.

The process model introduced in the paper based on dependency model offers a "lightweight" process modeling approach that supports collaborative definition and adaptation of the process compared to hard-coded or rigid processes that are hard to change after the process have been started. The business conversation manager (BCM) introduced in this paper builds on top of and allows users to utilize the existing systems that they are familiar with in their daily jobs such as MS SharePoint, and email. We introduce a minimum amount of abstractions in a simple and innovative manner to simplify the job of people in defining and managing people processes.

In terms of future work, we are planning to provide a catalog of conversations to users so that participants can find other related conversations within the project to that of their own conversations so that inter-conversation dependencies could be managed more efficiently. We are currently incorporating the capability to store the workplan of active conversations that are near conclusion as templates for future reuse. We are planning also to experimentally validate the system by having people use it in the context of service engagements.

Acknowledgement. Authors would like to thank Sujoy Basu and Susan Spence from HP Labs for their feedbacks and comments on earlier drafts of this paper.

References

1. Lee, J., Huynh, M.Q., Kwok, R.C., Pi, S.: IT outsourcing evolution—: past, present, and future. Commun. ACM 46(5), 84–89 (2003)
2. Hendriks, L., Carr, M.: ITIL: Best Practice in IT Service Management. In: Van Bon, J. (Hrsg.) The Guide to IT Service Management, London u. a, Band 1, pp. 131–150 (2002)
3. Hobson, S., Patil, S.: Public Disclosure versus Private Practice: Challenges in Business Process Management. In: Workshop on SOA, Globalization, People, & Work (SG-PAW 2009), Sweden (2009)
4. Project Management Institute, The Project Management Body of Knowledge, PMBOK (2000)
5. Jensen, K.: An Introduction to the Practical Use of Coloured Petri Nets. In: Reisig, W., Rozenberg, G. (eds.) APN 1998. LNCS, vol. 1492, pp. 237–292. Springer, Heidelberg (1998)
6. Choo, Y.: Hierarchical Nets: A Structured Petri Net Approach to Concurrency, Technical Report CaltechCSTR:1982.5044-tr-82, California Institute of Technology (1982)
7. Graupner, S., Motahari-Nezhad, H.R., Singhal, S., Basu, S.: Making Process from Best practices Frameworks Actionable. In: DDBP 2009: Second International Workshop on Dynamic and Declarative Business Processes, Auckland, New Zealand, September 1 (2009)
8. University of Karlsruhe, Semantic Media Wiki,
   http://semantic-mediawiki.org
9. TeleManagement Forum, Enhanced Telecom Operations Map (eTOM) – The Business Process Framework,
   http://www.tmforum.org/BestPracticesStandards/BusinessProces sFramework/1647/Home.html
10. Shangguan, Z., Gao, Z., Zhu, K.: Ontology-Based Process Modeling Using eTOM and ITIL. In: CONFENIS, vol. (2), pp. 1001–1010 (2007)
11. WS-BPEL Extensions for People—BPEL4People: A Joint White Paper by IBM and SAP (July 2005),
    http://www.sdn.sap.com/irj/servlet/prt/portal/prtroot/docs/ library/uuid/cfab6fdd-0501-0010-bc82-f5c2414080ed

12. Moran, T.P., Cozzi, A., Farrell, S.P.: Unified activity management: supporting people in e-business. ACM Commun. 48(12), 67–70 (2005)

13. Orbus Software, The iServer ITIL Solution,
    `http://www.orbussoftware.com/business-process-analysis/`
    `products/itil-solution/`

14. Dustdar, S.: Caramba — A Process-Aware Collaboration System Supporting Ad hoc and Collaborative Processes in Virtual Teams. Distrib. Parallel Databases 15(1), 45–66 (2004)

15. Reichert, M., Dadam, P.: ADEPT flex Supporting Dynamic Changes of Workflows Without Loosing Control. Journal of Intelligent Information Systems 10(2), 93–129 (1998)

16. Adams, M., ter Hofstede, A.H.M., Edmond, D., van der Aalst, W.M.P.: Implementing dynamic flexibility in workflows using worklets. Report BPM-06-06. BPMCenter.org (2006)

17. Volere. List of Requirement Managemnt Tools,
    `http://www.volere.co.uk/tools.htm`

18. HCi Journal, List of Document Management Software,
    `http://www.hci.com.au/hcisite3/journal/`
    `Listofdocumentmanagementsoftware.htm`

19. Dreiling, A.: Gravity – Collaborative Business Process Modelling within Google Wave, SAP Research (September 2009)

20. Itensil, Workflow-on-Wave (WoW), `http://itensil.com`

21. Dumas, M., van der Aalst, W.M., ter Hofstede, A.H.: Process Aware Information Systems: Bridging People and Software Through Process Technology. WileyBlackwell (2005)

22. Whittingham, K., Stolze, M., Ludwig, H.: The OpenWater Project - A substrate for process knowledge management tools, AAAI Technical Report SS-00-03 (2000)

23. Kreifelts, T., Hinrichs, E., Klein, K., Seuffert, P., Woetzel, G.: Experiences with the DOMINO office procedure system. In: Second Conference on European Conference on Computer-Supported Cooperative Work, The Netherlands, September 25 - 27 (1991)

Tools for Modeling and Generating Safe Interface Interactions in Web Applications

Marco Brambilla[1], Jordi Cabot[2], and Michael Grossniklaus[1]

[1] Politecnico di Milano, Dipartimento di Elettronica e Informazione
P.za L. Da Vinci, 32. I-20133 Milano - Italy
{mbrambil,grossniklaus}@polimi.it
[2] INRIA - École des Mines de Nantes
Rue Alfred Kastler, 4 B.P. 20722 - F-44307 NANTES Cedex 3 - France
jordi.cabot@inria.fr

Abstract. Modern Web applications that embed sophisticated user interfaces and business logic have rendered the original interaction paradigm of the Web obsolete. In previous work, we have advocated a paradigm shift from static content pages that are browsed by hyperlinks to a state-based model where back and forward navigation is replaced by a full-fledged interactive application paradigm, featuring undo and redo capabilities, with support for exception management policies and transactional properties. In this demonstration, we present an editor and code generator designed to build applications based on our approach.

1 Introduction

The Web has evolved from a platform for navigating hypertext documents to a platform for implementing complex business applications, where user interaction relies on richer interaction paradigms (RIA, AJAX). In this context, the original interaction paradigm of the Web, based on a simple navigation approach of moving from one page to another is too simplistic. Browsers themselves, that still provide the traditional features of *Back* and *Forward* page navigation along the browsing history, are inadequate for dealing with the complexity of current applications [1]. Depending on the bowser and the application, problems with the use of back and forward buttons include loss of data in pages with form fields, resetting the state of AJAX applications or repeatedly triggering a side effect of a link, e.g., the *Amazon bug*. The behaviour after exceptions and errors (e.g., session timeout) is also indeterministic.

These issues complicate the modelling of complex Web applications and hamper the user experience. State-based models are well suited for the specification of user interfaces and applications [2]. In previous work [3], we have therefore proposed a state-based modelling language to specify safe user interactions for Web applications, that is complementary to existing Web design methodologies, e.g., [4,5,6]. Our approach evolves the interaction paradigm by moving the Web from the browsing paradigm based on *Pages*, with related *Back* and *Forward* actions, to a full-fledged interactive application paradigm, based on the concept

B. Benatallah et al. (Eds.): ICWE 2010, LNCS 6189, pp. 482–485, 2010.

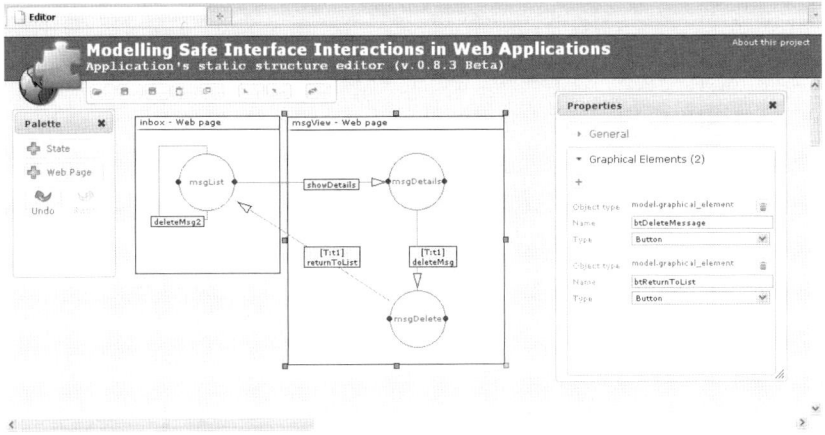

Fig. 1. Example Web interface model drawn with the online editor

of *State*, that features controlled *Undo* and *Redo* capabilities, exception management, and transactional properties.

This paper presents a toolset consisting in (1) a model editor to specify Web application interfaces, (2) an API that grants access to the model concepts, and (3) a code generator that automatically produces prototypical applications from the models that exploit our API at runtime for granting safe navigation.

2 Modelling Safe Interfaces for Web Applications

The first step of the development process is the specification of the interface and behaviour of the Web application. In our proposal Web applications are represented as state machines consisting of *states* (i.e., possible situations the application can be in) and *transitions* (i.e., changes from a state to another, triggered by an event). A single Web page can comprise several states, depending on the granularity chosen by the designer. Additional modeling primitives allow the definition of exception events and states (that model the response to unexpected situations) and the definition of *transaction* regions, i.e., a set of states that must be accomplished with all-or-nothing semantics. We also offer a set of predefined kinds of transitions between states (e.g., click button, list selection, ...) to facilitate the definition of the state machine.

As an example, Fig. 1 shows our model editor[1] at work, depicting a model for a Web email application. Page *inbox* shows an index of all available messages, that can be deleted or selected for visualization. The *msgView* page shows the details of the selected message, which can be deleted. The *deleteMsg* and *returnToList* transitions belong to the same *transaction T1*. If users undo the deletion, they are actually sent back to the previously deleted message, which is also restored (through a rollback operation) in the application state.

---

[1] Available at `http://home.dei.polimi.it/mbrambil/safeinterfaces.html`
    (beta version)

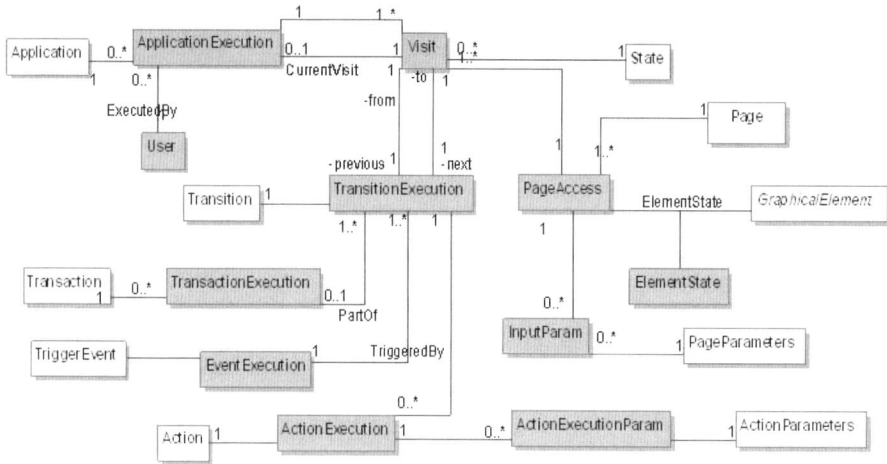

Fig. 2. Interaction metamodel

The online editor has been implemented as a Rich Internet Application, exploiting the OpenJacob Draw2D and Yahoo! User Interface libraries. The editor allows to save, load, edit, and validate models, and provides automatic generation of running prototypes from the models.

The modeling language used to describe the state machines is defined by the internal metamodel shown in Fig. 2 (white classes). Our metamodel is based on the the state machines sublanguage of the UML, adapted to the Web applications domain by adding concepts like Page, GraphicalElement, Transaction and so forth (as described above).

3   Run-Time Support for Safe Web Interactions

Our tool also helps to implement the modeled Web application by (1) automatically parsing the model information and passing it on to a predefined server component that acts as a controller for the application (MVC architecture) and (2) providing a run-time API that programmers can use to interact with the controller and easily manage all events involving state changes in the application and implement correct state behaviours (including undo and redo features) with little effort.

The data structures used internally by the controller to manage run-time dynamic information (current state the user is in, input parameters, user events) are shown in Fig. 2 (grey classes). For instance, every move of the application user to a state is recorded as a new *Visit*. Obviously, the same user can visit the same state several times. The visits trace is permanently stored to allow undo/redo computation.

Both the static and run-time information of the application can be accessed/ updated using our API. For reasons of space, we only present the main API functions. For instance, methods `getNext` and `getPrevious` can be used by

Table 1. API Methods (Excerpt)

Method	Remarks
`Visit::getNext(): Visit`	queries next visit
`Visit::getPrevious(): Visit`	queries previous visit
`ApplicationExecution::do (EventExec e, Parameter[] p): Visit`	moves to next visit and performs the corresponding actions
`ApplicationExecution::redo():Visit`	moves to the (previously visited) next visit
`ApplicationExecution::undo():Visit`	undoes the last transition and actions
`TransactionExecution::rollback()`	rollbacks the transaction

the application developer to query the next or previous visit in the history, respectively. Instead, the do and undo method are then used to actually perform a move to the next or previous visit and, thus, they manipulate the history records during the process. Note that the do method uses the parsed information from the state machine to know the state to go to according to the current user state and the event triggered by the user. The method redo re-visits a state that has already been visited.

4 Conclusion

We have sketched an approach for modeling Web application interfaces using extended state machines. A run-time API supports the implementation of applications and ensures safe and deterministic application behaviour even in the case of exceptions. As future work, we plan to validate the solution in industrial case studies and provide full coverage of transactionality of side effects.

References

1. Baresi, L., Denaro, G., Mainetti, L., Paolini, P.: Assertions to Better Specify the Amazon Bug. In: Proc. SEKE '02, pp. 585–592 (2002)
2. Draheim, D., Weber, G.: Modelling Form-based Interfaces with Bipartite State Machines. Interacting with Computers 17(2), 207–228 (2005)
3. Brambilla, M., Cabot, J., Grossniklaus, M.: Modelling Safe Interface Interactions in Web Applications. In: Proc. ER 2009, pp. 387–400 (2009)
4. Schwabe, D., Rossi, G., Barbosa, S.D.J.: Systematic Hypermedia Application Design with OOHDM. In: Proc. Hypertext '96, pp. 116–128 (1996)
5. Ceri, S., Fraternali, P., Bongio, A., Brambilla, M., Comai, S., Matera, M.: Designing Data-Intensive Web Applications. Morgan Kaufmann, San Francisco (2002)
6. Vdovják, R., Frăsincar, F., Houben, G.J., Barna, P.: Engineering Semantic Web Information Systems in Hera. Journal of Web Engineering 1(1-2), 3–26 (2003)

Linking Related Documents: Combining Tag Clouds and Search Queries

Christoph Trattner and Denis Helic

Graz Technical University of Technology
Inffeldgasse 21a/16c
A-8010 Graz
ctrattner@iicm.edu, dhelic@tugraz.at

Abstract. Nowadays, Web encyclopedias suffer from a high bounce rate. Typically, users come to an encyclopaedia from a search engine and upon reading the first page on the site they leave it immediately thereafter. To tackle this problem in systems such as Web shops additional browsing tools for easy finding of related content are provided. In this paper we present a tool that links related content in an encyclopaedia in a usable and visually appealing manner. The tool combines two promising approaches – tag clouds and historic search queries – into a new single one. Hence, each document in the system is enriched with a tag cloud containing collections of related concepts populated from historic search queries. A preliminary implementation of the tool is already provided within a Web encyclopaedia called Austria-Forum.

Keywords: query tags, tags, tag clouds, linking.

1 Introduction

Content in Web encyclopedias such as Wikipedia is mainly accessed through search engines. Typically, users with an interest in a certain encyclopedic topic submit a Google search, click on a Wikipedia document from the result list and upon reading the document they either go back to Google to refine their search, or close their browsers if they have already found the information they needed. Such a user behaviour on encyclopedia sites is traceable through a typical high bounce rate (see Alexa[1] for instance) of such sites. Essentially, users do not browse or search in Wikipedia to find further relevant content - they are rather using Google for that purpose.

In our opinion, Web encyclopedias lack simple and usable tools that involve users in explorative browsing or searching for related documents. In other Web systems, most notably Web shops, different approaches have been applied to tackle this situation. For example, one popular approach involves offering related information through collaborative filtering techniques as on Amazon. Google or Yahoo! apply a similar approach to offer related content by taking the users' search query history into account via sponsored links [4].

---

[1] http://www.alexa.com/siteinfo/wikipedia.org

B. Benatallah et al. (Eds.): ICWE 2010, LNCS 6189, pp. 486–489, 2010.

Recently, social bookmarking systems emerged as an interesting alternative to search engines for finding relevant content [3,7]. These systems apply the concept of social navigation [5] i.e. users browse by means of so-called tag clouds, which are collections of keywords assigned to different online resources by different users [2] driven by different motivations [8].

In this paper we introduce a novel approach of offering related content to users of Web encyclopedias. The approach is based on simple idea of integrating a tagging system into a Web encyclopedia and populating that system not only with user-generated tags but also with automatically collected Google query tags. In this way we combine two promising approaches successfully applied elsewhere into a single new one – the access to related content is granted not only through social trails left in the system by other users but also through search history of general user population. To test this idea a prototype tool has been implemented within a Web encyclopedia called Austria-Forum[2].

The paper is structured as follows. Section 2 presents the basic idea of this new approach and provides an analysis of its potentials. Section 3 shortly discusses the implementation of the idea within Austria-Forum. Finally, Section 4 concludes the paper and provides an outlook for the future work in this area.

2   Approach

The basic idea of this new approach is to combine provision of related documents as offered by social bookmarking sites and by e.g. Google search query history. On the one hand, tag clouds represent a usable and interesting alternative navigation tool in modern Web-based systems. Moreover, they are very close to the idea of explorative browsing [6], i.e. they capture nicely the intent of users coming to a system from a search engine - users have searched in e.g. Google and now they click on a concept in a tag cloud that reflects their original search intent. On the other hand, Google search query history, i.e. queries that are "referrers" to found documents are an invaluable source of information for refining user search in the system. It is our belief that an integration of such historical queries into a tag cloud user interface provides a promising opportunity to lead users to related documents.

To make this idea work the tag clouds need to be calculated in a context or resource-specific way, i.e. each resource in the system is associated with a special tag cloud. This resource-specific tag cloud captures the most important concepts and topics related to the current document and hence provides a useful navigational tool for exploration of related resources in the system. In addition to the user-generated tags the related concepts and topics are obtained from historic Google search queries leading to the resource in question.

To investigate the feasibility of this approach before implementing it, we conducted an analysis of tagging data automatically obtained from Google queries for Austria-Forum (AF). Thus, Google query tags have been collected over a period of four months and analyzed using the following metrics: number of tags

---

[2] http://www.austria-lexikon.at

Table 1. Growth of tagging set over time with user-generated and Google query tags

<table>
<tr><td colspan="5" align="center">(a) User Tags</td><td colspan="5" align="center">(b) Google Query Tags</td></tr>
<tr><td>day</td><td>#t</td><td>#t_new</td><td>#r</td><td>#r_new</td><td>day</td><td>#t</td><td>#t_new</td><td>#r</td><td>#r_new</td></tr>
<tr><td>-200</td><td>3,202</td><td>3,202</td><td>4,884</td><td>4,884</td><td>-60</td><td>3,906</td><td>3,906</td><td>1,698</td><td>1,698</td></tr>
<tr><td>-160</td><td>7,829</td><td>4,627</td><td>7,450</td><td>2,566</td><td>-50</td><td>7,020</td><td>3,114</td><td>3,160</td><td>1,462</td></tr>
<tr><td>-120</td><td>8,980</td><td>1,151</td><td>9,109</td><td>1,659</td><td>-40</td><td>10,018</td><td>2,998</td><td>4,710</td><td>1,550</td></tr>
<tr><td>-80</td><td>10,009</td><td>1,029</td><td>11,523</td><td>2,414</td><td>-30</td><td>12,772</td><td>2,754</td><td>6,245</td><td>1,535</td></tr>
<tr><td>-40</td><td>10,628</td><td>619</td><td>12,421</td><td>898</td><td>-20</td><td>15,615</td><td>2,843</td><td>8,055</td><td>1,810</td></tr>
<tr><td>now</td><td>11,097</td><td>469</td><td>12,871</td><td>450</td><td>-10</td><td>17,743</td><td>2,128</td><td>9,368</td><td>1,313</td></tr>
<tr><td></td><td></td><td></td><td></td><td></td><td>now</td><td>19,867</td><td>2,124</td><td>10,659</td><td>1,291</td></tr>
</table>

$\#t$, number of new tags $\#t_{new}$, number of resources $\#r$, and number of new resources $\#r_{new}$. The analysis observed the changes in these metrics over time.

Table 1 shows the potential of the Google query term approach. Over 10,659 AF resources were tagged during a period of 60 days by Google query tags. Compared to the user-generated tags in AF, which show an average increase of 399.35 tagged resources per 10 days for the last 200 days (see Table 1(a)), an average of around 1,500 new tags per 10 days for the last 60 days has been achieved with the query tags (see Table 1(b)). Thus, automatic tagging approach annotated four times more resources than the human approach within AF.

As the last step two tagging datasets have been combined. The combined dataset annotates 20,688 resources, which is an increase of nearly 100% in the number of annotated resource as compared to user-generated tags. Additionally, the combined dataset contains 27,824 unique tags (an increase of 150%). Similar results have been obtained by [1] for the `stanford.edu` domain.

3   Implementation

The first prototypical implementation of the tool consists of four modules.

Tag Collection Module: The module consists of two sub-modules: a client- and a server sub-module. The client collects HTTP-Referrer information of the users that comes from the Google search engine to a particular resource within Austria-Forum. The client sub-module is implemented via JavaScript AJAX. The server is a Web service that processes HTTP-Referrer headers sent over by the client sub-module. Firstly, the service identifies single query terms and denotes them as potential *tags*. Secondly, to filter out the noisy tags a stop word and a character filter is applied.

Tag Storage Module: This module stores the tags obtained by the collection module into a database. Currently, the tag database is hosted on a MySql server as a normalized tag database.

Tag (Cloud) Generation Module: To provide the access to related documents a resource-specific tag cloud is calculated by this module. This tag cloud is of the form $TC_r = (t_1, ..., t_n, r_1, ..., r_m)$, where $r_1, ..., r_m$ are the resources that have any of $t_1, ..., t_n$ tags in common. The calculated tag clouds are cached on the server-side to improve the performance of the system. For retrieving the tags and the corresponding resources this module provides a simple interface that consists of two functions: $GetTags(URL)$ (generates a XML representation of a tag cloud), and $GetLinks(URL, tag)$ (generates a XML representation of the resource list for a particular tag).

Tag Cloud Presentation Module: This modul is a client-side AJAX module implemented in JavaScript. It manipulates the browser DOM objects to render a tag cloud in a visually appealing fashion.

4 Conclusions

In this paper we presented an approach for exploring related resources in Web encyclopedias. The tool aims at offering additional navigational paths to related resources for users of such systems in general, and for users who come to these systems from a search engine such as Google. The future work will include development of a theoretical framework to compare this approach to other approaches aiming at a provision of related content in web-based information systems. In addition to theoretical investigations, a usability study to assess the acceptance and usefulness of the tool will be carried out.

References

1. Antonellis, I., Garcia-Molina, H., Karim, J.: Tagging with queries: How and why. In: ACM WSDM (2009)
2. Heymann, P., Paepcke, A., Garcia-Molina, H.: Tagging human knowledge. In: Proceedings of the Third ACM International Conference on Web Search and Data Mining, New York, NY, USA, pp. 51–61 (2010)
3. Mesnage, C.S., Carman, M.J.: Tag navigation. In: SoSEA '09: Proceedings of the 2nd International Workshop on Social Software Engineering and Applications, New York, NY, USA , pp. 29–32(2009)
4. Mehta, A., Saberi, A., Vazirani, U., Vazirani, V.: AdWords and generalized online matching. J. ACM 54(5), 22 (2007)
5. Millen, D., Feinberg, J.: Using social tagging to improve social navigation. In: Workshop on the Social Navigation and Community Based Adaptation Technologies, Dublin, Ireland (2006)
6. Sinclair, J., Cardew-Hall, M.: The folksonomy tag cloud: when is it useful? Journal of Information Science, 34–15 (2008)
7. Strohmaier, M.: Purpose Tagging - Capturing User Intent to Assist Goal-Oriented Social Search. In: SSM'08 Workshop on Search in Social Media, in conjunction with CIKM'08, Napa Valley, USA (2008)
8. Strohmaier, M., Koerner, C., Kern, R.: Why do Users Tag? Detecting Users' Motivation for Tagging in Social Tagging Systems. In: 4th International AAAI Conference on Weblogs and Social Media (ICWSM 2010), Washington, DC, USA, May 23-26 (2010)

GAmera: A Tool for WS-BPEL Composition Testing Using Mutation Analysis

Juan-José Domínguez-Jiménez, Antonia Estero-Botaro,
Antonio García-Domínguez, and Inmaculada Medina-Bulo

Dpt. Computer Languages and Systems,
University of Cádiz, Escuela Superior de Ingeniería,
C/Chile 1, CP 11003 Cádiz, Spain
{juanjose.dominguez,antonia.estero,antonio.garciadominguez,
inmaculada.medina}@uca.es

Abstract. This paper shows a novel tool, GAmera, the first mutant generation tool for testing Web Service compositions written in the WS-BPEL language. After several improvements and the development of a graphical interface, we consider GAmera to be a mature tool that implements an optimization technique to reduce the number of generated mutants without significant loss of testing effectiveness. A genetic algorithm is used for generating and selecting a subset of high-quality mutants. This selection reduces the computational cost of mutation testing. The subset of mutants generated with this tool allows the user to improve the quality of the initial test suite.

1 Introduction

The evolution of software towards Service-Oriented Architectures (SOAs) has led to the definition of a language that facilitates the composition of Web Services (WS): the OASIS Web Services Business Process Execution Language (WS-BPEL) 2.0 [1]. WS-BPEL allows us to develop new WS modeling more complex business processes on top of pre-existing WS. WS-BPEL is an XML-based language which specifies the behavior of a business process as a WS which interacts with other external WS independently of how they are implemented through message exchanges, synchronization and iteration primitives and fault or event handlers, among other constructs.

Mutation analysis has been validated as a powerful technique for testing programs and for the evaluation of the quality of test suites [2]. It generates mutants by applying mutation operators to the program to test. The resulting mutants contain a single syntactic change with respect to the original program. So, in order to apply this technique to programs in any language, we need a language-specific set of mutation operators and a tool for generating and executing the mutants.

Rice [3] lists the ten most important challenges in the automation of the test process. One of them is the lack of appropriate tools, as they are too costly to use or do not fit the tester's intention or the required environment. Several systems

B. Benatallah et al. (Eds.): ICWE 2010, LNCS 6189, pp. 490–493, 2010.

to generate mutants for programs written in various languages exist: Mothra [4] for Fortran, MuJava [5] for Java, ... GAmera is the first mutation testing tool for WS-BPEL.

One of the main drawbacks of mutation testing [2] is the high computational cost involved in the execution of the large number of mutants produced for some programs against their test suites. Most existing tools simply generate all the possible mutants. GAmera follows a different approach, generating only a subset of all the possible mutants. To select these mutants, GAmera incorporates a genetic algorithm (GA) that selects only high-quality mutants, reducing the computational cost.

In [6] we presented a set of specific mutation operators for WS-BPEL 2.0, in [7] we proposed a framework for the automatic generation of mutants for WS-BPEL based on GA and in [8] we showed the preliminary results of applying the new technique. The tool described in this paper is a direct consequence of these previous works. This paper shows the functionality and usefulness of GAmera after several improvements and the development of a graphical interface. This tool implements the GA integrated with the mutation operators defined for WS-BPEL in the previous works. This open-source tool is freely available at its official website[1].

2   Tool Design

The GAmera tool consists of three main components: the *analyzer*, the *mutant generator* and the *execution engine* that runs and evaluates the mutants.

Analyzer. It starts off the GAmera workflow by receiving as input the WS-BPEL composition under test and listing the mutation operators that can be applied to it. Figure 1 shows a screenshot of the application that interacts with the analyzer. The operators are displayed in separate tabs, depending on the category of the operator. The user can determine the set of mutation operators to use among all the available operators.

Mutant generator. Mutants are generated using the information received from the analyzer. The tool gives us the possibility of generating all the possible mutants, or selecting only a subset of them with the genetic algorithm.

The selection process uses two components. The first, called *mutant genetic search*, is a GA in which each individual represents a mutant. The GA is capable of automatically generating and selecting a set of mutants, using a fitness function that measures the quality of a mutant, depending on if there are or not test cases that kill it [7].

The second element is the *converter*, that transforms an individual of the GA into a WS-BPEL mutant. To perform this conversion, the tool uses a different XSLT stylesheet for each mutation operator.

Execution engine. The system executes the mutants generated against a test suite. Mutants are classified into three categories depending on their output:

---

[1] http://neptuno.uca.es/~gamera

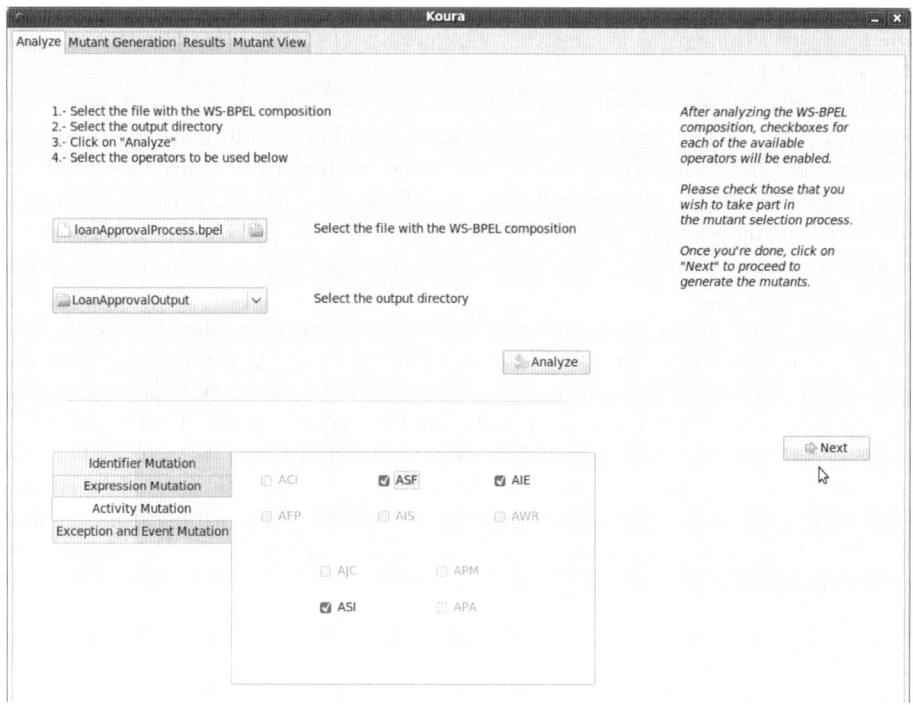

Fig. 1. Interaction with the analyzer

Killed. The output from the mutant differs from that of the original program for at least one test case.

Surviving. The output from the mutant and the original program are the same for all test cases.

Stillborn. The mutant had a deployment error and it could not be executed. Mutants in this state hint at how the design and implementation of the mutation operators should be revised.

For the execution of the original program and their mutants, GAmera uses the ActiveBPEL 4.1 [9] open-source WS-BPEL 2.0 engine and the BPELUnit [10], an open-source WS-BPEL unit test library which uses XML files to describe test suites.

2.1   Results

GAmera shows the results obtained in the execution of the mutants. It displays the total number of generated, killed, surviving and stillborn mutants. From these values, we can measure the quality of the initial test suite and let the user improve it by examining the surviving mutants and adding new test cases to kill them. For this purpose the tool includes a viewer which shows the differences between the original composition and the mutant.

3   Conclusions and Future Work

We have presented the first tool that automatically generates a set of high-quality mutants for WS-BPEL compositions using a genetic algorithm. This tool automates the test of WS-BPEL compositions and provides information that allows the user to improve the quality of the test suites.

The tool is useful both for developers of WS-BPEL compositions, since it automates the testing process, and for researchers in testing of WS-BPEL compositions that are intended to evaluate the quality of test suites.

We are currently working on the design of new mutation operators for providing various coverage criteria. A future line of work is adding a test case generator to the current tool. This generator will allow the user to enhance the quality of the test suite by generating automatically new appropriate test cases.

Acknowledgments

This paper has been funded by the Ministry of Education and Science (Spain) and FEDER (Europe) under the National Program for Research, Development and Innovation, Project SOAQSim (TIN2007-67843-C06-04).

References

1. OASIS: Web Services Business Process Execution Language 2.0 (2007), http://docs.oasis-open.org/wsbpel/2.0/OS/wsbpel-v2.0-OS.html
2. Offutt, A.J., Untch, R.H.: Mutation 2000: uniting the orthogonal. In: Mutation testing for the new century, pp. 34–44. Kluwer Academic Publishers, Dordrecht (2001)
3. Rice, R.: Surviving the top 10 challenges of software test automation. CrossTalk: The Journal of Defense Software Engineering, 26–29 (mayo 2002)
4. King, K.N., Offutt, A.J.: A Fortran Language System for Mutation-based Software Testing. Software - Practice and Experience 21(7), 685–718 (1991)
5. Ma, Y.S., Offutt, J., Kwon, Y.R.: MuJava: an automated class mutation system. Software Testing, Verification & Reliability 15(2), 97–133 (2005)
6. Estero-Botaro, A., Palomo-Lozano, F., Medina-Bulo, I.: Mutation operators for WS-BPEL 2.0. In: Proceedings of the 21th International Conference on Software & Systems Engineering and their Applications (2008)
7. Domínguez-Jiménez, J.J., Estero-Botaro, A., Medina-Bulo, I.: A framework for mutant genetic generation for WS-BPEL. In: Nielsen, M., Kucera, A., Miltersen, P.B., Palamidessi, C., Tuma, P., Valencia, F.D. (eds.) SOFSEM 2009. LNCS, vol. 5404, pp. 229–240. Springer, Heidelberg (2009)
8. Domínguez-Jiménez, J.J., Estero-Botaro, A., García-Domínguez, A., Medina-Bulo, I.: GAmera: An automatic mutant generation system for WS-BPEL. In: Proceedings of the 7th IEEE European Conference on Web Services, pp. 97–106. IEEE Computer Society Press, Los Alamitos (2009)
9. ActiveVOS: ActiveBPEL WS-BPEL and BPEL4WS engine (2008), http://sourceforge.net/projects/activebpel
10. Mayer, P., Lübke, D.: Towards a BPEL unit testing framework. In: TAV-WEB'06: Proceedings of the workshop on Testing, analysis, and verification of web services and applications, pp. 33–42. ACM, New York (2006)

Open, Distributed and Semantic Microblogging with SMOB*

Alexandre Passant[1], John G. Breslin[1,2], and Stefan Decker[1]

[1] Digital Enterprise Research Institute, National University of Ireland, Galway
firstname.lastname@deri.org
[2] School of Engineering and Informatics, National University of Ireland, Galway
john.breslin@nuigalway.ie

Abstract. This demo paper introduces SMOB, an open, distributed and semantic microblogging system using Semantic Web technologies (RDF(S)/OWL and SPARQL) and Linked Data principles. We present its ontology stack and related annotations, its distributed architecture, and its interlinking capabilities with the Linking Open Data cloud.

Keywords: Social Web, Semantic Web, Linked Data, Microblogging, Distributed Architectures.

1 Introduction

As many Web 2.0 services, microblogging[1] applications suffer from various limits. On the one hand, their close-world architecture strengthens the Web 2.0 data silo issues and makes microblog posts difficultly interoperable with other applications. On their other hand, their lack of machine-readable metadata and semantics entails that microblog posts cannot be fully exploited for advanced querying and reuse. In this demo paper, we present SMOB, an open-source framework for open, distributed and semantic microblogging that relies on Semantic Web technologies and Linked Data principles [1] to solve the aforementioned issues. It provides means to enable machine-readable description of microblog posts, and defines an open architecture where anyone can setup his own service, keeping control over his own data. To achieve this goal, SMOB relies on: (1) an ontology stack to represent microblogs (and their posts) combined with RDFa annotations to represent such data, (2) an open architecture based on distributed hubs that communicate and synchronise together using SPARQL/Update and its related HTTP protocol, and (3) interlinking components, so that microblog posts can be linked to existing resources from the Semantic Web, and especially from the Linking Open Data Cloud[2].

---

* The work presented in this paper has been funded in part by Science Foundation Ireland under Grant No. SFI/08/CE/I1380 (Líon 2).

[1] Microblogging consists in sharing short (generally under 140 characters) status update notifications. It notably became popular *via* Twitter — http://twitter.com.

[2] http://richard.cyganiak.de/2007/10/lod/

B. Benatallah et al. (Eds.): ICWE 2010, LNCS 6189, pp. 494–497, 2010.
© Springer-Verlag Berlin Heidelberg 2010

2   SMOB — Semantic MicROBlogging

2.1   The SMOB Ontologies Stack

In order to semantically-enhance microblogging services and microblog posts, there is a need for: (i) ontologies to represent users, their features (such as names, homepages, etc.) and their related social networking acquaintances; and (ii) ontologies to represent microblogging posts (and microblogs), including particular features such as hashtags (`#tag` patterns included in microblog posts to emphasise particular words), replies, and some contextual information (geographical context, presence status, etc.). Moreover, there is a need to link microblog posts to existing resources from the Semantic Web, in order to represent topics discussed in these posts without any ambiguity.

Regarding the first aspect, we naturally relied on FOAF — Friend of a Friend [3] — as it provides a simple way to define people, their attributes and their social acquaintances. Furthermore, SMOB users can reuse their existing FOAF profiles so that existing information about themselves is automatically linked from their posts. Then, to describe microblog posts and microblogs, we relied on and extended SIOC — Semantically-Interlinked Online Communities [2]. We also used OPO — Online Presence Ontology [5] — to describe users' presence information, such as geolocation. Finally, we relied on MOAT — Meaning Of A Tag [4] — to represent links between hashtags and Semantic Web resources.

Combined together, these ontologies form a complete stack for semantic microblogging, depicted in Fig. 1, and each microblog post generated with SMOB is provided in RDFa using these ontologies.

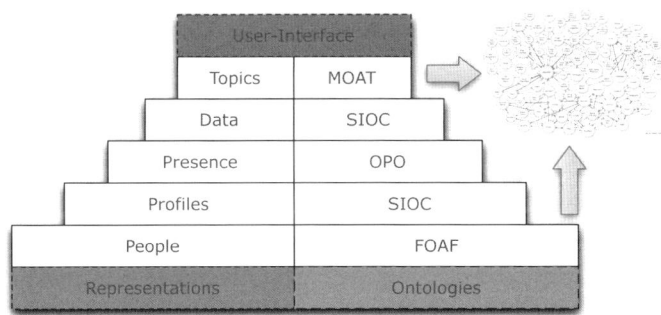

Fig. 1. The SMOB ontologies stack

2.2   A Distributed Architecture

The SMOB architecture is based on distributed hubs that act as microblogging clients and communicate each other to exchange microblog posts and following/followers notifications. That way, there is no centralised server but rather a set of hubs that contains microblog data and that can be easily replicated and extended, also letting users control and own their status updates. Hubs communicate each others and exchange content via HTTP using SPARQL/Update

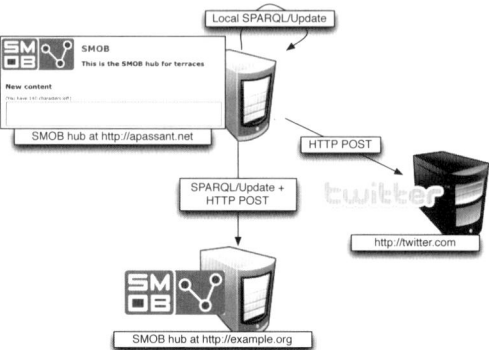

Fig. 2. Communication between SMOB hubs using SPARQL/Update

(the Update part of SPARQL, currently being standardised in the W3C[3]) and its `LOAD` clause (Fig. 2).

When a new microblog post is created, it is immediately stored in the user's hub and sent to the hubs of his followers using the aforementioned principles, so that they immediately receive and store it in their hub. SMOB also enables cross-posting to Twitter, and each hub provides its own SPARQL endpoint so that the data it contains can be easily queried and mashed-up with other data.

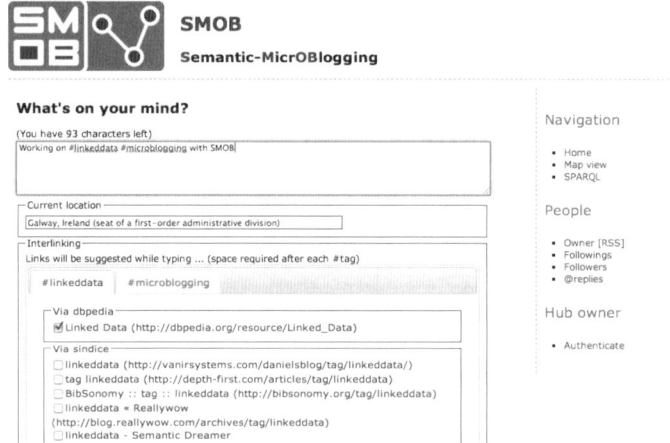

Fig. 3. The SMOB publishing interface and its interlinking components

2.3 Integrating Microblogging in the Linking Open Data Cloud

Finally, SMOB provides features for integrating microblog posts into the Linking Open Data cloud, in addition to the aforementioned reuse of FOAF profiles. On the one hand, it features a set of wrappers that automatically suggest URIs of

---

[3] http://www.w3.org/TR/sparql11-update/

existing resources for each hashtag used in microblog updates, relying on services such as Sindice[4] or DBpedia[5]. Once these suggestions are validated by the author, the mappings are modelled with MOAT and exposed in the post with RDFa. In addition, new wrappers can be deployed for integration with other sources, such as corporate knowledge bases. On the other hand, SMOB provides auto-completion features for geolocation information, relying on GeoNames[6] data. Fig. 3 depicts its user-interface, where these two interlinking components can be observed.

Using such interlinking, new features can be enabled, such as real-time geolocation or topic-based discovery of microblog data using SPARQL queries.

3   Conclusion

In this demo paper, we gave a short overview of SMOB, a system for Open, Distributed and Semantic Microblogging. SMOB is available at `http://smob.me` under the terms of the GNU/GPL license, and can be setup on any LAMP — Linux, Apache, MySQL, PHP — environment.

References

1. Berners-Lee, T.: Linked Data. Design Issues for the World Wide Web, World Wide Web Consortium (2006), `http://www.w3.org/DesignIssues/LinkedData.html`
2. Breslin, J.G., Harth, A., Bojārs, U., Decker, S.: Towards Semantically-Interlinked Online Communities. In: Gómez-Pérez, A., Euzenat, J. (eds.) ESWC 2005. LNCS, vol. 3532, pp. 500–514. Springer, Heidelberg (2005)
3. Brickley, D., Miller, L.: FOAF Vocabulary Specification. Namespace Document, FOAF Project (September 2, 2004), `http://xmlns.com/foaf/0.1/`
4. Passant, A., Laublet, P., Breslin, J.G., Decker, S.: A URI is Worth a Thousand Tags: From Tagging to Linked Data with MOAT. International Journal on Semantic Web and Information Systems (IJSWIS) 5(3), 71–94 (2009)
5. Stankovic, M.: Modeling Online Presence. In: Proceedings of the First Social Data on the Web Workshop. CEUR Workshop Proceedings, vol. 405. CEUR-WS.org (2008)

---

[4] `http://sindice.com`
[5] `http://dbpedia.org`
[6] `http://geonames.org`

The ServFace Builder - A WYSIWYG Approach for Building Service-Based Applications

Tobias Nestler[1], Marius Feldmann[2], Gerald Hübsch[2],
André Preußner[1], and Uwe Jugel[1]

[1] SAP Research Center Dresden
Chemnitzer Str. 48, 01187 Dresden, Germany
{tobias.nestler,andre.preussner,uwe.jugel}@sap.com
[2] Technische Universität Dresden, Department of Computer Science,
Institute for Systems Architecture, Computer Networks Group
{marius.feldmann,gerald.huebsch}@tu-dresden.de

Abstract. In this paper we present the ServFace Builder, an authoring tool that enables people without programming skills to design and create service-based interactive applications in a graphical manner. The tool exploits the concept of service annotations for developing multi-page interactive applications targeting various platforms and devices.

Keywords: Service Composition at the Presentation Layer, Model-driven Development, Service Frontends.

1 Background

Developing service-based interactive applications is time consuming and nontrivial. Especially, the development of user interfaces (UIs) for web services is still done manually by software developers for every new service. This is a very expensive task, involving many error-prone activities. Parameters have to be bound to input and output fields, operations to one or several application pages, such as forms or wizards, and operation invocations to UI-events like button-clicks.

This demo presents an authoring tool called *ServFace Builder* (Fig. 1). The tool leverages web service annotations [1] enabling a rapid development of simple service-based ineractive applications in a graphical manner. The tool applies the approach of service composition at the presentation layer, in which applications are build by composing web services based on their frontends, rather than application logic or data [2]. During the design process, each web service operation is visualized by a generated UI (called service frontend), and can be composed with other web service operations in a graphical manner. Thus, the user, in his role as a service composer and application designer, creates an application in WYSIWYG (What you see is what you get) style without writing any code.

The ServFace Builder aims to support domain experts. Domain experts are non-programmers that are familiar with a specific application domain. They need to understand the meaning of the provided service functionality without having a deep understanding of technical concepts like web services or service

B. Benatallah et al. (Eds.): ICWE 2010, LNCS 6189, pp. 498–501, 2010.

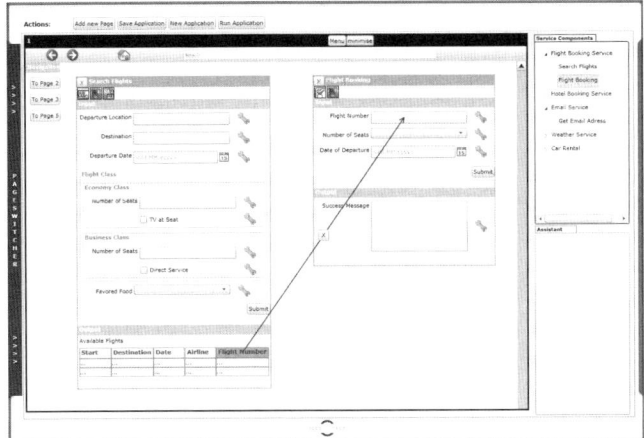

Fig. 1. The ServFace Builder

composition. Our demo shows, how we break down these aspects to the modeling of data-flow and page-flow to realize multi-page applications using our authoring tool. Finally, we show how the tool generates the executable application based on the designed models.

Technical details as well as the existing research challenges of the proposed approach are provided in [3] and [4].

2   The ServFace Builder - Components and Models

Components within the ServFace Builder are called *service frontends*, which are visualized via form-based UIs. They represent a single operation of a WSDL-described web service. The frontends consist of a nested container structure, which includes UI-elements like text fields or combo boxes that are bound to the corresponding service operation parameters. The ServFace Builder utilizes the advantages of the ServFace service annotations to visualize the service frontends already during design-time to give the user an impression of the resulting UI.

The annotations are created by the service developer to provide additional information about the web service, and present the knowledge transfered from the service developer to the service composer to facilitate the understanding and simplify the composition of web services. These annotations are reusable information fragments, which are usually not available for plain WSDL-based approaches, for which they might be non-formally defined in the service documentation. The ServFace annotations are stored in an annotation model that is based on a well-specified meta-model, which references to concrete elements within a WSDL-document. Service annotations provide extensive additional information covering visual, behavioral, and relational aspects. Visual aspects are, e.g., human readable *labels* for technical field names, *grouping* of parameters, definition of *enumerations* for predefined values. Aspects of the behavior of UI-elements are, e.g., client-side *validation* rules, *suggestion* of input values. And finally, Relation-annotations are used to express, e. g., *semantic data type relations*). An in-depth discussion of the annotation model, including examples and a description of the referencing mechanism that links annotations to service descriptions, can be found in [1].

Within the ServFace Builder, each web service is represented by a service model that is automatically inferred from the WSDL and additionally contains the corresponding annotation model. The service model bundles the necessary information about a service including their operations and parameters and holds a reference (Uniform Resource Identifier) to the WSDL file. The information stored in the service model serves as the foundation for the visualization of the frontends. Details about the service model and its elements can be found in [3].

The service model is part of the *Composite Application Model* (CAM). The CAM defines the overall application structure that is visualized within the ServFace Builder. Besides generally describing the integrated services, it describes inter-service connections as data flow, and the navigation flow of an application as page transitions coupled with operation invocations. The CAM is used as the serialization format for storing and loading the modeled interactive applications and serves as the input for the generation of executable applications. It is not bound to a specific platform, and thus can be reused for developing service-based interactive applications for various target technologies.

In the CAM, a composite application is represented by a set of pages. Every page is a container for service frontends and represents a dialog visible on the screen. The pages can be connected to each other to define a navigation flow. Adding a service operation to a page triggers the inference of the corresponding UI-elements necessary for using this operation. The CAM instance is continuously synchronized with user actions, when adding pages, integrating frontends, or modeling data flow. This instance can be serialized at any point in time as a set of Ecore-compliant XMI-files. These files are used as input for a model-to-code transformation process; the last step during the application development process. The CAM as well as the code generation process are described in [3].

3   Service Composition at the Presentation Layer

The ServFace Builder implements an approach coined *service composition at the presentation layer* [4] in order to combine services in a WYSIWYG manner. The user of the tool directly interacts with single UI-elements, entire service frontends and pages to model the desired application. No other abstraction layer is required to define data or control flows. In this demo, we present the graphical development process including the following steps to be accomplished by the user:

1. Select a target platform: As the initial step, the user has to select the target platform from a set of predefined platform templates. The size of the composition area is adjusted depending on the selection to reflect screen size limitations, and an initial CAM for the new application will be created.

2. Integrate a service frontend: After the platform selection, an initial blank page is created and the actual authoring process begins. The Service Component Browser offers all services provided via a connected service repository. To integrate a specific service operation into the application, the user has to drag it from the Service Component Browser to the Composition Area. The tool automatically creates the corresponding service frontend, which can be positioned on the page. An optional design step is to modify the configuration of the frontend, e.g., by hiding single UI elements.

3. Define a data-flow: To define a data-flow between two service frontends, the user has to connect UI-elements of both frontends to indicate the relationship. A connection is created by selecting the target UI-element that is to be filled with data and the source UI-element from which the data is to be taken. The data-flow is visualized in form of an arrow.

4. Define a page-flow: Frontends can be placed on one page or distributed over several pages to create a multi-page application. All pages are shown in a graph-like overview and can be connected to define the page flow. The definition of the page flow is done in a graphical way as well. The user has to connect two pages in order to create a page transition.

5. Deploy the application: After finishing the design process, the modeled application can be deployed according to the initial platform selected. Depending on the type of target platform, the application is either automatically deployed on a predefined deployment target (e.g., as a Spring-based web application) or is offered for download (e.g., as a Google Android application).

Acknowledgment

This work is supported by the EU Research Project (FP7) ServFace[1].

References

1. Janeiro, J., Preussner, A., Springer, T., Schill, A., Wauer, M.: Improving the Development of Service Based Applications Through Service Annotations. In: Proceedings of the WWW/Internet Conference (2009)
2. Daniel, F., Yu, J., Benatallah, B., Casati, F., Matera, M., Saint-Paul, R.: Understanding UI Integration: A Survey of Problems, Technologies, and Opportunities. IEEE Internet Computing 11(3), 59–66 (2007)
3. Feldmann, M., Nestler, T., Jugel, U., Muthmann, K., Hübsch, G., Schill, A.: Overview of an End User enabled Model-driven Development Approach for Interactive Applications based on Annotated Services. In: Proceedings of the 4th Workshop on Emerging Web Services Technology. ACM, New York (2009)
4. Nestler, T., Dannecker, L., Pursche, A.: User-centric composition of service frontends at the presentation layer. In: Proceedings of the 1st Workshop on User-generated Services, at ICSOC (2009)

---

[1] http://www.servface.eu

Extracting Client-Side Web User Interface Controls*

Josip Maras[1], Maja Štula[1], and Jan Carlson[2]

[1] University of Split, Croatia
[2] Mälardalen Real-Time Research Center, Mälardalen University, Västerås, Sweden
{josip.maras,maja.stula}@fesb.hr, jan.carlson@mdh.se

Abstract. Web applications that are highly dynamic and interactive on the client side are becoming increasingly popular. As with any other type of applications, reuse offers considerable benefits. In this paper we present our first results on extracting easily reusable web user-interface controls. We have developed a tool called Firecrow that facilitates the extraction of reusable client side controls by dynamically analyzing a series of interactions, carried out by the developer, in order to extract the source code and the resources necessary for the reuse of the desired web user-interface control.

1 Introduction

Web developers now routinely use sophisticated scripting languages and other active client-side technologies to provide users with rich experiences that approximate the performance of desktop applications [1]. Unfortunately, because of the very short time-to-market and fast pace of technology development, reuse is often not one of the primary concerns. When developers are building new applications, they often encounter problems that have already been solved in the past. Rather than re-inventing the wheel, or spending time componentizing the already available solution, they resort to pragmatic reuse [2]. This is especially true for web development [3], where basically all client side code is open.

In web development, a web page, whose layout is defined with HTML (HyperText Markup Language), is represented with the Document Object Model (DOM). All client side interactions are realized with JavaScript modifications of the DOM and the presentation of the web page is usually defined with CSS (Cascading Style Sheets). So, in order to be able to understand a web page and extract a reusable control, a developer has to be familiar with HTML, JavaScript and CSS, and has to be able to make sense of the interactions between the three separate parts that produce the end result. This is not a simple task, since there is no trivial mapping between the source code and the page displayed in the browser; the responsible code is often scattered between several files and is often intermixed with code irrelevant for the reuse task. For example, even extracting the required CSS styles is not a trivial process for the developer since CSS styles are inherited from parent elements, they are often defined in multiple files and

---

* This work was partially supported by the Swedish Foundation for Strategic Research via the strategic research centre PROGRESS, and the Unity Through Knowledge Fund supported by Croatian Government and the World Bank via the DICES project.

B. Benatallah et al. (Eds.): ICWE 2010, LNCS 6189, pp. 502–505, 2010.

can be dynamically changed while the application is executing (either with JavaScript or with CSS pseudo-classes). Required resources, such as images, can be included either directly as HTML nodes, or via CSS styles or dynamically with JavaScript code, and when transferring them the developer has to locate all used ones, copy them and adjust for the changed location (e.g. if some resources are included via absolute paths).

In this paper we propose a way of semi-automatically extracting reusable client-side controls. Using a tree-based DOM explorer the developer selects the HTML nodes that represent the desired user-interface (UI) control on the web page and executes the interactions that represent the behavior he/she wants to reuse. Based on the analysis done during the execution, all resources such as images, CSS styles, JavaScript and HTML code snippets required for the inclusion of the selected UI control are extracted.

The work presented in this paper is related to web application program slicing and extracting code responsible for the desired behavior of the selected web UI control. In the context of web engineering, Tonella and Ricca [4] define web application slicing as a process which results in a portion of the web application which exhibits the same behavior as the initial web application in terms of information of interest displayed to the user. There also exist two related approaches and tools that facilitate the understanding of dynamic web page behavior: Script InSight [5] and FireCrystal [6] that have a similar functionality of relating elements in the browser with the code responsible for them. However, they are both different to FireCrow in that they do not provide support for code extraction.

2   Method

If we set aside plugins, such as Java Applets, Flash or Silverlight, the client side of the web page is usually composed of three different parts: HTML code that defines the structure and the layout of the page, CSS code that defines the style and the presentation, and JavaScript code that is responsible for executing client-side business logic and interactive layout manipulation. Based on their mutual interactions and resources such as images, the browser engine renders the web page.

In order to extract the minimum amount of resources necessary for the reuse of the selected web UI control we have developed Firecrow, an extension to the Firebug[1] debugger, which supports a three-step extraction process (as shown in Figure 1).

In the fist step the developer chooses the part of the web page user interface that he/she wishes to reuse. Firecrow uses the Firebug's HTML DOM tree which makes it easy for the developer to choose HTML nodes that constitute the desired section of the user interface.

In the second step the developer starts the recording phase. This can be done in three modes: *(i)* Without tracking the execution of the JavaScript code — this mode is useful when the developer only wishes to reuse the layout and styles of the selected HTML nodes. No attempts for extracting the JavaScript code are made. *(ii)* Simple tracking of the executed JavaScript code — in this mode the tool keeps track of the executed JavaScript code. In that way, when the process is finished the tool extracts only the code

---

[1] Firebug is a plugin for the Firefox browser available for download from:
http://getfirebug.com

that was executed while in the recording phase. But, often web application code is not designed for reuse, so it is likely that the executed code will not only be concerned with the desired functionality. In that case the developer has to manually locate and remove all statements that could cause the extracted application to break (e.g. by accessing HTML nodes that are removed in the process of extraction). *(iii)* Advanced tracking of executed JavaScript code — in this mode the tool analyzes all executed JavaScript statements searching for statements that are in any way connected with HTML nodes that are removed in the process of extraction. After the process of extraction is completed, the tool annotates the source code with warnings that make it easier to locate those lines.

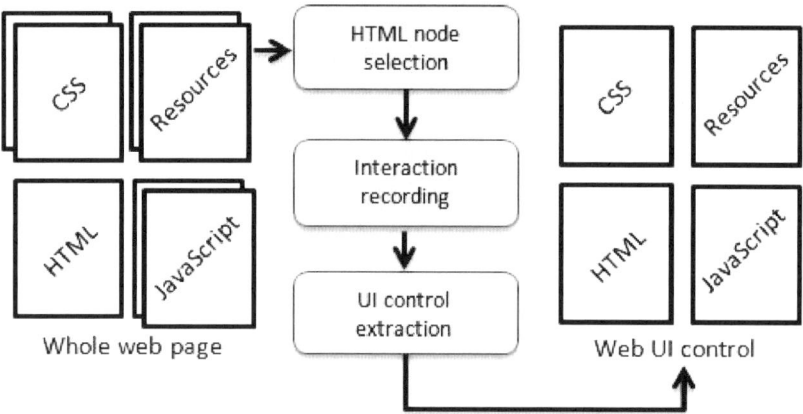

Fig. 1. The Firecrow extraction process

The reason for providing three different degrees of analysis is that each additional layer of analysis comes with a performance cost. For example, in the case of JavaScript intensive web applications, the last mode of advanced JavaScript code tracking could in some cases make the UI non-responsive.

The third step starts when the recording phase is finished — the tool, based on the selected resources, and the code executed while interacting with the UI control, extracts the necessary HTML code, CSS styles, resources and JavaScript code. This significantly simplifies the process of reuse, since the developer only has to include the extracted CSS styles and JavaScript code, and copy the extracted HTML code to the desired location.

2.1   Firecrow

Firecrow is realized as an extension to Firebug, which is a plugin for the Firefox browser. This allows taking advantage of the functionality provided by Firebug to analyze the CSS dependencies and the DOM of selected web page. Also, Firecrow is connected to the Firefox JavaScript engine in order to dynamically analyze the executed JavaScript code. An evaluation version of the tool, as well as a video showing how it can be used, is available from www.fesb.hr/~jomaras/Firecrow.

3   Limitations and Future Work

Firecrow is primarily designed to extract easily reusable client side web UI controls developed with a combination of JavaScript, HTML and CSS code. For those reasons there is currently no support for the extraction of Silverlight nor Flash UI controls. Also, since Firecrow is directly connected with the Firefox JavaScript engine, code specifically developed to be executed only in some other browser (e.g. Internet Explorer, Opera, Safari, etc.) will not be extracted.

An important remark is that, even though Firecrow does relatively simple dynamic analysis, in the case of web applications that are very JavaScript intensive there is a significant drop of performance. This is especially noticeable when doing advanced dynamic analysis. Currently we offer one way of tackling this problem: today many applications use JavaScript libraries which provide common infrastructure for easier development of JavaScript applications (out of Alexa Top 10000 Web sites 32,5% use at least one of JavaScript libraries [7]). These libraries are complex and are usually often reused as is. The developer can exclude those scripts from the dynamic analysis phase, in that way focusing only on the code that is realizing the desired functionality on a higher level of abstraction and not on the common library code that will most probably be reused as is.

So far, Firecrow only annotates the source code with warnings that show code statements that can potentially stop the application execution. As future work, we plan to expand this with the additional functionality of removing all source code statements that are not related with DOM modifications of the selected UI controls. Also, we will have to find a way to avoid performance problems when doing dynamic analysis in JavaScript intensive applications.

References

1. Wright, A.: Ready for a Web OS? ACM Commun. 52(12), 16–17 (2009)
2. Holmes, R.: Pragmatic Software Reuse. PhD thesis, University of Calgary, Canada (2008)
3. Brandt, J., Guo, P.J., Lewenstein, J., Klemmer, S.R.: Opportunistic programming: How rapid ideation and prototyping occur in practice. In: WEUSE '08: Proceedings of the 4th international workshop on End-user software engineering, pp. 1–5. ACM, New York (2008)
4. Tonella, P., Ricca, F.: Web application slicing in presence of dynamic code generation. Automated Software Engg. 12(2), 259–288 (2005)
5. Li, P., Wohlstadter, E.: Script insight: Using models to explore javascript code from the browser view. In: Proceedings of Web Engineering, 9th International Conference, ICWE 2009, San Sebastián, Spain, June 24-26, pp. 260–274 (2009)
6. Oney, S., Myers, B.: Firecrystal: Understanding interactive behaviors in dynamic web pages. In: VLHCC '09: Proceedings of the 2009 IEEE Symposium on Visual Languages and Human-Centric Computing (VL/HCC), pp. 105–108. IEEE Computer Society, Los Alamitos (2009)
7. BackendBattles: Javascript libraries (2010),
   http://www.backendbattles.com/JavaScript_Libraries

Applying Semantic Web Technology in a Mobile Setting: The Person Matcher

William Van Woensel, Sven Casteleyn, and Olga De Troyer

Vrije Universiteit Brussel, Pleinlaan 2, 1050 Brussel, Belgium
{William.Van.Woensel,Sven.Casteleyn,Olga.Detroyer}@vub.ac.be

Abstract. In a mobile setting, users use, handle and search for online information in a different way. Two features typically desired by mobile users are tailored information delivery and context awareness. In this paper, we elaborate a demo application that is built upon the existing SCOUT framework, which supports mobile, context-aware applications. The application illustrates the use of intrinsic mobile features, such as context- and environment-awareness, and combines them with the use of Semantic Web technologies to integrate and tailor knowledge present in distributed data sources.

Keywords: Mobile Web, Semantic Web, Context Awareness.

1 Introduction

With the up-to-par computing power and screen resolution of new generation smart phones, together with location awareness support (e.g., GPS), sensing possibilities (e.g., RFID) and the omnipresence of wireless internet access, an opportunity presents itself to deliver qualitative, context- and environment sensitive information. Due to the huge amount of information available from the user's surroundings, the main challenge thus becomes to filter and personalize this information, exploiting the aforementioned context data and the specific needs of the user.

SCOUT is a mobile application development framework that focuses on data acquisition from different, decentralized sources, with the aim of personalized and context-aware data delivery. It supports different sensing technologies to become aware of the surrounding environment, and is primarily based on Web technologies for communication and data delivery, and Semantic Web technologies for integrating and enriching the knowledge present in the decentralized data sources. In this demo, we demonstrate a mobile person matching application built on top of SCOUT. It automatically detects people in the vicinity, performs a detailed compatibility check based on their FOAF profiles and presents the results to the mobile user.

Tools already exist to find and visualize FOAF profiles; however, in contrast to FOAF visualization tools (e.g., WidgNaut [1]) or RDF search engines (e.g., [2]), the Person Matcher runs in a mobile setting, focuses on finding useful relations between two given FOAF profiles, and allows configurable weighting of relations.

B. Benatallah et al. (Eds.): ICWE 2010, LNCS 6189, pp. 506–509, 2010.

2   SCOUT in a Nutshell

The SCOUT framework consists of a layered architecture: each layer (from the bottom up) is shortly explained below. For a detailed description of SCOUT, see [3].

The *Detection Layer* is responsible for detecting identifiable physical entities in the vicinity of the user. The framework abstracts from the actual detection techniques employed, and only assumes the detected entity is able to communicate relevant information about itself (usually in the form of a URL pointing to an existing Website or RDF source). Our demo relies on two detection techniques: RFID and Bluetooth. Both techniques retrieve a URL from a nearby entity, which points to relevant information about the entity. In case of Bluetooth, a Bluetooth enabled device communicates this URL on request; in case of RFID, the URL is present on an RFID tag attached to the physical entity, which is then read by an RFID reader.

The *Location Management Layer* receives raw detection information from the Detection Layer, and conceptualizes it by creating positional relationships: when an entity is determined to be nearby, a positional relation is created; when the entity is no longer nearby, the positional relation is invalidated. Determining proximity (i.e., remoteness and nearness) is done using proximity strategies, which may differ depending on the available detection data and the specific detection technique used. A single proximity strategy is employed for both RFID and Bluetooth: as the detection range of the employed Bluetooth-enabled devices and RFID readers is relatively small, entities are considered nearby whenever they are in range, and no longer nearby when they move out of range.

The *Environment Layer* combines several models and services geared towards mobile context- and environment-aware application development. The Environment Model offers an integrated view on the data associated with (currently or past) nearby entities. It encompasses the User Model, which stores the user's characteristics, needs and preferences (in our demo, the User Model consists of the user's FOAF profile), and the Relation Model, which stores the (time-stamped) positional relationships provided by the Location Management Layer. In the Environment layer, Semantic Web technologies are exploited to represent and integrate data: RDF(S) / OWL to store and integrate the different data sources, and exploiting their reasoning capabilities to derive additional information; SPARQL to query the integrated models. The Environment Layer also provides mobile application developers with some basic services that provide access to these models: pull-based data retrieval, where arbitrary SPARQL queries are issued over the different models using the Query Service, and push-based data retrieval, where a Notification Service monitors changes in the environment and alerts registered applications of specific changes. Our demo application utilizes the Notification Service to be alerted of nearby entities. Furthermore, as only persons are relevant, a condition in the form of a SPARQL query ensures the application is only notified about entities of the type foaf:Person.

The SCOUT framework is written in JavaME. We employ the MicroJena API [4] to programmatically access and manipulate RDF data, and an external query server to handle SPARQL queries.

3 SCOUT Demo Application: The Person Matcher

As the mobile user is walking around, the Person matcher application is thus continuously provided with FOAF profiles of persons in his vicinity. The application subsequently calculates a "compatibility" score, based on a comparison between the FOAF profile of the nearby person and the user's own FOAF profile (stored in the User Model). The matching algorithm is grounded in establishing paths between the two FOAF profiles, and is based on a configurable weighting scheme.

The Person Matcher is likewise implemented in JavaME (MIDP 2.0, CLDC 1.1).

The weighting scheme

The user can employ the Matcher application for different reasons (e.g., finding colleagues to collaborate with; looking for new friends); depending on this reason, some (sequences of) FOAF properties will become more important in the matching process, while others become less important or even irrelevant. For this purpose, the Matcher application can be configured with different weighting schemes (specified in RDF format), identifying relevant FOAF property sequences and their importance (between 0 and 1). For our demo, we have provided a weighting scheme that is aimed at finding colleagues to collaborate with. For instance, two persons having created (foaf:made) documents with the same subject (foaf:topic) are potentially interested in collaborating, so the weight of the sequence "foaf:made <x> foaf:topic" will be high.

The matching algorithm

The matching algorithm looks for paths, consisting of properties identified in the weighting scheme, between the user's FOAF profile and the FOAF profile of the nearby person. These properties can be subdivided into two categories: "direct" linking properties, that link a person *directly* to another person (e.g., foaf:knows), and "indirect" linking properties, that connect a person to another person via a number of intermediary resources (the maximum amount is configurable). E.g., foaf:made links a person to a resource he made, to which other persons may also link using foaf:made.

The matching algorithm constructs a graph in a breadth-first manner, starting concurrently from both persons' FOAF profiles. The nodes in this graph correspond to Persons, edges to direct or indirect links. The algorithm is able to construct some links immediately, from data found in the initial two FOAF profiles. Subsequently, the algorithm retrieves the RDF sources of relevant resources (i.e., intermediary resources or linked persons), denoted by the rdfs:seeAlso property, and examines them for other relevant resources (i.e., similar to existing Semantic Web Crawlers, e.g. [5]), which are also added to the graph. During this process, the algorithm stops exploring link sequences if their total score (see below) falls below a certain threshold, and finishes once a predefined number of iterations is reached. A relevant "connection" is found when a link sequence connects the two initial Person nodes. Note that this algorithm combines data from a range of RDF sources, possibly resulting in new connections for which the data was not present in any single source.

The compatibility score between two FOAF profiles is the sum of the individual scores of connections found between the two graphs. Each connection's score is calculated as follows (j is the number of direct or indirect links in the connection):

$$\text{score}(connection) = \left(\prod_{0 < i \leq j}(weight_i) \right) \left(1 - \tfrac{j}{10} \right). \tag{1}$$

A connection's score equals the product of the weights of its contained links, while the last factor ensures that the score decreases with the length of the connection.

The user interface
The person matcher is a mobile application that, once started, continuously runs in the background. The following overviews are available: 1/ last 5 persons matched, 2/ best 5 matches of the day (figure 1a), 3/ the complete matching history. In these overviews, a detailed view of each match can be retrieved, which shows the total compatibility score, the name of the matched person, his profile picture (if available in the FOAF profile), and an overview of the connections linking the user with this person (figure 1b). Furthermore, the details on each connection can be obtained: i.e., the links of which it consists and the persons present in the connection (figure 1c).

Fig. 1. Person Matcher screenshots (a) (b) and (c)

4 Conclusion

This demo paper presents the Person Matcher. It relies on the SCOUT framework to detect and retrieve relevant semantic information from nearby persons, and calculates compatibility with these persons based on their FOAF profile. Both SCOUT and the Person Matcher are built utilizing Web technology for communication and content delivery, and Semantic Web technology for integrating and tailoring information.

References

1. WidgNaut, http://widgets.opera.com/widget/4037/
2. Ding, L., Finin, T., Joshi, A., Pan, R., Cost, R.S., Peng, Y., Reddivari, P., Doshi, V., Sachs, J.: Swoogle: a search and metadata engine for the semantic web. In: CIKM '04 (2004)
3. Van Woensel, W., Casteleyn, S., De Troyer, O.: A Framework for Decentralized, Context-Aware Mobile Applications Using Semantic Web technology. In: OTM Workshops (2009)
4. MicroJena, http://poseidon.elet.polimi.it/ca/?page_id=59
5. Biddulph, M.: Semantic web crawling. XML Europe (2004)

Syncro - Concurrent Editing Library for Google Wave

Michael Goderbauer, Markus Goetz, Alexander Grosskopf,
Andreas Meyer, and Mathias Weske

Hasso-Plattner-Institute, Potsdam 14482, Germany

Abstract. The web accelerated the way people collaborate globally distributed. With Google Wave, a rich and extensible real-time collaboration platform is becoming available to a large audience. Google implements an operational transformation (OT) approach to resolve conflicting concurrent edits. However, the OT interface is not available for developers of Wave feature extensions, such as collaborative model editors. Therefore, programmers have to implement their own conflict management solution.

This paper presents our lightweight library called syncro. Syncro addresses the problem in a general fashion and can be used for Wave gadget programming as well as for other collaboration platforms that need to maintain a common distributed state.

1 Introduction

Collaboration is the basis for joint value creation. With increasing complexity of tasks to conduct in business, science, and administration, efficient team collaboration has become a crucial success factor.

Software tools for collaboration typically support the exchange of data and centralized storage of information, for instance in shared workspaces. Only in recent years, technology became available allowing real-time collaboration via the web-browser. This enables team partners to edit a document concurrently and let others see changes instantly.

Google Wave is a new technology that promises to bring real-time collaboration to a large audience. A wave is a hosted conversation that allows multiple users to work on a text artifact at the same time. Editing conflicts are resolved using operational transformation (OT) [1]. Furthermore, a wave's functionality can be extended by gadgets. Even though gadgets live inside of waves, they are not per se collaborative because the OT interface implemented by Google is not available to third-party developers. For that reason, gadget programmers have to find their own answer to handle conflicting edits.

In Section 2, we discuss OT approaches as underlying concept for conflict resolution. We explain the syncro library, which allows conflict resolution in Google Wave gadgets in Section 3. Implementation details, demo-gadgets and further work are discussed in Section 4 before we conclude the paper in Section 5.

B. Benatallah et al. (Eds.): ICWE 2010, LNCS 6189, pp. 510–513, 2010.

2   Operational Transformation for Conflict Resolution

When multiple users edit a single artifact at the same time, conflicts are inevitable and therefore, sophisticated algorithms are needed to handle those conflicting edits. A very popular text-based approach to do this is operational transformation, which is implemented in several collaborative text editing tools, such as MoonEdit[1] or Etherpad[2]. OT relies on Lamport's notion of total ordering [2] to define a unique sequence of events in a distributed system (more details in Section 3). Moreover, OT performs transformations to apply remote modifications, such as insert or delete, on the local client [1]. In addition to the mentioned online text editing tools, this approach was also implemented in a generic OT library for Windows [3]. Linked with the code, the library turns most single-user desktop applications, such as Word, PowerPoint, or Maya, into a collaborative version.

Following the trend towards online collaboration, Google has also implemented OT for collaborative text editing inside Google Wave. However, as their OT interface is not available inside of gadgets, programmers have to implement their own conflict management. One possible approach would be the reimplementation of OT for gadgets. This would require serializing the artifact edited in the gadget to text before OT transformations are applied. From an engineering point of view the indirection of converting artifacts to text is far too complicated. A more convenient way would be to apply the modifications directly to the artifact.

3   Syncro - A Distributed Command Stack for Google Wave

Syncro is a library that provides a generally applicable solution for concurrent editing conflicts of complex artifacts in platforms like Google Wave. It's based on the command pattern [4], a software design pattern in which each modification to an artifact is encapsulated in a command object. Using this concept, syncro shares the commands among all participants of a wave.

Since Google Wave gadgets have no interface to send or receive commands, syncro stores this information in the gadget state, a key value store synchronized among all participants of a wave. To make sure that no command is overwritten, each command has to be stored at a unique key.

Furthermore, to assure that each participant has the same view on the edited artifact, the commands need to be applied in the same consistent order on all clients. The challenge of ordering events in a distributed system has been addressed by Leslie Lamport in his paper from 1978 [2]. He introduces an algorithm for synchronizing a distributed system of logical clocks, which can be used to order events consistently among all participants. These logical clocks are called Lamport clocks and their value is increased whenever an event occurs. In Google Wave the events are the commands created independently by each participant. Lamport's algorithm attaches a logical clock value and a unique sender ID to each command to establish a strict total order among all commands. Therefore,

---
[1] http://moonedit.com/
[2] http://etherpad.com/

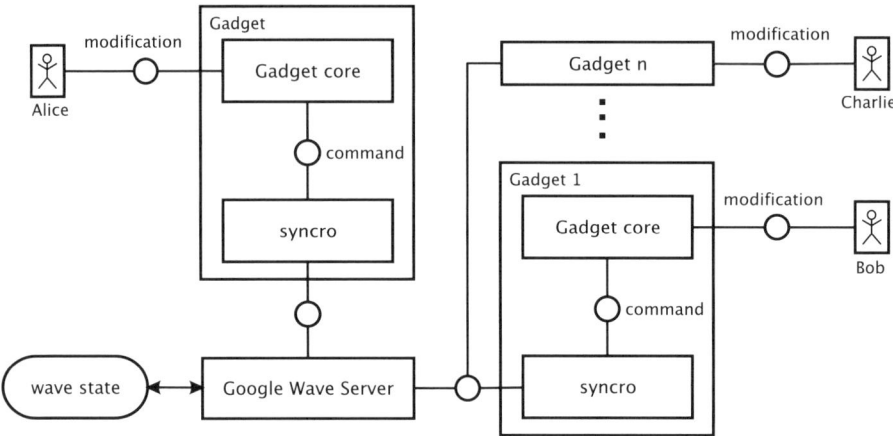

Fig. 1. Architecture of syncro

each combination of logical clock value and sender ID is unique. In our context, we can use the globally unique Google Wave user ID (e. g. alice@googlewave.com) as sender ID.

Based on this, we implemented a JavaScript library named syncro. It can be used by gadget programmers to solve the problems related to concurrent editing. The system architecture of syncro is shown in Figure 1. When Alice interacts with the gadget, each interaction creates a command. The gadget then pushes the command to its local syncro instance. Syncro attaches the current Lamport timestamp consisting of the clock value and Alice's Google Wave user ID to the command. Afterwards, the extended command is stored in the gadget state using its Lamport timestamp as unique key. This state modification invokes a callback in the syncro instances of the other Wave participants, here Bob and Charlie. The callback function retrieves the new command from the wave state and inserts it chronologically correct into the local command stack using the attached Lamport timestamp. Syncro then informs the gadget about the new command.

Obviously, Bob can also interact with his gadget while Alice's command is still in transmission. When Alice's command arrives at Bob's client and syncro decides that her command happened before Bob's, the syncro library calls the gadget with the commands to be reverted and the commands to be applied. In our case, the gadget reverts Bob's command before applying Alice's. Afterwards, Bob's command is reapplied. This guarantees that each gadget applies the commands in exactly the same order ensuring that all wave participants have the same view of the edited artifact in the gadget.

4   Library Implementation and Applications

The library presented in Section 3, the source code and a demo application is publicly available[3]. The demo shows the functionality of a simple collaborative

---

[3] http://bitbucket.org/processwave/syncro/

graphic editing tool and can be tried out in a public wave[4]. More technical details and further links can be found on our blog[5]. For programmers, syncro solves the problem of managing conflicting edits in Google Wave gadgets and its underlying concepts can be used in a wide variety of collaboration platforms. We implemented syncro as a framework component for a collaborative business process modeling tool we are integrating into Google Wave. Our modeling gadget will build on the open source project Oryx[6], a model editor running inside the browser. Oryx supports various process modeling languages (e.g. BPMN, EPCs, PetriNets), UML Class diagrams, xForms editing and much more. For example, Figure 1 was modeled with Oryx. A Wave-aware version of Oryx using syncro is planned by June 2010 and we will demonstrate it at the conference.

5   Conclusion

This paper addresses the challenge of editing artifacts collaboratively in Google Wave gadgets. To solve the conflicts related to those concurrent edits, we have developed a lightweight JavaScript library named syncro. Syncro is based on Lamport clocks [2] and programmers implementing the command pattern can use the library to enable collaborative editing. For us, syncro is a base technology to integrate the Oryx model editor into Google Wave.

Acknowledgements

We thank the processWave.org team for their development and design activities. Namely, Michael Goderbauer, Markus Goetz, Marvin Killing, Martin Kreichgauer, Martin Krueger, Christian Ress and Thomas Zimmermann.

References

1. Sun, C., Ellis, C.: Operational Transformation in Real-Time Group Editors: Issues, Algorithms, and Achievements. In: Proceedings of the 1998 ACM Conference on Computer Supported Cooperative Work, pp. 59–68. ACM, New York (1998)
2. Lamport, L.: Time, Clocks, and the Ordering of Events in a Distributed System. Communications of the ACM 21(7) (1978)
3. Sun, C., Xia, S., Sun, D., Chen, D., Shen, H., Cai, W.: Transparent Adaptation of Single-User Applications for Multi-User Real-Time Collaboration. ACM Trans. Comput.-Hum. Interact 13(4), 531–582 (2006)
4. Gamma, E., Helm, R., Johnson, R., Vlissides, J.: Design Patterns: Abstraction and Reuse of Object-Oriented Design. In: Nierstrasz, O. (ed.) ECOOP 1993. LNCS, vol. 707, pp. 406–431. Springer, Heidelberg (1993)

---

[4] http://tinyurl.com/demowave
[5] http://www.processwave.org/
[6] http://www.oryx-project.org/

An Eclipse Plug-in for Model-Driven Development of Rich Internet Applications[*]

Santiago Meliá, Jose-Javier Martínez, Sergio Mira,
Juan Antonio Osuna, and Jaime Gómez

Universidad de Alicante, IWAD, Campus de San Vicente del Raspeig,
Apartado 99 03080 Alicante, Spain
{santi,jmartinez,smira,jaosuna,jgomez}@dlsi.ua.es

Summary. Rich Internet Applications (RIAs) have recently appeared in the Internet market offering a rich and efficient User Interface similar to desktop applications. However, these applications are rather complex and their development requires design and implementation tasks that are time-consuming and error-prone. In this paper, we present a tool called OIDE (OOH4RIA Integrated Development Enviroment) aimed at accelerating the RIAs development through the OOH4RIA approach which establishes a RIA-specific model-driven process.

1 Introduction

In the last few years, a new type of sophisticated Web applications called Rich Internet Applications (RIAs) are breaking through the Internet market offering better responsiveness and a more extended user experience than the traditional Web applications. In essence, RIAs are client/server applications that are at the convergence of two competing development cultures: desktop and Web applications. They provide most of the deployment and maintainability benefits of Web applications, while supporting a much richer client User Interface (UI).

However, RIAs are complex applications, with time-consuming and error-prone design and implementation tasks. They require designing a rich user interface based on the composition of Graphical User Interface (GUI) widgets and event-based choreography between these widgets.

Therefore, RIA development requires new design methods and tools to represent this complex client/server architecture and to increase the efficiency of the development process through code generation techniques able to accelerate it and reduce errors. To achieve this goal we propose a seamless and domain-specific RIA development approach called OOH4RIA [4], which proposes a model-driven development process based on a set of models and transformations to obtain the implementation of RIAs. This approach specifies an almost complete Rich Internet Application (RIA) through the extension of the OOH server-side models (i.e. domain and navigation) and with two new platform-specific RIA presentation

---

[*] This work is supported by the Spanish Ministry of Education under contract TIN2007-67078 (ESPIA).

B. Benatallah et al. (Eds.): ICWE 2010, LNCS 6189, pp. 514–517, 2010.

models (i.e. presentation and orchestration). In order to give support to this approach, we have implemented a Rich Client Platform tool called OOH4RIA Integrated Development Environment (OIDE) [6] defined by a set of model-driven frameworks developed in Eclipse. Currently, this approach has been extended by introducing artifacts that gather new concerns: a RIA quality model and the technological and architectural RIA variability [5].

For an adequate comprehension of the OIDE tool, next section 2 presents a general perspective of OIDE development process which implements partially the OOH4RIA approach process. And finally, section 3 shows the most important technological features of OIDE.

2 An Overview of the OIDE Development Process

OIDE implements partially the OOH4RIA development process specifying an almost complete Rich Internet Application (RIA) through the extension of the OOH [2] server-side models (i.e. domain and navigation) and with two new RIA presentation models (i.e. presentation and orchestration).

This OIDE process starts by defining the OOH domain model, which is based on the UML class diagram notation, to represent the domain entities and the relationships between them. To do so, we have defined an extended domain MOF metamodel to obtain an Object-Relational Mapping without ambiguities. To improve the quality of the server side, the domain model has introduced several fundamental adjustments: (1) defining a topology of different operations such as create, delete, relationer, unrelationer, etc. to generate the CRUD (i.e. Create, Read, Update and Delete) functionality of a data-intensive server-side. (2) Dealing with a complete collection datatypes such as set, bag, list, etc. (3) Introducing concepts to remove the ambiguity of the Object-Relational Mapping (ORM) such as the object identifier to obtain the primary key, mapping from UML datatypes to database datatypes, database aliases in class (to name the table), in attribute (to name the column) and in roles (to name the foreign keys).

After specifying the domain model, the developer must design the OOH Navigation Model to define the navigation and visualization constraints. The navigation model – a DSL model – uses a proprietary notation defined by the OOH method formalized by a MOF metamodel. This model establishes the most relevant semantic paths through the information space filtering the domain elements available in the RIA client-side. It also introduces a set of OCL filters permitting to obtain information from the domain model.

At this point, the UI designer begins the definition of the RIA client-side establishing the complete layout by means of a structural representation of different widgets (e.g. dimensions x, y, position), panels (e.g. horizontal, vertical, flow, etc.) and style (e.g. colour, background, fonts, etc). There are many RIA frameworks, each with a different set of widgets with their own properties and events. For this reason, we have defined a platform-specific model called Presentation Model, which can be instantiated for each supported RIA framework (currently, OIDE gives support to Google Web Toolkit and Silverlight) thus obtaining similarity with the look and feel of a RIA UI.

To complete the information needed by an interactive UI, OOH4RIA incorporates a platform-dependent model called Orchestration Model, which helps introduce the dynamic behavior of the RIA UI. This model does not have a graphical representation in the OIDE tool, being defined by a set of Event-Condition-Action (ECA) rules defined by a property form. They determine how different widgets receive the events from users and if an OCL condition is accomplished they invoke a set of Actions that correspond to one or more widget methods or to one or more services offered by the server-side business logic. These events and methods offered by the OIDE tool are RIA framework specific (i.e. Silverlight or GWT) permitting the user to specify the proper arguments accurately.

The last step consists in executing the model-to-text transformation to obtain the RIA implementation. The process defines a transformation that generates the RIA server-side from the domain and navigation model, and a second transformation obtains the RIA client-side from the Presentation and Orchestration models.

3   The OIDE Design and Technological Features

The OIDE is a tool developed like an Eclipse plug-in based on other open-source relevant tools in the Eclipse Modeling Project: the Graphical Modeling Framework (GMF) [3] to represent domain-specific models and the Xpand language of the Model to Text (M2T) project [1] to develop the transformations that carry out the OIDE development process. This tool is a user-friendly IDE which specifies easily and seamless the server and the client side of a RIA. Its main characteristics are: (1) a WySWyG UI presentation model which emulates the same appearance and the layout spatial distribution than a specific RIA design tool (see Fig. 1). (2) An intuitive ECA rules Tree which permits to define the UI behavior selecting events and methods of a RIA framework widgets. (3) An OCL checking which avoids introducing invalid model configurations. (4) A rapid prototyping that helps reducing the development iterations. (5) The integration of the model transformation editor which allows developers to modify them for a specific application. To do so, the tool generates a set of default transformation rules for each new OIDE project, thus allowing developers to manipulate these transformations introducing exceptions or a specific code for this project. OIDE integrates the following transformation languages provided by the Eclipse Modeling Model to Text project [1]: Xpand, a model-to-text transformation language, Xtend, a model-to-model transformation language and Check to represent OCL-like constrains. Specifically, Xpand provides two interesting features: On the one hand, a polymorphism rules invocation that allows developers to introduce new rules without having to modify the default rules provided by the tool. And the other hand, an incremental generation facility that detects which parts of a model have changed since the last generation process and determines which files need to be generated due to that change and which are unaffected by it.

To implement the OIDE DSLs, we have defined an EMOF metamodel (specifically an EMF metamodel) which establish the metaclasses, attributes and relationships between the elements of the OOH4RIA models. At this point, we use the GMF framework to generate automatically the graphical editor for each different models and the EMF to produce a set of Java classes that enable us to view and edit this metamodel. Next, we must establish a correspondence between the relevant MOF metaclasses and a graphical element (a node or a link) using a set of XML files. This generates a graphical editor (canvas and tool) based on GEF.

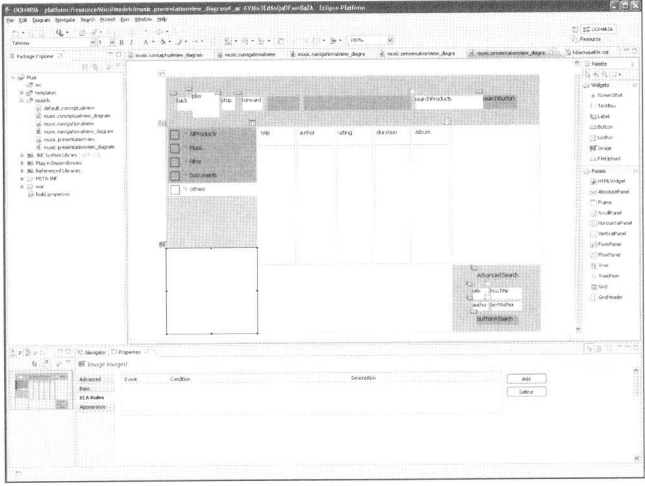

Fig. 1. A snapshot of the Presentation Model in OIDE tool

Currently, the OIDE tool permits to generate the final implementation in two alternative RIA frameworks: GWT and Silverlight. GWT is a DOM-based RIA framework developed by Google, permitting developers to program with a Java API by generating a multi-browser (DHTML and Javascript) code. Thus, OIDE integrates a GWT plug-in which permits to prototype the RIA client-side in the Eclipse IDE. On the other hand, Microsoft Silverlight is a very promising plugin-based RIA framework that integrates multimedia, graphics, animations and interactivity into a single runtime environment.

References

1. Eclipse Modeling Model to Text Project,
   http://www.eclipse.org/modeling/m2t/
2. Gómez, J., Cachero, C., Pastor, O.: Conceptual Modeling of Device-Independent Web Applications. IEEE Multimedia 8(2), 26–39 (2001)
3. Graphical Modeling Framework (GMF),
   http://www.eclipse.org/modeling/gmf/
4. Meliá, S., et al.: A Model-Driven Development for GWT-Based Rich Internet Applications with OOH4RIA. In: ICWE '08, pp. 13–23. IEEE Computer Society, USA (2008)
5. Meliá, S., et al.: Facing Architectural and Technological Variability of Rich Internet Applications. IEEE Internet Computing 99, 30–38 (2010)
6. OOH4RIA Integrated Development Enviroment (OIDE),
   http://suma2.dlsi.ua.es/ooh4ria/

A Cross-Platform Software System to Create and Deploy Mobile Mashups

Sandra Kaltofen, Marcelo Milrad, and Arianit Kurti

Centre of Learning and Knowledge Technologies (CeLeKT)
School of Computer Science, Physics and Mathematics
Linnaeus University, Sweden
{sandra.kaltofen,marcelo.milrad,arianit.kurti}@lnu.se

Abstract. Changes in usage patterns of mobile services are continuously influenced by the enhanced features of mobile devices and software applications. Current cross-platform frameworks that allow the implementation of advanced mobile applications have triggered recent developments in relation to end-user mobile services and mobile mashups creation. Inspired by these latest developments, this paper presents our current development related to a cross-platform software system that enables the creation of mobile mashups within an end-user programming environment.

Keywords: End-user programming, Mobile Mashup, Cross-platform development.

1 Introduction

Mobile devices are getting more powerful in terms of computational power and functionalities. GPS sensors, high definition cameras and Internet access are perceived as standard features of today's mobile devices. The market for mobile applications is continuously growing in direct relation to these latest developments. The more powerful these devices will get, the more powerful mobile applications will become. One of the drawbacks in the development of mobile applications is that developers have to face with the restrictions in the mobile phone hardware and the devices' specifications [1]. Furthermore, they need to have knowledge about different programming languages because the Software Development Kits (SDKs) released by the platform creators are usually tied to a specific language. All these developments generate the need of having software systems that provide end-users with a simple way for the creation of mobile applications [2]. Such systems should be able to integrate different features to create cross-platform mobile applications in such a way that they would not be bound to a particular mobile platform. Another central point of focus is *mobile mashups* and the lack of possibilities to create mashups optimized for mobile usage. In order to address these specific issues, we have designed and developed a system that enables the creation and deployment of customized cross-platform mobile mashups using an end-user programming environment.

B. Benatallah et al. (Eds.): ICWE 2010, LNCS 6189, pp. 518–521, 2010.

2 Requirements and Proposed Solution

Requirements engineering for web applications follows the general guidelines used for software applications. According to [3], the main functional requirements of web applications need to incorporate organization, application domain, navigation and interaction requirements. Guided by these principles and the outcomes of our literature survey, we have identified a number of features that a mobile mashup system should have: 1. Mobile mashups should not be bound to device specifications and should work cross-platform [1]; 2. The mashups needs to access social networks and other Web 2.0 services to exchange data [4]; 3. Previous programming skills should not be required to create mobile mashups [5]; 4. The functionalities for the mashups are offered as customizable components [2]; 5. The environment needs to allow sharing the created mashups with other users [2, 5]; 6. The environment needs to work with visual programming concepts and technologies [6].

Proposed solution: Based on the above-mentioned features, we have developed a system as described in figure 1. Different components are used in order to support enhanced customization of mobile mashups. *Service components* are used to provide core functionalities for a mashup; like the access to server site features and important Web APIs and/or their combination. These components are used to allow the user to receive existing content or to create new content by entering personal data. *Layout components* are used to present information and content such as text, images or videos. *Device components* are used to access the advanced features of today's mobile devices like the GPS sensor data, activation of the photo camera and access to the device content. *Additional components* are proposed to provide enhanced extensibility to the software system.

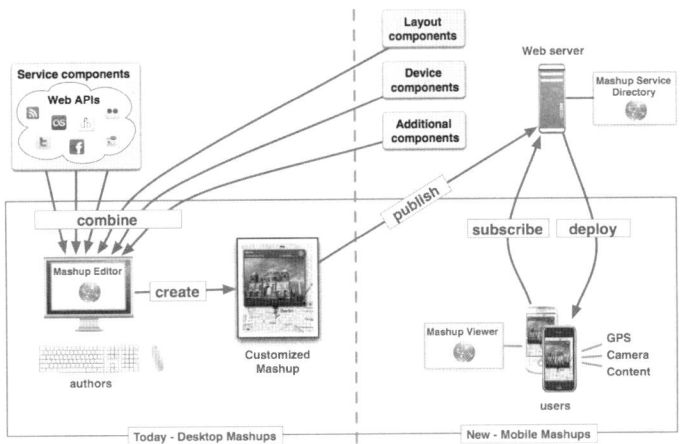

Fig. 1. Overview of the proposed solution

Implementation: Our software solution consists of two applications: an end-user programming editor to create the mobile mashups and a viewer to deploy them on a mobile device. The *Mobile Mashup Editor* was implemented as a Rich Internet

Application (RIA) using the *Google Web Toolkit (*http://code.google.com/webtoolkit) framework. The editor provides a GUI that enables to combine and to configure mashup components using visual programming concepts and techniques. Using this editor, mashups can be published to a service directory on a web server to make them accessible to other users. The *Mobile Mashup Viewer* runs on the mobile device as a native application. The viewer was implemented using the cross-platform mobile framework *Titanium Mobile* (www.appcelerator.com) and it provides access to the remote mashup service directory. In contrast to desktop mashups, mobile mashups have to run within the viewer application and not within a web browser. This is necessary to be able to access the sensor data and the content of mobile devices within a mashup. For that reason, the editor creates and publishes the mashup not as a mobile web application but rather as description of the created mashup. This description contains the mashup structure and settings, its components and their configuration. The viewer can interpret this description and deploys the mobile mashup application out of it. The proposed software solution is cross-platform; both on the editor and the viewer site.

3   Cross-Platform Mobile Mashup Example

A mobile mashup can contain different pages and components that are added to each specific application. When a component is added to a page it can be configured, resized and positioned on it. A prototype of the *Mobile Mashup Editor* is shown in figure 2 and contains an example with three pages. The example below has three different components: 1) A *service component* for a Google Map that shows either an address the author can configure or a location information received through a Web API, 2) A *layout component* with a label and an entered mailing address text and 3) A *device component* for calling a configured telephone number. Within the *Mobile Mashup Viewer* application, published mashups can be deployed. The deployed example mashup has been tested on the iPhone and Android platforms as shown in figures 3 & 4. After the deployment, the user can navigate through the different mashup pages. The layout of the pages and the functionalities are equal to the designed pages and the configuration the author made in the editor.

Fig. 2. Mobile Mashup Editor   **Fig. 3.** iPhone   **Fig. 4.** Android

4 Conclusion and Future Work

The focus of the prototype implementation was to create the basis for an extensible software system capable of creating and deploying mobile mashups. Initial tests of our current prototype on the iPhone and Android platforms have proven the validity of this approach. The system offers new ways and possibilities for the creation of mobile applications that will be accessible from several mobile platforms. Moreover, the end-user programming environment, the *Mobile Mashup Editor*, provides an easy to use development platform, as an author can make use of visual programming concepts and techniques. Aspects related to privacy and trusts are beyond the scope of this paper.

Innovative features and benefits: The proposed solution provides the following unique features compared to existing approaches or systems: 1. The installed viewer application on the mobile device allows an easy access to the published mashups. Developers can therefore provide mobile mashup applications without the restrictions of an application store or licensing; 2. The use of cross-platform technologies on the editor and viewer site does not bind the solution to particular platforms and devices. As a result, the potential benefits of the proposed software system can be described as follows: 1. The deployed mobile applications have the layout and functionalities that were designed for and can use the features of the mobile device like GPS, camera and the device content; 2. The software system is component-based and therefore extensible through the integration of *additional components*. In our future development, we plan to integrate database components to our system. This will allow the *Mobile Mashup Viewer* to provide offline working capabilities and synchronisation with a remote database. Another line of exploration is the specification and development of *additional components* that can be used in different application domains.

References

1. Chaisatien, P., Tokuda, T.: A Web-Based Mashup Tool for Information Integration and Delivery to Mobile Devices. In: Proceedings of the 9th International Conference on Web Engineering. LNCS, vol. 5648, pp. 489–492. Springer, Berlin (2009)
2. Bosch, J.: From software product lines to software ecosystems. In: Proceedings of the 13th International Software Product Line Conference, San Francisco, California, August 24-28, pp. 111–119. ACM Press, New York (2009)
3. Casteleyn, S., Florian, D., Dolog, P., Matera, M.: Engineering Web Applications: Data Centric Systems and Applications. Springer, Berlin (2009)
4. Sheth, A.: Citizen Sensing, Social Signals, and Enriching Human Experience. IEEE Internet Computing, 87–92 (2009)
5. Jensen, C.S., Vicente, C.R., Wind, R.: User-Generated Content: The Case for Mobile Services. IEEE Computer 41, 116–118 (2008)
6. Trevor, J.: Doing the mobile mashup, pp. 104–106. IEEE Computer Society, Los Alamitos (2008)

A Blog-Centered IPTV Environment for Enhancing Contents Provision, Consumption, and Evolution

In-Young Ko, Sang-Ho Choi, and Han-Gyu Ko

Dept. of Computer Science, Korea Advanced Institute of Science and Technology,
335 Gwahangno, Yuseong-gu, Daejeon, 305-701, Republic of Korea
{iko,shchoi9,kohangyu}@kaist.ac.kr

Abstract. There have been some efforts to take advantages of the Web for the IPTV domain to overcome its limitations. As users become the center of the contents creation and distribution, motivating user participation is the key to the success of the Web-based IPTV. In this paper, we propose a new IPTV framework, called a blog-centered IPTV, where personal blogs are the first-class entities that represent user interests in IPTV contents. An IPTV blog provides a user with a set of interfaces for finding, accessing and organizing IPTV contents based on their needs, and becomes an active entity to join communities and to participate in making community contents evolved. We have implemented a prototype of the blog-centered IPTV, and showed that users can easily find and access their desired contents and successfully build community-based contents.

Keywords: IPTV, Web-based IPTV, Blog, Community, Social Networks.

1 Introduction

The Web is now being considered as an important platform for new IPTV (Internet Protocol Television). The essential characteristics of the Web such as openness, variety, and accessibility bring IPTV users with facilities to promote the content creation and distribution by allowing users to collaboratively participate in creating, organizing and sharing their contents [1]. There have been some efforts to take advantages of the Web for the IPTV domain such as Joost, Babelgum, and Metacafe[1]. To effectively motivate user participation, which is the key to the success of Web-based IPTV, more user-centric functions and facilities are needed than simply sharing of video contents [2].

In this paper, we propose a new IPTV framework, called a blog-centered IPTV, where personal Web logs (blogs) are the first-class entities that represent user interests in IPTV contents. An IPTV blog provides a user with a set of interfaces for finding, accessing and organizing IPTV contents based on their needs, and becomes an active entity to join communities and to participate in making community contents evolved. Although it has similar features with Blogosphere [3], as Fig. 1 depicts, the

---

[1] http://www.joost.com/, http://www.babelgum.com/, http://www.metacafe.com/

B. Benatallah et al. (Eds.): ICWE 2010, LNCS 6189, pp. 522–526, 2010.

blog-centered IPTV covers the entire lifecycle of IPTV communities and supports user activities of content consumption, evolution, syndication, and provision. In addition, it automatically identifies potential IPTV communities by analyzing the social and personal characteristics of users, and recommends a user with an existing community to join or a potentially useful community to create [4].

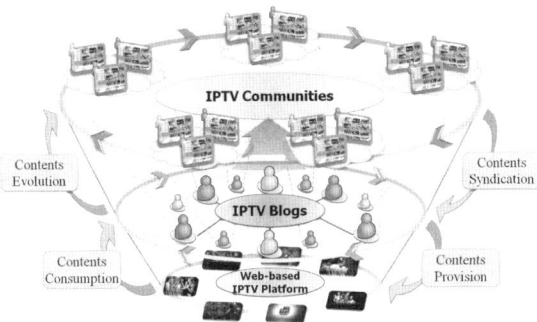

Fig. 1. Overview of the Blog-centered IPTV Environment

For an IPTV community, the IPTV blogs proactively contribute to accumulate and organize various contents that are relevant to the community. Community contents are then consumed by each blog in a personalized way. Users can easily locate and access the contents that meet their interests and needs via their IPTV blogs.

The rest of this paper is organized as follows. In Section 2, we explain the architecture of the blog-centered IPTV environment including essential components and their relations. Section 3 presents a blog-centered IPTV prototype and the paper ends in Section 4 with the conclusion and future work.

2   Architecture of the Blog-Centered IPTV Environment

As shown in Fig. 2, the architecture of the blog-Centered IPTV environment consists of three main layers: (a) Media Layer, (b) IPTV Blog Layer, and (c) IPTV Community Layer. The media layer stores and provides various resources including media contents and their metadata, and user-related information. The media contents are accessed via the *media enabler* that locates and streamlines the contents. The *semantic enabler* provides an ontology-based model and reasoning methods to represent and manage semantic metadata of IPTV contents. The *social connector* accesses user-related data and extracts social network information to be used for recommending potential blog communities.

In the IPTV blog layer, the *IPTV blog handler* provides a basic set of functions to create or customize a blog by using a template, to define the profile information of a blog, and to create channels that integrate and deliver blog contents to other users. The *semantic tagging* facilities allow users to put ontology-based semantic tags on multimedia contents. The semantic tags are automatically collected, monitored and analyzed to identify user preferences and to determine potential communities to be recommended to the users. The *semantic media syndication* component provides functions to enable blogs to subscribe for each others' contents based on their needs represented in an ontology-based semantic model. The *contents mash-up engine* supports users with a set of methods to integrate various types of contents such as texts, Web services as well as regular video contents, and make them delivered via the blog channels. The *semantic search engine* allows blogs to be equipped with semantically-based search capability to produce more relevant search results based on user preferences and interests.

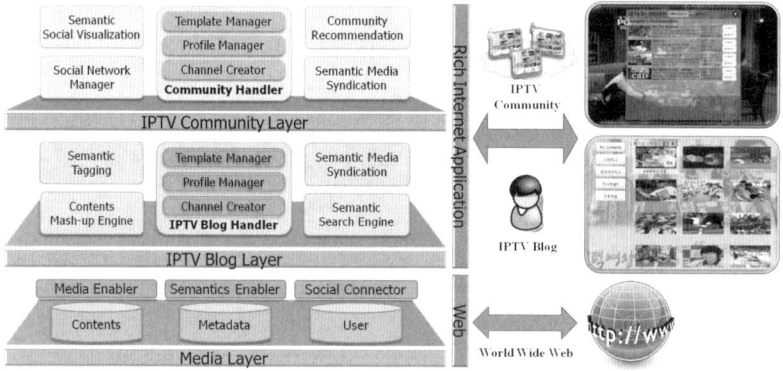

Fig. 2. Architecture of the blog-centered IPTV environment

Similar to the IPTV blog handler, the *IPTV community handler* at the IPTV community layer provides functionality of creating and managing IPTV communities. The *social network manager* identifies a social network of users centered on the owner of a blog based on their preferences and social relationship information. The *semantic social visualization* module visualizes a map of users in a social network and allows users to browse through the map based on user interests. The *community recommendation* module recommends users to create a new IPTV community or to join an existing one by comparing profiles of blogs and by grouping relevant blogs based on their semantic properties. A blog can subscribe for a community channels by using the functions provided by the *semantic media syndication* component.

3   Prototype and Application

To show the effectiveness of our blog-centered IPTV, we implemented a prototype. For IPTV clients, we used Silverlight[2], which is one of the popular RIA (Rich Internet Application) technologies [5]. We used IIS (Internet Information Services) for the Web server and WMS (Windows Media Services) for the media server. The right side of Fig. 2 shows the screen shots of an IPTV blog and a community page. In a blog, metadata of relevant contents are arranged in a chronicle order with

---

[2] http://www.silverlight.net/

comments, rankings, tags and other annotations. An IPTV community page shows an aggregated view of information about the most relevant contents from the blogs in the community.

Users can find relevant contents through the semantic search capability, and as shown in Fig. 3 (left), the search result is displayed in a radar view, by which users can easily identify semantic closeness of the contents found toward their needs. Users can also browse through different categories of contents by centering a different content, and narrow down into a more specific set of contents by applying various semantic filters such as genres, time, location, emotions, etc. that are automatically enabled based on the context of the contents and users. By using the social network browser (the right-side screen shot in Fig. 3), users can navigate through a group of users who have similar interests, and access contents organized by other users. In addition, the users can create a new IPTV community by inviting other users in the social network.

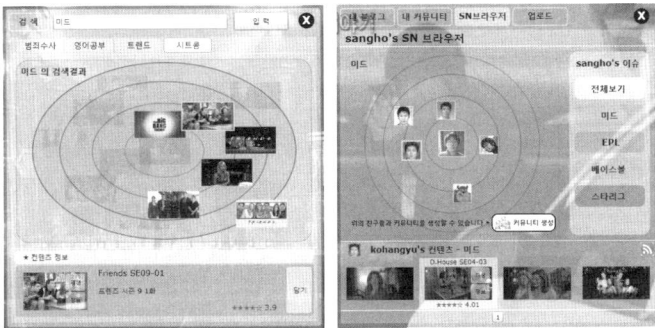

Fig. 3. Semantic search interface (left), and a social network browser (right)

4   Conclusion

In this paper, we proposed a new IPTV environment called a blog-centered IPTV, which provides facilities of semantically-based content search and visual browsing, and IPTV community recommendation and management. In this approach, personal blogs are the first class entities that organize IPTV contents in a personalized way, and actively find and join potential communities. We believe that this framework will enhance the provision, consumption, and evolution of IPTV contents by motivating more user participations. We are currently extending our approach by adding features to monitor community activities and to manage the lifecycle of communities according to them.

Acknowledgments. This work was supported by the IT R&D program of MKE/ KEIT. [2008-S-006-02, Development of Open-IPTV (IPTV2.0) Technologies for Wired and Wireless Networks]

References

1. O'Reilly, T.: What Is Web 2.0: Design Patterns and Business Models for the Next Generation of Software. Published on O'Reilly, Communications & Strategies (1), 17 (First Quarter, 2007)
2. Koh, J., et al.: Encouraging participation in virtual communities. Communications of the ACM 50(2), 69–73 (2007)
3. Herring, S.C., Kouper, I., Paolillo, J.C., Scheidt, L.A., Tyworth, M., Welsch, P., Wright, E., Yu, N.: Conversation in the Blogosphere: An Analysis From the Bottom Up. In: Proceedings of the 38th Annual Hawaii International Conference on System Sciences (HICSS'05), Big Island, Hawaii, January 3-6 (2005)
4. Ko, H.-G., Choi, S.-H., Ko, I.-Y.: A Community Recommendation Method based on Social Networks for Web 2.0-based IPTV. In: 16th International Conference on Digital Signal Processing, Santorini, Greece, July 5-7 (2009)
5. O'Rourke, C.: A Look at Rich Internet Applications. Oracle Magazine 18(4), 59–60 (2004)

Factic: Personalized Exploratory Search in the Semantic Web

Michal Tvarožek and Mária Bieliková

Institute of Informatics and Software Engineering, Faculty of Informatics
and Information Technologies, Slovak University of Technology
Ilkovičova 3, 842 16 Bratislava, Slovakia
Name.Surname@fiit.stuba.sk

Abstract. Effective access to information on the Web requires constant improvement in existing search, navigation and visualization approaches due to the size, complexity and dynamic nature of the web information space. We combine and extend personalization approaches, faceted browsers, graph-based visualization and tree-based history visualization in order to provide users with advanced information exploration and discovery capabilities. We present our personalized exploratory browser Factic as a unique client-side tool for effective Semantic Web exploration.

1   Introduction and Related Work

The Web has become an almost ubiquitous virtual information space where information is stored, published and shared in many different forms. To address problems associated with information access on the Web (e.g., the navigation problem, complex [semantic] query construction, information overload), several directions and initiatives are being pursued such as the Semantic Web [1], Adaptive web-based systems [2] or Exploratory search [3].

In our previous work, we have demonstrated the practicality of adaptation in a faceted browser by applying personalization principles to a faceted semantic browser thus improving overall user experience when working with an ontological repository [4]. While our previous approach provided good personalized navigation and query construction support, it offered limited support for visual information discovery, e.g., using different widgets to render topic-, time- and location-based information as demonstrated by VisGets [5].

Furthermore, it had to be configured for a specific domain model and thus was not seamlessly applicable across multiple (open) domains at the same time. Consequently, we focused on improving two aspects of our solution – *exploratory search enhancement* and *automatic user interface generation* based on domain metadata in order to improve *end user experience* when accessing information in the Semantic/Deep web, which is typically inaccessible via search engines.

2   Personalized Exploratory Search Browser

Our original personalized faceted browser *Factic* [4] performed faceted browsing and personalization to adapt facets, i.e. order, hide and annotate them while also

B. Benatallah et al. (Eds.): ICWE 2010, LNCS 6189, pp. 527–530, 2010.

recommending restrictions based on user characteristics. To further improve user experience, we completely reworked *Factic* to facilitate two new features:

- *Multi-paradigm exploration* – combined view-based (faceted), keyword-based, content-based exploration with multiple adaptive result overviews and collaborative resource annotation (see Fig. 1), graph-based visualization and browsing of information (see Fig. 2) and custom/external content rendering.
- *User interface generation* – (semi)automated generation of the browser's user interface based on semantic metadata from the domain ontology to facilitate use in multiple or changing information domains such as the Web. We primarily focus on *facet generation*, i.e. what data correspond to facets, how data sources are queried, how facets are visualized and what interaction options users can employ; *result overview generation*, i.e. what attributes of results and how they are visualized; *graph view generation*, i.e. which nodes and how are visualized, what attributes to show and what layout to employ.

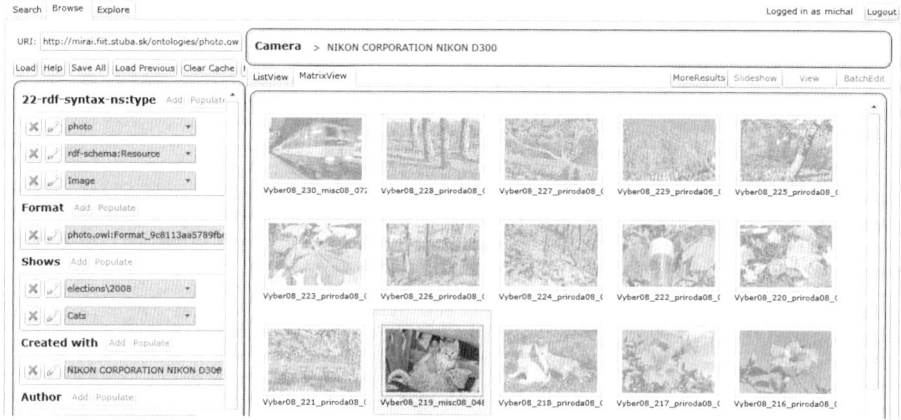

Fig. 1. Example of a faceted query result overview in a matrix (right) with one selected result whose attributes are edited in the annotation pane (left). The facets used for querying are hidden below the annotation pane, which is normally invisible.

We developed *Factic* as a client-side Silverlight application running inside a web browser (see Fig. 2). This simplifies deployment and enables it to process and store information on the client device. *Factic* is primarily centered around end-user specific functionality such as visualization, personalization, user modeling and profile management. We also process end-user specific data (e.g., activity logs and the derived user models) on the client thus reducing unwanted privacy exposure (with optional sharing with the server-side).

Consequently, *Factic* works as an intelligent front-end to (multiple) search and/or information providers thus effectively delegating querying, indexing and crawling services to third-party providers (e.g., in the future Google, Sindice, DBPedia). Our modular architecture allows us to easily incorporate new views

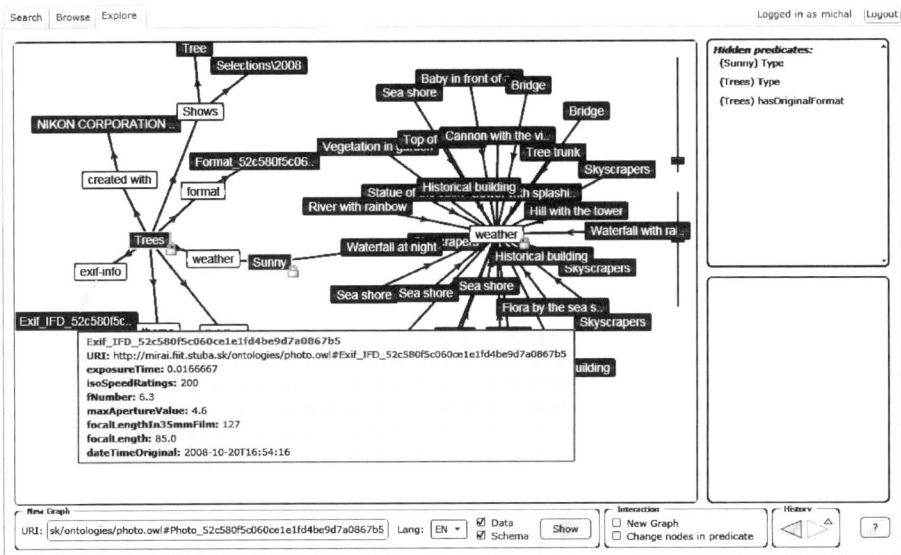

Fig. 2. The graph-based view shows different resources (e.g., photos, authors) as nodes. Users can expand nodes or navigate by centering the visible window on new nodes. Hovering over nodes shows additional annotations and associated thumbnails.

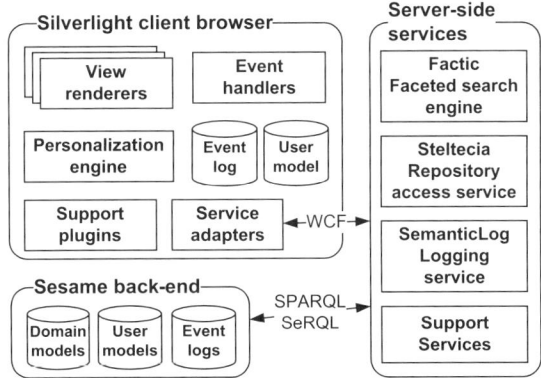

Fig. 3. The client renders the user interface and provides personalization support (top). The server includes web (WCF) services for faceted search (Factic), ontological repository access (Steltecia), and event logging (SemanticLog) for global statistics tracking (right). All services store their data in a common ontological repository in Sesame.

and also gives us the flexibility to add new data sources provided that they contain enough semantic metadata to generate user interface widgets (see Fig. 3).

Our solution is primarily suitable for digital libraries or other (semi)structured domains which contain semantic metadata (we also experimented with publications and job offers). We evaluate our browser via user studies in the image domain where our data set contains roughly 8 000 manually and semi-automatically annotated images. Our results have shown the viability of adaptive interface generation in addition to user experience improvements in terms of better orientation, revisitation support and shorter task times.

3   Summary and Future Work

Our exploratory search browser *Factic* enables *adaptive Semantic Web exploration* by empowering end-users with access to semantic information spaces with:

- Adaptive view generation.
- Interactive multi-paradigm exploratory search.
- Personalized recommendation for individual users.

Our client-side personalization and user modeling focus enables us to address privacy and identity issues as personally identifiable data never have to leave a user's system. Furthermore, we see the extension of our approach to accommodate legacy web resources as one possible direction of future work ultimately leading toward a *Next Generation Web Browser* for seamless exploration of semantic and legacy web content as outlined in [6].

Acknowledgment. This work was partially supported by the grants VEGA 1/0508/09, KEGA 028-025STU-4/2010 and it is the partial result of the Research & Development Operational Programme for the project Support of Center of Excellence for Smart Technologies, Systems and Services, ITMS 26240120005, co-funded by the ERDF.

References

1. Shadbolt, N., Berners-Lee, T., Hall, W.: The semantic web revisited. IEEE Intelligent Systems 21(3), 96–101 (2006)
2. Brusilovsky, P., Kobsa, A., Nejdl, W. (eds.): The Adaptive Web: Methods and Strategies of Web Personalization. LNCS, vol. 4321. Springer, Berlin (June 2007)
3. Marchionini, G.: Exploratory search: from finding to understanding. Comm. of the ACM 49(4), 41–46 (2006)
4. Tvarožek, M., Bieliková, M.: Personalized faceted navigation in the semantic web. In: Baresi, L., Fraternali, P., Houben, G.-J. (eds.) ICWE 2007. LNCS, vol. 4607, pp. 511–515. Springer, Heidelberg (2007)
5. Dörk, M., Carpendale, S., Collins, C., Williamson, C.: Visgets: Coordinated visualizations for web-based information exploration and discovery. IEEE Transactions on Visualization and Computer Graphics 14(6), 1205–1212 (2008)
6. Tvarožek, M., Bieliková, M.: Reinventing the web browser for the semantic web. In: WI-IAT '09: Proc. of the 2009 IEEE/WIC/ACM Int. Conf. on Web Intelligence and Intelligent Agent Technology, pp. 113–116. IEEE CS, Los Alamitos (2009)

Takuan: A Tool for WS-BPEL Composition Testing Using Dynamic Invariant Generation

Manuel Palomo-Duarte, Antonio García-Domínguez, Inmaculada Medina-Bulo,
Alejandro Alvarez-Ayllón, and Javier Santacruz

Department of Computer Languages and Systems. Universidad de Cádiz, Escuela
Superior de Ingeniería, C/Chile 1, CP 11003 Cádiz, Spain
{manuel.palomo,antonio.garciadominguez,inmaculada.medina@uca.es,
alejandro.alvarez,javier.santacruz}@uca.es

Abstract. WS-BPEL eases programming in the large by composing web
services, but poses new challenges to classical white-box testing techniques. These have to be updated to take context into account and cope
with its specific instructions for web service management. Takuan is an
open-source system that dynamically generates invariants reflecting the
internal logic of a WS-BPEL composition. After several improvements
and the development of a graphical interface, we consider Takuan to be a
mature tool that can help find both bugs in the WS-BPEL composition
and missing test cases in the test suite.

Keywords: Web services, service composition, WS-BPEL, white-box
testing, dynamic invariant generation.

1 Introduction

The OASIS WS-BPEL 2.0 standard allows the user to develop advanced web
services (WS) by composing others. However, it presents a challenge [1] for traditional white-box testing techniques, firstly due to its dependency on context,
and secondly because of the inclusion of WS-specific instructions not found in
most programming languages (like those for fault and compensation handling).

Automatic dynamic invariant generation [2] has proved to be a successful
technique to assist in white-box testing of programs written in imperative languages. Let us note that, throughout this work, the term *dynamic invariant* (or
likely invariant) is considered, as in most related works, in its broadest sense: a
property that a program holds for a specified test suite.

Takuan [3] is an open-source system that dynamically generates invariants
reflecting the internal logic of a WS-BPEL composition. After several improvements [4,5] Takuan can be easily used from a graphical interface to support
WS-BPEL composition testing, helping find bugs in the composition code and
improve the test suite.

2 Dynamic Invariant Generation for WS-BPEL

Dynamic invariant generation process is not based on a formal analysis of the
composition, but on information collected from several executions. This way, if

B. Benatallah et al. (Eds.): ICWE 2010, LNCS 6189, pp. 531–534, 2010.

we use a good test suite as input, all of the features of the environment and the complex internal logic of the composition (loops, fault handling, etc.) will be reflected in the execution logs, and the generator will infer true invariants. On the other hand, if we use an incomplete or biased test suite, we will get false invariants. These are due to an incomplete test suite that needs additional test cases which will falsify those invariants.

There are different ways to use dynamically generated invariants to help program testing [2]. We will focus on two of them, as shown in figure 1: debugging a program and improving a test suite.

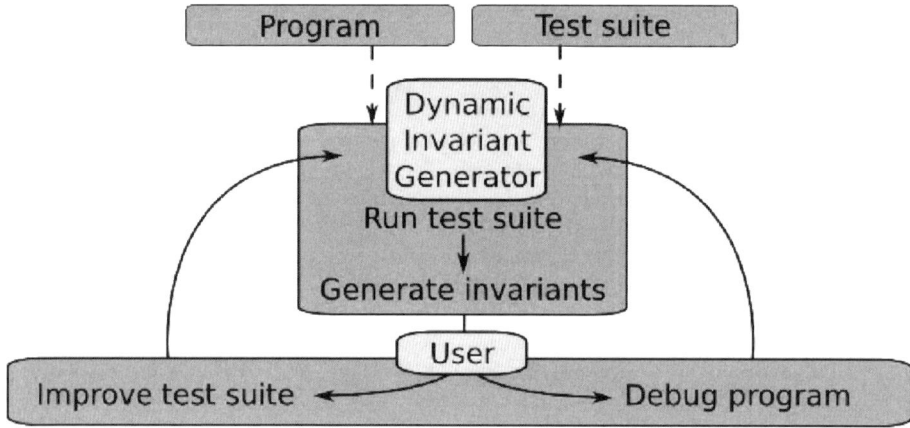

Fig. 1. Dynamic invariant generation feedback loop

The usual dynamic invariant generation workflow starts with the user providing an initial test suite and a program. After running the program against the test suite, unexpected invariants may be obtained. In that case, the user needs to check if that difference was caused by a bug in the code. After fixing the bug, the user could run the new (corrected) program against the same test suite and check if those invariants are not inferred.

If the user considers the program to be correct, the fault may lie on the test suite, which is lacking test cases that disprove those invariants. By appropriately extending the test case, they would not be inferred again in the next run. This cycle can be run as many times as needed, fixing bugs in the program code and improving the test suite until the expected invariants are obtained.

3   Takuan

Takuan integrates our own code with several well-tested open-source systems [6]: the BPELUnit unit testing library, the ActiveBPEL engine and the Daikon dynamic invariant generator. All of them have been modified to create a dynamic invariant generation workflow that includes several optimizations to reduce running time and improve the invariants for the WS-BPEL language [4,5].

Takuan is available for free download under the terms of the GNU/GPL license. In its home site [7] we provide both source code for automatic script-guided compilation and a ready-to-use virtual machine.

Takuan can be invoked directly from a shell or through a graphical plugin for the NetBeans IDE. It can connect with a Takuan instance, both locally and remotely through its IP address. Figure 2 shows the plugin running.

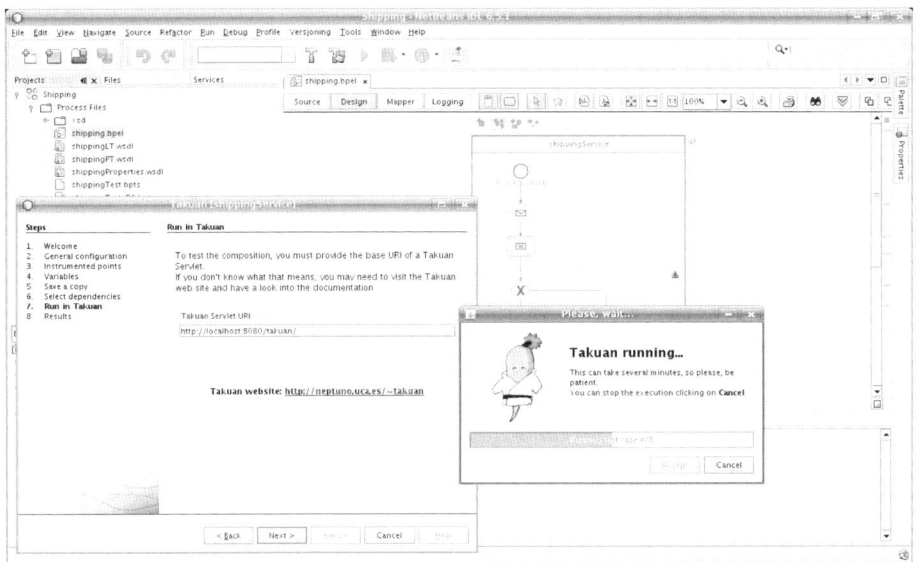

Fig. 2. Takuan NetBeans Plugin running

By default, Takuan logs every variable value before and after every sequence instruction in the composition to check for invariants. The user can choose which variables to consider for invariants and in which instructions.

Additionally, Takuan can replace some of the external services with *mock-ups*: dummy services which will act according to user instructions. It is only recommended when not all external services are available for testing, or to define *what-if* scenarios under certain predefined WS behavior. In any of these cases, the behavior provided by each mockup in each test case (a SOAP answer or a certain fault) will be part of the test suite specification. The user will be responsible for providing suitable values for them.

4 Conclusions and Future Work

Takuan is an open-source dynamic invariant generator for WS-BPEL compositions. It takes a WS-BPEL 2.0 process definition with its dependencies and a test suite, and infers a list of invariants. In this paper we have shown two ways to use Takuan in a feedback loop: highlighting bugs in the composition code and helping to improve a test suite.

We are currently developing a new graphical user interface, named Idig-inBPEL, that will ease massive systematic Takuan use [7]. Our future work will be to extend Takuan to infer non-functional properties that can be included in composition specification, so it can be used as automatically as possible inside a Service-Oriented-Architecture-specific developing methodology [8].

Acknowledgments

This paper has been funded by the Department of Education and Science (Spain) and FEDER funds under the National Program for Research, Development and Innovation. Project SOAQSim (TIN2007-67843-C06-04).

References

1. Bucchiarone, A., Melgratti, H., Severoni, F.: Testing service composition. In: Proceedings of the 8th Argentine Symposium on Software Engineering, ASSE '07 (2007)
2. Ernst, M.D., Cockrell, J., Griswold, W.G., Notkin, D.: Dynamically discovering likely program invariants to support program evolution. IEEE Transactions on Software Engineering 27(2), 99–123 (2001)
3. Palomo-Duarte, M., García-Domínguez, A., Medina-Bulo, I.: Takuan: A dynamic invariant generation system for WS-BPEL compositions. In: ECOWS '08: Proceedings of the 2008 Sixth European Conference on Web Services, Washington, DC, USA, pp. 63–72. IEEE Computer Society, Los Alamitos (2008)
4. Palomo-Duarte, M., García-Domínguez, A., Medina-Bulo, I.: Improving Takuan to analyze a meta-search engine WS-BPEL composition. In: SOSE '08: Proceedings of the 2008 IEEE International Symposium on Service-Oriented System Engineering, Washington, DC, USA, pp. 109–114. IEEE Computer Society Press, Los Alamitos (2008)
5. Palomo-Duarte, M., García-Domínguez, A., Medina-Bulo, I.: Enhancing WS-BPEL dynamic invariant generation using XML Schema and XPath information. In: Gaedke, M., Grossniklaus, M., Díaz, O. (eds.) ICWE '09. LNCS, vol. 5648, pp. 469–472. Springer, Heidelberg (2009)
6. Palomo-Duarte, M., García-Domínguez, A., Medina-Bulo, I.: An architecture for dynamic invariant generation in WS-BPEL web service compositions. In: Proceedings of ICE-B 2008 - International Conference on e-Business, Porto, Portugal. INSTICC Press (July 2008)
7. SPI&FM Group: Official Takuan home site, http://neptuno.uca.es/~takuan
8. García Domínguez, A., Medina Bulo, I., Marcos Bárcena, M.: Hacia la integración de técnicas de pruebas en metodologías dirigidas por modelos para SOA. In: Actas de las V Jornadas Científico-Técnicas en Servicios Web y SOA, Madrid (October 2009)

Author Index

Printing: Mercedes-Druck, Berlin
Binding: Stein+Lehmann, Berlin